POWER MANAGEMENT TECHNIQUES FOR INTEGRATED CIRCUIT DESIGN

POWER MANAGEMENT TECHNIQUES FOR INTEGRATED CIRCUIT DESIGN

Ke-Horng Chen
National Chiao Tung University, Taiwan

WILEY

This edition first published 2016
© 2016 John Wiley & Sons Singapore Pte. Ltd.

Registered Office
John Wiley & Sons Singapore Pte. Ltd., 1 Fusionopolis Walk, #07-01 Solaris South Tower, Singapore 138628.

For details of our global editorial offices, for customer services and for information about how to apply for permission to reuse the copyright material in this book please see our website at www.wiley.com.

All Rights Reserved. No part of this publication may be reproduced, stored in a retrieval system or transmitted, in any form or by any means, electronic, mechanical, photocopying, recording, scanning, or otherwise, except as expressly permitted by law, without either the prior written permission of the Publisher, or authorization through payment of the appropriate photocopy fee to the Copyright Clearance Center. Requests for permission should be addressed to the Publisher, John Wiley & Sons Singapore Pte. Ltd., 1 Fusionopolis Walk, #07-01 Solaris South Tower, Singapore 138628, tel: 65-66438000, fax: 65-66438008, email: enquiry@wiley.com.

Wiley also publishes its books in a variety of electronic formats. Some content that appears in print may not be available in electronic books.

Designations used by companies to distinguish their products are often claimed as trademarks. All brand names and product names used in this book are trade names, service marks, trademarks or registered trademarks of their respective owners. The Publisher is not associated with any product or vendor mentioned in this book. This publication is designed to provide accurate and authoritative information in regard to the subject matter covered. It is sold on the understanding that the Publisher is not engaged in rendering professional services. If professional advice or other expert assistance is required, the services of a competent professional should be sought.

Limit of Liability/Disclaimer of Warranty: While the publisher and author have used their best efforts in preparing this book, they make no representations or warranties with respect to the accuracy or completeness of the contents of this book and specifically disclaim any implied warranties of merchantability or fitness for a particular purpose. It is sold on the understanding that the publisher is not engaged in rendering professional services and neither the publisher nor the author shall be liable for damages arising herefrom. If professional advice or other expert assistance is required, the services of a competent professional should be sought.

Library of Congress Cataloging-in-Publication Data

Names: Chen, Ke-Horng, author.
Title: Power management techniques for integrated circuit design / Ke-Horng Chen.
Description: Chichester, UK ; Hoboken, NJ : John Wiley & Sons, 2016. |
  Includes bibliographical references and index.
Identifiers: LCCN 2016002873 (print) | LCCN 2016007188 (ebook) | ISBN
  9781118896815 (cloth) | ISBN 9781118896822 (pdf) | ISBN 9781118896839 (epub)
Subjects: LCSH: Voltage regulators—Design and construction. | Power
  semiconductors–Design and construction. | Integrated circuits—Design and
  construction. | Electric power—Conservation.
Classification: LCC TK2851 .C5138 2016 (print) | LCC TK2851 (ebook) |
  DDC 621.39/5—dc23
LC record available at http://lccn.loc.gov/2016002873

Cover image: Photoslash/iStockphoto

Set in 10/12pt Times by SPi Global, Pondicherry, India

1   2016

To my respected parents, Li-Yun Wu and He-Nan Chen, and my wife, Hsin-Hua Pai

Contents

About the Author	xii
Preface	xiii
Acknowledgments	xv

1 Introduction — 1
  1.1 Moore's Law — 1
  1.2 Technology Process Impact: Power Management IC from 0.5 micro-meter to 28 nano-meter — 1
    1.2.1 *MOSFET Structure* — 1
    1.2.2 *Scaling Effects* — 7
    1.2.3 *Leakage Power Dissipation* — 9
  1.3 Challenge of Power Management IC in Advanced Technological Products — 14
    1.3.1 *Multi-V_{th} Technology* — 14
    1.3.2 *Performance Boosters* — 15
    1.3.3 *Layout-Dependent Proximity Effects* — 19
    1.3.4 *Impacts on Circuit Design* — 20
  1.4 Basic Definition Principles in Power Management Module — 22
    1.4.1 *Load Regulation* — 22
    1.4.2 *Transient Voltage Variations* — 23
    1.4.3 *Conduction Loss and Switching Loss* — 24
    1.4.4 *Power Conversion Efficiency* — 25
  References — 25

2 Design of Low Dropout (LDO) Regulators — 28
  2.1 Basic LDO Architecture — 29
    2.1.1 *Types of Pass Device* — 31
  2.2 Compensation Skills — 34
    2.2.1 *Pole Distribution* — 34

		2.2.2 Zero Distribution and Right-Half-Plane (RHP) Zero	40

- 2.3 Design Consideration for LDO Regulators — 42
  - 2.3.1 Dropout Voltage — 43
  - 2.3.2 Efficiency — 44
  - 2.3.3 Line/Load Regulation — 45
  - 2.3.4 Transient Output Voltage Variation Caused by Sudden Load Current Change — 46
- 2.4 Analog-LDO Regulators — 50
  - 2.4.1 Characteristics of Dominant-Pole Compensation — 50
  - 2.4.2 Characteristics of C-free Structure — 56
  - 2.4.3 Design of Low-Voltage C-free LDO Regulator — 62
  - 2.4.4 Alleviating Minimum Load Current Constraint through the Current Feedback Compensation (CFC) Technique in the Multi-stage C-free LDO Regulator — 66
  - 2.4.5 Multi-stage LDO Regulator with Feedforward Path and Dynamic Gain Adjustment (DGA) — 75
- 2.5 Design Guidelines for LDO Regulators — 79
  - 2.5.1 Simulation Tips and Analyses — 81
  - 2.5.2 Technique for Breaking the Loop in AC Analysis Simulation — 82
  - 2.5.3 Example of the Simulation Results of the LDO Regulator with Dominant-Pole Compensation — 85
- 2.6 Digital-LDO (D-LDO) Design — 93
  - 2.6.1 Basic D-LDO — 94
  - 2.6.2 D-LDO with Lattice Asynchronous Self-Timed Control — 96
  - 2.6.3 Dynamic Voltage Scaling (DVS) — 100
- 2.7 Switchable Digital/Analog-LDO (D/A-LDO) Regulator with Analog DVS Technique — 110
  - 2.7.1 ADVS Technique — 110
  - 2.7.2 Switchable D/A-LDO Regulator — 113
- References — 120

3 Design of Switching Power Regulators — 122
- 3.1 Basic Concept — 122
- 3.2 Overview of the Control Method and Operation Principle — 125
- 3.3 Small Signal Modeling and Compensation Techniques in SWR — 131
  - 3.3.1 Small Signal Modeling of Voltage-Mode SWR — 131
  - 3.3.2 Small Signal Modeling of the Closed-Loop Voltage-Mode SWR — 135
  - 3.3.3 Small Signal Modeling of Current-Mode SWR — 150
- References — 169

4 Ripple-Based Control Technique Part I — 170
- 4.1 Basic Topology of Ripple-Based Control — 171
  - 4.1.1 Hysteretic Control — 173
  - 4.1.2 On-Time Control — 176

		4.1.3	Off-Time Control	179

- 4.1.3 Off-Time Control — 179
- 4.1.4 Constant Frequency with Peak Voltage Control and Constant Frequency with Valley Voltage Control — 182
- 4.1.5 Summary of Topology of Ripple-Based Control — 183
- 4.2 Stability Criterion of On-Time Controlled Buck Converter — 185
  - 4.2.1 Derivation of the Stability Criterion — 185
  - 4.2.2 Selection of Output Capacitor — 197
- 4.3 Design Techniques When Using *MLCC* with a Small Value of R_{ESR} — 201
  - 4.3.1 Use of Additional Ramp Signal — 202
  - 4.3.2 Use of Additional Current Feedback Path — 204
  - 4.3.3 Comparison of On-Time Control with an Additional Current Feedback Path — 254
  - 4.3.4 Ripple-Reshaping Technique to Compensate a Small Value of R_{ESR} — 256
  - 4.3.5 Experimental Result of Ripple-Reshaped Function — 262
- References — 269

5 Ripple-Based Control Technique Part II — 270

- 5.1 Design Techniques for Enhancing Voltage Regulation Performance — 270
  - 5.1.1 Accuracy in DC Voltage Regulation — 270
  - 5.1.2 V^2 Structure for Ripple-Based Control — 271
  - 5.1.3 V^2 On-Time Control with an Additional Ramp or Current Feedback Path — 275
  - 5.1.4 Compensator for V^2 Structure with Small R_{ESR} — 277
  - 5.1.5 Ripple-Based Control with Quadratic Differential and Integration Technique if Small R_{ESR} is Used — 283
  - 5.1.6 Robust Ripple Regulator (R3) — 294
- 5.2 Analysis of Switching Frequency Variation to Reduce Electromagnetic Interference — 297
  - 5.2.1 Improvement of Noise Immunity of Feedback Signal — 298
  - 5.2.2 Bypassing Path to Filter the High-Frequency Noise of the Feedback Signal — 299
  - 5.2.3 Technique of PLL Modulator — 302
  - 5.2.4 Full Analysis of Frequency Variation under Different v_{IN}, v_{OUT}, and i_{Load} — 304
  - 5.2.5 Adaptive On-Time Controller for Pseudo-Constant f_{SW} — 313
- 5.3 Optimum On-Time Controller for Pseudo-Constant f_{SW} — 321
  - 5.3.1 Algorithm for Optimum On-Time Control — 322
  - 5.3.2 Type-I Optimum On-Time Controller with Equivalent V_{IN} and $V_{OUT,eq}$ — 323
  - 5.3.3 Type-II Optimum On-Time Controller with Equivalent V_{DUTY} — 331
  - 5.3.4 Frequency Clamper — 333
  - 5.3.5 Comparison of Different On-Time Controllers — 333
  - 5.3.6 Simulation Result of Optimum On-Time Controller — 335
  - 5.3.7 Experimental Result of Optimum On-Time Controller — 335
- References — 343

6 Single-Inductor Multiple-Output (SIMO) Converter — 345
- 6.1 Basic Topology of SIMO Converters — 345
  - 6.1.1 Architecture — 345
  - 6.1.2 Cross Regulation — 347
- 6.2 Applications of SIMO Converters — 348
  - 6.2.1 System-on-Chip — 348
  - 6.2.2 Portable Electronics Systems — 350
- 6.3 Design Guidelines of SIMO Converters — 351
  - 6.3.1 Energy Delivery Paths — 351
  - 6.3.2 Classifications of Control Methods — 359
  - 6.3.3 Design Goals — 363
- 6.4 SIMO Converter Techniques for Soc — 364
  - 6.4.1 Superposition Theorem in Inductor Current Control — 364
  - 6.4.2 Dual-Mode Energy Delivery Methodology — 366
  - 6.4.3 Energy-Mode Transition — 367
  - 6.4.4 Automatic Energy Bypass — 371
  - 6.4.5 Elimination of Transient Cross Regulation — 372
  - 6.4.6 Circuit Implementations — 376
  - 6.4.7 Experimental Results — 387
- 6.5 SIMO Converter Techniques for Tablets — 397
  - 6.5.1 Output Independent Gate Drive Control in SIMO Converter — 397
  - 6.5.2 CCM/GM Relative Skip Energy Control in SIMO Converter — 405
  - 6.5.3 Bidirectional Dynamic Slope Compensation in SIMO Converter — 415
  - 6.5.4 Circuit Implementations — 420
  - 6.5.5 Experimental Results — 427
- References — 441

7 Switching-Based Battery Charger — 443
- 7.1 Introduction — 443
  - 7.1.1 Pure Charge State — 447
  - 7.1.2 Direct Supply State — 448
  - 7.1.3 Plug Off State — 448
  - 7.1.4 CAS State — 448
- 7.2 Small Signal Analysis of Switching-Based Battery Charger — 449
- 7.3 Closed-Loop Equivalent Model — 454
- 7.4 Simulation with PSIM — 461
- 7.5 Turbo-boost Charger — 465
- 7.6 Influence of Built-In Resistance in the Charger System — 470
- 7.7 Design Example: Continuous Built-In Resistance Detection — 472
  - 7.7.1 CBIRD Operation — 473
  - 7.7.2 CBIRD Circuit Implementation — 476
  - 7.7.3 Experimental Results — 480
- References — 482

8 Energy-Harvesting Systems — 483
- 8.1 Introduction to Energy-Harvesting Systems — 483
- 8.2 Energy-Harvesting Sources — 486
  - *8.2.1 Vibration Electromagnetic Transducers* — 487
  - *8.2.2 Piezoelectric Generator* — 490
  - *8.2.3 Electrostatic Energy Generator* — 491
  - *8.2.4 Wind-Powered Energy Generator* — 492
  - *8.2.5 Thermoelectric Generator* — 494
  - *8.2.6 Solar Cells* — 496
  - *8.2.7 Magnetic Coil* — 498
  - *8.2.8 RF/Wireless* — 501
- 8.3 Energy-Harvesting Circuits — 502
  - *8.3.1 Basic Concept of Energy-Harvesting Circuits* — 502
  - *8.3.2 AC Source Energy-Harvesting Circuits* — 505
  - *8.3.3 DC-Source Energy-Harvesting Circuits* — 511
- 8.4 Maximum Power Point Tracking — 514
  - *8.4.1 Basic Concept of Maximum Power Point Tracking* — 514
  - *8.4.2 Impedance Matching* — 515
  - *8.4.3 Resistor Emulation* — 516
  - *8.4.4 MPPT Method* — 518
- References — 523

Index — 527

About the Author

Ke-Horng Chen received his B.S., M.S., and Ph.D. degrees in electrical engineering from the National Taiwan University, Taipei, Taiwan in 1994, 1996, and 2003, respectively.

From 1996 to 1998, he was a part-time IC Designer at Philips, Taipei, Taiwan. From 1998 to 2000, he was an Application Engineer at Avanti Ltd., Taiwan. From 2000 to 2003, he was a Project Manager at ACARD Ltd., where he was engaged in designing power management ICs. He is currently Director of the Institute of Electrical Control Engineering and a Professor with the Department of Electrical and Computer Engineering, National Chiao Tung University, Hsinchu, Taiwan, where he has organized a Mixed-Signal and Power Management IC Laboratory. He is the author or coauthor of more than 200 papers published in journals and conferences, and also holds several patents. His current research interests include power management ICs, mixed-signal circuit designs, and display algorithm and driver designs of liquid crystal display (LCD) TVs.

Dr. Chen has served as an Associate Editor of *IEEE Transactions on Power Electronics* and *IEEE Transactions on Circuits and Systems – Part II: Express Briefs*. He is also an Associate Editor of *IEEE Transactions on Circuits and Systems – Part I*. He joined the Editorial Board of *Analog Integrated Circuits and Signal Processing* in 2013. He is on the IEEE Circuits and Systems (CAS) VLSI Systems and Applications Technical Committee, and the IEEE CAS Power and Energy Circuits and Systems Technical Committee. He belongs to the Society for Information Display (SID) and International Display Manufacturing Conference (IDMC) Technical Program Sub-committees. He is Tutorial Co-Chair of IEEE Asia Pacific Conference on Circuits and Systems (APCCAS) 2012 and Track Chair of Integrated Power Electronics, IEEE International Conference on Power Electronics and Drive Systems (PEDS) 2013. He is Technical Program Co-Chair of IEEE International Future Energy Electronics Conference (IFEEC) 2013. He has served as CAS Taipei Section Chair since 2015. He is also Technical Program Committee Member, European Solid-State Circuits Conference (ESSCIRC) 2014–present.

Preface

Over the past three decades, power management technology has become more important as portable and wearable electronics have become part of our daily lives. It is important to realize the detailed design of power management circuits, including low dropout (LDO) regulators, switching power converters (SWRs), switched-capacitor designs among others, if battery usage lifetime and power-conversion efficiency need to be extended. Although some circuits can be found in analog or power electronics books, the reader cannot get an overall understanding of power management designs. Thus, I have written this book to collect useful material related to power management designs in recent years.

Power management IC designs use low-voltage (LV) and high-voltage (HV) devices. The specialty of this book is including LV and HV power management designs. Moreover, the objective of the book is to let the reader understand the process trend and demand of today's applications from the first. The mathematical analysis in the book is simplified, because in my opinion the reader needs to have the ability to understand the function of power management circuits. After that, the reader can analyze the whole power system and derive the complicated mathematical results. Thus, I have used many easy-to-understand figures in the book to let the reader realize why and how power management should be implemented. Although the reader can understand this via derived equations in some similar books, they can have the fun of thinking about and implementing their own designs if they study the circuits in this book by inspection rather than by equations. Moreover, digital and analog design techniques are introduced because a combination of digital and analog skills can give maximum performance of power management in system-on-chip (SoC) applications.

I have taught most of the material in this book both at the National Chiao Tung University, Hsinchu, Taiwan and in Taiwan industry. The order, the format, and the content are all carefully polished when I deliver the material to readers. It is a pity that much material is not included in this book. However, I encourage the reader to apply the concepts to similar power management designs. I have included some design guidelines in this book to let the reader realize the objective of each design.

Chapter 1 provides the reader with knowledge of LV and HV device characteristics and structure in different advanced technologies for learning the material in this book.

Chapter 2 describes the general design of an LDO regulator used in many power management circuits. Compensation skills are introduced to let the reader realize how to ensure power stability in case of any disturbance from input, output, and loading. A digital LDO regulator is also included for LV applications.

Chapter 3 includes the design guidelines of voltage-mode and current-mode switching power regulators. Compensation skills are also introduced to quantify the behavior of basic pulse-width-modulation (PWM) SWRs by inspection.

Chapter 4 introduces the ripple-based control technique for some applications that demand the features of fast transient response, low power consumption, and compact size solution. In particular, fast transient response is the trend for SWR designs to improve the performance of dynamic voltage/frequency scaling techniques and/or reference tracking techniques.

Chapter 5 shows some ripple-based control techniques to improve the performance of basic designs. Even if parasitic effects become large, the techniques presented here can still have excellent performance. Readers can train themselves by using the circuits in this book, proved for silicon, to implement useful power management circuits.

Chapter 6 shows state-of-the-art single-inductor multiple-output (SIMO) converters used in SoC to minimize the power module size. The power stage design and controller design are included in this chapter. We use the design concepts introduced in Chapters 2–5. The reader can obtain advanced training in power management designs here.

Chapter 7 shows the switching-based battery charger to complete the full function of power management in SoC designs. The basic stability proved by some behavior simulators can let the reader know how to model and increase the whole battery charger system.

Chapter 8 includes some energy-harvesting techniques to let the reader realize the possibility of obtaining energy from the environment. How to convert and how to improve efficiency are shown in this chapter.

Acknowledgments

This book has benefited from the recent research results of my Master and Ph.D. students. Many experts in both this research field and industry contributed much useful material to this book. Among them are Shen-Yu Peng (National Chiao Tung University, Hsinchu, Taiwan), Meng-Wei Chien (RealTek Corporation, Hsinchu, Taiwan), and Ying-Wei Chou (MediaTek Inc., Hsinchu, Taiwan). I say "thank you" to them.

I thank Yu-Huei Lee (RichTek Inc., Hsinchu, Taiwan), Yi-Ping Su (NovaTek Inc., Hsinchu, Taiwan), Wei-Chung Chen (MediaTek Inc., Hsinchu, Taiwan), Te-Fu Yang (Phison Electronics Corporation, Hsinchu, Taiwan), and Tzu-Chi Huang (MediaTek Inc., Hsinchu, Taiwan) for their contributions to this book.

My wife, Hsin-Hua, has made some contributions to this book. She encouraged me to complete the whole book using a range of useful circuits proved for silicon. She collected much useful material, including simulation and experimental results.

The book's production was made possible with the cooperation of staff at John Wiley. I thank James Murphy, Preethi Belkese, Maggie Zhang, Gunalan Lakshmipathy, Revathy Kaliyamoorthy, and Clarissa Lim. Without their help, there would be no book.

1

Introduction

1.1 Moore's Law

Over the past few decades, the number of transistors per square inch on integrated circuits (ICs) has doubled every 18 months, which is the forecast of Moore's law and is a continuing condition. However, a physical limitation appears when the transistor size shrinks to 28 nm. Several technology performance boosters, for example dual stress liner (DSL) technology, strained silicon techniques, and the stress memorization technique (SMT), are required to retain the performance of transistors. The industry has failed to keep to the trend predicted by Moore's law. Figure 1.1 depicts how the rate of transistor size scaling has slowed down and is likely to break Moore's law by the end of 2015.

1.2 Technology Process Impact: Power Management IC from 0.5 micro-meter to 28 nano-meter

1.2.1 MOSFET Structure

The voltage stress issue of metal–oxide–semiconductor field-effect transistors (MOSFETs) in drivers and power MOSFETs needs careful consideration. The evolution of MOSFETs and their applications are based on different input supply voltage (Figure 1.2). In advanced processes (i.e., 40, 28, and 22 nm), core MOSFETs with characteristics of small silicon size and high speed are used in low-voltage applications. Moreover, conventional low-voltage MOSFETs are applied for low supply voltage conditions in normal processes, such as 22 nm, 0.18 μm, 0.25 μm, and 0.5 μm. Nevertheless, the drain-to-source voltage, V_{DS} of low-voltage MOSFETs cannot tolerate a high voltage and punches, and will break the MOSFET when the input supply voltage increases. Therefore, double-diffused metal–oxide–semiconductors (DMOSs), vertical

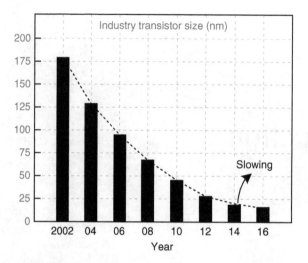

Figure 1.1 Transistor size scaling rate has slowed down

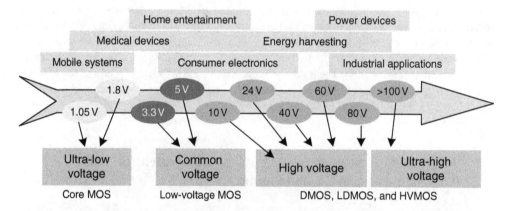

Figure 1.2 Evolution of MOSFETs and applications with different input supply voltages

double-diffused metal–oxide–semiconductors (VDMOSs), and laterally diffused metal–oxide–semiconductors (LDMOSs) are applied to bear a high V_{DS}. However, the gate-to-source voltage, V_{GS} of such MOSFETs cannot endure a high voltage, which will also damage the MOSFET. A high-voltage metal–oxide–semiconductor (HVMOS) solves the problem here, because its structure can tolerate a high voltage of both V_{DS} and V_{GS}.

The structures and characteristics of low-voltage MOSFETs, core MOSFETs, DMOSs, VDMOSs, LDMOSs, and HVMOSs are introduced in the following subsections, followed by a comparison of these MOSFETs.

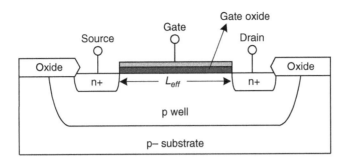

Figure 1.3 Structure of typical n-channel low-voltage MOSFET

1.2.1.1 Low-Voltage MOSFET

The structure of a typical n-channel low-voltage MOSFET is shown in Figure 1.3. Compared with LDMOSs and HVMOSs, the simple structure of a low-voltage MOSFET has the advantages of small silicon area and longest effective channel length (L_{eff}), which is defined as the contact area between the p well and the gate in the n-channel low-voltage MOSFET. Moreover, a thin-gate oxide is designed to achieve the high-speed on-and-off switching of the MOSFET. However, this thin-gate oxide cannot bear the high voltage stress of the V_{GS}. Moreover, the V_{DS} only operates in low-voltage stress conditions, because the drift region of drain is too small to tolerate a high voltage of V_{DS}.

1.2.1.2 Core MOSFET

The integrated technique of system-on-chip (Soc) has improved. A core MOSFET with small silicon size reduces the silicon area and increases the operating speed of the Soc [1, 2]. Moreover, the supply voltage evaluates to 1.8 V, 1.05 V, or lower voltages to reduce the system's power dissipation. Therefore, the voltage stress of a core MOSFET cannot bear a conventional supply voltage, such as 3.3 or 5 V, because the oxide layer of the core MOSFET is thinner than that of a low-voltage MOSFET. Conventional supply voltages damage the thinner oxide layer.

1.2.1.3 Double-Diffused MOSFET

Figure 1.4 shows a DMOS structure [3, 4]. The effective channel length is produced by p-type diffusion and gate oxide. Moreover, the n-type substrate is very lightly doped in this structure. Light doping provides enough space for expansion of the depleted region between the p-type diffusion and the n+ drain contact regions. Therefore, the breakdown voltage between drain and source is enlarged. This structure can endure a high voltage of V_{DS} but not a high voltage of V_{GS}, because of its thin gate oxide.

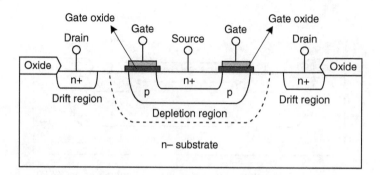

Figure 1.4 Structure of DMOS

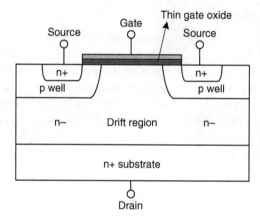

Figure 1.5 Structure of VDMOS

1.2.1.4 Vertical Double-Diffused MOSFET

The VDMOS structure combines the concepts of vertical power structures and lateral double diffusion (Figure 1.5) [5]. The drain voltage is vertically supported by the n− layer. Moreover, current flows laterally from the source through the channel, which is parallel to the silicon surface, and then turns at a right angle to flow vertically down through the n− drain layer to the n+ substrate and the drain contact. An effective channel is formed, if a sufficiently positive gate voltage is applied, and the extra drift region of the n− layer can tolerate a high voltage of V_{DS}. However, the thin gate oxide cannot bear a high voltage of V_{GS}.

1.2.1.5 Laterally Diffused MOSFET

LDMOS is also applied to solve the problem of high voltage V_{DS}. The structure of a typical n-channel LDMOS is similar to that of a low-voltage MOSFET, as shown in Figure 1.6 [6, 7]. The difference is that the LDMOS extends the drain drift region by adding an n-well

Introduction

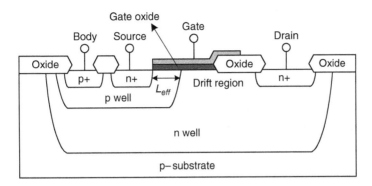

Figure 1.6 Structure of typical n-channel LDMOS

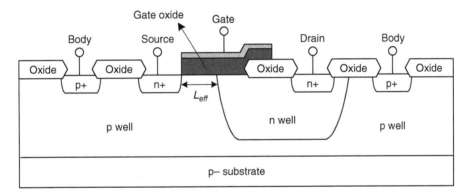

Figure 1.7 Structure of n-channel asymmetric HVMOS

layer to achieve the high-voltage stress tolerance of V_{DS}. Although this structure solves the high-voltage problem of V_{DS}, it has several disadvantages. The effective channel length defined by the p well underneath the gate is needed because of the drift extension for the wide drift region. Therefore, the structure causes a significant gate-to-drain overlap region and extends the silicon area, which is proportional to the cost. Moreover, the LDMOS is non-symmetric in the drain and source regions because the extended drift region and the overlap region are only designed on the drain side. However, the thickness of the LDMOS gate oxide is similar to that of a low-voltage MOSFET and cannot tolerate the high voltage of V_{GS}.

1.2.1.6 Asymmetric High-Voltage MOSFET

Asymmetric HVMOSs are applied for high-voltage processes to solve the high-voltage issue with both V_{GS} and V_{DS}. The structure of an n-channel asymmetric HVMOS is illustrated in Figure 1.7 [8, 9]. Similarly, an extra n well is used to stretch the drift region of the drain to bear the high voltage of V_{DS}. Therefore, a reduced effective channel length and gate-to-drain overlay region also occur in the HVMOS structure. Moreover, the thickness of the asymmetric HVMOS gate oxide is increased to extend the high-voltage tolerance of V_{GS}. However, the thicker gate oxide causes decreased on-and-off switching speed of the asymmetric HVMOS and results in longer system delay times. Furthermore, the silicon area of the asymmetric HVMOS is larger than that of the low-voltage MOSFET, DMOS, and LDMOS because of the extended drift region of the drain and the thicker gate oxide. This asymmetric HVMOS is also non-symmetric in the drain and source regions because only the drain has extended drift region. Consequently, the asymmetric HVMOS has many disadvantages, which decrease system efficiency and increase cost. However, the asymmetric HVMOS is necessary and the aforementioned disadvantages cannot be avoided when the system is in high V_{DS} and high V_{GS} conditions. Unfortunately, asymmetric HVMOSs cannot be used in cascade schemes because asymmetric HVMOSs cannot endure the high voltage of V_{SB}.

1.2.1.7 Symmetric High-Voltage MOSFET

In a symmetric high-voltage MOSFET, the source region does not have an extra n well for the drift region compared with the drain region as shown in Figure 1.7. That is, the source-to-body region cannot tolerate the high-voltage condition. Thus, the symmetric HVMOS structure is preferred to extend the voltage stress tolerance of V_{SB}, as shown in Figure 1.8 [8, 9], and overcome this problem. The difference with this structure is that the source region of the symmetric HVMOS copies the same method as the drain region to solve the high-voltage issue of V_{SB}. Both the drain and source regions have an extra n well to endure the high voltage of V_{DS} and V_{SB}. Moreover, the gate oxide in the symmetric HVMOS is as thick as that in the asymmetric HVMOS for the high voltage of V_{GS}. Although the symmetric HVMOS can tolerate all high voltages of V_{DS}, V_{GS}, and V_{SB}, its structure needs a large silicon area, which increases the cost.

1.2.1.8 Comparison

The advantages and disadvantages of low-voltage MOSFETs, core MOSFETs, DMOSs, LDMOSs, and HVMOSs are trade-offs in any design. Figure 1.9 shows different types of

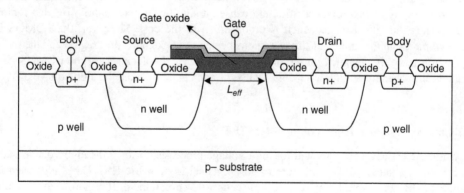

Figure 1.8 Structure of n-channel symmetric HVMOS

Introduction

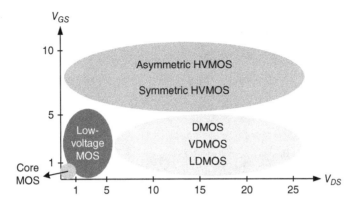

Figure 1.9 Different types of MOSFETs with different values of V_{GS} and V_{DS}

Table 1.1 Comparison of different MOSFETs

	Silicon area	High-voltage V_{DS}	High-voltage V_{GS}	High-voltage V_{SB}
Low-voltage MOSFET	Small	Damage	Damage	Damage
Double DMOSFET	Medium	Safe	Damage	Damage
Vertical DMOSFET	Medium	Safe	Damage	Damage
Lateral DMOSFET	Medium	Safe	Damage	Damage
Asymmetric HVMOSFET	Large	Safe	Safe	Damage
Symmetric HVMOSFET	Large	Safe	Safe	Safe

MOSFETs with different values of V_{GS} and V_{DS}. Choosing a suitable device under different supply voltage conditions from the different MOSFETs listed in Table 1.1 can achieve high efficiency and/or minimize the cost of the silicon area.

1.2.2 Scaling Effects

Given that complementary metal–oxide–semiconductor (CMOS) technology processes are downscaling, more transistors can be fabricated to reduce the cost of ICs. Electrical oxide thickness (EOT), parasitic capacitance, and nominal supply voltage also decrease with each new technology. Therefore, transistor operating speeds are increased and active power consumption is lowered. However, the threshold voltage (V_{th}) cannot be scaled down simultaneously with the scaling down of the MOSFET gate length because the leakage current would be extremely large. Static power is increasing enormously, and cannot be neglected because the CMOS technology is continuously scaling down as shown in Figure 1.10. Furthermore, several scaling effects appear in short-channel devices, such as velocity saturation, hot carrier injection, large device mismatch, threshold voltage variations, and increased leakage current. In addition, the reduction of voltage rails, decreased dynamic range, and increased power density are challenges for designers – especially in analog circuit designs [10].

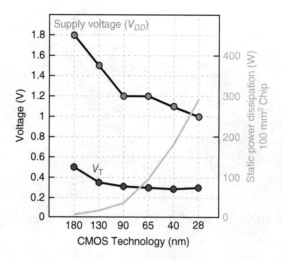

Figure 1.10 Voltage decreasing and static power increasing vs. CMOS technology scaling down

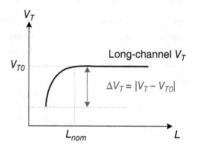

Figure 1.11 V_{th} decreases with channel length; V_{th} roll-off effect

The distance from the source to the drain in short-channel devices is reduced, and thus the depletion regions from the drain and the source are closer. Therefore, as depicted in Figure 1.11, V_{th} decreases exponentially if the channel is shorter than a certain gate length (L_{nom}), and the leakage current becomes unacceptable. This condition is known as the V_{th} roll-off effect, which circuit designers want to avoid. A first-order model describes that the effect of V_{th} roll-off is a function of EOT and substrate doping [2]. Therefore, increasing the substrate doping density to narrow the source/drain depletion region and scaling the EOT can minimize ΔV_{th} in short-channel devices.

Both the vertical electric field (V_{GS}) and the horizontal electric field (V_{DS}) can change the channel potential in short-channel devices. Figure 1.12(a) shows that the potential energy of a long-channel device is independent of the drain voltage. The potential barrier of a long-channel device is only controlled by its gate voltage. As Figure 1.12(b) shows, the potential energy of a short-channel device is also affected by its drain voltage. Both gate/source and drain/source voltages can affect the drain current. Given that the x–y electric fields are coupled, the transistor can conduct electrons even if the gate voltage is smaller than V_{th}. Therefore, V_{th}

Introduction

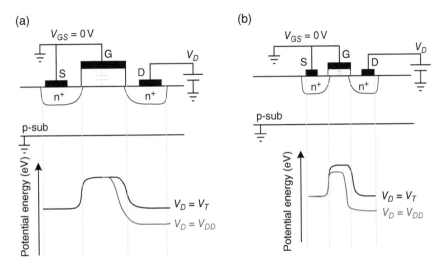

Figure 1.12 (a) Potential energy of long-channel device. (b) Potential energy of short-channel device

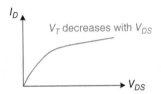

Figure 1.13 Drain current increases with V_{DS} in DIBL effect

decreases as the drain bias voltage increases. This phenomenon is known as the drain-induced barrier lowering (DIBL) effect. It is defined by Eq. (1.1) and shown in Figure 1.13:

$$DIBL \equiv \frac{|\Delta V_{th}|}{|\Delta V_{DS}|} \tag{1.1}$$

1.2.3 Leakage Power Dissipation

Deep sub-micrometer technology integrates high density and diversity of circuit functions on the same chip. This integration has changed the concept of power consumption compared with that in conventional long-channel transistors. The power dissipation trend shows that leakage power accounts for almost one-third of total power consumption among processes below 130 nm, as shown in Figure 1.14. With the continuous scaling down of technology, the supply voltage decreases to meet the requirements of power density and system reliability, but the leakage current increases exponentially.

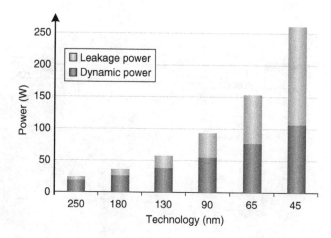

Figure 1.14 Trend of dynamic and leakage power dissipation [11]

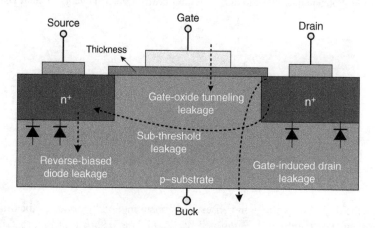

Figure 1.15 Leakage current path in short-channel devices

A major concern in scaling channel size effects is that the gradual increase of leakage current leads to large static power [12, 13]. Static power needs to be carefully considered, because its value affects operation and performance. Figure 1.15 shows the leakage current in short-channel devices. Given the threshold voltage V_{th} scaling the non-ideal off-state characteristics of a transistor, sub-threshold current is constantly drawn from the power supply to the ground, even when a transistor operates in the cut-off region. Furthermore, a thinner gate-oxide structure is necessary to control the short-channel effect (SCE) and maintain the transistor driving strength under low supply voltage. The structure also increases the gate leakage current. Thus, the primary leakage currents are sub-threshold and gate-oxide tunneling leakage currents in deep sub-micrometer CMOS circuits. The leakage current includes the gate direct-tunneling leakage current (I_G), the gate-induced drain leakage current (I_{GIDL}), the reverse-biased junction leakage current (I_{REV}), and the sub-threshold or weak-inversion leakage current (I_{SUB}). I_{GIDL} is

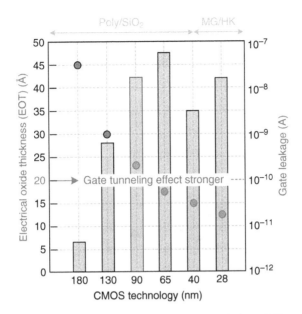

Figure 1.16 Gate leakage increases with CMOS technology trends from 0.18 μm to 65 nm when the oxide thickness scales down

caused by a large horizontal electric field in the overlapping region between drain and gate. It occurs from drain to substrate and increases with higher drain voltage. I_{REV} occurs because p–n junctions are heavily doped in short-channel devices. It is also known as band-to-band tunneling and increases as the transistor feature size decreases continuously [14].

The leakage current of an off-state transistor (I_{OFF}) indicates $V_{GS} = 0$ V and $I_G = 0$. Thus, its value is equal to the summation of $I_{GIDL} + I_{REV} + I_{SUB}$. Figure 1.16 shows that the gate leakage increases with CMOS technology trends from 0.18 μm to 65 nm because the oxide thickness continuously scales down. However, the leakage current decreases significantly for a 40 nm CMOS process, because of several performance boosters. However, the current's value increases continuously if the channel length scales down constantly from 40 to 28 nm. High-k/metal gate (HK/MG) technology was introduced in 2006; they use high-k dielectric materials, such as TiO_2 and Ta_2O_5, and reduce the gate-oxide leakage current. The reason for this is that a high-k dielectric allows a less aggressive reduction in gate dielectric thickness and obtains the required gate overdrive even under low supply voltage operation in deep sub-micron CMOS technologies. The HK/MG is an effective technique to reduce leakage currents and has excellent gate control compared with conventional silicon-dioxide gate insulators. Figure 1.17(b) shows that the quantum tunneling effect of electrons/holes occurs and causes I_G to become larger than that in Figure 1.17(a) without quantum tunneling effect, when the EOT is smaller than 20Å. The quantum tunneling effect is a major source of leakage current in deep sub-micron CMOS technologies. The leakage current is still a serious factor, which cannot be ignored in deep sub-micron CMOS technologies, although the introduction of HK/MG technology can reduce I_G.

As depicted in Figure 1.18, the gate-oxide tunneling leakage current is composed of leakage from the gate to the channel, including gate-to-source leakage current (I_{GCS}) and gate-to-drain

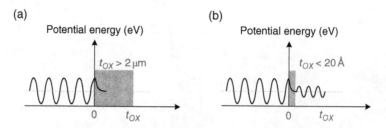

Figure 1.17 (a) Tunneling effect in long-channel devices. (b) Tunneling effect in short-channel devices

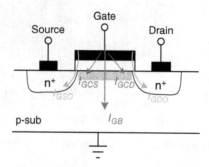

Figure 1.18 Composition of gate-oxide leakage current

leakage current (I_{GCD}), leakage from gate oxide to source (I_{GSO}) or drain overlapping regions (I_{GDO}), and leakage from gate to substrate (I_{GB}) [15]. Aggressive scaling down of the gate-oxide thickness increases the electric field oxide across the gate insulator and results in electron tunneling from gate to substrate or vice versa. The resulting current is called the gate-oxide tunneling current, which is the major leakage current in nanometer CMOS technology. Two mechanisms are responsible for this phenomenon. The first is called the Fowler–Nordheim (FN) tunneling mechanism, in which electrons tunnel into the conduction band of the oxide layer. Direct tunneling is more dominant than the FN tunneling mechanism. Here, the electrons tunnel directly to the gate through the forbidden energy gap of the silicon-dioxide layer, which appears when the gap is less than 3–4 nm thick. The resulting current is called the gate direct-tunneling leakage current, and it flows from the gate through the oxide insulation to the substrate or vice versa. Mechanisms for direct tunneling include electron tunneling in the conduction band, electron tunneling in the valence band, and hole tunneling in the valence band. The dominant source of leakage is the direct tunneling of electrons through the gate oxide [16]. This leakage current depends exponentially on the oxide thickness and the supply voltage, as expressed in Eq. (1.2), where: E_{ox} is the electric field across the oxide; W and L are the effective transistor width and length, respectively; $A = q^3/16\pi^2 h\Phi_{ox}$, $B = 4\pi\Phi_{ox}^{1.2}(2m_{ox})^{0.5}/3hq$. m_{ox} is the effective mass of the tunneling particle, Φ_{ox} is the tunneling barrier height, h is $\pi/2$ times Planck's constant, and q is the electron charge. Moreover, the tunneling current exists in both the on-state and the off-state of the MOSFET:

$$I_{gox} = W \cdot L \cdot A \cdot E_{ox}^2 \cdot e^{-B/E_{ox}} \tag{1.2}$$

Introduction

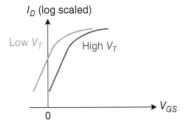

Figure 1.19 I_D vs. V_{GS} under different V_{th} shows the trade-off between operation speed and leakage current for circuit designers

The dominant leakage current among three parts of I_{OFF} is I_{SUB}, resulting in drain/source current for a MOSFET in weak inversion. Sub-threshold leakage is the drain/source current of a transistor during operation in weak inversion, where transistors continue to turn on despite the gate-source voltage being below the threshold voltage. Sub-threshold conduction occurs, unlike for a strong inversion region, in which the drift current dominates because of the diffusion current of the minority carriers in the channel of a MOSFET device. This condition occurs for several reasons. First is the weak inversion effect. Carriers move by diffusion along the surface, similar to when the gate voltage is below V_{th}, allowing charge transport across the base of the bipolar transistors. The weak inversion current becomes significant when the gate-to-source voltage is lower than, but close to, the threshold voltage of the device. Second is the DIBL effect, which is the reduction of the threshold voltage of the transistor at higher drain voltages. I_{SUB} increases exponentially as V_{th} decreases and/or as T increases, Eq. (1.3). η is a factor depending on the channel length, and is between 1 and 2; q is the charge on an electron; k is Boltzmann's constant; T is temperature; W and L are the width and length of the MOSFET, respectively [2]:

$$I_{DS}(\text{nA}) = 100 \cdot \frac{W}{L} \cdot e^{q(V_{GS}-V_{th})/\eta kT} \tag{1.3}$$

I_{SUB} is strongly sensitive to process, supply voltage, and temperature variation. Circuit designers aim for a low V_{th} to achieve a fast transient response and a high drain/source current in strong inversion, as depicted in Figure 1.19. In contrast, a high V_{th} is required for a low leakage current off-state. A trade-off occurs between operation speed and leakage current, which explains why V_{th} cannot be reduced as the gate length continues to scale down.

Parasitic diodes exist between the source/drain diffusion region and the substrate in one transistor, as depicted in Figure 1.15. These parasitic-diode p–n junctions must be reverse-biased for proper transistor operation. This condition results from minority carrier diffusion and drift near the edge of depletion regions, and also from the generation of electron–hole pairs in the depletion regions of reverse-bias junctions. This results in the leading transistor consuming power in the form of a reverse-bias current I_{REV}, which is drawn from the power supply. The current magnitude depends on the area of source/drain diffusion and the current density, which in turn is determined by the doping concentration. Highly doped shallow junctions and halo doping, necessary to control SCEs in nanometer devices, have escalated the leakage

Table 1.2 Transistor leakage in short-channel devices

Leakage current	Influence	Existence	Solutions
I_G	Large	On-state and off-state	HK/MG
I_{SUB}	Large	Off-state	Higher V_{th}
I_{REV}	Very small	On-state and off-state	Dopant profile optimization
I_{GIDL}	Small	Off-state	Dopant profile optimization

current. Thus, the electron tunneling effect across the p–n junction causes junction leakage. The currents can be expressed as in Eqs. (1.4) and (1.5) [17]:

$$I_{BD} = A \cdot J_s \cdot \left(e^{q \cdot V_{BD}/kT} - 1\right) \tag{1.4}$$

$$I_{BS} = A \cdot J_s \cdot \left(e^{q \cdot V_{BS}/kT} - 1\right) \tag{1.5}$$

where A is the area of the junction, J_s is the reverse saturation current density, V_{BD} and V_{BS} are the reverse bias voltage across the junction, and kT/q is the thermal voltage. Reverse-biased diode leakage further becomes crucial if the source/drain region continues to be heavily doped.

I_{GIDL} is caused by a high field effect in the drain junction of MOS transistors. The silicon surface in an NMOS transistor has almost the same potential as the p-type substrate, and acts like a p region because of the heavier doping than the substrate when the gate is biased to form an accumulation layer in the silicon surface under the gate. A dramatic increase in some effects, such as avalanche multiplication and band-to-band tunneling, can be observed when the gate is at zero or negative voltage, and the drain is at the supply voltage level. Minority carriers underneath the gate are swept to the substrate to form the I_{GIDL} path, as shown in Eq. (1.6) [18]. A higher supply voltage and a thinner oxide lead to an increased I_{GIDL}, as shown in Eq. (1.6), where A is a pre-exponential parameter, B (typically 23–70 MV/cm) is a physically based exponential parameter, and E_s is the transverse electric field at the surface depending on operating voltage and gate-oxide thickness:

$$I_{GIDL} = A \cdot E_s \cdot e^{-B/E_s} \tag{1.6}$$

The steep doping profile that occurs at the drain edge increases the band-to-band tunneling currents, especially as the drain/bulk voltage increases. Therefore, a thinner oxide and a higher supply voltage increase I_{GIDL} drastically. Controlling the doping concentration in the drain of the transistor is the best way to control I_{GIDL}. Table 1.2 summarizes the contribution of leakage currents in short-channel devices. Corresponding solutions are also listed.

1.3 Challenge of Power Management IC in Advanced Technological Products

1.3.1 Multi-V_{th} Technology

The power density increases continuously with the scaling down of the gate length. Thus, multiple-threshold voltage (multi-V_{th}) ICs have been widely used to optimize power and maintain

Introduction

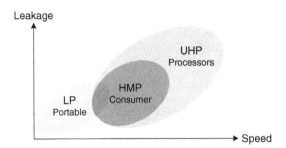

Figure 1.20 Trade-off between operation speed and leakage current

Table 1.3 Performance optimization strategies

Requirement	Chosen devices	Design methodology
Ultra-high performance	U-LVT/LVT/RVT	1. Simulated with U-LVT and LVT 2. Replaced by RVT on non-critical path
High performance medium power	LVT/RVT	1. Simulated with LVT 2. Replaced by RVT on non-critical path
Low power	HVT/RVT	1. Simulated with HVT 2. Replaced by RVT on large-timing path

operation speed in various industries. Long-channel devices in previous CMOS technologies utilized a V_{th} implant to dominate the threshold voltage. Today, multi-threshold devices in deep sub-micron CMOS technologies are fabricated with different channel length and halo implantation optimization. The area cost and layout blueprint are similar to those for high-threshold voltage (HVT) and low-threshold voltage (LVT) devices. This convenience enables circuit designers to implement multi-V_{th} ICs. As illustrated in Figure 1.20, a trade-off exists between operating speed and leakage current among different products. For example, in ultra-high-performance applications, processors in mobile phones are dominated by ultra-low-threshold voltage (U-LVT) devices with selective LVT devices or regular-threshold voltage (RVT) devices on non-critical paths. Table 1.3 summarizes the choice of several devices with corresponding requirements and design methodologies.

1.3.2 Performance Boosters

In Figure 1.21, the transistor on-state current density (I_{ON}) continues to increase because strained-silicon technologies are introduced around a 90 nm node. Continuing I_{ON} improvement is related to a growing range of performance boosters, as listed in Table 1.4. Some techniques, such as the contact-etch-stop-liner (CESL) technique, SMT, and embedded silicon-germanium (eSiGe) technologies for improving the PMOS transistor, are used to enhance 90 and 65 nm processes. The CESL technique provides tension and compression to NMOS and PMOS transistors, respectively, to significantly improve mobility [19]. For CMOS

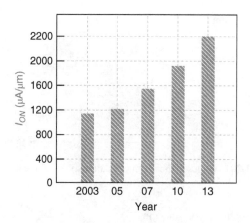

Figure 1.21 Trends of I_{ON} in NMOS

Table 1.4 Performance boosters vs. technology node

Performance boosters	90 nm	65 nm	40 nm	28 nm
CESL	✓	✓	✗	✗
SMT	✗	✗	✓	✓
eSiGe for PMOS	✓	✓	✓	✓
SiC for NMOS	✗	✗	✓	✓
DSL	✗	✗	✓	✓
HK/MG technology	✗	✗	✓	✓

technologies below 65 nm, SMT is introduced and widely used to boost the performance of NMOS transistors. SMT can result in longitudinal tensile stress, which is good for NMOS mobility [20]. A brief description of SMT formation is provided in Figure 1.22(a)–(c) [21]. Tensile nitride is formed after the S/D implantation as SiN deposition. SMT formation can be achieved after the removal of SiN. Moreover, Figure 1.23 shows trends of NMOS I_{ON} with SMT. However, SMT cannot be used in metal gates, which is an important trend in relation to the technology below the 40 nm node.

In further advanced technologies, such as 40 and 28 nm, SiC for improving the NMOS transistor, the DSL technique, and the HK/MG technique are employed to improve device performance. Moreover, Figure 1.24(a), (b) shows the dimension in 2D and the cross-section of the device, respectively. PMOS and NMOS have each aimed for a uniaxial stress direction to improve the performance, as shown in Table 1.5.

Moreover, an embedded source/drain stressor for the PMOS transistor in a 90 nm process is introduced and known as eSiGe S/D technology, as shown in Figure 1.25(a). This technology provides longitudinal compressive stress for PMOS mobility but has no influence on NMOS. However, it has drawbacks, such as a larger junction leakage [23]. Thus, a Si_3N_4 cap layer is utilized in the 40 nm process to enhance NMOS mobility. The Si_3N_4 cap layer can provide longitudinal tensile stress. Figure 1.25(b) shows the NMOS transistor with its corresponding stressor. The DSL technology in Figure 1.26 is used to boost the performance of a different type

Introduction

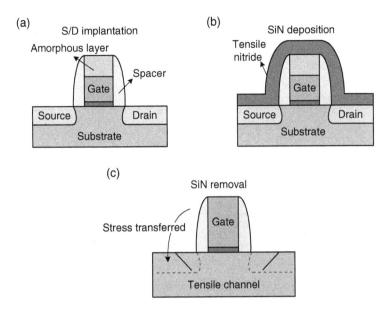

Figure 1.22 Brief description of SMT formation: (a) S/D implantation; (b) SiN deposition; (c) SiN removal

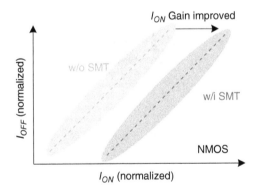

Figure 1.23 Trends of NMOS I_{ON} with SMT

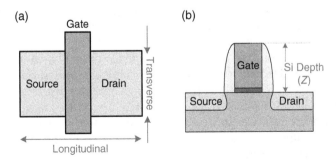

Figure 1.24 Direction of stress to transistor: (a) longitudinal and transverse; (b) Z-direction [22]

Table 1.5 Desired stress in CMOS process [22]

Direction	NMOS	PMOS
Longitudinal	Tensile	Compressive
Transverse	Tensile	Tensile
Si-depth (Z)	Compressive	Tensile

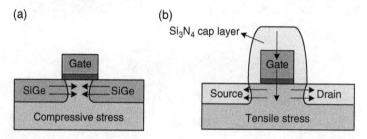

Figure 1.25 MOSFET with each corresponding stressor: (a) compressive stress for PMOS transistor; (b) tensile stress for NMOS transistor

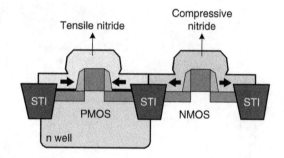

Figure 1.26 DSL technology for both NMOS and PMOS transistors

of MOSFET, because this DSL technology applies a SiN film to create tensile stress on the NMOS transistor and compressive stress on the PMOS transistor.

Gate leakage has become a major problem when the EOT is scaled below 20Å. Moreover, dopant penetration and dielectric degradation occur [24]. For a continuously scaled down EOT, the integration of HK/MG technology in a 40/28 nm node significantly improves the performance in terms of gate leakage reduction. HK/MG technology utilizes a high-k dielectric to increase the gate-oxide capacitance (C_{OX}), although it has already been increased by scaling down the EOT. A higher on-state driving current with increased C_{OX} ensures improved performance. Thus, HK/MG technology is widely used in ultra-high-performance products. There are two main integration methods for HK/MG technology: gate-first metal-inserted-poly-Si gate (MIPS) and gate-last replaced metal gate (RMG). The main difference between these two methods is that the metal electrode is deposited either before or after the high-temperature activation anneals during fabrication. Both methods are under continuing development. A comparison of the two methods is presented in Table 1.6 [25–30].

Introduction

Table 1.6 Comparison table for two HK/MG approaches [25–30]

	MIPS	RMG
High-k dielectric	First	First or last
Metal gate	First	Last
Thermal budget	High	Low
EOT	Thick	Thin
Mobility	Low	High
Process complexity	Low	High
Cost	Low	High

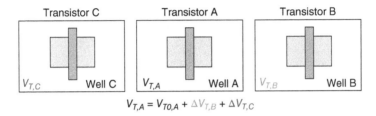

Figure 1.27 Explanation of layout-dependent proximity effect

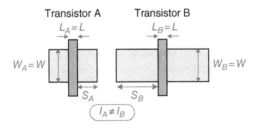

Figure 1.28 Explanation of LOD effect

1.3.3 Layout-Dependent Proximity Effects

Given the nanometer CMOS technology, a growing number of transistors have been fabricated on the same well. Transistors are very close to each other and induce several layout-dependent proximity effects. These effects are unintentional and contrary to the intentional stress effect introduced in Section 1.2.2, because transistors interact with each other. Figure 1.27 shows a simple example in which the threshold voltage can be changed by neighboring transistors. Proximity effects can degrade the on-state current and shift the threshold voltage.

The length of diffusion (LOD) effect is caused by mechanical compressive stress induced by shallow trench isolation (STI), which is used to isolate devices and produce compressive stress as the wafer cools down [31]. The LOD effect with different diffusion length can cause a current difference even if transistors have the same length and width, as shown in Figure 1.28. However, the well proximity effect (WPE), as shown in Figure 1.29, is induced because

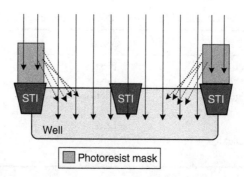

Figure 1.29 Explanation of WPE effect

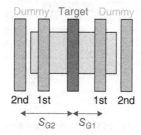

Figure 1.30 Explanation of PSE effect

high-energy dopant ions scatter into the well edge from the edge of the photoresist mask [32]. Therefore, transistors exhibit a difference in threshold voltage V_{th} because different distances form a well edge. $V_{th,A}$ is affected by the variations of $V_{th,B}$ and $V_{th,C}$. A polyspace effect (PSE) is caused by the first or second neighboring polylines, as shown in Figure 1.30. Transistors with minimum poly-to-poly spacing have a large variation in V_{th} and on-state current. Designers today have to focus on minimizing layout-dependent proximity effects during layout and post-simulation to ensure chip quality in nanometer technologies.

1.3.4 Impacts on Circuit Design

Nanometer devices cause impacts on physical limitations and IC design, especially analog IC design. As mentioned in Section 1.2.1, the major problem is gate tunneling leakage as transistors are continuously scaling down. The gate-leakage tunneling current can introduce a low-frequency pole (f_{gate}) and degrade system stability, transient response time, and voltage variation.

The degradation of the loop filter of a phase-lock-loop (PLL) can enlarge the frequency variation [33]. A test circuit diagram of gate tunneling current is shown in Figure 1.31(a), and its small signal model is shown in Figure 1.31(b), where the gate-leakage tunneling current (i_C) is through C_{OX} and depends on the input signal frequency. The total gate leakage is composed of i_{leak} and i_C, where i_{leak} is the gate leakage that is not caused by the tunneling effect. Therefore, two current components have the same magnitude at one certain frequency (f_{gate}). f_{gate} can be

Introduction

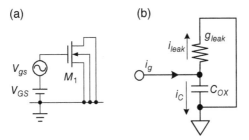

Figure 1.31 Test diagrams of f_{gate} because of gate leakage current: (a) circuit diagram; (b) small signal equivalent model

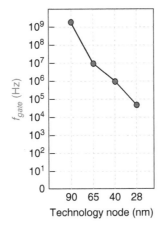

Figure 1.32 Trends of f_{gate} vs. technology node

estimated as expressed in Eq. (1.7) [34]. Figure 1.32 explains that f_{gate} is lowered because of the enlarged gate-tunneling leakage current and has to be considered because the technology node scales down continuously. Moreover, Figure 1.33 shows that the gate-tunneling leakage also enlarges the mismatch among current mirrored transistors:

$$f_{gate} = \frac{1}{2\pi C_{IN}(g_{Leakage})} \quad (1.7)$$

Another major change is that the characteristics of transistor symmetry break down in deep sub-micron technologies. For example, all poly gates of core devices in a 28 nm process must be in the vertical direction, and a single transistor can be influenced by other transistors several micrometers away. A simple description of a broken-down symmetry is shown in Figure 1.34.

Analog designers widely utilize certain digital circuits or circuit techniques to minimize device mismatch and solve the mismatch problems that accompany process scaling. For example, digital circuits, such as trimming logic and auto-calibration circuits, are widely inserted into critical analog systems or feedback loops. The requirements of a digital

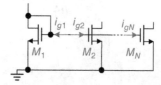

Figure 1.33 Current mirror mismatch because of gate current leakage

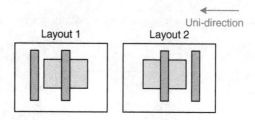

Figure 1.34 Symmetry breaking down in deep sub-micron technology

temperature sensor are increased, because temperature variation is a significant factor in scaling effects. Moreover, a circuit technique, such as stack devices, is proposed to minimize the leakage current. Therefore, the trends of scaling devices continue to challenge circuit designers in complex ways.

1.4 Basic Definition Principles in Power Management Module

Some commonly used design specifications are defined to evaluate the performance of power management modules. These design specifications include both transient and steady-state performance.

1.4.1 Load Regulation

Load regulation, as defined in Eq. (1.8), is one of the steady-state performances that influence output voltage accuracy. Different load conditions indicate the voltage regulation ability of the power management module. A smaller value of the load regulation ensures higher precision of the output voltage. Load regulation is related to the open-loop gain of the converter. Therefore, higher open-loop gain is required to ensure better regulation, but the system stability worsens. Compensation techniques need a relationship between input variation and output voltage regulation. A small value of the line regulation indicates better immunity in case of input voltage variation. Consequently, good line regulation performance leads to robust operation against the input supply variation:

$$Line\ regulation = \frac{\Delta v_{OUT}}{\Delta v_{IN}}(V/V) \tag{1.8}$$

Introduction

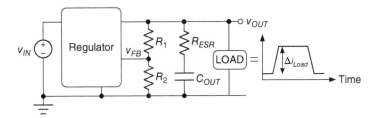

Figure 1.35 Illustration of time domain analysis for load transient response

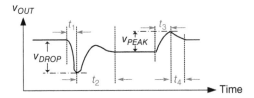

Figure 1.36 Output voltage waveform of load transient response in regulator

1.4.2 Transient Voltage Variations

Distinct operating modes in Soc applications cause different transient responses. The power management module needs to handle a versatile load transient response and ensure high performance of the Soc. A good transient response implies a small voltage transient variation and a fast settling time at the output voltage when a sudden load change occurs. System bandwidth determines the performance of the transient response. That is, a wide system bandwidth indicates a short response time contributed by a fast control loop in the power management module. Frequency domain and time domain analyses are performed to further achieve transient response.

In the frequency domain analysis of a power management module, the position of the poles and zeros leads to distinct responses when the transient load occurs. A large system bandwidth and a fast response time would be obtained if the system dominant pole is set at a higher frequency region; however, stability is difficult to guarantee. A slow transient response time is derived once the dominant pole is set at low frequencies; nevertheless, stable operation can be ensured easily. To correctly refer to the frequency domain analysis, an illustration of time domain analysis for the load transient response is shown in Figure 1.35. The output node usually contains a feedback voltage divider (R_1 and R_2), an output capacitor (C_{OUT}), and its equivalent series resistance (R_{ESR}). The output voltage waveform of the load transient response is shown in Figure 1.36. When the load transient response occurs in case of a light-to-heavy load change, the transient period t_1 and the drop voltage v_{DROP} are determined by the system bandwidth and the output capacitor C_{OUT} with its equivalent series resistance R_{ESR}. The output capacitor functions as a current source to sustain the request for the output load current. The voltage drop v_{DROP} in the load transient response can be formulated as:

$$v_{DROP} = v_{ESR} + v_{cap} = \Delta i_{OUT} R_{ESR} + \frac{\Delta i_{OUT} \cdot t_1}{C_{OUT}} \tag{1.9}$$

The transient period t_2 is determined by the time during which the high-side power switch can charge the output capacitor back to its regulated voltage. The phase margin of the control loop and any pole/zero doublets affect the transient settling time. The total period, containing t_1 and t_2, is known as the transient recovery time. The voltage variation v_{PEAK} occurs when the output load suddenly changes from heavy to light load. The determination of the transient periods, t_3 and t_4, is similar to that of the transient periods, t_1 and t_2, respectively. The voltage v_{PEAK} in the load transient response can also be formulated as:

$$v_{PEAK} = v_{ESR} + v_{cap} = \Delta i_{OUT} R_{ESR} + \frac{\Delta i_{OUT} \cdot t_3}{C_{OUT}} \quad (1.10)$$

1.4.3 Conduction Loss and Switching Loss

Given the large driving current flowing through the power stage of the power management module, the conduction loss obviously affects the power conversion efficiency. The conduction loss of the high-side power switch in the buck converter can be expressed as in Eq. (1.11). The on-resistance of the high-side power switch R_{onp} is inversely proportional to the power transistor size:

$$P_{conH} = (i_{LOAD})^2 \cdot R_{onp} \cdot \frac{v_{OUT}}{v_{IN}} \quad (1.11)$$

The conduction loss of the low-side power switch in the buck converter can be shown as in Eq. (1.12), with the low-side power switch on-resistance R_{onn}:

$$P_{conL} = (i_{LOAD})^2 \cdot R_{onn} \cdot \frac{v_{IN} - v_{OUT}}{v_{IN}} \quad (1.12)$$

A trade-off is observed between the conduction loss and the active silicon area. The utilization of large power switches directly reduces the conduction loss at the power stage, but the cost also increases because of the large silicon area. The switching loss results from the switching interval of the power switches. As depicted in Figure 1.37, the switching loss is proportional to the input voltage v_{IN} and the load current i_{OUT} in the buck converter. When the power switches turn on, the driving current increases, whereas the voltage across the drain and source of the power switch decreases. That is, the intersection of the conducting current and the voltage drop

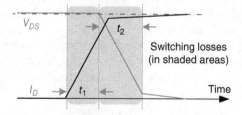

Figure 1.37 Illustration of switching loss of power switch

across the power switch causes power loss during the switching period. The switching loss during the period t_1 in the switching operation of the power switch can be obtained as:

$$P_{swt1} = \frac{1}{2}(v_{IN} \cdot i_{LOAD}) \cdot t_1 \qquad (1.13)$$

When the conducting current rises to its target value, the current is sustained, whereas the drop in voltage across the power switch decreases. Similarly, the switching loss derived during the period t_2 in the switching operation is shown as:

$$P_{swt2} = \frac{1}{2}(v_{IN} \cdot i_{LOAD}) \cdot t_2 \qquad (1.14)$$

The switching loss in one switching operation of one power switch is the summation of P_{swt1} and P_{swt2}. The total switching loss of one power switch with a constant switching frequency f_{sw} in the power management module can be expressed as:

$$P_{sw} = P_{swt1} + P_{swt2} = (v_{IN} \cdot i_{LOAD}) \cdot (t_1 + t_2) \cdot f_{sw} \qquad (1.15)$$

1.4.4 Power Conversion Efficiency

Power conversion efficiency is the most important design issue, especially in the battery-operated power management module. Power conversion efficiency is defined as the ratio of the provided output power to the total power received from the input supply. It can be formulated as:

$$\eta = \frac{P_{OUT}}{P_{IN}} = \frac{P_{OUT}}{P_{OUT} + P_{con} + P_{sw} + P_Q + P_{others}} \qquad (1.16)$$

The total power received from the input supply needs to provide the following items: the output power (P_{OUT}), the conduction loss at the power stage (P_{con}), the switching loss at the power stage (P_{sw}), the quiescent current power of the control circuits (P_Q), and the power loss caused by parasitic elements in realistic silicon fabrication (P_{others}). Therefore, minimizing the power losses can lead to high power conversion efficiency to enhance the competitiveness of a power management module.

References

[1] Kim, K. (2015) Silicon technologies and solutions for the data-driven world. *IEEE International Solid-State Circuits Conference (ISSCC), Digest of Technical Papers*, San Francisco, CA, February 22–26, pp. 1–7.
[2] Hu, C.C. (2010) *Modern Semiconductor Devices for Integrated Circuits*. Prentice-Hall, Upper Saddle River, NJ.
[3] Lin, H.C. and Jones, W.N. (1973) Computer analysis of the double-diffused MOS transistor for integrated circuits. *IEEE Transactions on Electron Devices*, **20**(3), 275–283.

[4] Yang, S., Sheu, G., Guo, J., and Tasi, J.R. (2011) Application of multi-lateral double diffused field ring in ultra-high-voltage device MOS transistor design. *International Conference on Electronic Measurement & Instruments (ICEMI)*, August 16–19, 2011, pp. 85–88.

[5] Anghel, C. (2004) High voltage devices for standard MOS technologies—characterisation and modelling. PhD thesis, EPFL, Lausanne.

[6] Radhakrishna, U., DasGupta, A., DasGupta, N., and Chakravorty, A. (2011) Modeling of SOI-LD MOS transistor including impact ionization, snapback, and self-heating. *IEEE Transactions on Electron Devices*, **58**(11), 4035–4041.

[7] Ma, Y., Jeng, M.-C., and Liu, Z., *HVMOS and LDMOS Modeling Review*. Cadence Design Systems, Inc., 2007.

[8] Tien, W.W.-Y. and Chen, F.-H. (2008) Recessed drift region for HVMOS breakdown improvement. US Patent 20080246083 A1, October 9, 2008.

[9] Tien, W.W.-Y. and Chen, F.-H. (2012) Recessed drift region for HVMOS breakdown improvement. US Patent 8138559 B2, March 20, 2012.

[10] Jain, S., Khare, S., Yada, V., et al. (2012) A 280 mV-to-1.2 V wide-operating-range IA-32 processor in 32 nm CMOS. *IEEE International Solid-State Circuits Conference (ISSCC), Digest of Technical Papers*, San Francisco, CA, February 19–23, pp. 66–68.

[11] Synopsys Inc. (2008) Low Power Trends and Methodology. http://www.synopsys.com/Solutions/EndSolutions/advanced-lowpower/Documents/lp_trend08_godwin.pdf (accessed November 14, 2015).

[12] Lee, D., Blaauw, D., and Sylvester, D. (2004) Gate oxide leakage current analysis and reduction for VLSI circuits. *IEEE Transactions on Very Large Scale Integration (VLSI) Systems*, **12**(2), 155–166.

[13] Choi, C.-H., Nam, K.-Y., Yu, Z. and Dutton, R.W. (2001) Impact of gate direct tunneling current on circuit performance: A simulation study. *IEEE Transactions on Electron Devices*, **48**(12), 2823–2829.

[14] Takeda, E., Matsuoka, H., Igura, Y., and Asai, S. (1988) A band to band tunneling MOS device (B/sup 2/T-MOSFET)—A kind of 'Si quantum device.' *International Electron Devices Meeting (IEDM), Technical Digest*, December 11–14, 1988, pp. 402–405.

[15] Agarwal, A., Kim, C.H., Mukhopadhyay, S., and Roy, K. (2004) Leakage in nano-scale technologies: Mechanisms, impact and design considerations. *Proceedings of the Design Automation Conference*, July 7–11, 2004, pp. 6–11.

[16] Lee, W. and Hu, C. (2001) Modeling CMOS tunneling currents through ultrathin gate oxide due to conduction- and valence-band electron tunneling. *IEEE Transactions on Electron Devices*, **48**(7), 1366–1373.

[17] Allen, P.E. and Holberg, D.R. (2002) *CMOS Analog Circuit Design*. Oxford University Press, New York, pp. 531–571.

[18] Lindert, N., Yoshida, M., Wann, C., and Hu, C. (1996) Comparison of GIDL in p+-poly PMOS and n+-poly PMOS devices. *IEEE Electron Device Letters*, **17**(6), 285–287.

[19] Lin, C.-T., Fang, Y.-K., Yeh, W.-K., et al. (2007) Impacts of notched-gate structure on contact etch stop layer (CESL) stressed 90-nm nMOSFET. *IEEE Electron Device Letters*, **28**(5), 376–378.

[20] Pandey, S.M., Liu, J., Hooi, Z.S., et al. (2011) Mechanism of stress memorization technique (SMT) and method to maximize its effect. *IEEE Electron Device Letters*, **32**(4), 467–469.

[21] Ortolland, C., Okuno, Y., Verheyen, P., et al. (2009) Stress memorization technique—fundamental understanding and low-cost integration for advanced CMOS technology using a nonselective process. *IEEE Transactions on Electron Devices*, **56**(8), 1690–1697.

[22] Ghani, T., Armstrong, M., Auth, C., et al. (2003) A 90 nm high volume manufacturing logic technology featuring novel 45 nm gate length strained silicon CMOS transistors. *IEEE Electron Devices Meeting (IEDM), Technical Digest*, December 8–10, 2003, pp. 11.6.1–11.6.3.

[23] Mistry, K., Armstrong, M., Auth, C., et al. (2004) Delaying forever: Uniaxial strained silicon transistors in a 90 nm CMOS technology. *Proceedings of the IEEE Symposium on VLSI Circuits*, June 2004, pp. 50–51.

[24] Ang, K.-W., Chui, K.-J., Bliznetsov, V., et al. (2004) Enhanced performance in 50 nm N-MOSFETs with silicon-carbon source/drain regions. *IEEE Electron Devices Meeting (IEDM), Technical Digest*, December 13–15, 2004, pp. 1069–1071.

[25] Ragnarsson, L.-A., Li, Z., Tseng, J., et al. (2009) Ultra-low EOT gate first and gate last high performance CMOS achieved by gate-electrode optimization. *IEEE Electron Devices Meeting (IEDM), Technical Digest*, December 7–9, 2009, pp. 663–666.

[26] Auth, C., Cappellani, A., Chun, J.-S., et al. (2008) 45 nm high-k + metal gate strain-enhanced transistors. *Proceedings of the IEEE Symposium on VLSI Circuits*, June 2008, pp. 128–129.

[27] Mistry, K., Allen, C., Auth, C., *et al.* (2007) A 45 nm logic technology with high-k+metal gate transistors, strained silicon, 9 Cu interconnect layers, 193 nm dry patterning, and 100% pb-free packaging. *IEEE Electron Devices Meeting (IEDM), Technical Digest*, December 5–8, 2007, pp. 247–250.

[28] Henson, K., Bu, H., Na, M.H., *et al.* (2008) Gate length scaling and high drive currents enabled for high performance SOI technology using high-k/metal gate. *IEEE Electron Devices Meeting (IEDM), Technical Digest*, December 15–17, 2008, pp. 645–648.

[29] Tomimatsu, T., Goto, Y., Kato, H., *et al.* (2009) Cost-effective 28-nm LSTP CMOS using gate-first metal gate/high-k technology. *Proceedings of the IEEE Symposium on VLSI Circuits*, June 2009, pp. 36–37.

[30] Choi, K., Jagannathan, H., Choi, C., *et al.* (2009) Extremely scaled gate-first high-k/metal gate stack with EOT of 0.55nm using novel interfacial layer scavenging techniques for 22nm technology node and beyond. *Proceedings of the IEEE Symposium on VLSI Circuits*, June 2009, pp. 138–139.

[31] Sheu, Y.-M., Chang, C.-S., Lin, H.-C., *et al.* (2003) Impact of STI mechanical stress in highly scaled MOSFETs. *Proceedings of the IEEE Symposium on VLSI Circuits*, June 2003, pp. 269–272.

[32] Drennan, P.G., Kniffin, M.L., and Locascio, D.R. (2006) Implications of proximity effects for analog design. *Proceedings of the IEEE Custom Integrated Circuits Conference (CICC)*, September 2006, pp. 169–176.

[33] Chen, J.-S. and Ker, M.-D. (2009) Impact of gate leakage on performances of phase-locked loop circuit in nanoscale CMOS technology. *IEEE Transactions on Electron Devices*, **56**(8), pp. 1774–1779.

[34] Soumya, P., Chittaranjan, M., and Amit, P. (2014) *Nano-Scale CMOS Analog Circuits: Models and CAD Techniques for High-Level Design Citation*. CRC Press, Boca Raton, FL.

2

Design of Low Dropout (LDO) Regulators

Low dropout (LDO) regulators are widely used in portable electronic devices because they occupy small chip and printed circuit board (PCB) areas. The performance advantages of these regulators include low quiescent current and wide bandwidth (BW), which result in fast transit response. Unlike switching regulators (SWRs), LDO regulators convert voltage through a linear operation, and thus a well-regulated output voltage can be derived without the occurrence of output voltage ripples. However, LDO regulators suffer from an inherent disadvantage, namely poor power conversion efficiency (PCE) if the ratio of the output voltage to the input voltage is small, because a large voltage stress over the pass transistor causes significant power loss. When considering the small silicon area, compact PCB, and reduced discrete component costs, an LDO regulator is one candidate that can provide a conversion function for regulated and scaled-down voltages. LDO regulators are frequently utilized as post-regulators in series with SWRs to suppress voltage ripples from the switching operation of SWRs caused by their large open-loop gain. In general, the combination of an SWR in series with an LDO regulator can be viewed as a simple and efficient power management module if the ratio of the output voltage to the input voltage, as well as the open-loop gain, can remain large in the LDO regulator. An LDO regulator is regarded as a voltage buffer for decreasing voltage ripples at the cost of slightly reduced PCE.

LDO regulators have structures that are defined according to their controlling methods and compensated skills. As illustrated in Figure 2.1, LDO regulators may either be analog-low dropout (A-LDO) regulators or digital-low dropout (D-LDO) regulators. The controller of the former is designed with analog circuits, whereas that of the latter is designed with digital circuits. A-LDO regulators have two major categories, namely dominant pole compensation, which involves a large compensation capacitor C_{out} located at the output node, and a capacitor-free (*C-free*) structure, which involves a dominant pole generated by Miller compensation capacitance.

Power Management Techniques for Integrated Circuit Design, First Edition. Ke-Horng Chen.
© 2016 John Wiley & Sons Singapore Pte Ltd. Published 2016 by John Wiley & Sons Singapore Pte Ltd.

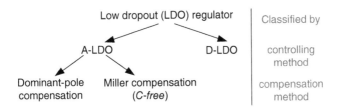

Figure 2.1 Classification of LDO regulators

Applications determine the suitable type of LDO to use. Various multimedia and portable devices demand Soc integration. Soc typically requires a compact-size, fast-transient, low-noise supply voltage. A *C-free* structure can be made compact at the expense of a large quiescent current and transient voltage variation. An A-LDO regulator with dominant pole compensation improves the transient voltage variation and reduces the quiescent current at the cost of a large output capacitor. When considering a D-LDO regulator as the post-regulator, an improvement in transient response and regulator size leads to the feasibility of integration; however, a trade-off occurs between regulation performance and quiescent current.

In this chapter, the basic LDO regulator is first introduced. Then, concerns over compensation for loop stability are presented to develop dominant pole compensation and *C-free* structures. Design flow and tips are also illustrated to understand the specifications and performance in A-LDO regulators. The characteristics of A-LDO and D-LDO regulators are then discussed and compared. LDO regulators with specific features are introduced to satisfy the requirements of various applications. Furthermore, low-power techniques, including novel dynamic-voltage scaling (DVS) and analog dynamic-voltage scaling (ADVS), are introduced for compatible utilization in Soc applications.

2.1 Basic LDO Architecture

In general, the voltage drop across power transistors in LDOs results in unavoidable heat in the power transistors. LDOs are appropriate for low-power applications because of heat dissipation. A simple LDO consists of an error amplifier (EA) and power transistors to eliminate ripples and regulate output voltage. LDO regulators have the smallest layout and footprint area, and a low quiescent current to provide small volume and high current efficiency. Figure 2.2 illustrates the architecture of a basic LDO regulator. The dropout voltage across the pass transistors is defined as the voltage difference between the input supply voltage and the output voltage. A small dropout voltage can lead to satisfactory PCE. However, decreasing the dropout voltage of an LDO may decrease the gain of the last stage, and thus deteriorate the regulation performance. A trade-off occurs between PCE and regulation performance. Designers generally use large power transistors to reduce the dropout voltage effectively and take advantage of its low on-resistance. However, large power transistors increase the silicon area, and thus the chip cost also increases. Moreover, the transient response is slowed down by the limited slew rate designed in the EA when driving at the gate of the power transistor. Given the differences between the reference voltage V_{REF} and the feedback voltage V_{FB}, the output voltage of the EA

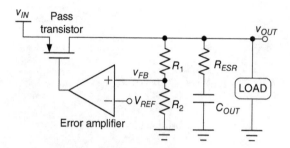

Figure 2.2 Structure of a basic LDO regulator

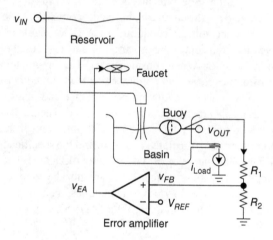

Figure 2.3 Operation of LDOs explained using an emulated reservoir and basin system controlled by a buoy with closed-loop system

controls the gate voltage of the pass transistors, and thus modulates the current flowing through the pass transistors.

Given the negative feedback control, LDO regulators can regulate the output voltage irrespective of load current and input voltage variations if the loop gain is sufficiently large. Any switching disturbance from a pre-regulator, such as SWRs, can be effectively reduced by an effective loop gain. However, an increased loop gain may cause stability deterioration because high-frequency poles will be higher than the crossover frequency. Hence, another trade-off occurs between the regulation performance and system stability.

As illustrated in Figure 2.3, the operation of LDOs can be explained by an emulated reservoir and basin system, which is controlled by a buoy with a closed-loop system. The input voltage V_{IN} is emulated by the reservoir. The power transistor functions as a faucet, and the charge stored in an output capacitor can be regarded as the water stored in a basin. The pipe aperture of the faucet implies driving capability. The height of the water in the basin represents the level of the output voltage. The load current can be viewed as water flowing out at the bottom of a basin. To control the water level at a constant height, similar to a regulated voltage level labeled

V_{OUT} in LDOs, a closed-loop system should be formed using the buoy to detect the water level. The detected water level is scaled down and compared with the predefined reference voltage V_{REF} to obtain an error control signal V_{EA}. That is, V_{EA} can be used to determine how tight or loose the faucet is. Water flowing out of the faucet can correspond to the value of V_{EA}. Under a dynamic equilibrium condition, the water flowing into and out of a basin remains in dynamic balance, which is maintained by the negative feedback system. In the emulated reservoir and basin system, only the water level that is similar to the output voltage level in LDOs is monitored. Thus, the system is called a voltage-mode control system.

How fast the faucet can be turned tightly or loosely determines how fast the water level can reach the desired level in case of a sudden change in output load current. In the emulated reservoir and basin system, the water level is controlled by the negative feedback closed-loop system, wherein the water level is sensed by the buoy. A large pipe aperture of the faucet can deliver high current to the basin, but the faucet experiences driving difficulty of the EA output. Thus, turning the faucet to an adequate position requires good driving capability of the EA. Furthermore, the water level is easily maintained at a constant level by a basin with a large volume, because a sudden considerable change in loading current cannot instantly disturb the water level of a large basin even if the faucet has not been turned to an adequate position. Moreover, if the output load current changes from high to low, then the water level will remain high for a short period because of the absence of an extra path for surplus water to flow out of the basin. To reach the desired water level, an extra path where water can rapidly flow out of the basin is required, which leads to wastage of valuable water. Evidently, this approach is not energy efficient and not environmentally friendly. If cost and energy efficiency are considered, a bulky basin is not the appropriate design to keep the water level constant.

In summary, two concerns are worth mentioning. First, the driving capability of the EA depends on the faucet size, and affects the rapid delivery of the current from the reservoir to the basin. Second, the basin size determines the volume of water stored in it and influences the maintenance of the water level in case of any sudden change in output load current. These two issues can be mapped into similar LDO designs, and thus two possible large capacitances should be considered carefully in LDO design. The first large capacitance is located at the output node and the second is located at the output of the EA, which can be mapped into two capacitances at V_{OUT} and V_{EA} in Figure 2.3, respectively. The values of these capacitances determine the types of suitable LDO topology. Therefore, application requirements have to be considered first. One of the possible topologies has to be selected to satisfy all specifications. In the following sections, compensation and advanced techniques are introduced to enhance LDO performance.

2.1.1 Types of Pass Device

The faucet is the power transistor. Thus, the type of power transistor suitable for LDO designs should be identified. All possible pass transistor designs are shown in Figure 2.4. Considering the type of bipolar transistor (BJT) as pass transistor, NPN Darlington, NPN, and PNP shown in Figure 2.4(a)–(c), respectively, can be used to obtain the advantage of high driving capability because of the high current gain of BJTs. However, BJTs have two obvious disadvantages. First, the dropout voltage required by the voltage headroom, which consists of base/emitter and collector/emitter voltages, to keep the BJT working in the active region

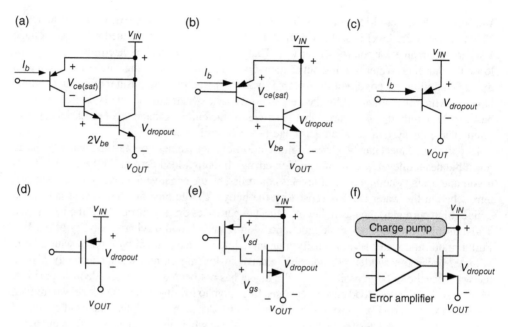

Figure 2.4 Pass devices include (a) NPN Darlington, (b) NPN, (c) PNP, (d) N-MOSFET, (e) P-MOSFET, and (f) N-MOSFET with one charge pump circuit

is large. In Figure 2.4(a)–(c), the dropout voltage is $V_{ce(sat)} + 2V_{be}$, $V_{ce(sat)} + V_{be}$, and $V_{ec(sat)}$, respectively. Thus, in low-quiescent and low-power Soc applications, Darlington NPN and NPN configurations are unsuitable because of their large dropout voltage. By contrast, the PNP configuration is the most appropriate configuration because its minimum dropout voltage is only $V_{ec(sat)}$, which is typically 0.4 V. The second critical disadvantage is the large leakage current at the base terminal. The base current is basically proportional to the value of I_C/β_n or I_C/β_p, where I_C is the collector current and β_n, β_p are the current gains of the NPN and PNP transistors, respectively. In general, β_n and β_p are within the ranges of 50–100 and 5–10, respectively. The leakage current caused by the non-zero base current is strongly affected by current gain. If a LDO voltage is required, then the most appropriate BJT is the PNP transistor; however, this transistor suffers from a large leakage current because of its low β_p compared with the NPN transistor, which has a large β_n.

A MOSFET is selected to replace the BJT as the power transistor to eliminate these two critical and obvious disadvantages, as shown in Figure 2.4(d)–(f). In general, no DC flows into or out of the gate terminal of the MOSFET. Thus, the leakage current problem can be solved completely and low quiescence is guaranteed. However, the driving capability of a MOSFET is considerably smaller than that of a BJT, and thus a large aspect ratio of the MOSFET is required to generate a similar driving current to that provided by a BJT. An obvious disadvantage is the large silicon area occupied by the MOSFET.

In Figure 2.4(d) the p-channel MOSFET (P-MOSFET) with large aspect ratio has a lower dropout voltage than its counterpart PNP transistor, which has a minimum dropout voltage limited to approximately 0.2 V. That is, if the LDO regulator requires LDO voltage and low quiescent current, then the P-MOSFET is the best choice among the various configurations.

Table 2.1 Comparison of different pass devices

	NPN Darlington	NPN	PNP	NMOS	PMOS
I_{Load}	High	High	High	Medium	Medium
I_q	Medium	Medium	Large	Low	Low
$V_{dropout}$	$V_{ce(sat)} + 2V_{be}$	$V_{ce(sat)} + V_{be}$	$V_{ec(sat)}$	$V_{gs} + V_{ds(sat)}$	$V_{sd(sat)}$
Speed	Fast	Fast	Slow	Medium	Medium
Compensation	Easy	Easy	Complex	Easy	Complex

However, the P-MOSFET also has several disadvantages. For example, it requires a larger silicon area to compensate for the lower mobility μ_p than in the n-channel MOSFET (N-MOSFET). Moreover, the P-MOSFET works as a common source (CS) stage in an LDO regulator. The gate-to-drain capacitance C_{gd} is amplified by the Miller effect and increases the difficulty of loop compensation. By contrast, the N-MOSFET shown in Figure 2.4(e) works as a source follower (SF) stage that functions as a buffer stage in an LDO regulator. Thus, the compensation complexity of an LDO regulator with N-MOSFET is considerably lower than that of an LDO regulator with P-MOSFET. Considering the dropout voltage of an N-MOSFET, a large voltage headroom equal to $V_{sd} + V_{gs}$ is the critical disadvantage, which is similar to that in the BJT. To address this problem in commercial products, a charge pump circuit can be used to decrease the dropout voltage, as shown in Figure 2.4(f). The boosted supply voltage for the EA can minimize the dropout voltage, but an LDO regulator suffers from switching noise caused by the charge pump circuit. A trade-off appears to occurs between noise suppression and dropout voltage. An off-chip low-pass filter is inserted between the control signal and the gate of the N-MOSFET, because an on-chip low-pass filter is difficult to apply. Consequently, the disadvantages are extra off-chip components and a large PCB area.

The products listed in Table 2.1 use P-MOSFETs as power transistors because of their low quiescence and LDO voltage at the cost of complex compensation skills. In the following section, compensation skills are introduced by assuming that a P-MOSFET is used as the power transistor. Before ending this subsection, the I–V curve of the power P-MOSFET is observed, as shown in Figure 2.5. The P-MOSFET works similarly to the faucet shown in Figure 2.3. Controlling the value of the source-to-gate voltage V_{SG}, which is similar to the tightness or looseness of the faucet, can determine the current flowing through the power P-MOSFET. At light loads (e.g., i_{D1} in Figure 2.5), a small V_{SG} indicates less current flowing out of the input supply voltage V_{IN}. By contrast, increasing V_{SG} at heavy loads (e.g., i_{D2}) can loosen the faucet to force a high current to flow through the power P-MOSFET. In case of a load change from light to heavy (e.g., from i_{D1} to i_{D2}), a large $V_{dropout}$ can guarantee that the power P-MOSFET will work in the saturation region for high gains in CS configuration, where $V_{dropout}$ is the source-to-drain voltage of the P-MOSFET.

A high gain in the system loop leads to a well-regulated output voltage. However, as mentioned previously, a large $V_{dropout}$ implies poor PCE of an LDO regulator. Ensuring that the power P-MOSFET at the saturation region has a constant $V_{dropout}$ and a large $V_{dropout}$ is required at heavy loads, but redundant and power inefficient at light loads because of the small V_{SG}. If $V_{dropout}$ can be adjusted according to the output load current conditions, then the PCE can be kept high over a wide load range, which is referred to as the ADVS technique. ADVS and digital-dynamic voltage scaling (DDVS) are introduced later.

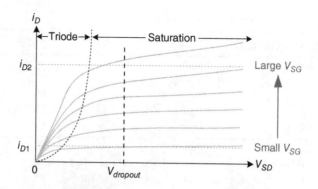

Figure 2.5 *I–V* curve of the power P-MOSFET

2.2 Compensation Skills

As illustrated in Figure 2.2, the regulation of an LDO regulator is structured with a negative feedback loop, and thus the output voltage can be regulated at a desired voltage level. Therefore, the frequency response should be designed carefully not only for stability, but also for transient response. In general, compensation skills should ensure that an LDO regulator has various capabilities, including low quiescent current at light loads, high current efficiency, low voltage operation, and high driving capability. Different compensation skills present various challenges that correspond to different specifications of systems and applications. In this section the characteristics of poles and zeros provide the concept of compensation skills for LDO regulators. Pole distribution is introduced first.

2.2.1 Pole Distribution

First, in the design of the LDO regulator shown in Figure 2.6(a), the existence of poles can be clearly identified at P_0 and P_1, located at the LDO output and the first stage output, respectively. Without considering the Miller effect on the gate-to-drain capacitance and by adding output capacitors at the output node, P_1 is below P_0, as shown in Figure 2.6(b), because a large parasitic capacitance C_{par} and a large output resistance $r_{OUT(ea)}$ are observed at the gate of the power MOSFET and at the output of the EA, respectively. $r_{OUT(ea)}$ is large because of the low quiescent current requirement. Given that the power P-MOSFET should be large to provide high driving capability, the gate-to-drain capacitance C_{GD} is sufficiently large to be a Miller capacitance with the gain contributed by the power P-MOSFET. Thus, P_1 and P_0 are split by the Miller capacitance C_{GD}; that is, P_0 moves toward high frequencies as a new high-frequency pole P_0' while P_1 moves toward the origin as a new low-frequency pole P_1'. The root locus of the pole-splitting effect that results from a large C_{GD} is illustrated in Figure 2.6(b). Consequently, system stability can be ensured if two poles can be widely split by C_{GD}. That is, a low-frequency pole P_1' at the gate of the power MOSFET can be regarded as the system-dominant pole without being affected by other low-frequency poles because the first non-dominant pole P_0' is now located at high frequencies.

Design of Low Dropout (LDO) Regulators

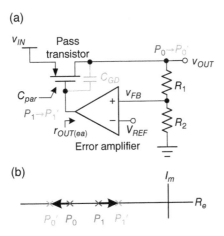

Figure 2.6 Basic LDO regulator. (a) Schematic. (b) Root locus showing the pole-splitting effect with Miller capacitance C_{GD}

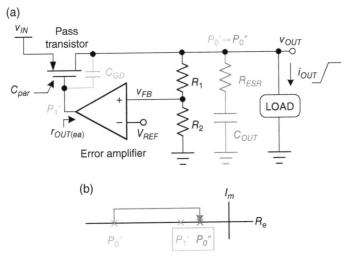

Figure 2.7 Basic LDO regulator with a large output capacitor to deal with significant sudden loading current change. (a) Schematic. (b) Root locus

When checking the requirements of LDO regulators, experiencing the sudden load changes required for system operation is necessary. As previously mentioned, a sudden load change cannot immediately change the water level of a big basin in the emulated reservoir and basin system. Thus, connecting a large output capacitor at the output to sustain a well-regulated output voltage is easy, as shown in Figure 2.7(a). A large output capacitor solves the sudden load change problem, but the stability decreases because the large output capacitor forces the output pole toward the origin with the role of dominant pole. Meanwhile, the pole at the gate of the power P-MOSFET is not the dominant pole anymore. The position of the output pole changes from P_0' to P_0'', where P_0'' is considerably lower than P_1' because of the large output capacitor. P_0'' becomes the new dominant pole of the system when a large output capacitor is used. However, we should remember the subsequent problem that two low-frequency poles, namely P_0'' and P_1' (the output pole and the pole at the gate of the power P-MOSFET, respectively) exist in the LDO regulator. Consequently, the phase margin (PM) is deteriorated by two low-frequency poles, as shown in Figure 2.7(b). Given that the output pole is now the dominant pole, the previous non-dominant pole is located at the gate of the power MOSFET and should move toward high frequencies to increase the system stability. Obviously, we should determine how to push P_1' toward high frequencies or add a compensation zero to cancel its effect. If this problem is reviewed carefully, we can determine that the output resistance of EA is large because of the low quiescent current. The resistor/capacitor (RC) time, which is constant at the gate of the power P-MOSFET, is too large. Thus, the EA cannot directly drive large parasitic capacitance at the gate of the power P-MOSFET if a low quiescent current is still required.

A useful technique to alleviate large RC time constants occurring at the gate of the power P-MOSFET is to add a buffer stage between the output of the EA and the gate of the power MOSFET, as illustrated in Figure 2.8(a). Consequently, the driving capability and the PM can be simultaneously enhanced by the inserted buffer stage. The buffer stage splits a low-frequency pole P_1' into two high-frequency poles, namely P_1^* and P_1^{**}, as shown in Figure 2.8(b). P_1^* and P_1^{**} have small RC time constants, that is $C_{par}{}^*r_{OUT(buf)}$ and $C_{IN(buf)}{}^*r_{OUT(ea)}$, respectively, where $C_{IN(buf)}$ and $r_{OUT(buf)}$ are the input capacitance and output resistance of the buffer stage, respectively. $C_{IN(buf)}$ and $r_{OUT(buf)}$ are small because of the characteristics of the voltage buffer. The large RC time constant stops increasing when the buffer stage is used. P_1^* is below P_1^{**} because of the large parasitic capacitance C_{par} at the gate of the power P-MOSFET. P_1^* becomes the first non-dominant pole after the buffer stage is inserted.

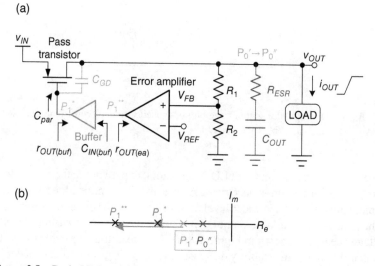

Figure 2.8 Basic LDO regulator with a buffer stage. (a) Schematic. (b) Root locus

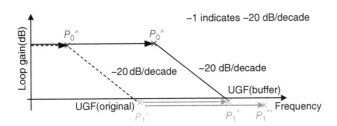

Figure 2.9 Relationship between the dominant pole and non-dominant poles in the frequency domain before and after the buffer stage is inserted

After the buffer stage is inserted, the low-frequency pole P_1^* gains a small RC time constant. If P_1^* is set at the unity gain and no low-frequency pole exists below P_1^*, excluding the output pole P_0'', a line with slope -20 dB/decade can be drawn to determine the highest position of P_0'', as illustrated in Figure 2.9. Similarly, before the buffer stage is inserted, the original position of P_0'' can be determined by drawing a line with slope -20 dB/decade from the location of P_1', which is also found at unity gain. Figure 2.9 shows the possible working range of P_0''. In conclusion, if P_0'' is located at the highest position, then the RC time constant of P_0'' can be reduced because P_1^* is above P_1'. The advantages of a smaller RC time constant of P_0'' are twofold. First, the LDO regulator can tolerate a high loading current, which indicates a small equivalent output resistance. Second, the LDO regulator can remain stable if a small output capacitor is used at medium loads instead of the conventional design without using the buffer stage.

Basically, the buffer can be a simple SF, as depicted in Figure 2.10(a), with an equivalent output resistance of $1/g_{m1}$. However, $1/g_{m1}$ is not sufficiently small to push the pole toward high frequencies. Thus, the design target is to reduce the equivalent output resistance. The analysis can involve a test voltage v_t applied at source M_1 when the equivalent output resistance is measured, as depicted at the top left of Figure 2.10(b). The small signal current i_t ($\approx g_{m1}v_t$) is redirected to the base terminal of an additional BJT Q_1 to generate an amplified current βi_t. The advantage of this technique is that the small signal i_t is enlarged by the current gain β of the BJT. Given the local negative feedback, the equivalent output resistance of the buffer can be attenuated by a factor of $(\beta + 1)$, which can be expressed as:

$$r_{OUT(buf)} \approx \frac{1}{g_{m1}} \cdot \frac{1}{(\beta+1)} \quad (2.1)$$

In general, only lateral PNP or NPN BJT is provided in the standard CMOS process. Thus, the designer can use a bipolar-CMOS-DMOS (BCD) process to facilitate design at the expense of cost and silicon area. To solve this issue, the function of the BJT device can be substituted by combining MOSFETs from M_{U1} to M_{U4}, as shown in Figure 2.10(c). A new equivalent output resistance is expressed in Eq. (2.2), which indicates that the output resistance is reduced by the $(g_{m2}r_{o1} + g_{m3}g_{m4}r_{o1}r_{o4})$ contributed by the local negative feedback control of MOSFETs from M_{U1} to M_{U4} without any additional process:

$$r_{OUT(buf)} \approx \frac{1}{g_{m1}} \cdot \frac{1}{(g_{m2}r_{o1} + g_{m3}g_{m4}r_{o1}r_{o4})} \quad (2.2)$$

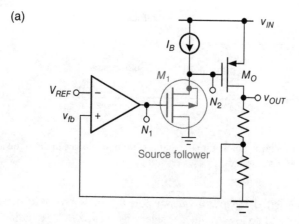

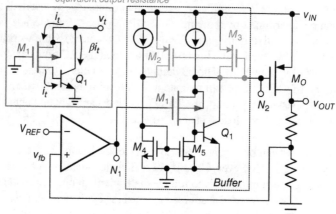

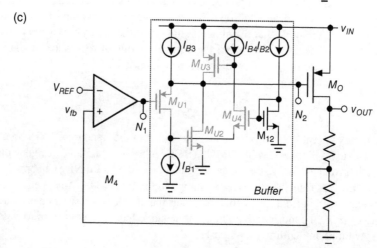

Figure 2.10 Structure of the LDO regulator with a buffer stage, which can be implemented with (a) an SF configuration, (b) an SF with a local negative feedback implemented by BJT, and (c) an SF with a reduced output resistance implemented by CMOS devices

As previously mentioned, the size of the output capacitor can be reduced if a buffer stage is used. The output capacitor cannot be implemented on the same silicon because P_1^* in Figure 2.9 should be pushed higher than P_1', because the acceptable size of an on-chip capacitor is within the range of several pico-farads compared with an external output capacitor that measures several micro-farads. That is, the advantage of the buffer stage is that it allows the external output capacitor to be smaller than that of the conventional LDO design without a buffer stage. Consequently, the external output capacitor remains too large to become an on-chip capacitor even if a buffer stage is used.

To obtain a compact-sized solution for the LDO regulator, the output capacitor shown in Figure 2.7 is used because the large charge stored in a large output capacitor can deal with instant load changes. If this large output capacitor is removed, then the power transistor should be rapidly turned on or off because the small charge stored in the parasitic output capacitance cannot satisfy the load change requirements of the system. If the system bandwidth is sufficiently large, the LDO regulator has the capability to handle a sudden load change even when the large output capacitor is removed. The Miller compensation skill, which is completely contrary to the dominant-pole compensation skill, can be applied to LDO regulators without adding any large external output capacitor. The Miller compensation skill can have a large bandwidth if a gain-stage amplifier is inserted between the EA and the power transistor. Thus, a high bandwidth is achieved but Miller capacitors are necessary. The inherent gate-to-drain capacitance C_{GD} can be regarded as a Miller capacitor. Only one Miller capacitor C_m is necessary, as shown in Figure 2.11(a). The characteristics and features of the two aforementioned compensation skills should be examined to identify which skill is more suitable for Soc applications.

Figure 2.11(a) shows that a Miller capacitor C_m is added and a large output capacitor C_{OUT} is removed. Consequently, Figure 2.11(b) shows that the pole at V_{OUT} moves toward high frequencies from P_0' to P_0''', while the pole at the output of the EA moves toward low frequencies from P_1' to P_1''. P_1'' replaces P_0' in the dominant-pole compensation skill as the new dominant pole because of the Miller effect. The advantage of this process is that the Miller capacitor C_m is considerably smaller than the original C_{OUT}, because the Miller effect amplifies the equivalent capacitance. Moreover, another Miller capacitance C_{GD} can split P_2 and P_0''' into high

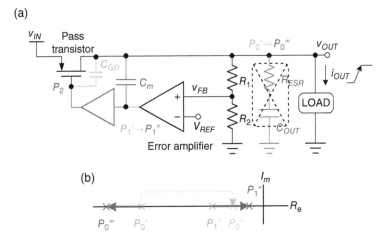

Figure 2.11 Basic LDO regulator with Miller compensation. (a) Schematic. (b) Root locus

frequencies. However, without a large C_{OUT}, any load transient response will cause a larger output transient voltage variation than the dominant-pole compensation skill if two compensation skills are assumed to have the same unity gain frequency (UGF), because the small basin cannot accommodate large water system requirements if the bandwidth does not increase.

When the load changes from heavy to light, the equivalent output impedance decreases, and then P_0''', which is the pole at V_{OUT}, moves toward a lower frequency. To maintain the performance of the transient response, extending the UGF is a challenge, particularly under ultra-light load conditions.

2.2.2 Zero Distribution and Right-Half-Plane (RHP) Zero

LDO regulators experience a load change from Soc applications. Thus, the bandwidth of LDO regulators should be sufficiently high to have a good transient load response. To extend the bandwidth further, a left-half-plane (LHP) zero is required to cancel the effect of the first non-dominant pole. We provide several common techniques to generate LHP zero, without being affected by right-half-plane (RHP) zero. The first well-known technique, as shown in Figure 2.12(a), is to use an equivalent series resistance at the output capacitor to form an LHP zero according to the equivalent output impedance:

$$Z_{OUT} = R_{ESR} + \frac{1}{sC_{OUT}} = \frac{1 + sR_{ESR}C_{OUT}}{sC_{OUT}} \tag{2.3}$$

The ESR value is temperature dependent, and thus determining an exact value in the design is difficult. Moreover, in case of a loading current change, any current flowing through the ESR will cause a large IR-drop voltage. The transient voltage variation becomes large, which is not desired in power management design. Thus, low ESR output capacitors are favored in converter designs.

In general, in a CS configuration, the Miller capacitor C_m will contribute an RHP zero, which is not desired in regulator designs. RHP zero is caused by a feedforward path through C_m from v_i to v_o. RHP zero significantly deteriorates the system stability, because it contributes +20 dB and −90° per decade. Thus, removing the RHP zero effect is necessary. As shown in Figure 2.12(b), the second technique is to convert RHP zero into LHP zero through the null resistance R_Z in series with C_m. If $1/g_{m1}$ is considerably smaller than R_Z, then the expression of LHP zero (ω_Z) is shown as:

$$\omega_Z \approx \frac{1}{C_m\left(g_{m1}^{-1} - R_Z\right)} \tag{2.4}$$

where g_{m1} is the transconductance of MOSFET M_1. Furthermore, given that the implementation of R_Z occupies a large silicon area, conventional MOSFET resistance working in the triode region can be used. However, its equivalent resistance value suffers from perturbations resulting from output voltage variation. Figure 2.12(c) shows an effective technique to remove RHP zero by adding a SF in series with a Miller capacitor C_m on the feedback path from v_o to v_i, which indicates that the voltage difference across C_m remains at $v_i - v_o$. However, the input capacitance observed at v_i is the series equivalent capacitance generated by C_m and C_{gs2}. Consequently, the equivalent capacitance becomes smaller than that of the original design. That

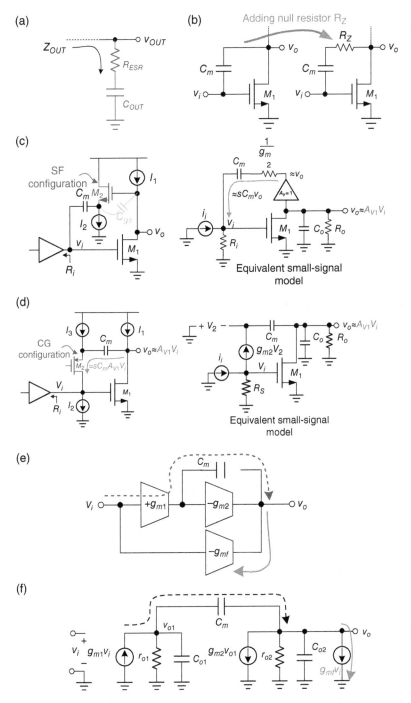

Figure 2.12 LHP zero generation methods. (a) ESR on the output capacitor. (b) Miller capacitor with a null resistor R_Z. (c) SF configuration used to remove RHP zero. (d) CG configuration used to remove RHP zero. (e) Feedforward path forms parallel architecture to obtain one LHP zero. (f) Small signal model of (e) showing the effect of the feedforward path

is, RHP zero moves toward infinity, and thus an LHP zero is generated. After calculation, the transfer function is derived as:

$$\frac{v_o}{i_i} = \frac{-g_{m1}R_oR_i(1+C_m/g_{m2}s)}{R_oC_oC_m(1/g_{m2}+R_i)s^2 + [(1/g_{m2}+g_{m1}R_oR_i)C_C + R_oC_o]s + 1} \quad (2.5)$$

The generated LHP zero is expressed in Eq. (2.6), which is determined by controlling the transconductance value of M_2:

$$\omega_Z = -\frac{g_{m2}}{C_m} \quad (2.6)$$

The RHP zero disappears in the circuit, but a disadvantage must be pointed out in this case. The maximum allowable output voltage must be higher than $V_{OD(current\ source\ I2)} + V_{GS2}$, which indicates that the output voltage cannot be too low. Thus, another technique can be used to improve the solution to this problem. The circuit configuration is shown in Figure 2.12(d), where transistor M_2 works as a common gate (CG) configuration. The output voltage signal is converted into a current signal sC_mv_o, which is injected into the v_i node and summed with the input current i_i. The transfer function from i_i to v_o is expressed in Eq. (2.7) with one generated LHP zero, as shown in Eq. (2.8):

$$\frac{V_o}{i_i} = \frac{-g_{m1}R_iR_o\left(1+C_m/g_{m2}s\right)}{R_o/g_{m2}C_oC_ms^2 + \left[(1+g_{m1}R_i)R_oC_m + C_m/g_{m2} + R_oC_o\right]s + 1} \quad (2.7)$$

$$\omega_Z = -\frac{g_{m2}}{C_m} \quad (2.8)$$

Moreover, the feedforward path can generate an LHP zero via an active signal path, because it can alleviate the effect of RHP zero through C_m. Figure 2.12(e) shows a two-stage amplifier with the Miller compensation capacitor C_m. The feedforward amplifier with transconductance g_{mf} can provide a feedforward signal path that is used to move RHP zero toward infinity, thus transforming it into LHP. That is, if g_{mf} is equal to g_{m1}, then no net current signal out of the output exists, and thus the output voltage is zero when ω approaches infinity, as shown in Figure 2.12(f). After the transfer function is derived, setting $g_{mf} = g_{m1}$ can move zero to LHP and equal to $-g_{m2}/C_{o1}$, which is above the UGF, where C_{o1} is the output capacitance of the first stage. RHP zero is effectively removed from the system, even if LHP zero is not well defined in this case.

An explanation of LHP and RHP zeros helps in the following LDO designs. Before providing the detailed designs, the specifications and design considerations of the LDO regulator are carefully addressed in the following section.

2.3 Design Consideration for LDO Regulators

Several specifications should be mentioned in LDO designs. Such specifications can be classified into two categories, namely steady-state and dynamic characteristics. Steady-state specifications include dropout voltage $V_{dropout}$, line regulation, load regulation, temperature coefficient (TC), and maximum/minimum load current, among others. Meanwhile, dynamic specifications include line/load transient voltage variations, transient recovery time, pole/zero doublets, PM, and adaptive quiescent operating current, among others. The following subsections explain each specification in the LDO designs.

2.3.1 Dropout Voltage

As shown in Figure 2.13, when a certain desired V_{OUT} is assumed under certain loading conditions, different V_{IN} refer to various operating regions, including the off region, dropout region, and regulation region. Different operating regions lead to varying V_{OUT} performances. In the regulation region, V_{OUT} can be regulated within the allowable dropout voltage $V_{dropout}$. This voltage can be estimated by the voltage difference between the input voltage and the output voltage when V_{OUT} deviates from its nominal regulated voltage of approximately 2%. That is, the voltage across the power transistor that is defined as $V_{dropout}$. When $V_{dropout}$ increases with decreasing V_{IN}, the power transistor eventually loses its gain and V_{OUT} gradually deviates from its desired value. As soon as the power transistor turns to function as a switch instead of an analogy gain stage, the system loop gain drastically decreases and consequently, the LDO regulator loses its capability to regulate V_{OUT}. The V_{OUT} value is determined according to V_{IN}, I_{Load}, and R_{on}, where R_{on} is the equivalent resistance of the power transistor. Under this situation, the operation enters the dropout region. The voltage across the power transistor can basically be expressed as:

$$V_{dropout} = I_o R_{on} \qquad (2.9)$$

As V_{IN} decreases further, R_{on} increases continuously because the input voltage that drives the power transistor is too low. Under this situation, V_{OUT} is decreased to the ground and the operation enters the off region.

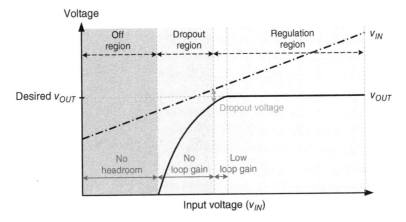

Figure 2.13 Characteristic curve of input voltage and output voltage that defines the dropout voltage $V_{dropout}$

Operation in the regulation region is necessary for LDO regulators to provide regulated performance and driving capability. When a small $V_{dropout}$ and a large I_{Load} are desired, LDO regulators should be out of the regulation region. The feedback controller can adjust the R_{on} that corresponds to $V_{dropout}$ to regulate V_{OUT} when operation occurs in the regulation region. The minimum $V_{dropout}$ is the worst-case scenario in the regulation region, when LDO regulators consider the largest I_{Load} and the minimum R_{on}. To ensure that LDO regulators are operating within the regulation region, increasing the size of the power transistors can reduce R_{on} and then ease the limitation of $V_{dropout}$. However, this process increases the complexity of the compensation network.

LDO regulators are typically utilized for two purposes. The first purpose is to function as the voltage converter to provide step-down supply voltage. In this case, the requirements are simple architecture, small silicon area, and low cost with minimal requirement for passive components such as inductors and capacitors. However, a large $V_{dropout}$ is inevitable because LDO regulators are used for step-down voltage conversion. The PCE of LDO regulators is sacrificed because of the considerable power loss across power transistors compared with SWRs.

The second purpose of LDO regulators is to work as post-regulators to provide a noise-suppression function for low-noise voltage in sensitive analogy circuits. $V_{dropout}$ is expected to be minimal to prevent power loss. In modern LDO regulator designs, the trade-off between efficiency and the silicon area occupied by power transistors is considered; thus, $V_{dropout}$ is designed within the range of 200 mV.

2.3.2 Efficiency

PCE is defined as the ratio of the output power P_{OUT} to the input power P_{IN}, as expressed in Eq. (2.10). The output power P_{OUT} is equal to the product of the output voltage and the output loading current. The input power is equal to the product of the input voltage and the input current I_{IN}:

$$\eta = \frac{P_{OUT}}{P_{IN}} = \frac{V_{OUT} I_{Load}}{V_{IN} I_{IN}} \qquad (2.10)$$

In Eq. (2.11), I_{IN} includes the loading current I_{Load} and the quiescent current I_q that consists of all the biasing currents in the EA, the bandgap reference, feedback resistance network, and so on:

$$I_{IN} = I_{Load} + I_q \qquad (2.11)$$

Overall, the PCE in Eq. (2.10) can be modified to Eq. (2.12), which is a product of voltage conversion efficiency and current efficiency as respectively defined in Eqs. (2.13) and (2.14):

$$\eta = \frac{I_{Load} V_{OUT}}{(I_{Load} + I_q) V_{IN}} = \eta_V \cdot \eta_I \qquad (2.12)$$

Design of Low Dropout (LDO) Regulators

where

$$\eta_V = \frac{V_{OUT}}{V_{IN}} \times 100\% \quad (2.13)$$

and

$$\eta_I = \frac{I_{Load}}{I_{Load} + I_q} \times 100\% \quad (2.14)$$

Under heavy loads, the PCE in Eq. (2.12) is approximately equal to the ratio of V_{OUT} to V_{IN}, because I_{Load} is considerably larger than I_q. This finding indicates that the dropout voltage $V_{dropout}$ limits the efficiency performance. Meanwhile, the current efficiency η_I cannot be ignored in expressing the overall PCE under light loads. The standby and light-load modes occupy most of the time in portable devices. That is, a high η_I becomes an important factor in determining battery usage time. Thus, to obtain high current efficiency, the quiescent current I_q should be adaptive to the variation in loading current I_{Load} to obtain a high η_I over a wide load range. Under light loads, a low quiescent current requirement is typically set to maintain a high η_I. Having a low I_q if the system loading current is low in standby mode has become a challenge. Under heavy loads, however, a high η_I can still be maintained if I_q is increased to correspond to the loading current. Even if I_q can be adjusted adaptively by the loading current, maintaining the performance of the LDO regulator remains a challenge because of the limited bandwidth and slew rate of the operational amplifier (OPA). The required performance of LDO designs is first reviewed. Then, possible techniques to satisfy the specifications are discussed later in this book.

2.3.3 Line/Load Regulation

Line and load regulations are two important steady-state performances of LDO regulators because they indicate the percentage of output voltage variation in case a perturbation occurs from the input voltage and the output loading current, respectively. Line regulation is defined as the output voltage variation caused by an input line voltage variation, as shown in Figure 2.14. Line regulation can be expressed as Eq. (2.15), where L_o is the loop gain of the LDO regulator; g_m and r_o are the transconductance and the output resistance of the power MOSFET, respectively; β is the feedback factor. To obtain good line regulation, a high loop gain is required but

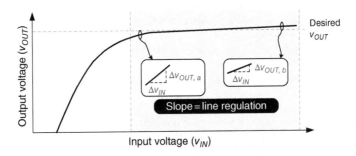

Figure 2.14 Definition of line regulation

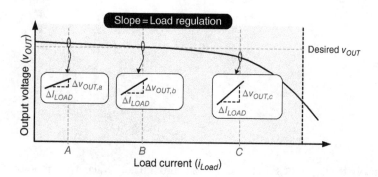

Figure 2.15 Definition of load regulation

stability may deteriorate. Thus, a trade-off occurs between regulation performance/accuracy and stability:

$$\text{Line regulation} = \frac{\Delta V_{OUT}}{\Delta V_{IN}} \approx \frac{g_m r_o}{L_o} + \frac{1}{\beta} \cdot \left(\frac{\Delta V_{REF}}{\Delta V_{IN}} \right) \qquad (2.15)$$

When V_{IN} decreases, the dropout voltage also decreases. In Figure 2.14, the line regulation is drastically degraded when V_{IN} is reduced to a certain level. Given the constraint of dropout voltage, the loop gain of the LDO regulator is degraded because of the reduced V_{IN}. That is, $\Delta V_{OUT,a} > \Delta V_{OUT,b}$, assuming that the input variation ΔV_{IN} is the same as that in Figure 2.14.

The load regulation is defined by the output voltage variation in case of a change in loading current, as shown in Figure 2.15. To obtain good load regulation, a high loop gain is required, but the stability worsens and the compensation becomes complex. The expression of load regulation is given in Eq. (2.16), where L_o is the loop gain of the LDO regulator and r_o is the output resistance of the power MOSFET:

$$\text{Load regulation} = \frac{\Delta V_{OUT}}{\Delta I_{OUT}} = -\frac{r_o}{1 + L_o} \qquad (2.16)$$

If a P-MOSFET is selected as the power transistor, then the loop gain of the LDO regulator depends strongly on the load conditions. A heavy load condition typically reduces gain, and thus the load regulation is degraded gradually as shown in Figure 2.15. When the load current change (ΔI_{Load}) is the same, the change in output voltage is large under heavy load conditions. That is, $\Delta V_{OUT,c} > \Delta V_{OUT,b} > \Delta V_{OUT,a}$ (Figure 2.15), because the loading current at point C is significantly higher than that at point A.

2.3.4 Transient Output Voltage Variation Caused by Sudden Load Current Change

The transient response can be used to test the dynamic performance of LDO regulators. This response can be categorized into two types. The first type, called load transient response, is caused by load current variation. The second type, called line transient response, is caused by line voltage variation.

Design of Low Dropout (LDO) Regulators

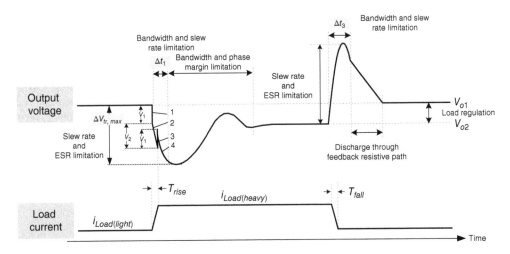

Figure 2.16 Load transient response in case of a load current change

As shown in Figure 2.16, when the LDO regulator experiences a load current step change, the output voltage requires time to be regulated by the negative feedback control. During a light-to-heavy load current change, the transient voltage variation can be divided into four parts. The first segment of the output voltage variation is contributed by the equivalent series inductance (ESL) in the output capacitor, because the voltage variation is equal to $V_1 = ESL^* di/dt$, which can also be expressed as $V_1 = ESL^*(I_{Load(heavy)} - I_{Load(light)})/T_{rise}$. Similarly, the ESR in the output capacitor also expresses the voltage drop V_2 as segment 2. When the load current is set to $I_{Load(heavy)}$, the ESL expresses the upward effect as segment 3. Finally, the output voltage experiences a transient response that is determined by the dynamic behavior of the converter and by the applied control method in segment 4. The ESR can sometimes be regarded as the current sensor to let the output ripple as the pulse width modulation (PWM) ramp in ripple-based control. The utilization of ESR as current sensor will be explained in detail in the following sections.

In case of a load current change, the pass device cannot immediately supply sufficient load current because this device works similarly to a faucet and requires time to be turned on to obtain a large current. Therefore, the output voltage experiences a voltage drop. The drop period Δt_1 depends on the closed-loop bandwidth BW_{cl} and the slew rate at the power MOSFET gate terminal. The response time can be approximated using Eq. (2.17), where C_{par} is the parasitic capacitance at the power MOSFET gate terminal and I_{sr} is the biasing current of the EA:

$$\Delta t_1 \approx \frac{1}{BW_{cl}} + t_{sr} = \frac{1}{BW_{cl}} + C_{par}\frac{\Delta V}{I_{sr}} \qquad (2.17)$$

Here, BW_{cl} is the bandwidth at light loads.

Similarly, Δt_3 can be expressed as in Eq. (2.18). However, the values of Δt_1 and Δt_3 are different, because the bandwidths under light and heavy load conditions vary:

$$\Delta t_3 \approx \frac{1}{BW_{cl}} + t_{sr} = \frac{1}{BW_{cl}} + C_{par}\frac{\Delta V}{I_{sr}} \qquad (2.18)$$

where BW_{cl} is the bandwidth at heavy loads.

In this case, we can derive the maximum transient voltage variation $\Delta V_{tr,max}$ during transient response. To simplify the expression, the ESL effect is first ignored. Thus, in case of light-to-heavy and heavy-to-light load changes, as shown in Figures 2.17 and 2.18, respectively, the power MOSFET delivers current that is either too low or too high, respectively, to the output load. Considering the energy difference between the input voltage source and the output loading, the output capacitor works as the buffer to keep the output voltage well regulated before a dynamic equilibrium is established. That is, the output capacitor can deliver energy to the output first before the gate-to-source voltage power MOSFET is adjusted to an adequate voltage (Figure 2.17). By contrast, additional energy is injected into the output capacitor before the power MOSFET is turned off to a suitable operating position (Figure 2.18). That is, the maximum transient output voltage variation $\Delta V_{tr,max}$ can be derived using Eq. (2.19) to represent the undershoot/overshoot voltage:

$$\Delta V_{tr,\max} = \frac{I_{Load(heavy)} - I_{Load(light)}}{C_{OUT}} \cdot \Delta t_1 \text{ (or } \Delta t_3) + \Delta V_{ESR} \qquad (2.19)$$

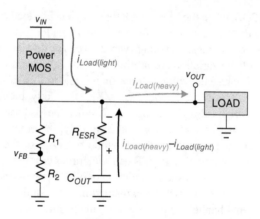

Figure 2.17 Output voltage drops in case of a light-to-heavy load change

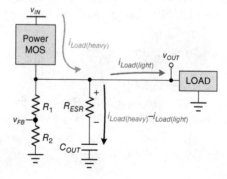

Figure 2.18 Output voltage overshoots in case of a heavy-to-light load change

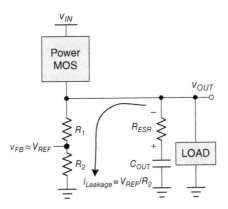

Figure 2.19 Output overshoot voltage is dissipated by the feedback divider resistors

The first term in Eq. (2.19) indicates that the output capacitor C_{OUT} functions as the buffer before the power MOSFET can work properly after the period Δt_1 or Δt_3, which is the time during which the power MOSFET starts to react to a light-to-heavy or a heavy-to-light load change, respectively. Simultaneously, current flows into or out of the output capacitor. The IR-drop voltage caused by the ESR can be represented by the second term in Eq. (2.19). Under both light-to-heavy and heavy-to-light load changes, the IR-drop voltage caused by the ESR will result in significant output transient voltage variation. Thus, ESR is undesirable in LDO regulators. In currently available commercial products, capacitors with low ESR are selected. For example, a multi-layer ceramic capacitor (MLCC) is frequently used in LDO regulators to ensure low output voltage ripples.

The overshoot voltage shown on the right of Figure 2.16 can be dissipated by the leakage current at the feedback divider resistors, as shown in Figure 2.19. Thus, the decreasing slope of the output voltage depends on the value of the divider resistors. However, a large overshoot voltage may damage the next Soc stage, which is implemented in an advanced nanometer process with low voltage characteristics. Thus, in conventional LDO regulators, a dummy resistive load can be connected to the output to dissipate additional energy. However, this process is applied at the cost of efficiency.

The other dynamic performance is line transient response. When the input voltage steps down to a smaller value, the power MOSFET will deliver a reduced load current, and thus cause a voltage dip. This response is similar to the load transient response in case of a light-to-heavy load current change. By contrast, when the input voltage steps up to a larger value, the power MOSFET will deliver a higher load current than the required output load, and thus produce an overshoot voltage. This response is similar to the load transient response in case of a heavy-to-light load current change.

Based on the preceding discussion, a conclusion can be drawn on which general methods can be used to improve the dynamic transient response. These methods include applying a wide bandwidth for a closed loop, implementing a fast slew rate at the power MOSFET gate terminal, using a large output capacitor, and reducing the ESR. These methods are similar to those presented in the discussion of the emulated reservoir and basin system. A large output capacitor can sustain the voltage variation in terms of temperature, but limits the bandwidth. With a large output capacitor, the compensation becomes too complex to be able to extend the bandwidth.

By contrast, when a large output capacitor is discarded, a wider bandwidth is required and the complexity increases. Moreover, a fast slew rate is required to compensate for the degraded transient response caused by the absence of a large output capacitor. This condition is the reason why the quiescent current is typically higher when a *C-free* structure is applied compared with when dominated pole compensation is implemented. In conclusion, a trade-off between performance and cost occurs in these methods.

2.4 Analog-LDO Regulators

As discussed in Sections 2.2 and 2.3, the design strategy in A-LDO regulators can be classified into two basic categories according to compensation skill. That is, the location of the dominant pole is first determined. In the first category, the dominant pole is located at the output because a large output capacitor is used. In the second category, called a *C-free* or capacitor-less LDO regulator, the dominant pole is generated by the Miller capacitance without using a large output capacitor.

Conventional LDO regulators use a large output capacitor with several microfarads to set the dominant pole at the output node. Considering that a large physical capacitor is used, the compensation method is called the dominant-pole compensation skill. An obvious disadvantage of this method is the use of a large off-chip capacitor, which increases the PCB area.

By contrast, the dominant pole can be generated by an amplified capacitance through the Miller effect. Using a large physical off-chip capacitor to form the dominant pole is unnecessary. The *C-free* LDO regulator got its name from the removal of the large off-chip capacitor, which was replaced with a small on-chip Miller capacitor to form the dominant pole and increase the system stability. When considering highly integrated LDO regulators for Soc applications, the *C-free* LDO regulator exhibits the advantages of reducing the effects of bonding wires and decreasing the silicon area occupied by the input/output (I/O) pads to connect off-chip passive components.

According to the following subsections, the specifications of LDO regulators for portable electronics should include three basic requirements: low quiescent current, wide input range operation, and improved regulation performance. Based on these requirements, we examine the advantages and disadvantages of the two categories of LDO regulators to identify the differences between these designs. The LDO regulators with the dominant-pole compensation skill are examined first.

2.4.1 Characteristics of Dominant-Pole Compensation

Figure 2.20 shows the LDO regulator with the dominant-pole compensation skill. In general, the dominant-pole compensation skill is utilized when the equivalent output capacitance of the next stage circuit is undetermined. Inserting a large output capacitor C_{OUT} can ensure that the dominant pole is located at the output node even if the input capacitance of the next stage circuit is unknown. Moreover, it can deal with the large load step occurring at the next stage circuit.

A basic LDO regulator includes an EA, feedback divider resistors R_1 and R_2, and a power P-MOSFET. In this regulator, a large gate-to-drain capacitance C_{gd} can be expected across the gate and drain terminals of the P-MOSFET. That is, the total parasitic capacitance C_{par} at the gate of the P-MOSFET is determined by the Miller capacitance $C_{gd}^{*}(1 + A_{v(Mp)})$, where $A_{v(Mp)}$ is the voltage gain of the P-MOSFET in the CS configuration. Moreover, the output resistance

Design of Low Dropout (LDO) Regulators

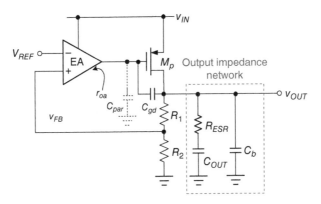

Figure 2.20 Conventional LDO regulator

r_{oa} of the EA is large because of its low quiescent current in portable electronics. The pole at the gate of the P-MOSFET, which is called the first non-dominant pole P_{1st_non}, will be at low frequencies and expressed through the following equation:

$$P_{1st\_non} = \frac{1}{r_{oa}C_{par}} \tag{2.20}$$

The dominant-pole compensation skill connects a suitable output impedance network to the output node. The constitution of the output impedance network includes two capacitors: the large output capacitor C_{OUT} and the bypass capacitor C_b, which has a smaller value than C_{OUT}. C_b can generate a pole at high frequencies to promote decade loop gain and reduce high-frequency noise. Moreover, the ESR resistance R_{ESR} of C_{OUT} should be considered in stability analysis, whereas the ESR resistance of C_b can be disregarded because C_{OUT} is larger than C_b.

In this case, all poles and zeros are examined if the output impedance network is used. The equivalent output impedance Z_{OUT} can be expressed as in Eq. (2.21), where r_{out} is the equivalent output resistance:

$$\begin{aligned}Z_{OUT} &= r_{out}||(R_1+R_2)||\frac{(1+sR_{ESR}C_{OUT})}{sC_{OUT}}||\frac{1}{sC_b}\\ &= \frac{(r_{op}||(R_1+R_2))\cdot(1+sR_{ESR}C_{OUT})}{s^2(r_{op}||(R_1+R_2))R_{ESR}C_{OUT}C_b + s\left[(r_{op}||(R_1+R_2))+R_{ESR}\right]C_{OUT} + s\left[r_{op}||(R_1+R_2)\right]C_b + 1}\end{aligned} \tag{2.21}$$

The denominator in Eq. (2.21) is a second-order polynomial that indicates the two poles contributed by C_{OUT} and C_b, as expressed in Eqs. (2.22) and (2.23), respectively:

$$P_0 = \frac{1}{r_{out}C_{OUT}}, \text{ if } (R_1+R_2) >> r_{op} >> R_{ESR} \tag{2.22}$$

$$P_{2ndhskip-1pt\_non} = \frac{1}{R_{ESR}C_b} \tag{2.23}$$

Meanwhile, the numerator in Eq. (2.12) is a first-order polynomial. A zero, which is contributed by R_{ESR}, is located at low frequencies:

$$Z_{ESR} = \frac{1}{R_{ESR}C_{OUT}} \qquad (2.24)$$

In this case, the relationship of three poles and one zero is illustrated in Figure 2.21(a). Two low-frequency poles are located below the UGF. Thus, ESR zero can alleviate the effect of the first non-dominant pole P_{1st_non} while ensuring that the second non-dominant pole P_{2nd_non} is above the UGF to satisfy the stability requirement. That is, the stability requirement sets an ESR stable region in the design of LDO regulators. If ESR zero is found at an extremely low frequency, a high gain will cause P_{2nd_non} to be located below the UGF, which reduces the system stability. By contrast, if ESR zero is located at high frequencies, then the effect of P_{2nd_non} will not be alleviated before the UGF, and thus the PM is insufficient. As shown in Figure 2.21(b), the curve of load current versus ESR value shows the ESR stable region set by the stability requirement.

Furthermore, different load currents will change the shape of the frequency response, as shown in Figure 2.21(c). According to Eq. (2.25), the dominant pole P_0 will move toward high frequencies if the load current increases. P_0 is regarded as a load-dependent pole:

$$P_0 = \frac{1}{r_{out}C_{OUT}} \propto I_{Load}, \text{ where } r_{out} \propto \frac{1}{I_{Load}} \qquad (2.25)$$

Similarly, the loop gain L_O and the UGF can respectively be derived from Eqs. (2.26) and (2.27):

$$L_O = \beta A_{EA} A_{v(Mp)} \propto g_{mp} r_{OUT} \propto \sqrt{I_{Load}} \times \frac{1}{\lambda I_{Load}} = \frac{1}{\sqrt{I_{Load}}} \qquad (2.26)$$

where $\beta = \frac{R_2}{R_1 + R_2}$, A_{EA} is the error amplifier gain, and $g_{mp}(\propto \sqrt{I_{Load}})$ is the transconductance of the power MOSFET;

$$UGF = L_O \cdot P_0 \propto \sqrt{I_{Load}} \qquad (2.27)$$

Although the loop gain L_O decreases when the load current increases, the UGF still moves toward high frequencies because P_0 moves toward such frequencies more drastically than the decrease in L_O. However, a further decrease in L_O will degrade the regulation performance without benefiting the system stability.

Figure 2.22 shows the output voltage waveforms in the LDO regulator with dominant-pole compensation. The PM degrades as the load increases, because the UGF moves toward high frequencies, and thus high-frequency poles appear below the UGF. When the PM degrades to a value that is smaller than 45°, the transient performance suffers from an obvious damping voltage variation (ΔV_{OUT}) and long settling time (t_s). When the PM degrades to a value that is smaller than 0° under an ultra-heavy load, the feedback loop becomes a positive feedback and the output voltage experiences oscillation.

Design of Low Dropout (LDO) Regulators

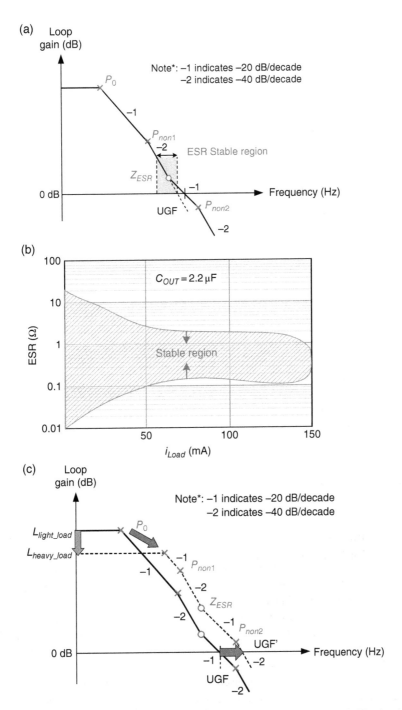

Figure 2.21 Analysis of the frequency response of the LDO regulator with dominant-pole compensation. (a) ESR stable region. (b) Relationship of load current vs. ESR. (c) Frequency response under different load current conditions

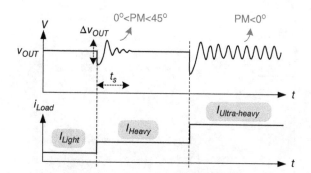

Figure 2.22 In dominant-pole compensation, the unstable phenomenon of the LDO regulator is caused by load current limitation as the load gradually changes from light to heavy

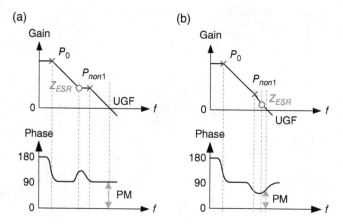

Figure 2.23 If the second non-dominant pole can be disregarded, then (a) a large ESR results in a robust and stable system, whereas (b) a small ESR can achieve a small recovery transient time

Moreover, when ESR zero is utilized to compensate for the first non-dominant pole, it can be designed at a frequency lower or higher than P_{non1}, as shown in Figure 2.23(a), (b), respectively. When P_{non2} is considerably higher than the UGF and can be disregarded, the case presented in Figure 2.23(a) is more robust because the frequency of Z_{ESR} is lower than P_{non1} and can increase the PM to approximately 90°. However, the LDO regulator with 90° PM exhibits a stable transient response but a long recovery time. Moreover, a low-frequency Z_{ESR} implies a large R_{ESR} value, which seriously deteriorates the dip voltage of the transient response. By contrast, the case shown in Figure 2.23(b) can achieve an improved transient response with an adequate PM of approximately 60°.

Furthermore, the variation in R_{ESR} caused by the temperature change leads to a difficulty in achieving system stability. A variable R_{ESR} causes different Z_{ESR} frequencies, which in turn result in a varying UGF, even if the load condition does not change. The PM can be deteriorated by P_{non2} when the UGF draws closer to P_{non2}.

Consequently, the ESR compensation skill can hardly ensure stability under different load conditions because of the non-constant unit gain frequency. That is, no simple rule can define the ESR compensation. The best means to ensure stability is to use a large ESR in the worst-case scenario. However, an undesirably large R_{ESR} may cause an unintended voltage dip when the load changes. As such, other techniques to generate an extra zero without using a large ESR for the dominant-pole compensation skill can be utilized. Moreover, inserting a buffer stage can cause the non-dominant pole that is moving toward higher frequencies to alleviate the UGF constraint, as discussed in Section 2.2.

2.4.1.1 Considering the Power MOSFET at the Triode Region

In the preceding discussion, the characteristic of frequency response is based on the power MOSFET that is operating in the saturation region. In general, when the voltage conversion ratio of the LDO regulator is large, $V_{dropout}$ is also large, which enables the power MOSFET to operate easily in the saturation region with an acceptable silicon area. By contrast, when $V_{dropout}$ is small or when the load range is large, keeping the power MOSFET operating within the saturation region is difficult. That is, the power MOSFET will operate in the triode region. In Soc applications, for example, the battery voltage and the conversion that supplies voltage to core devices are decreased simultaneously. Consequently, $V_{dropout}$ is reduced and the power transistor operates in the triode region. The characteristics of frequency response are different from the results derived in Eq. (2.27). The characteristics of the dominant pole and loop gain can be derived from Eqs. (2.28) and (2.29), respectively. In addition, the UGF expressed in Eq. (2.30) is constant and independent of load current conditions:

$$P_{don} = P_0 = \frac{1}{r_{out}C_{OUT}} \propto \sqrt{I_{Load}}, \text{where } r_{out} \propto \frac{1}{\sqrt{I_{Load}}} \qquad (2.28)$$

$$L_o = \beta A_{oa} A_p = \beta g_{ma} r_{oa} g_{mp} r_{out} \propto \frac{1}{\sqrt{I_{Load}}} \qquad (2.29)$$

with $\begin{cases} \beta, g_{ma}, \text{and } r_{oa} \text{ constant} \\ g_{mp} = \text{constant, when } V_{DS} \text{ is constant} \\ r_{out} \propto \frac{1}{\sqrt{I_{Load}}} \end{cases}$

$$\text{UGF} = L_o \cdot P_{don} \propto \frac{1}{\sqrt{I_{Load}}} \cdot \sqrt{I_{Load}} = \text{constant} \qquad (2.30)$$

Furthermore, for the power MOSFET that is operating in the triode region, the gain contributed by the power transistor degrades considerably. To maintain good output voltage regulation, enhancing the EA gain is necessary. Therefore, designing a gain-enhanced EA and an adequate pole/zero distribution is an important concern in the following discussion.

2.4.2 Characteristics of C-free Structure

An A-LDO regulator with Miller compensation can be used to supply energy to the Soc if the output capacitance from the next stage is known. No extra output capacitor is required, and thus this regulator is called the *C-free* LDO regulator. The advantage of this regulator is its high integration in the entire Soc, with an embedded power management system, because of the absence of an output capacitor. Given that the output capacitance consists of parasitic capacitance, the transient response of *C-free* LDO regulators will be considerably different from that of dominant-pole compensation, even if only one small bypass capacitor C_b is used to prevent a large dip voltage from occurring in case of load current change.

As shown in the emulated reservoir and basin system illustrated in Figure 2.3, the basin is too small to experience any sudden load current change. Thus, the only means to solve the sudden change in load current is to turn on the faucet rapidly to provide sufficient water to the output. Consequently, the *C-free* LDO regulator should be designed as a multi-stage architecture to enable it to attain a high DC gain, which can extend the bandwidth. Furthermore, a small Miller capacitor is typically used to guarantee system stability even when the DC gain is increased by a multi-stage architecture.

However, removing the output capacitor also results in several disadvantages in the regulated output voltage. Without a large output capacitor, the output voltage easily suffers from any noise disturbance or load transient variation. A noise disturbance at the output node is difficult to suppress because of the absence of a suitable bypass path. Such a path is formed by a large output capacitor in the dominant-pole compensation technique to generate high-frequency noise signals. When a heavy-to-light/light-to-heavy load transient occurs, redundant or insufficient energy will cause the voltage to overshoot or undershoot, respectively. To suppress the transient voltage variation, a large UGF is necessary to control the power transistor immediately and provide adequate power delivery. When a designer aims to obtain a large UGF for fast transient response, all non-dominant poles should be moved to frequencies that are considerably higher than the UGF. A multi-stage EA should be carefully designed to suppress any noise disturbance effectively.

Figure 2.24 shows the simplest LDO regulator designed using a three-stage OPA compensated with single Miller compensation (SMC) with a Miller capacitor C_m [1, 2]. By considering a large gate-to-drain capacitance C_{gd}, the LDO regulator has an inherent nested Miller compensation (NMC) structure [3–6]. The first stage consists of the transconductance g_{m1} of differential pairs to provide high gain. The second stage comprises positive g_{m2} as a single positive gain-boosting stage. The third stage involves the power P-MOSFET with its output impedance. Given its three stages, this structure has at least three poles. The dominant pole P_{non1} at the first stage output, which is contributed by a Miller capacitor C_m, is located at low frequencies. The first non-dominant pole P_{non2} appears at the gate of the power P-MOSFET because of its large size. The second non-dominant pole P_0, which is contributed by the output impedance because of its parasitic capacitance, is carried by the power line of the sub-supplied circuit and the equivalent output resistance.

Figure 2.25 shows a transistor-level example of the LDO regulator compensated by a single Miller capacitor C_m. The first stage, which is implemented by a differential amplifier, consists of transistors M_{P1}–M_{P3} and M_{N4}, M_{N5}. g_{m1} is the transconductance of the input differential pairs M_{P2} and M_{P3}. Transistors M_{N6} and M_{P7}, M_{P8} form the second stage to provide a positive gain-boosting effect, where g_{m2} is the transconductance of M_{N6}. The power

Design of Low Dropout (LDO) Regulators

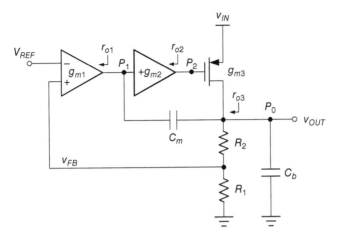

Figure 2.24 SMC for the three-stage LDO regulator

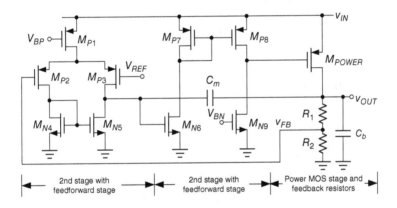

Figure 2.25 SMC capacitor for the three-stage LDO regulator

MOSFET and its output impedance form the third stage, where g_{m3} is the transconductance of M_{POWER}.

Figure 2.26 shows the simplified open-loop structure when considering all equivalent parasitic capacitances to analyze the frequency response, where g_{m1}, g_{m2}, and g_{m3} are the transconductances of the first, second, and third stages, respectively. r_{o1}, r_{o2}, and r_{o3} are the equivalent output resistance of each stage. C_1, C_2, and C_b are the equivalent output capacitance of each stage. In particular, C_{gd} is included because of the large capacitance contributed by the power P-MOSFET.

The small signal model shown in Figure 2.26 is illustrated further in Figure 2.27. Based on Kirchhoff's current law (KCL) and Kirchhoff's voltage law (KVL) theorems, the transfer function from input to output can be given as in Eq. (2.31), where A_0 is the open-loop

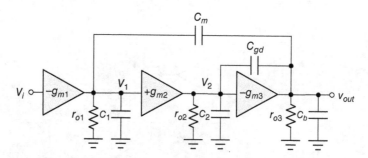

Figure 2.26 Analysis of the LDO regulator with SMC skill

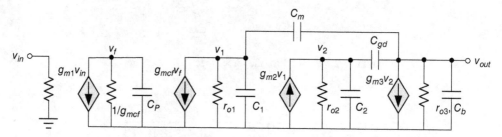

Figure 2.27 Equivalent small signal model for the basic structure of the three-stage *C-free* LDO regulator

DC gain and P_{dom} (the dominant pole of the system) are expressed in Eqs. (2.32) and (2.33), respectively:

$$\frac{v_{out}}{v_i} = A_o \frac{\left(1 + s\dfrac{C_{gd}}{g_{m3}} - s^2 \dfrac{C_m C_{gd}}{g_{m2} g_{m3}}\right)}{\left(\dfrac{s}{P_{dom}} + 1\right)\left[s^2 \dfrac{C_{gd} C_b}{g_{m2} g_{m3}} + s\dfrac{C_{gd}(g_{m3} - g_{m2})}{g_{m2} g_{m3}} + 1\right]} \quad (2.31)$$

$$A_0 = g_{m1} r_{o1} g_{m2} r_{o2} g_{m3} r_{o3} \quad (2.32)$$

$$P_{dom} = \frac{1}{r_{o1}\left[C_m(g_{m2} r_{o2} g_{m3} r_{o3})\right]} \quad (2.33)$$

Based on the numerator of Eq. (2.31), the second-order polynomial contains two zeros, which are separately located in RHP and LHP. The two zeros can be disregarded because they are located at high frequencies compared with the UGF. As mentioned in Section 2.2.2, if one LHP zero is required, then one null resistance can be in series with the Miller capacitor to convert RHP zero to LHP zero. One interesting result that can be observed is the UGF given in Eq. (2.34), which shows that its value is independent of the output capacitance because all non-dominant poles are pushed above the UGF:

$$\text{UGF} = A_0 \cdot P_{dom} = \frac{g_{m1}}{C_m} \quad (2.34)$$

The root locus of the *C-free* LDO regulator is shown in Figure 2.29 when the load current decreases. Under heavy loads, two non-dominant poles are separated at LHP. Two high-frequency non-dominant poles, namely P_{non1} and P_{non2}, can be expressed in Eqs. (2.35) and (2.36), respectively. These poles are contributed by the gate-to-drain capacitance C_{gd} of the power MOSFET and the output capacitance C_3, respectively:

$$P_{non1} \approx \frac{g_{m2}}{C_{gd}} \tag{2.35}$$

$$P_{non2} \approx \frac{g_{m3} C_{gd}}{C_2 C_b} \tag{2.36}$$

P_{non2} is load dependent because it is located at the output node. That is, P_{non2} moves toward high and low frequencies at heavy and light loads. In case of decreasing load current, two separated poles will run into each other to become complex poles, as shown at point B in Figure 2.28. In this case, these poles remain located at LHP, and thus the system is stable but the PM decreases continuously. Once the load current decreases further to light or ultra-light load conditions, these two poles may move from LHP to RHP. The system becomes unstable, as shown at point C in Figure 2.28.

The natural frequency ω_0 and the damping factor Q are derived from Eqs. (2.37) and (2.38), respectively:

$$\omega_0 = \sqrt{\frac{g_{m2} g_{m3}}{C_2 C_3}} \tag{2.37}$$

$$Q = \sqrt{\frac{C_2 C_b}{g_{m2} g_{m3}}} \frac{g_{m2} g_{m3}}{(g_{m3} - g_{m2}) C_{gd}} \tag{2.38}$$

Similarly, we can explain the scenario of complex poles under light loads at different Q values. Given the decrease in g_{m3} under light loads, the denominator will gradually decrease and even become negative, which indicates that a high Q effect becomes serious under light loads.

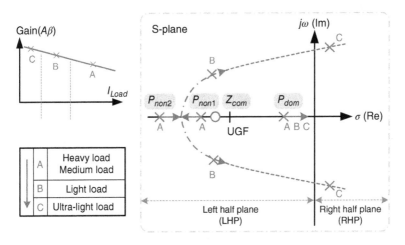

Figure 2.28 Root locus of the *C-free* LDO regulator when the load current decreases

As soon as g_{m3} becomes smaller than g_{m2}, the complex poles move from LHP to RHP. Then, the system becomes unstable. Figure 2.29(a), (b) shows the positions of the two non-dominant poles under heavy and light loads, respectively. Given the frequency peaking that occurs in Figure 2.29(b) caused by the high Q effect, the gain margin and PM are insufficient to guarantee system stability.

As illustrated in Figure 2.30, the output voltage gradually becomes unstable because of the variation in complex poles and the increase in Q, which is an expected phenomenon when the load current decreases continuously to ultra-light load conditions. Output voltage ringing and long settling time (t_s) during the transient period are caused by insufficient gain margin and PM.

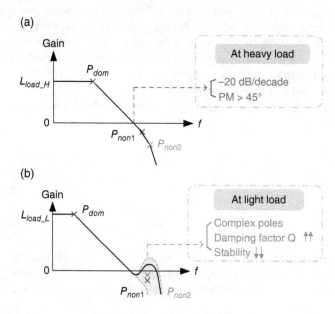

Figure 2.29 Frequency response of the *C-free* LDO regulator at (a) heavy loads and (b) light loads

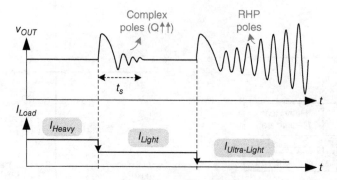

Figure 2.30 Load transient response in case of different load current changes in the *C-free* LDO regulator

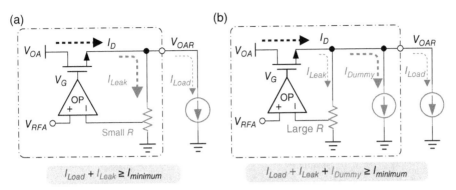

Figure 2.31 Conventional solution to alleviate minimum load limitation in the *C-free* LDO regulator with (a) an increased quiescent current or (b) a dummy loading current

Moreover, the slewing problem affects the transient response time of the *C-free* LDO regulator. Without the assistance of an output capacitor during the load transient period, the transient response can be improved by a slightly higher quiescent current to achieve a low impedance and a high slew rate (SR). However, the current efficiency is poor under light loads. Moreover, the biasing current of the second stage should be sufficiently large to drive the power MOSFET and satisfy the transient requirements. Consequently, a high quiescent current increases the values of g_{m2} and Q. That is, the minimum load current constraint will become larger than the expected value because increasing g_{m2} will cause the denominator to decrease according to Eq. (2.45). Compared with the corresponding LDO design with dominant-pole compensation, the *C-free* LDO regulator exhibits poor transient performance if the quiescent current is limited to the same value in both compensation techniques.

In designing *C-free* LDO regulators, the minimum load current requirement constraint seriously degrades the PCE under light-load or no-load conditions. An extra minimum loading current at V_{OUT} of tens or hundreds of microamperes is necessary. That is, to avoid an unstable light-load scenario, the load current should be larger than the minimum load current constraint. The *C-free* LDO regulator exhibits the disadvantage of minimum load current constraint, which may cause the efficiency to degrade under considerably light loads. In battery-powered electronics, such a regulator will shorten the battery usage time. The basic minimum load current can be established by the leakage current generated by the feedback divider resistors. A dummy loading current is sometimes consumed at the output. Consequently, efficiency is restricted because of the increased leakage current (I_{Leak}) or dummy load current (I_{Dummy}) payment to satisfy the minimum load current requirements, as shown in Figure 2.31(a), (b), respectively. Although an ultra-light load is required when a Soc system enters standby mode, the current wasted through the power MOSFET remains large and results in substantial power loss. Thus, the battery usage time of portable electronics is shortened.

2.4.2.1 Comparison between *C-out* LDO and *C-free* LDO

The preceding discussion presents the characteristics of dominant-pole compensation and the *C-free* structure. Table 2.2 provides a comparison of the LDO regulator design with different compensation methods. Considering Soc applications, the *C-free* LDO regulator is widely

Table 2.2 Comparison between dominant-pole compensation LDO and *C-free* LDO

	Dominant-pole compensation LDO	*C-free* LDO
Dominant pole	Output capacitor	Miller capacitor
UGB to load condition	1. Saturation-region power MOSFET → Dependent 2. Linear-region power MOSFET → Independent	Independent
Load range	Limitation of maximum load	Limitation of minimum load
Quiescent current	1. Increase for good SR and transient response	1. Increase for good SR 2. Increase for moving non-dominant pole to higher frequency 3. Increase for stability at ultra-light load
Defense of voltage variation	Excellent	Poor
DVS speed	Slow	Fast

utilized compared with the dominant-pole compensation LDO regulator. However, based on the table, the *C-free* LDO controls the degraded performance because of the discarded large output capacitor. To improve performance and achieve competitive specifications, the following subsections introduce the design methodologies for the *C-free* LDO regulator.

2.4.3 Design of Low-Voltage C-free *LDO Regulator*

For the Soc applications fabricated in an advanced process, the voltage stress that can be sustained by core devices is drastically scaled down. Implementing a high gain in conventional designs via a cascode structure is difficult, because of the decreasing voltage headroom caused by the decreasing input voltage. Given the decreasing voltage headroom, designing a multi-stage *C-free* LDO regulator under a low-input-voltage operation is useful. Under low-voltage operation, a conventional LDO regulator with dominant-pole compensation suffers from low DC gain, and thus the regulation performance deteriorates drastically. In the currently available nanometer CMOS process, the Soc requires the local LDO regulator to provide regulation performance and high power supply rejection (PSR). That is, a multi-stage *C-free* LDO regulator with a capacitor-less structure and a high DC gain is one suitable design for low-input-voltage operation. In this study, we present several design techniques for *C-free* LDO regulators to improve the regulation performance.

Figure 2.32 shows the schematic of a multi-stage *C-free* LDO with a compensation-enhancement multi-stage amplifier (CEMA) [7]. To obtain high gain, the structure must consist of three gain stages. Transistors M_7 and M_8 constitute the first gain stage. The second stage is composed of transistor M_{11} with a current mirror to obtain a positive gain. Meanwhile, the

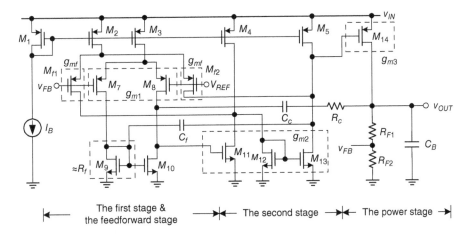

Figure 2.32 Schematic of the multi-stage *C-free* LDO regulator with CEMA

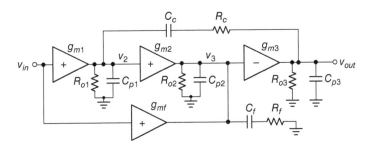

Figure 2.33 Small signal model of the CEMA structure in an open loop

power MOSFET M_{14} constitutes the third gain stage. The feedforward stage consists of transistors M_{f1} and M_{f2} to generate one compensation zero. The equivalent resistance R_f is composed of the diode-connected transistor M_9, and thus has an equivalent resistance of $1/g_{m9}$. This structure can be supplied by sub-1 V to overcome the small voltage headroom in the analog design and achieve on-chip compensation by simultaneously utilizing two small compensation capacitors. Given the cascaded structure, the voltage gain of the multi-stage amplifier can be increased even under low-voltage operation. A high gain and a large UGF can still be maintained by the multi-stage amplifier to achieve good regulation and a fast transient response, respectively.

Figure 2.33 shows the equivalent circuit of the multi-stage *C-free* LDO design, where C_c, C_f, R_c, and R_f are the compensation components. $L_O(s)$ expressed in Eq. (2.39) is the transfer function of the open loop. After simplifying Eq. (2.39), the DC gain and the positions of poles and zeros are shown in Eq. (2.40), which includes three poles (ω_{pL2}, ω_{ph1L2}, and ω_{ph2L2}) and two zeros (ω_{zL2} and ω_{zhL2}):

$$L_O(s) \approx \frac{K[1+a_1 s + a_2 s^2]}{(1+sC_c g_{m2} g_{m3} R_{o1} R_{o2} R_{o3})[1+b_1 s + b_2 s^2]} \quad (2.39)$$

where

$$K = g_{m1}g_{m2}g_{m3}R_{o1}R_{o2}R_{o3}$$

$$a_2 = \frac{g_{m1}g_{m2}R_{o1}R_cR_fC_cC_f}{g_{m1}g_{m2}R_{o1}+g_{mf}}, \quad a_1 = R_cC_c + \frac{g_{mf}R_{o1}C_c}{g_{m1}g_{m2}R_{o1}+g_{mf}} + R_fC_f$$

$$b_1 = \frac{(C_f+C_{p2})(C_c(R_{o1}+R_{o3}+R_c)+C_{p3}R_{o3})}{C_cg_{m2}g_{m3}R_{o1}R_{o3}} + \frac{C_{p3}(R_{o1}+R_c)}{g_{m2}g_{m3}R_{o1}R_{o2}}$$

$$b_2 = \frac{C_{p3}(C_f+C_{p2})(R_{o1}+R_c)}{g_{m2}g_{m3}R_{o1}}$$

$$L_O(s) \approx \frac{K\left(1+\dfrac{s}{\omega_{zL2}}\right)\left(1+\dfrac{s}{\omega_{zhL2}}\right)}{\left(1+\dfrac{s}{\omega_{pL2}}\right)\left(1+\dfrac{s}{\omega_{ph1L2}}\right)\left(1+\dfrac{s}{\omega_{ph2L2}}\right)} \qquad (2.40)$$

According to Miller's theorem, pole ω_{pL2}, which is provided by Eq. (2.41) using a small on-chip capacitor C_c within the range of several pico-farads, is regarded as the system dominant pole:

$$\omega_{pL2} = \frac{1}{C_cg_{m2}g_{m3}R_{o1}R_{o2}R_{o3}} \qquad (2.41)$$

The compensation zero ω_{zL2}, as expressed in Eq. (2.42), is generated by the feedforward gain stage in CEMA. Moreover, the on-chip compensation resistor R_c can further push ω_{zL2} toward low frequencies to achieve adequate PM:

$$\omega_{zL2} \approx \frac{g_{m1}g_{m2}}{C_c(g_{mf}+R_cg_{m1}g_{m2})} \qquad (2.42)$$

The generated low-frequency pole/zero pair, namely ω_{pL2} and ω_{zL2}, can guarantee system stability. Considering the other two non-dominant poles contributed by the second-order polynomial of the denominator of Eq. (2.39), the stability will deteriorate when complex poles are formed. To avoid complex poles, the criteria presented through Eq. (2.43) should be held:

$$b_1^2 \geq 4b_2$$

$$\Rightarrow \left(\frac{(C_f+C_{p2})(C_c(R_{o1}+R_{o3}+R_c)+C_{p3}R_{o3})}{C_cg_{m2}g_{m3}R_{o1}R_{o3}} + \frac{C_{p3}(R_{o1}+R_c)}{g_{m2}g_{m3}R_{o1}R_{o2}}\right)^2 \geq 4 \cdot \frac{C_{p3}(C_f+C_{p2})(R_{o1}+R_c)}{g_{m2}g_{m3}R_{o1}}$$

(2.43)

In particular, satisfying the aforementioned inequality becomes difficult if the value of C_{p2} is small when the LDO regulator with a small power MOSFET requires working under low-power operation. Consequently, a compensation capacitor C_f is utilized to satisfy Eq. (2.43) easily. The pole-splitting technique is achieved for two non-dominant poles by providing the low impedance of $1/g_{mf}$ through a short path with C_f at high frequencies. That is, the pole

Design of Low Dropout (LDO) Regulators

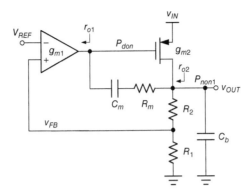

Figure 2.34 Two-stage *C-free* LDO regulator

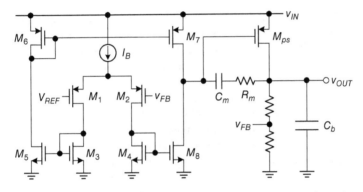

Figure 2.35 Circuit implementation of the two-stage *C-free* LDO regulator

at the second stage output can be moved toward higher frequencies. To understand the advantage of the multi-stage *C-free* LDO further, Figure 2.34 illustrates a simplified *C-free* LDO with two-stage structure for detailed comparison.

Figure 2.35 provides an example to illustrate the two-stage *C-free* LDO regulator.

The transfer function can be derived from Eq. (2.44) according to the small signal model shown in Figure 2.36:

$$L_O'(s) \approx A_0 \frac{\left(1 + \dfrac{s}{z_1}\right)}{\left(\dfrac{s}{P_{dom}} + 1\right)\left(\dfrac{s}{P_{non1}} + 1\right)} \tag{2.44}$$

where

$$A_0 = g_{m1} r_{o1} g_{m2} r_{o2}$$

$$P_{dom} \approx \frac{1}{r_{o1}\left[(C_m + C_{gd}) g_{m2} r_{o2}\right]}$$

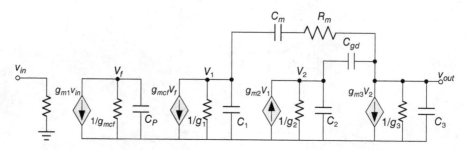

Figure 2.36 Equivalent small signal model of the two-stage *C-free* LDO regulator

$$P_{non1} = \frac{1}{r_{o2}C_b}, Z_1 = \frac{1}{R_m C_m}$$

In Eq. (2.44), the dominant pole P_{dom} is contributed by the Miller capacitor and C_{gd} at the node of the gate of the power P-MOSFET. Given that the power MOSFET is sufficiently large, C_{gd} behaves as the Miller capacitance without requiring an extra physical capacitor. By contrast, the only non-dominant pole P_{non1} is generated at the output node. To extend the UGF further, zero Z_1, which is provided by C_m and R_m, can be used to cancel the effect of P_{non1}. However, the two-stage *C-free* LDO cannot achieve a high gain. Moreover, the UGF is limited by P_{non1} under light loads because Z_1 cannot compensate for the PM over a wide load range.

Figure 2.37 presents a comparison of the frequency responses of the LDO regulators with a two-stage structure, a three-stage structure, and a three-stage structure with CEMA. The $L_O'(s)$ from the two-stage *C-free* LDO has a DC voltage gain that is significantly smaller than 40 dB, which cannot guarantee a regulated output driving voltage in the power management module. Meanwhile, the $L_O(s)$ from the three-stage structure with CEMA can effectively provide a high DC gain even under a low supply voltage operation of 1 V. That is, the DC voltage gain can be raised higher than 80 dB to achieve good regulation performance. In addition, the compensation zero enhancement and non-dominant pole splitting in CEMA are indicated. The location of the output filter pole of the DC/DC converter is also provided in Figure 2.37. The system PM varies from 55° to 80° under different load conditions.

2.4.4 Alleviating Minimum Load Current Constraint through the Current Feedback Compensation (CFC) Technique in the Multi-stage C-free LDO Regulator

In general, the LDO regulator should guarantee high PSR and fast transient response. Hence, these two factors should be simultaneously considered along with minimum load current. In this case, the multi-stage LDO with the current feedback compensation (CFC) technique can be used to alleviate the minimum load constraint for the *C-free* LDO regulator while achieving high PSR and fast transient response.

The CFC structure consists of three stages, as shown in Figure 2.38. The first stage, which involves M_1–M_6, converts differential in-signals to single-end out-signals and achieves high gain. The second stage involves M_7–M_{12} and R_B. During this stage, transistors M_7–M_{10} form

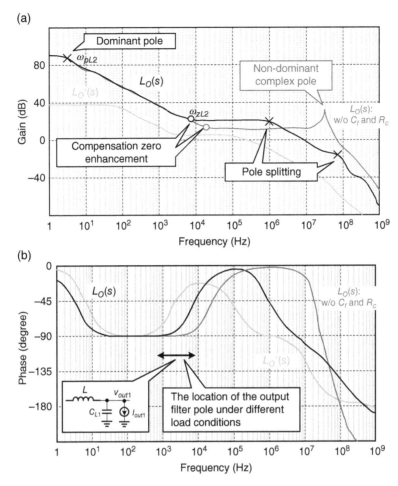

Figure 2.37 Comparison of three different EAs in frequency response. (a) Gain. (b) Phase

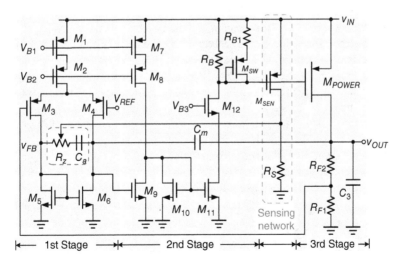

Figure 2.38 Architecture of the CFC *C-free* LDO regulator

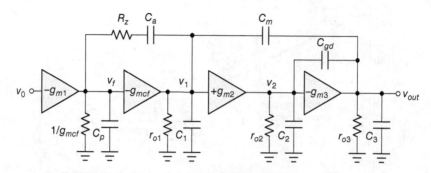

Figure 2.39 Equivalent small signal circuit of the CFC *C-free* LDO regulator

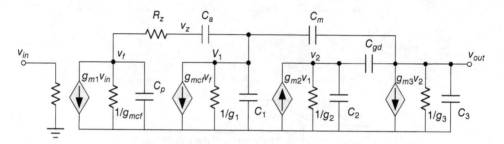

Figure 2.40 Complete small signal model of the CFC *C-free* LDO regulator

a wideband stage and create a ground reference. Meanwhile, transistors M_{11} and M_{12} form a CS stage with resistive load R_B to achieve high PSR performance. The third stage is structured by the power P-MOSFET with CS configuration. The feedback divider resistors R_{F1} and R_{F2} form the shunt feedback to regulate the output voltage. The Miller capacitor is connected to both the ground reference node, which is the output of the first stage, and the output node to avoid noise that is directly passing from the supply source to the output. The compensation network C_a and the resistance R_Z are used to control the gain and the phase dynamically under different loads during the first stage. The value of the resistance R_Z can be adjusted by the current sensing network, which is composed of M_{SEN} and R_S.

The analysis structure of the CFC *C-free* LDO regulator is illustrated in Figure 2.39. g_{m1}, g_{m2}, and g_{m3} are the equivalent transconductances of each stage. r_{o1}, r_{o2}, and r_{o3} are the equivalent output reactances of each stage. C_1, C_2, and C_3 are the lumped parasitic capacitors of each stage. Given that the value of the gate-to-drain capacitor of the power P-MOSFET is large, the parasitic capacitor should be considered and is represented as C_{gd}.

Based on the KVL and KCL theorems, the transfer function of Figure 2.40 from the input to the output can be derived from Eq. (2.45), where A_o is the open-loop DC gain as expressed in Eq. (2.46). The dominant pole of this system is P_{dom}:

$$\frac{v_{out}}{v_0} = -A_{vo} \frac{(sC_aR_z+1)(a_2s^2+a_1s-1)}{\left(1+\dfrac{s}{P_{dom}}\right)\left(1+\dfrac{s}{P_{non1}}\right)} \cdot \frac{1}{(b_2s^2+b_1s+1)} \qquad (2.45)$$

Design of Low Dropout (LDO) Regulators

where

$$a_2 = \frac{C_m C_2}{g_{m2} g_{m3}}, \quad a_1 = \frac{g_{m2} C_{gd}}{g_{m2} g_{m3}}, \quad b_2 = \frac{C_2 C_3}{g_{m2} g_{m3}}$$

$$b_1 = \frac{\left(\frac{1}{g_{mcf}} + R_z\right)\left(C_3/r_{o2} + C_2/r_{o3} + (g_{m3} - g_{m2})C_{gd}\right)C_a C_m + (C_m + 2C_a)C_2 C_3}{\left(C_3/r_{o2} + C_2/r_{o3} + (g_{m3} - g_{m2})C_{gd}\right)(C_m + 2C_a) + 2g_{m2}C_{gd}C_a + \left(\frac{1}{g_{mcf}} + R_z\right)g_{m2}g_{m3}C_a C_m}$$

$$A_{vo} = g_{m1} r_{o1} g_{m2} r_{o2} g_{m3} r_{o3} \tag{2.46}$$

The system consists of four poles and three zeros. One LHP zero Z_{LHP} for dynamic compensation is contributed by the capacitor C_a and the resistor R_Z:

$$Z_{LHP} = \frac{1}{C_a R_Z} \tag{2.47}$$

The two zeros provided by the second-order polynomial are each located in RHP and LHP. These zeros can be disregarded because they are considerably higher than the UGF. The dominant pole in Eq. (2.48), which is proportional to the root of I_{Load}, is decided by C_m and the output resistance in the first stage:

$$P_{dom} = \frac{g_1 g_2 g_3}{g_{m2} g_{m3} C_m} \propto \frac{1}{\sqrt{I_{Load}} \times \frac{1}{I_{Load}}} = \sqrt{I_{Load}} \tag{2.48}$$

The second and third non-dominant poles, P_{non2} and P_{non3}, are respectively generated by the gate-to-drain capacitance C_{gd} at the gate of the power P-MOSFET and the capacitor C_3 at the output node. Given that P_{non3} depends strongly on I_{Load}, the two poles from the second-order polynomial in the denominator of Eq. (2.45) can be expressed in different forms under various load conditions for a clear analysis. P_{non3} moves toward high frequencies under heavy loads, and thus P_{non2} and P_{non3}, which are roughly expressed as in Eq. (2.49), are far from each other. By contrast, P_{non2} and P_{non3} are close to each other when P_{non3} moves toward low frequencies under light loads.

$$P_{non2} = \frac{g_{m2}}{C_{gd}}, \quad P_{non3} = \frac{g_{m3} C_{gd}}{C_2 C_3} \quad \text{under heavy loads} \tag{2.49}$$

To discuss the effect of complex poles, the natural frequency ω_0 and a damping factor featured in P_{non2} and P_{non3} can be derived as follows:

$$\omega_0 = \sqrt{\frac{g_{m2} g_{m3}}{C_2 C_3}} \quad \text{and}$$

$$Q = \sqrt{\frac{C_2 C_3}{g_{m2} g_{m3}}} \frac{(g_2 C_3 + g_3 C_2 + (g_{m3} - g_{m2})C_{gd})(C_m + 2C_a) + 2g_{m2}C_{gd}C_a + \left(\frac{1}{g_{mcf}} + R_z\right)g_{m2}g_{m3}C_a C_m}{\left(\frac{1}{g_{mcf}} + R_z\right)(g_2 C_3 + g_3 C_2 + (g_{m3} - g_{m2})C_{gd})C_a C_m + (C_m + 2C_a)C_2 C_3}$$

$$\tag{2.50}$$

under light loads

To alleviate the influence of a high Q value, which results from the two non-dominant poles under light loads, an extra pole P_{non1}, expressed in Eq. (2.51), is inserted into the frequency around the UGF:

$$P_{non1} = \frac{g_{m2}g_{m3}C_m}{\left(\frac{1}{g_{mcf}}+R_Z\right)g_{m2}g_{m3}C_aC_m + \left(C_2/r_{o3}+g_{m3}C_{gd}\right)(C_m+2C_a)} \propto \frac{1}{k_1+k_2\sqrt{I_{Load}}} \quad (2.51)$$

where

$$k_1 = C_a\left(\frac{1}{g_{mcf}}+R_Z\right) \text{ and } k_2 = \frac{C_2(C_m+2C_a)}{g_{m2}C_m}$$

Figure 2.41 illustrates the root locus of the poles and zeros when the load current increases from light to heavy. The UGF is nearly constant in the aforementioned *C-free* structure. According to Eqs. (2.48) and (2.51), P_{dom} moves to a high frequency, whereas P_{non1} moves to a low frequency. The lower frequency of P_{non1} deteriorates the PM, and thus the compensated zero with a large resistor R_Z can dynamically move to a lower frequency under heavy loads to guarantee stability. Moreover, the complex poles will separate into two poles as the load current increases. This phenomenon implies that the effects of P_{non2} and P_{non3} can be disregarded under heavy loads.

Compared with the Q value of the basic *C-free* LDO regulator shown in Eq. (2.38), the CFC technique uses the C_a, R_Z, and dynamic output impedance of the second stage to introduce other factors and maintain a low Q value over a wide range of load conditions, as indicated in Eq. (2.50). Moreover, this technique realizes the minimum load constraint by an existing LHP pole even under ultra-light loads.

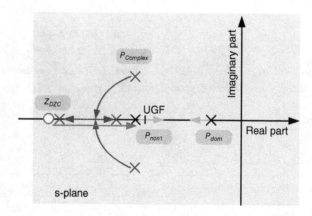

Figure 2.41 Poles and zero locations of the CFC LDO as the load increases

2.4.4.1 Ultra-Light Load Condition with Small R_Z

Under light loads, P_{non2} and P_{non3} form complex poles with a three-stage LDO design. In this case, Eq. (2.50) can be simplified to Eq. (2.52) and Q can be reduced by the second and third terms of the denominator, that is $C_2C_3(C_m+2C_a)/((1/g_{mcf})+R_Z)$ and $g_2C_aC_mC_3$, to decrease its value:

$$Q = \sqrt{\frac{C_2C_3}{g_{m2}g_{m3}}} \frac{g_{m2}g_{m3}C_aC_m}{(g_{m3}-g_{m2})C_aC_mC_{gd}+g_2C_aC_mC_3+\frac{C_2C_3(C_m+2C_a)}{\left(\frac{1}{g_{mcf}}+R_Z\right)}} \quad (2.52)$$

The compensated resistor R_Z should be small to achieve a low Q value. Based on direct observation, R_Z increases the equivalent resistance at the output node through the path of C_m at high frequencies. That is, the equivalent resistance is R_Z in series with $1/g_{mcf}$. The sensing network can adjust R_Z to have a small value under ultra-light loads. Q and P_{non1} can be derived from Eqs. (2.53) and (2.54), respectively, as follows:

$$P_{non1} = \frac{g_{mcf}}{C_a} \quad (2.53)$$

$$Q = \sqrt{\frac{C_2C_3}{g_{m2}g_{m3}} \cdot \frac{\frac{1}{g_{mcf}}g_{m2}g_{m3}C_aC_m}{(C_m+2C_a)C_2C_3}} = \sqrt{\frac{C_2C_3}{g_{m2}g_{m3}} \cdot \frac{\frac{1}{g_{mcf}}g_{m2}g_{m3}C_m}{\left(\frac{C_m}{C_a}+2\right)C_2C_3}} \quad (2.54)$$

Although R_Z is helpful in reducing Q, the peaking of Q still affects stability. To improve the constraint of minimum load current further, pole P_{non1} is set at a frequency that is near the UGF. Figure 2.42 illustrates the contribution of P_{non1}, which accelerates the decay of the gain after surpassing the frequency of the UGF.

To maintain the PM at approximately 60°, P_{non1} must be placed twice higher than the UGF, as shown in Eq. (2.55). This setup results in a low Q of 5.

$$P_{non1} > 2 \cdot \text{UGF} = 2\frac{g_{m1}}{C_m} \quad (2.55)$$

To avoid complex poles and an unstable system, the first non-dominant pole must be set to at least half the natural frequency, as shown in Eq. (2.56). Given that the magnitude rolls off with a slope of −20 dB/decade after the UGF and a slope of −40 dB/decade after the first non-dominant pole, a gain margin of at least 18 dB exists for complex poles and low Q approximations.

$$P_{non1} < \frac{1}{2}\omega_o \quad (2.56)$$

The compensated zero is located at a relatively high frequency compared with the UGF, because R_Z is small. Consequently, the PM of the overall system can be determined by Eq. (2.57), which is near 60°:

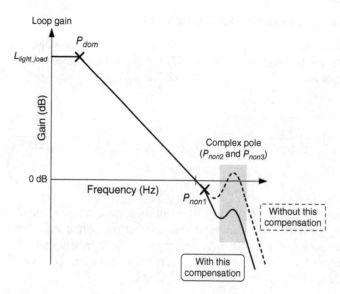

Figure 2.42 CFC capacitor-free LDO under a light load condition

$$PM = 180° - \tan^{-1}\left(\frac{UGF}{p_{don}}\right) - \tan^{-1}\left(\frac{UGF}{p_{non1}}\right) - \tan^{-1}\left(\frac{\frac{UGF}{\omega_o}}{Q\left[1 - \left(\frac{UGF}{\omega_o}\right)^2\right]}\right)$$

$$= 90° - \tan^{-1}\left(\frac{UGF}{p_{non1}}\right) - \tan^{-1}\left(\frac{\frac{UGF}{\omega_o}}{Q\left[1 - \left(\frac{UGF}{\omega_o}\right)^2\right]}\right) = 90° - 26.56° - 2° \approx 60°$$

(2.57)

Based on the preceding analysis, compensation capacitors C_m and C_a can be derived from Eqs. (2.58) and (2.59), respectively, as follows:

$$C_m = 2\frac{g_{m1}}{g_{mcf}} C_a = 4g_{m1}\sqrt{\frac{C_2 C_3}{g_{m2} g_{m3}}} \tag{2.58}$$

$$C_a = 2g_{mcf}\sqrt{\frac{C_2 C_3}{g_{m2} g_{m3}}} \tag{2.59}$$

2.4.4.2 Light-to-Medium Load Condition

As the load increases from light to medium, that is 1–10 mA, P_{non1} moves toward lower frequencies because R_Z increases. As the R_Z value increases, P_{non1}, ω_0, and Q can be derived from

Eqs. (2.60) to (2.62), respectively. In this case, a low Q can also be maintained, and Eq. (2.63) still holds true because of the slight probability of R_Z:

$$P_{non1} = \frac{1}{\left(\frac{1}{g_{mcf}} + R_Z\right)C_a} \tag{2.60}$$

$$\omega_o = \sqrt{\frac{g_{m2}g_{m3}}{C_2 C_3}} \tag{2.61}$$

$$Q = \sqrt{\frac{C_2 C_3}{g_{m2}g_{m3}}} \cdot \frac{g_{m2}}{C_{gd}} \tag{2.62}$$

$$P_{non1} > 2 \cdot \text{UGF} \tag{2.63}$$

The PM of the overall system can be determined by P_{dom} and P_{non1}, as shown in Eq. (2.64):

$$\begin{aligned} \text{PM} &= 180° - \tan^{-1}\left(\frac{\text{UGF}}{P_{dom}}\right) - \tan^{-1}\left(\frac{\text{UGF}}{P_{non1}}\right) \\ &= 90° - \tan^{-1}\left(\frac{\text{UGF}}{P_{non1}}\right) \\ &\approx 60° \end{aligned} \tag{2.64}$$

2.4.4.3 Heavy Load Condition with Large R_Z

As the load increases further from 10 to 100 mA, P_{non1}, as shown in Eq. (2.65), will move to a lower frequency, because both the output reactance g_3 and R_Z increase. The low-frequency zero Z_{DZC}, expressed in Eq. (2.66), will alleviate the P_{non1} effect:

$$P_{non1} = \frac{g_{m2}g_{m3}C_m}{\left(\frac{1}{g_{mcf}} + R_Z\right)g_{m2}g_{m3}C_a C_m + g_3 C_2(C_m + 2C_a)} \tag{2.65}$$

$$Z_1 = \frac{1}{R_Z C_a} \tag{2.66}$$

The second-order polynomial in Eq. (2.45) can be simplified to Eq. (2.67) under heavy loads. If the discriminant function of the second-order polynomial in Eq. (2.67) is smaller than zero, then a pair of complex poles exists in the system. However, these complex poles are located at high frequencies and have no effect on the system stability. If the discriminant function of the second-order polynomial in Eq. (2.67) is larger than zero, then two separate poles exist in the system. Finally, if a second non-dominant pole exists, then the locations of the poles are shown in Eq. (2.68), which are found at high frequencies and have negligible effects on the system stability:

$$s^2 \frac{C_2 C_3}{g_{m2} g_{m3}} + s \frac{g_3 C_2 + g_{m3} C_{gd}}{g_{m2} g_{m3}} + 1 \approx s^2 \frac{C_2 C_3}{g_{m2} g_{m3}} + s \frac{g_{m3} C_{gd}}{g_{m2} g_{m3}} + 1 \qquad (2.67)$$

$$P_{non2} = \frac{g_{m2} g_{m3}}{g_3 C_2 + g_{m3} C_{gd}} \text{ and } P_{non3} = \frac{g_3 C_2 + g_{m3} C_{gd}}{C_2 C_3}$$

$$\Rightarrow P_{non2} = \frac{g_{m2}}{C_{gd}} \text{ and } P_{non3} = \frac{g_{m3} C_{gd}}{C_2 C_3} \qquad (2.68)$$

The overall system contains two low poles and one low zero. The system stability can be determined as follows:

$$\begin{aligned} \text{PM} &= 180° - \tan^{-1}\left(\frac{\text{UGF}}{p_{don}}\right) - \tan^{-1}\left(\frac{\text{UGF}}{p_{1st\text{-}non}}\right) + \tan^{-1}\left(\frac{\text{UGF}}{z_{LHP}}\right) \\ &= 90° - \tan^{-1}\left(\frac{\text{UGF}}{p_{1st\text{-}non}}\right) + \tan^{-1}\left(\frac{\text{UGF}}{z_{LHP}}\right) \\ &\approx 60° \end{aligned} \qquad (2.69)$$

2.4.4.4 Summary of CFC Capacitor-Free LDO Regulators

The frequency responses of CFC capacitor-free LDO regulators under different load conditions are summarized in Figure 2.43. Under ultra-light load conditions smaller than 1 mA, the dominant pole contributes a 90° phase shift. The first non-dominant pole contributes a near 30° phase shift, and the complex poles contribute minimal phase shift. The overall system stability can be maintained at 60°. Under light to medium load conditions, that is approximately 1–10 mA, the dominant pole contributes the same phase shift. The first non-dominant pole moves to lower frequencies and contributes a 30° phase shift. The complex poles move to higher frequencies and do not contribute any phase shift. Thus, the system stability is maintained at 60°. Finally, under medium to heavy load conditions, approximately 10–100 mA, the dominant pole contributes the same phase shift. The first non-dominant pole moves further

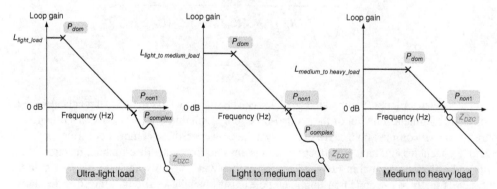

Figure 2.43 Variations of poles and zero locations in the CFC *C-free* LDO regulator over a wide load range

toward even lower frequencies, and the dynamic zero also moves toward lower frequencies to alleviate the pole effect. Therefore, the PM of the CFC *C-free* LDO regulator can be maintained at 60° within the entire load range.

2.4.5 Multi-stage LDO Regulator with Feedforward Path and Dynamic Gain Adjustment (DGA)

With the development of advanced nanometer processes in recent years, the supply voltage has been continuously scaled down. Consequently, LDO regulators are required to operate under sub-1 V. To understand the design in detail, an example implemented in a 65 nm CMOS technology is presented here. The design specifications are described as follows. The input voltage V_{IN} can range from 0.9 to 1.2 V; the latter is the voltage stress of the 65 nm core device. The dropout voltage is approximately 200 mV, with a maximum load current of 100 mA. To achieve stable operation, a minimum load of 50 μA is required.

Figure 2.44 shows a schematic of the four-stage LDO with feedforward path and dynamic gain adjustment (DGA). Moreover, the feedforward gain stage contributed by M_{13} is included to accelerate the transient response at the gate of the power P-MOSFET M_P. As mentioned earlier, the feedforward path contributes LHP zero in the frequency domain to improve stability. The DGA mechanism, which consists of M_{14} and R_S, is used to sense the load current and adjust the resistance value of R_p to guarantee stability even at ultra-light loads.

Figure 2.45 shows the structure of the sub-1 V multi-stage LDO design. Considering its cascade structure, the voltage gain of the multi-stage amplifier can be increased under low-voltage operation. A high-voltage gain is basically composed of three gain stages: g_{m1}, which includes g_{md}, g_{m2}, and g_{mL}. Moreover, a feedforward gain stage g_{mf} accelerates the slew rate at the gate of the pass element and generates LHP zero to improve LDO stability. In addition, the compensation capacitor C_c is amplified by Miller's theorem to determine the dominant pole in the system. C_f is inserted to split high-frequency complex poles under light loads and prevent

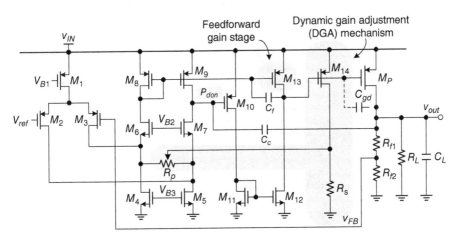

Figure 2.44 Schematic of the sub-1 V four-stage LDO with feedforward gain stage and DGA mechanism

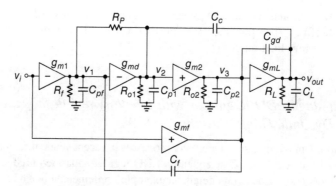

Figure 2.45 Structure of the sub-1 V multi-stage LDO regulator

non-dominant complex poles from moving toward RHP to ensure stability. Furthermore, resistance R_p is proportionally adjusted to I_{Load} by the DGA mechanism. Hence, R_p can be used to adjust the DC voltage gain and the UGF to avoid phase deterioration under ultra-light loads. The transfer function is given in Eq. (2.70) according to Figure 2.45. The exact analysis of the multi-stage LDO can be illustrated through three different output load conditions, as discussed in the following subsections.

$$\frac{v_{out}}{v_{in}} \approx \frac{g_{m1}g_{m2}g_{mL}R_{o1}R_{o2}R_L \frac{R_f(g_{md}R_p-1)}{(R_p+g_{md}R_{o1}R_f)} \left[1+s\left(R_f C_f + \frac{g_{mf}R_{o1}C_c}{g_{m1}g_{m2}R_{o1}+g_{mf}}\right)\right]}{(1+sC_c g_{m2}g_{mL}R_{o1}R_{o2}R_L)\left[1+s\frac{(R_p+R_f)(g_{mL}(C_{gd}+C_f)-g_{m2}(C_c+C_L)+g_{mf}(C_f+C_2+C_c))}{g_{m2}g_{mL}} + s^2 \frac{C_L(C_{gd}+C_f+C_2)}{g_{m2}g_{mL}}\right]} \quad (2.70)$$

2.4.5.1 Medium to Heavy Load Currents (g_{mL} Significantly Larger than g_{m1} and g_{m2})

As the load current increases from 10 to 100 mA, the transfer function in Eq. (2.70) is approximated as:

$$\frac{v_{out}}{v_{in}} \approx \frac{g_{m1}g_{m2}g_{mL}R_{o1}R_{o2}R_L\left(1+s\frac{g_{mf1}C_c}{g_{m1}g_{m2}}\right)}{(1+sC_c g_{m2}g_{mL}R_{o1}R_{o2}R_L)\left[1+s\frac{(C_{gd}+C_f+C_2)(R_p+R_f)}{g_{m2}} + s^2\frac{C_L(C_{gd}+C_f+C_2)}{g_{m2}g_{mL}}\right]} \quad (2.71)$$

The dominant pole ω_{don}, a feedforward LHP zero ω_{zf}, and a system UGF are given respectively as:

$$\omega_{don} = \frac{1}{C_c g_{m2}g_{mL}R_{o1}R_{o2}R_L} \quad (2.72)$$

$$\omega_{zf} = \frac{g_{m1}g_{m2}}{C_c g_{mf1}} \quad (2.73)$$

$$\text{UGF} = \frac{g_{m1}}{C_c} \quad (2.74)$$

The first non-dominant pole at the gate of the power MOSFET is close to the feedforward zero ω_{zf}. Moreover, the second non-dominant pole at the output node is located at high frequencies because of the large g_{mL}. Consequently, a one-pole system is achieved with a theoretical PM of 90°. Thus, the stable operation of the multi-stage LDO is derived under medium to heavy load conditions because of the split high-frequency LHP non-dominant poles.

2.4.5.2 Light to Medium Output Load Currents (g_{mL} Slightly Larger than g_{m1} and g_{m2})

As the load current increases from a light to a medium load range, that is approximately 1–10 mA, the non-dominant pole at the output node moves toward the origin, which results in the occurrence of complex poles, with the first non-dominant pole at the gate of the power MOSFET. Based on the transfer function of Eq. (2.70), the relationship among non-dominant complex poles can be exhibited as follows:

$$\text{UGF} = \frac{g_{m1}}{C_c} \quad (2.75)$$

Under this condition, complex poles are located at high frequencies, but will have an influence on the phase shift. The complex pole frequency ω_o and Q are respectively indicated by:

$$\omega_o = \sqrt{\frac{g_{m2}g_{mL}}{C_L(C_{gd} + C_f + C_2)}} \quad (2.76)$$

$$Q = \sqrt{\frac{C_L(C_{gd} + C_f + C_2)}{g_{m2}g_{mL}}} \\ \times \left[\frac{g_{m2}g_{mL}}{(R_p + R_f)\left(g_{mL}(C_{gd} + C_f) - g_{m2}(C_c + C_L) + g_{mf}(C_f + C_2 + C_c)\right)} \right] \quad (2.77)$$

Increasing g_{mL} and C_f can push complex poles toward higher frequencies, and thus decrease the Q value to improve the system stability. Thus, implementing C_f can minimize the unintended phase shift caused by complex poles under light to medium load conditions.

2.4.5.3 Ultra-Light Load Current (g_{mL} Smaller than g_{m1} and g_{m2})

Under ultra-light loads (i.e., below 1 mA) a small g_{mL} will lead to an unstable LDO system. As mentioned in the discussion on light to medium load conditions, non-dominant complex poles with a high Q factor reduce the PM. Consequently, a higher Q factor is derived with decreasing

g_{mL}. Non-dominant complex poles will move toward the RHP to deteriorate system stability. The boundary condition between stable and unstable operations can be expressed as:

$$g_{mL}(C_{gd}+C_f) - g_{m2}(C_c+C_L) + g_{mf}(C_f+C_2+C_c) > 0$$
$$\Rightarrow g_{mL} > \frac{g_{m2}(C_c+C_L) - g_{mf}(C_f+C_2+C_c)}{(C_{gd}+C_f)} \quad (2.78)$$

RHP poles can be eliminated when the s term of the second-order function in Eq. (2.70) is positive. Thus, a minimum value of g_{mL} must be contained to ensure stability. A minimum load current of the multi-stage LDO can be obtained to maintain a minimum g_{mL} value.

Nevertheless, non-dominant complex poles under ultra-light loads still affect the phase shift around the UGF. A load-dependent resistance R_p, proportional to the output loading current, is used to adjust the DC voltage gain in the DGA mechanism. Given the high voltage gain at ultra-light loads, decreasing the DC voltage gain and the UGF does not obviously influence the transient response. The DGA mechanism can minimize the effect caused by non-dominant complex poles, and thus ensure LDO stability.

Figure 2.46 shows the frequency response under an ultra-light load condition of 50 μA with and without a flying capacitor C_f and the DGA mechanism. In the case without both C_f and the DGA mechanism, non-dominant complex poles appear near the UGF and result in a poor PM. With the implementation of C_f, the Q factor decreases but an inadequate PM remains. The DGA mechanism can slightly decrease the DC voltage gain and the UGF of the LDO to enhance the system PM. Thus, operation under an ultra-light load can be realized.

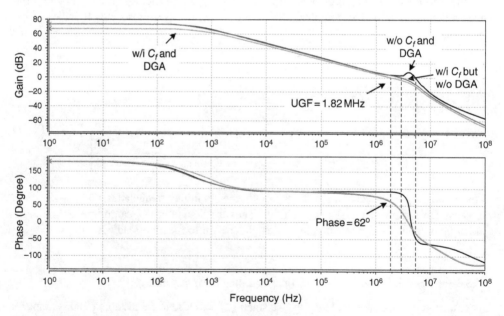

Figure 2.46 Comparison of frequency responses under a 50 μA load condition at different implementations

Design of Low Dropout (LDO) Regulators

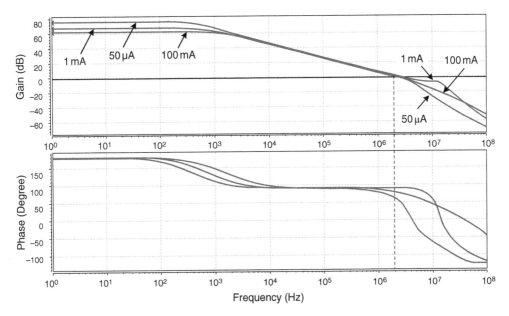

Figure 2.47 Frequency responses under different output load conditions

Figure 2.47 shows the frequency responses under different load conditions, ranging from 50 μA to 1 mA, and then to 100 mA. With the DGA mechanism, which can dynamically adjust the DC voltage gain through load detection, the correct operating function is achieved under all load conditions.

Figure 2.48 shows the load transient response under an input voltage of 1 V. In Figure 2.48(a), the output voltage is regulated to 0.8 V with undershoot and overshoot voltages of approximately 52 and 85 mV, respectively, when the load changes from 50 μA to 100 mA, and vice versa, within a load step of 1 μs. The transient recovery time is approximately 0.6 μs. By contrast, the transient response under an output voltage of 0.6 V is presented in Figure 2.48(b). The undershoot and overshoot voltages are 43 and 70 mV, respectively, with a load step ranging from 50 μA to 100 mA, and vice versa. The transient recovery time is approximately 0.6 μs.

2.5 Design Guidelines for LDO Regulators

In this section of the book we provide typical guidelines for designing LDO regulators. The crucial design parameters that should be considered carefully before implementing a circuit are as follows:

- reading the specifications carefully
- load considerations
- output capacitor considerations
- on-chip capacitor considerations
- MOSFET-cap considerations
- system compensation considerations

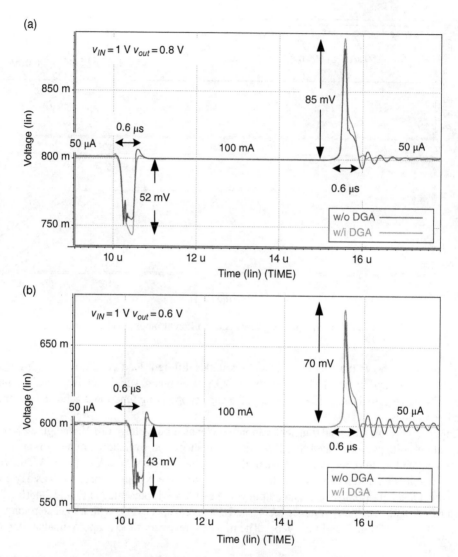

Figure 2.48 Load transient responses under 50 μA to 100 mA, and vice versa. (a) $V_{IN} = 1$ V, $V_{OUT} = 0.8$ V. (b) $V_{IN} = 1$ V, $V_{OUT} = 0.6$ V

- voltage range
- power MOSFET size estimation.

Before starting, the specifications should be read carefully. Compensation methods and techniques have both advantages and limitations, that is, the actual design exhibits trade-off with performance. Consequently, some important conditions provided in the specifications can help designers determine an adequate structure.

For example, load considerations include the desired driving capability and the load range. The driving capability suggests that designers should provide a wide bandwidth, a large SR, or a large output capacitance to sustain V_{OUT} regulation during instances of changing loads. To satisfy the load range requirement, a designer should be careful with the maximum/minimum load limitation when using dominant-pole compensation or a *C-free* structure. In some applications, such as power for PA, the required large capacitance with nano- to micro-farads at the node of the supply voltage is unavoidable. Such a large capacitance can provide low frequency. For the desired gain, PM, and UGF, a pole can obstruct frequency response compensation regardless of whether dominant-pole compensation or a *C-free* structure is applied. By contrast, on-chip capacitances are utilized for the desired poles or zeros to provide adequate frequency response. In this case, the capacitance value should be determined carefully when considering the cost of the silicon area. On-chip capacitance can be realized using different types of device, such as poly–insulator–poly (PIP), metal–insulator–metal (MIM), or MOSFET-capacitance. In particular, the cross voltage can directly vary the capacitance when MOSFET-capacitance is used. Thus, the designer should ensure that cross voltages are sufficiently large under all conditions with varying loads, input voltages, and output voltages. In general, the specifications of load consideration and output capacitance loading provide sufficient information for the designer to plan the compensation strategy. Meanwhile, the operating ranges of the input or output voltage are related to the frequency response and stability because different cross voltages can vary the gain, small signal resistance, and operating region. Once a power transistor operates in the linear and saturation regions at different input/output voltages, the gain and small signal resistance of the transistor significantly change and seriously influence the frequency response. In this case, compensation should be adjusted adequately for different situations. Finally, the size of the power transistor is estimated. The desired dropout voltage and driving capability can basically determine the minimum size of a power transistor. Hence, the designed size should be larger than the minimum size. By contrast, the maximum size of a power transistor should be limited not only because of concerns about using unnecessary silicon area, but also because an extremely large power transistor can degrade the DC operation of the EA under light loads. For example, to drive a large p-type power transistor under a light load, the driving voltage level should be extremely high. This voltage may be the upper voltage headroom of the last EA stage, and thus the gain is decreased drastically. Consequently, a power transistor with adequate size should be designed.

2.5.1 *Simulation Tips and Analyses*

After circuit implementation, performance can be checked through the following simulations. AC analysis is used to analyze the frequency response, including gain, PM, and UGF. Moreover, checking the slew rate in case of a large signal operation is also important. DC analysis is employed to check the operating region of each MOSFET, and DC specifications such as line and load regulations. Transient analysis is utilized to check dynamic performance during the transient period and power-on sequence. Planning the power-on sequence with an adequate soft start is important to ensure safe operation, without any damage caused by overcurrent, overvoltage, and so on, before quiescent operation is established.

Finally, full-chip simulation should be carefully executed. In this step, designers should consider possible parasitic components existing on the silicon and the PCB. Such components include the capacitance of the back-end circuit; on-resistance through the power delivery path; and R, L, and C in bond wires, among others. Moreover, designers should consider process, voltage, and temperature (PVT) variations. However, considering numerous factors draws from the simulation time. As previously stated, appropriate design and simulation considerations can effectively guarantee the yield rate after fabrication:

- AC analysis
- DC analysis
- transient analysis
- power-on sequence and soft-start design
- full-chip simulation
- consideration of PVT variations.

Designers should know how to utilize simulation tools to analyze circuit performance, as discussed in the preceding paragraphs. These tools can also be used to verify the original functions and yield rate after fabrication. Therefore, the authors strongly encourage designers to anticipate all possible situations that chips may encounter in practice. Designers should first learn to establish simulation environments. Understanding design issues and practical system operating environments is helpful for this purpose. However, checking all possible considerations is impossible because such a process not only increases the design complexity but also requires considerable simulation time. Therefore, the trade-off between robust simulation and efficient design procedure should be balanced carefully.

2.5.2 Technique for Breaking the Loop in AC Analysis Simulation

An open-loop structure without a feedback path is applied to reduce complexity or alleviate the BW constraint, while the regulated voltage is varied significantly because of the loading effect. To guarantee the quality of the regulated voltage, a closed-loop structure is used as the basis to design power regulators. The stability of the closed-loop structure should be designed carefully; such stability is generally confirmed by the gain margin and PM. However, a closed-loop structure experiences difficulties in AC simulation because the processes of injecting the stimulus and obtaining the output signal should be determined. The characteristics of a closed-loop system can be obtained from an open-loop analysis whose bode plot reflects its corresponding gain margin and PM. That is, the analysis is simplified by conducting an open-loop analysis, which is achieved by breaking the loop. Thus, determining how to break the loop adequately is important to maintain simulation accuracy without being affected by the loading effect. Two methods are recommended for this purpose, and these are introduced as follows.

Figure 2.49 illustrates a typical circuit block that represents a power converter with a feedback loop. Block A provides the open-loop gain, whereas block β provides the path to feed the output voltage V_{OUT}, which enables the reference voltage V_{REF} to regulate the output voltage.

When breaking the loop, the input and output impedances should be considered at the broken node to build an accurate model. In Figure 2.50, the output impedance $Z_{o,\beta}$ of block β is added to the front end of the breaking node while the input impedance $Z_{i,A}$ of block A is added to the back end of the breaking node. In Figure 2.51, the test AC signal source v_{ac}, a test inductor L, and a test capacitor C are used to extract the transfer function during simulations. The test inductor L conducts DC information to maintain the DC operating voltage in a steady state. Meanwhile, L blocks AC information, and thus the circuit forms as an open loop in view of the small signal aspect. By contrast, the test capacitor C conducts only the test AC signal

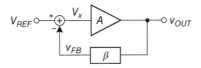

Figure 2.49 Model of a typical converter circuit

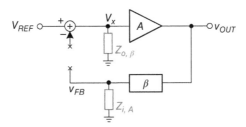

Figure 2.50 Basic principle for breaking a feedback loop

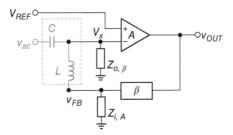

Figure 2.51 General method for breaking a feedback loop

v_{ac}, but blocks any DC information. Consequently, the DC voltage of V_X will not be influenced by the AC signal source. Given that L and C are expected to be ideal components with infinite values, their values should be high, on the order of several megas. For example, L can be 100 MH and C can be 100 MF. The transfer function can be obtained by observing the value of v_{FB}/v_{ac}.

However, deriving accurate input/output impedance values is not easy or convenient. As such, one possible method to break a loop is to select a high impedance node as the breaking node to reduce the complexity of the analysis. For example, a MOSFET gate is typically chosen as the breaking node because of its high impedance. Consequently, the input impedance $Z_{i,A}$ can be disregarded. Moreover, the output impedance $Z_{o,\beta}$ can also be ignored. The AC and DC components at V_x can be determined directly by v_{ac} and v_{FB} through C and L, respectively, to derive the transfer function. This procedure does not provide an exact solution to the transfer function, because the assumption is high impedance at the gate of the power MOSFET. Furthermore, designers may experience a situation in which the high impedance node is difficult to find in the loop. When the breaking node is not a high impedance node, the corresponding impedance $Z_{i,A}$ should be considered; however, this action increases the complexity of analysis.

Figure 2.52 presents another approach to establish an accurate model related to the original circuit. Test capacitors, C_1 and C_2, and test inductors, L_1 and L_2, provide adequate AC and DC paths for the same purpose mentioned earlier. The values of these test passive components are set on the order of several megas. When considering the DC domain, the DC operating voltage

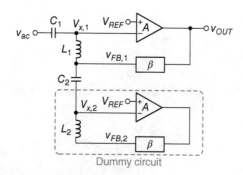

Figure 2.52 Accurate model for breaking a feedback loop

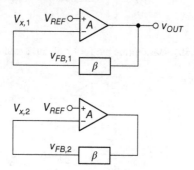

Figure 2.53 DC operating points correctly established

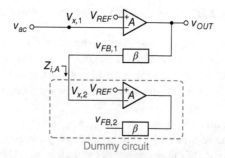

Figure 2.54 In the AC domain, the circuit topology reflects an exact solution to the transfer function without being affected by the selection of the breaking node

of each circuit block can be established as shown in Figure 2.53, because of the short paths provided by the inductors L_1 and L_2, and the open paths provided by C_1 and C_2. By contrast, when considering the AC domain, Figure 2.54 shows that the capacitors C_1 and C_2 are shorter and the inductors L_1 and L_2 are opened, and thus the circuit block exhibits one open loop and the dummy circuit is utilized to provide $Z_{i,A}$. Consequently, the AC and DC characteristics are confirmed, and thus this approach can reflect an accurate model without being affected by the selection of the breaking node. Similarly, the transfer function can be obtained from the value of $v_{FB,1}/v_{ac}$.

2.5.3 Example of the Simulation Results of the LDO Regulator with Dominant-Pole Compensation

In this subsection, we provide basic simulation methods to verify the performance and function of the LDO regulator whose specifications are listed in Table 2.3. These will guide the reader to perform the simulations step by step. The verified performances include stability, dropout voltage, line regulation, load regulation, overshoot/undershoot voltage, and settling time during transient response.

In AC analysis, Figure 2.55 illustrates the closed loop in the LDO regulator with dominant-pole compensation. The first step is to identify the possible location of the breaking node in the entire LDO regulator. In general, the input impedance of the EA is assumed to be infinitely large because the first stage is implemented by a CS amplifier. Thus, Figure 2.56(a) shows that the breaking node is selected on the path from V_{FB} to the non-inverting terminal of the EA. Figure 2.56(b) shows that L_{ideal} and C_{ideal} are added to provide the paths for conducting DC and AC signals, respectively, to determine the quiescent operating points and the transfer function, respectively. The conducting path for DC signals can maintain the function of DC operation with adequate values of V_{IN}, V_{OUT}, V_{REF}, and I_{Load}. Based on DC operation, the conducting path for AC signals can determine the frequency response. Readers may believe that this process is an alternative method to connect DC voltage sources directly to the breaking node as quiescent operating points. However, a high loop gain in the LDO regulator will result in a severe error if a slight error caused by the deviation of quiescent operating points occurs.

Table 2.3 Specifications of the LDO regulator

Parameter	Value	Unit
Compensation	Dominant-pole compensation	
Technology	0.25 μm CMOS (5 V device)	
Input voltage (V_{IN})	3.0–4.5	V
Output voltage (V_{OUT})	2.7	V
Reference voltage (V_{REF})	1.2	V
Load current (I_{Load})	1–100	mA
Output capacitor (C_{OUT})	4.7	μF
ESR of output capacitance (R_{ESR})	5	Ω

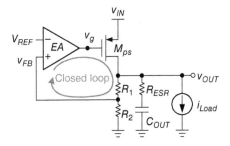

Figure 2.55 Illustration of the LDO regulator for AC analysis

Given the simple single-stage design, the poles and zeros of the EA, which are located at high frequencies, can be disregarded. Therefore, two poles, P_1 and P_2, which are located at V_{OUT} and v_g, respectively, are considered for stability analysis. Moreover, ESR zero Z_1, which is contributed by R_{SER} and C_{OUT} at V_{OUT}, can have a phase boost of 90°. If dominant-pole compensation is used, Figure 2.57 shows the frequency response when V_{IN}, V_{OUT}, and I_{Load} are 3 V, 2.8 V, and 100 mA, respectively. P_1 is located at approximately 1 kHz and zero Z_1 is designed at approximately 4 kHz to cancel the effect of P_2. Compared with those in Figure 2.57, the frequency responses in Figures 2.58 and 2.59 show the different locations of zero Z_1 and provide different performances. In Figure 2.58, the UGF and PM increase and decrease, respectively, because a large R_{ESR} (50 Ω) causes Z_1 to be located at lower frequencies compared with P_2. In Figure 2.59, the UGF and PM decrease and increase, respectively, because a smaller R_{ESR}

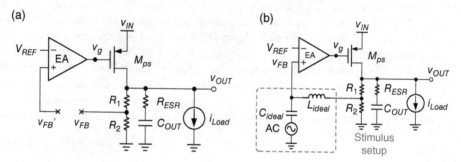

Figure 2.56 (a) Conceptual approach to break the loop at the high impedance node. (b) Setup of the stimulus

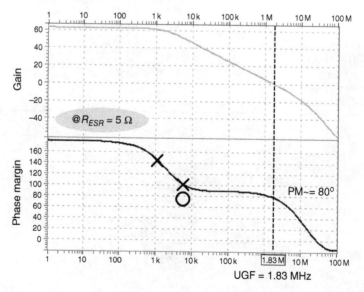

Figure 2.57 Frequency response with R_{ESR} = 5 Ω and I_{Load} = 100 mA

Design of Low Dropout (LDO) Regulators

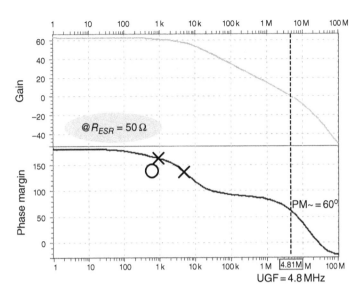

Figure 2.58 Frequency response with $R_{ESR} = 50\ \Omega$ and $I_{Load} = 100$ mA

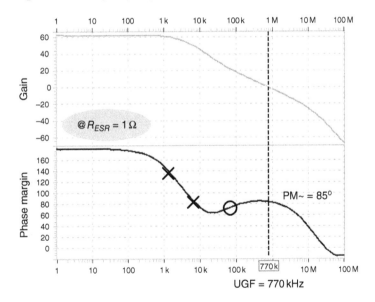

Figure 2.59 Frequency response with $R_{ESR} = 1\ \Omega$ and $I_{Load} = 100$ mA

(1 Ω) causes zero Z_1 to be located at higher frequencies compared with P_2. Consequently, $R_{ESR} = 1\ \Omega$ can provide adequate frequency compensation.

In Figure 2.60 the frequency response is verified under different loading conditions, including 1 and 100 mA. A decreasing loading condition leads to higher gain ($\text{Gain}_{(1\ mA)} > \text{Gain}_{(100\ mA)}$), lower frequency of P_1 ($P_{1(1\ mA)} > P_{1(100\ mA)}$), and smaller UGF ($\text{UGF}_{(1\ mA)} < \text{UGF}_{(100\ mA)}$). Moreover, when checking the simulation conditions, a complete verification should include

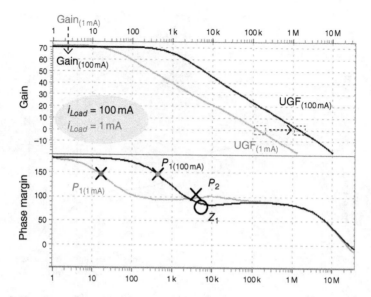

Figure 2.60 Comparison of frequency responses under $I_{Load} = 1$ mA and $I_{Load} = 100$ mA

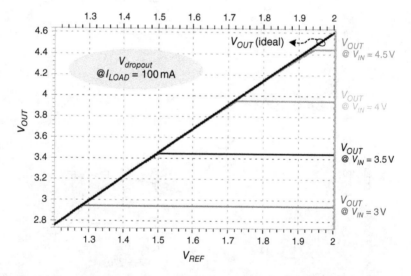

Figure 2.61 Dropout voltage with different V_{IN} values ranging from 3 to 4.5 V

all combinations of different V_{IN}, V_{OUT}, I_{Load}, temperature, and so on, to ensure system stability even under the worst operating conditions (Figure 2.60).

DC analysis can verify the dropout voltage, load regulation, and line regulation. First, the dropout voltage $V_{dropout}$ is observed if V_{IN} and V_{OUT} increase continuously, as shown in Figure 2.61. If I_{Load} is set to its maximum value of 100 mA, V_{OUT} is observed in case of increasing V_{IN} and V_{REF}. The ideal output value $V_{OUT(ideal)}$ can track the increasing trend of V_{REF}

because of the negative feedback control. When V_{IN} is set to 3 V, increasing V_{REF} causes V_{OUT} to also increase. As shown in Figure 2.62(a), where V_{IN} is set to 3 V, the loop gain becomes weak, and thus the dropout voltage can be determined as approximately 140 mV when the curve of V_{OUT} deviates from $V_{OUT(ideal)}$. Similarly, Figure 2.62(b)–(d) shows $V_{dropout}$ when V_{IN} is 3.5, 4, and 4.5 V, respectively.

Figures 2.63 and 2.64 show the load regulation and the line regulation, respectively. In Figure 2.63, the load regulation is verified when V_{IN} is 3 V, V_{OUT} is 2.7 V, and I_{Load} ranges from 1 to 100 mA. The variation at V_{OUT} is 2 mV, which indicates that the performance of load regulation is approximately 20.2 mV/A. In Figure 2.64, I_{Load} is set to its maximum condition of

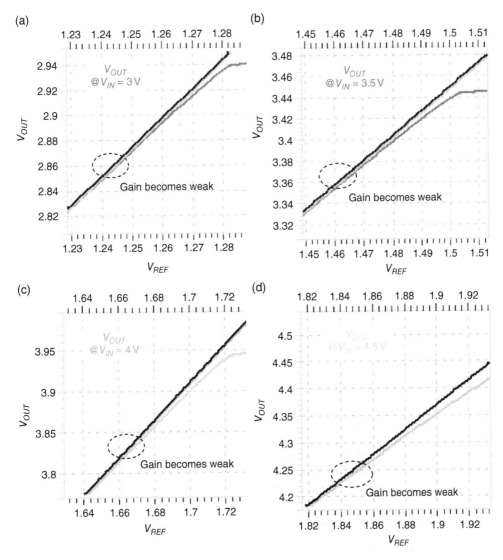

Figure 2.62 Dropout voltage with (a) V_{IN} = 3 V, (b) V_{IN} = 3.5 V, (c) V_{IN} = 4 V, and (d) V_{IN} = 4.5 V

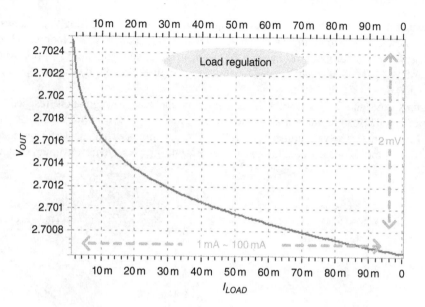

Figure 2.63 Load regulation when I_{Load} ranges from 1 to 100 mA

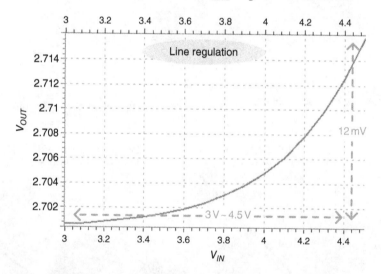

Figure 2.64 Line regulation when V_{IN} ranges from 3 to 4.5 V

100 mA, because this condition will lead to the worst loop-gain case. Line regulation is verified when V_{IN} ranges from 3 to 4.5 V, V_{OUT} is 2.7 V, and I_{Load} is 100 mA. The variation at V_{OUT} is 12 mV, which indicates that the performance of load regulation is approximately 8 mV/V.

Figure 2.65 shows the simulation results of transient response. In the beginning, V_{IN} is set to increase from 0 to 3 V within 50 μs, which emulates the power-on situation in real cases. After the power-on period of 50 μs, the system has not yet settled down because V_g and I_{MPS} still have varying values, although V_{OUT} is regulated at 2.7 V. V_{OUT} consists of the cross voltages of C_{OUT} and R_{ESR}. This variable indicates that C_{OUT} has not yet been charged to the desired voltage of

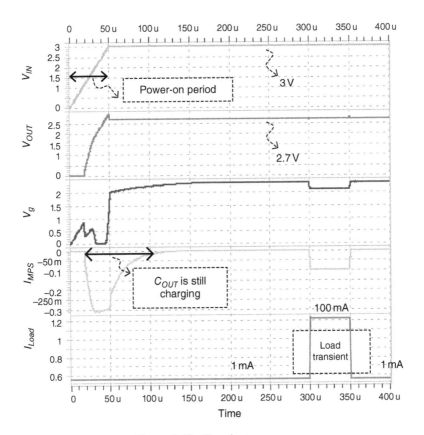

Figure 2.65 Transient response

2.7 V. To analyze the load transient response, designers should avoid setting any loading change during this unstable period.

During the transient response period, the loading current changes from 1 to 100 mA and from 100 to 1 mA at 300 and 350 μs, respectively. The slew rate of the loading current is 1 μs. Figure 2.66 provides detailed zoomed-in images of waveforms that occur during the load transient period. When the loading current changes from 1 to 100 mA, the EA controls V_g to adjust the corresponding driving capability of the power P-MOSFET. In particular, V_g experiences a sudden voltage change from 2.34 to 2 V, which incurs a large signal variation and indicates that the large-signal characteristics of the EA should be carefully designed. The slew rate of the EA always affects the transient response, even if the PM is adequate. Figures 2.67 and 2.68 show the voltage undershoot, voltage overshoot, and settling time. Moreover, the performance of load regulation can be verified from V_{OUT} in a steady state under different loading current conditions, as shown in Figure 2.69.

If the load transient performance cannot satisfy the specifications, designers can modify the design or the compensation to recheck the simulation results. Notably, regardless of whether AC analysis or transient response analysis is checked first, the other should always be checked after. The cross reference between AC analysis and transient response analysis should be consistent.

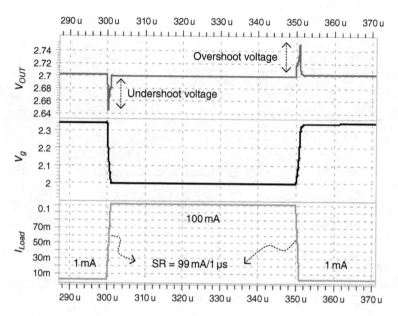

Figure 2.66 Transient response when I_{Load} changes from 1 to 100 mA, and vice versa

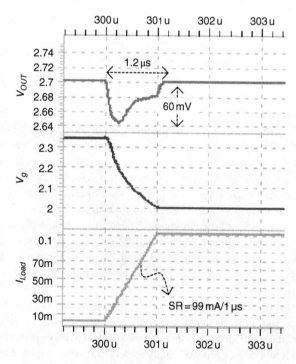

Figure 2.67 Load transient from light to heavy loads

Design of Low Dropout (LDO) Regulators

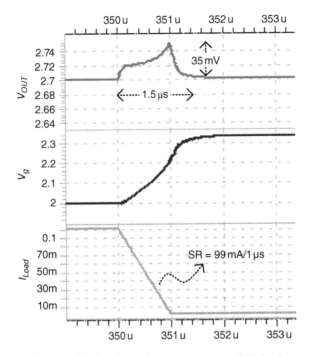

Figure 2.68 Load transient from heavy to light loads

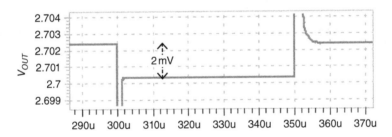

Figure 2.69 Load regulation when I_{Load} changes from 1 to 100 mA, and vice versa

2.6 Digital-LDO (D-LDO) Design

To obtain a high PCE, several approaches have been developed to maintain stability and consume a low quiescent current, even under no-load conditions. However, a low quiescent current induces difficulty in performance enhancement for most LDO regulators. Moreover, for adequate PM and transient response, a structure becomes more complex and the design cost increases correspondingly. Thus, D-LDO regulators have been proposed; the controllers of these regulators are implemented by digital circuits, to turn on/off the power MOSFET array fully [8–10]. Considering their digital implementation, D-LDO regulators can feature fast transit response, a simplified compensation network, and they can work under ultra-low supply voltages. We first introduce the concept of D-LDO regulators and then provide an advanced controller to improve the performance further.

2.6.1 Basic D-LDO

Figure 2.70 illustrates the basic structure of a D-LDO regulator, which consists of a power MOSFET array, a comparator, and a digital controller. In the design of the power MOSFET array, a large power MOSFET is divided into several sub-MOSFET units (SMUs). Each unit is driven by an n-bit digital control signal from the digital controller. Instead of a complex and noise-sensitive analog amplifier, a comparator is used to monitor the output voltage V_{OUT} and compare it with the reference voltage V_{REF}. That is, the logic high or low of the comparator output signal is fed into the digital controller to determine the number of turning-on SMUs. Different numbers of turning-on SMUs can be regarded as several on-resistances that are parallel to an equivalent on-resistance of the power MOSFET. Consequently, different driving capabilities that correspond to various loading current conditions can be adjusted. For example, under heavy loading current conditions, many SMUs are turned on to reduce the equivalent on-resistance and obtain high driving capability.

In general, a simple digital controller can be realized by an up/down counter or a shift register that is controlled by a one-bit compact or output. Moreover, an additional clock signal can be used to synchronize all control signals. The size of the SMUs in the MOSFET array can be designed in binary-code form. As shown in Figure 2.71 [8], the digital controller can be realized by a shift register, wherein the SMUs in the MOSFET array are designed with uniform size.

Compared with the power MOSFET that is controlled by an analog amplifier, a digital control method fully turns on/off each SMU in the power MOSFET array. From a digital perspective, fully turning a system on and off indicates high noise immunity. In a steady state, the number of turning-on power MOSFETs indicates the loading current condition. The shift register will have a dynamic equilibrium under a bistable condition that oscillates between two coding words. That is, the number of turning-on SMUs increases and decreases back and forth if equilibrium is established. However, the back-and-forth operation incurs voltage ripples, as shown in Figure 2.72; these ripples are similar to the switching voltage ripples in SWRs. Unlike

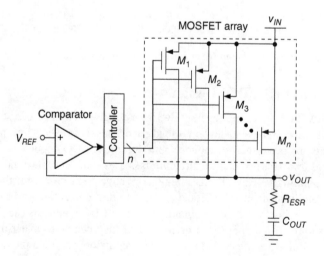

Figure 2.70 Basic structure of a D-LDO regulator

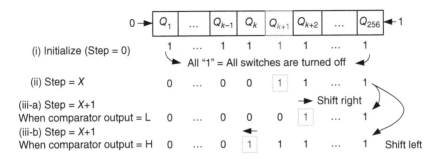

Figure 2.71 Shift register used to control on/off switching of the power MOSFET array

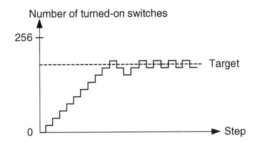

Figure 2.72 Up/down counting control method in D-LDO regulators causes undesired voltage ripples

A-LDO regulators, the undesired voltage ripples will degrade the advantages of the ripple-free characteristic of LDO regulators. The resolution of a switching MOSFET array determines how large a voltage ripple is. A high resolution results in a low output ripple. Furthermore, the transient response time is determined by the frequency of a clock signal and the resolution of the switching MOSFET array. A high frequency results in a fast response and a short recovery time. Nevertheless, the operating frequency cannot be increased without limitation because a high-frequency operation will result in insubstantial power consumption, thus worsening the PCE. That is, the bandwidth of D-LDO regulators is determined by the operating frequency and the resolution of the power MOSFET array when the digital control method is used.

A fast transient response can be achieved more easily in the digital control method than in the analog control method if the operating frequency is simply increased without considering the PCE. Moreover, the digital control method has another advantage, that is, only the pole at the output should be considered because full-swing signals can push poles and zeros to high frequencies. The system can be simplified into a single-pole system, alleviating the constraint of the loading current range, particularly for the minimum light load requirement in the *C-free* LDO regulator.

Given the advantages of fast transient and wide load range operation, D-LDO regulators are more suitable for DVS-based applications. Moreover, a wide operating voltage ranges from the device threshold voltage V_{th} to nearly the highest supply voltage (even if only the minimum biasing current is used to ensure voltage regulation). When considering an advanced nanometer process, D-LDO regulators are also among the candidates for point-of-load (POL) power converters. The following subsections introduce several advanced control methods for D-LDO regulators. For example, the lattice asynchronous self-timed control (LASC) technique is introduced to effectively reduce the power consumption required for synchronous clock control and output voltage caused by a bi-stable operation. Expanding the concept of the DVS technique from the task layer to the instruction layer, the instruction cycle-based dynamic voltage scaling (iDVS) technique is introduced to achieve the advantage of effective power saving. Considerable power can be saved at both the architectural and the circuit levels if the iDVS technique is used.

2.6.2 D-LDO with Lattice Asynchronous Self-Timed Control

A basic D-LDO regulator synchronously adjusts the driving capability with a predefined reference clock. Although a fast transient response and low-voltage operation can be obtained with D-LDO regulators compared with A-LDO regulators, the power consumption remains constrained by the synchronous clock. To improve the performance, a LASC control for *C-free* D-LDO regulators, as illustrated in Figure 2.73, has been developed without any clock signal because of its asynchronous control. The operation of the LASC is similar to that of a clock-free bidirectional shift register that determines the activity of each SMU. Without a synchronous clock, asynchronous control realizes the hand-shaking operation between adjacent SCUs. Therefore, the trade-off between the PCE and the transient response speed can be compromised. That is, in case of a load change, the equivalent clock

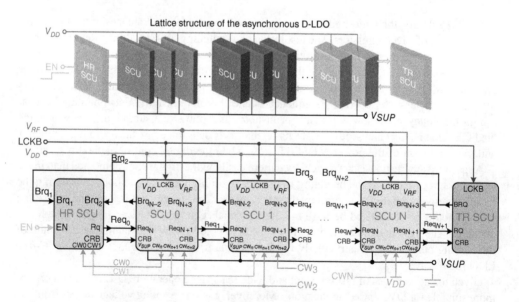

Figure 2.73 Implementation of an asynchronous D-LDO regulator with the LASC technique [11]

frequency is effectively increased to achieve a fast transient response. By contrast, the equivalent clock frequency is automatically decreased for energy/power saving in standby mode. This condition indicates that the D-LDO regulator with the asynchronous control technique can simultaneously exhibit the advantages of high PCE and fast transient response.

A driven source is an event, and thus the problems of clock skew and synchronous surging current never occur. Figure 2.74 shows that the LASC D-LDO regulator consists of an SCU, an SR-latch comparator, a heading reflector (HR), and a terminal reflector (TR). As shown in Figure 2.74(a), the SCU contains a Muller C-gate, an SR-latch comparator, a power switch, a path multiplexer, and control logics to modulate power switches to ensure the regulated output voltage V_{SUP}. In Figure 2.74(b), the SR-latch comparator is triggered by an active high enable signal EN, which is controlled by the forward request pulse of the previous stage. The dynamic comparator compares V_{SUP} with V_{RF} to generate signals CW_0–CW_N to turn on/off the corresponding power switch.

The path multiplexer determines the forward request signal Req_{n+1}, either from the prior stage or the later stage backward request signal Brq_{n+3} according to the results CW_n, CW_{n+1}, and CW_{n+2}. The table in Figure 2.74(a) shows the overall operating principle of the SCU-based Muller C-gate self-timed control. As shown in Figure 2.74(c), HR ensures that all SCUs in the LASC D-LDO regulator will return to their initial states. HR also guarantees that power switches are turned off by setting the signal current-ripple based (CRB) to low when the EN signal is forced to be low. The Muller C-gate shown in Figure 2.74(d) is a basic component of asynchronous circuits. The behavior of an n-input Muller C-gate changes the output state to 3high if all inputs are high and to low if all inputs are low. Otherwise, the n-input Muller C-gate maintains the same output as that in the previous state. As shown in Figure 2.74(e), the rising edge detector (RED) circuit generates a single pulse to trigger the HR circuit to pump the first request pulse, and thus activate the LASC D-LDO regulator when EN changes from low to high. To address the boundary condition, the TR circuit, as depicted in Figure 2.74(f), helps the forward request signal reflected from the termination when V_{SUP} cannot acquire a sufficient power supply during the final SCU stage. Furthermore, HR prevents the backward request signal from disappearing when V_{SUP} derives an overcharge load during the first SCU stage.

Figure 2.75(a), (b) shows the timing diagram of a single SCU stage operation under different conditions that correspond to the circuit in Figure 2.74(a). When V_{SUP} is smaller than the reference voltage V_{RF} in an SCU stage that is triggered by the signal Req_n from the prior stage, the level-active SR-latch comparator outputs low-signal CW_n to turn on the power switch. Thus, the voltage for V_{SUP} can be increased to track V_{RF}. The forward request signal Req_{n+1} is generated by the self-time control mechanism after a deterministic delay, that is "delay X + delay Y," when the next SCU stage performs a shift-right operation, which activates additional power switches to regulate V_{SUP}. If V_{SUP} is greater than V_{RF}, then the control signal CW_n will be pulled high to turn off the power switch at this stage. The backward request signal Brq_{n-2} will be triggered by the self-time mechanism after a deterministic delay, that is "delay X + delay Y," when the prior SCU stage performs a shift-left operation to reduce the driving capability of V_{SUP}. Figure 2.75(c) shows the operation timing diagram of the LASC D-LDO regulator. First, EN is pulled low and then, the LCKB signal is forced to high during the power-on reset state. All SCU stages are initialized to turn off all power switches. Once the power-on sequence of the

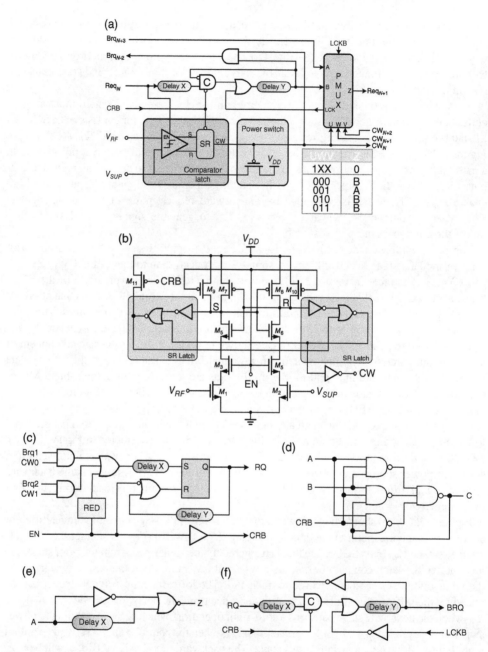

Figure 2.74 Implementations of (a) SCU, (b) SR-latch comparator, (c) HR, (d) Muller C-gate, (e) rising edge detector, and (f) TR

Design of Low Dropout (LDO) Regulators

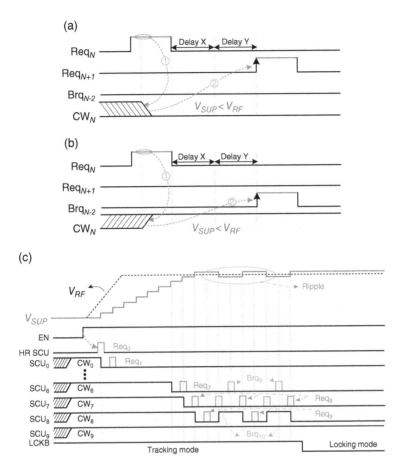

Figure 2.75 Timing diagrams. (a) Single SCU operation when V_{SUP} is smaller than V_{RF}. (b) Single SCU operation when V_{SUP} is larger than V_{RF}. (c) LASC operation when EN is activated

processor is complete and the EN signal is forced to high, the HR SCU pumps the first request signal Req_0 into the LASC controller, and thus the asynchronous D-LDO regulator output voltage V_{SUP} can start tracking the reference voltage V_{RF} according to the required power. During the up-tracking period, the LASC functions as the shift-right operation to increase the number of turning-on power switches by shifting the control signals from CW_0 to CW_N. When V_{SUP} reaches its target value of V_{RF}, backward request signals are generated to stop the delivery of supplementary power to V_{SUP}. When the LASC operation converges to the adjacent SCU stages, or when the present supply voltage is adequate for correct operation, the signal LCKB is cleared to change the operating state from the tracking mode and return to the locking mode. The operation of the LASC D-LDO regulator ends with the indication of the signal LCKB. Thus, output voltage ripples are eliminated in the LASC D-LDO regulator because all SCUs are in a steady state. Therefore, all devices are also in a steady state and the current consumption closely approaches the 0.18 μm process core device leakage current, which is approximately

80 nA because of the fully digital D-LDO regulator. The LASC D-LDO regulator achieves a fast response and an ultra-low static current consumption simultaneously.

2.6.3 Dynamic Voltage Scaling (DVS)

2.6.3.1 Introduction to Basic DVS

Portable electronics are essential products in our daily lives, covering a wide range of applications, including devices and gadgets used for entertainment, communication, and biomedicine. These electronic products contain processors, such as digital signal processors (DSPs), advanced reduced instruction set computing machines (ARMs), and microcontroller units (MCUs), as core components. Therefore, designing low-power processors to save as much power as possible, and extend the battery life of portable devices, is critical.

Figure 2.76 shows the hierarchical processor architecture. The figure also demonstrates that programs are executed from the high-level operating system (OS) layer to the lowest transistor component layer. The program stored in the memory is accessed by the OS to dispatch and schedule different priority tasks. The basic unit of a task is the individual instruction.

After the processor decodes the instructions, the logic gate circuits are activated to perform specific computations. The corresponding layer then accepts the logical control signals to enable or disable millions of transistors. Hence, various techniques have been presented to reduce the power consumption of processors according to the hierarchical processor architecture. In Figure 2.76, state-of-the-art multiple-threshold voltage and body-bias adjustment techniques are employed to reduce the power consumption down to the lowest level of the architecture, that is, the transistor component layer [12]. For simplicity, the clustered-voltage-scaling (CVS) technique at the logic gate layer is adopted to reduce power, because

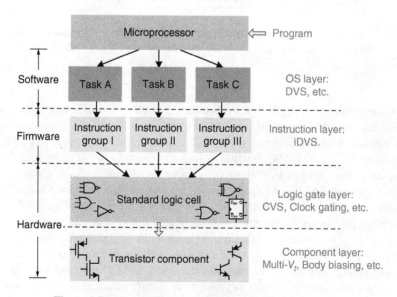

Figure 2.76 Low-power management strategy for processors

of the distinct CVS of the characteristics of each function block [13]. These techniques exhibit limited power-reduction capabilities and require specific processes supported by the foundries, or careful layout placement of a logical cell with multi-power grids. By contrast, the DVS technique in the OS layer is an effective means to reduce the power consumption, because dynamic power consumption depends on a quadratic function of the supply voltage V and the clock frequency f, as shown in Eq. (2.79), where C is the equivalent dynamic operating capacitance:

$$P \propto CV^2 f \qquad (2.79)$$

The DVS technique is appropriate for low-power DSP designs fabricated using the standard CMOS process [14–16]. The conventional DVS task-based control circuit presented in [17] and illustrated in Figure 2.77 uses a closed loop to ensure that the clock frequency ($f_{DESIRED}$) satisfies the desired processor operating clock frequency, which is assigned by the OS and stored in the frequency register for a specific task execution. If peak performance is unnecessary, then the operating clock frequency can be decreased to save considerable power. To evaluate the operating clock frequency, the ring oscillator is provided with a real-time supply voltage V_{SUP} that is generated by an inductor-based SWR to determine the digital numerical clock frequency (f_{CLOCK}). Moreover, the supply-determined f_{CLOCK} is compared with $f_{DESIRED}$ to identify the digital frequency error signal (f_{ERROR}) and produce the control signal by using a digital filter. Finally, the drivers located after the digital loop filter turn the power MOSFETs on/off to modify the output voltage V_{SUP}. Given the closed loop, V_{SUP} can be adjusted to be sufficiently high to guarantee that f_{CLOCK} is close to $f_{DESIRED}$.

Therefore, the dynamic frequency scaling (DFS) technique is implemented by the rapid processor clock frequency according to the minimal and dynamic scaling supply voltage. If high efficiency is considered, then the suitable power supply regulator is an inductor-based converter with a low transient response. Consequently, the DVS tracking speed is restricted from a few microseconds to a few milliseconds. The delay caused by voltage tracking decreases efficiency and processor performance. Various fast voltage tracking methods for high-performance DVS response have been reported to satisfy the DVS requirement [18, 19].

The conventional task-based DVS technique allows all tasks in a scheduler to complete just-in-time operations. The OS depends on run-time workload to adjust the supply voltage dynamically, which leads to a substantial power saving [20, 21]. However, the supply voltage V_{DD} of

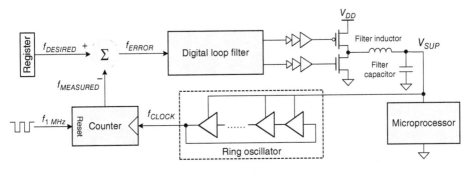

Figure 2.77 Conventional task-based DVS control circuit

the conventional task-based DVS is limited by the highest power instruction in a task operation, as illustrated in Figure 2.78(a). That is, V_{DD} is kept high without being scaled down to save power. Thus, the conventional task-based DVS with a conservative scheduler will fail to scale down V_{DD} because the processor has no slack time. Alternatively, the conventional task-based DVS technique can change the processor operating clock frequency to facilitate a voltage scaling operation; however, this process deteriorates its performance.

A rapidly changing processor clock frequency induces several problems when controlling peripheral modules. These problems include control of signal timing errors and a missing communication data latch in peripheral devices such as synchronous dynamic random access memory (SDRAM), inter-integrated circuit (I^2C), analog-to-digital converter (ADC), digital-to-analog converter (DAC), universal asynchronous receiver/transmitter (UART), and flash memory peripheral interface. The reason for such a problem is the dependence of peripheral devices in Soc on a constant clock and predictable control signals. An alternative iDVS technique [22] employs the DVS technique on the instruction layer to overcome the aforementioned drawbacks. One task consists of numerous instructions, as shown in Figure 2.78(b). If the DVS technique is applied on the instruction layer, then more slack time slots can be derived in the iDVS technique. This technique can reduce the power consumption more than the conventional task-based DVS technique.

To further understand this, the processor is regarded as a dynamic loading and emulated by an adjustable resistor that is controlled by instructions, as shown in Figure 2.79(a). The iDVS technique, which is based on different instructions, does not change or stall the processor operating clock frequency. Thus, the processor performs with a minimum power supply to save power. Through the task-based DVS technique, power consumption can be reduced if the operating tasks have low power consumption. Low-power DSP designs can have a powerful energy-saving capability if the DVS technique is applied on the instruction layer. Figure 2.79(b) shows an example of DSP with the iDVS technique to save power dissipation, wherein the LASC D-LDO regulator is used as the voltage buffer to improve the transient response. The SWR cannot satisfy the slew rate requirement of the DSP core. Thus, the LASC D-LDO regulator can exhibit fast reference tracking to satisfy the voltage slew rate requirement. Consequently, the DSP performance and efficiency can be maintained simultaneously. Moreover, at the DSP core side, the iDVS controller requires the adaptive instruction-cycle control (AIC) circuit to guarantee fast voltage tracking speed and high operating frequency. Moreover, an additional design flow for the iDVS technique, with the help of an automatic computer-aided design (CAD) tool, is also required to maximize power-saving performance.

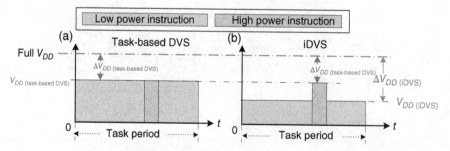

Figure 2.78 (a) Conventional task-based DVS limited by high power instruction. (b) iDVS effectively reduces power consumption

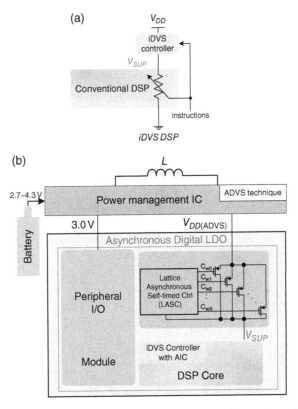

Figure 2.79 (a) Concept of iDVS operation. (b) DSP core with implementation of the iDVS technique

2.6.3.2 Instruction-Cycle-Based Dynamic Voltage Scaling (iDVS)

Processors are designed to provide versatile application programs. When listening to a Moving Picture Experts Group audio layer III (MP3) while simultaneously browsing a picture from a flash memory device, the OS dispatches file-system access and the MP3/Joint Photographic Experts Group (JPEG) decoding algorithm. Similarly, when taking a phone call, the OS of an embedded system launches the key-scan service and the speech compressing/decompressing code-excited linear prediction (CELP) algorithm once mobile phone buttons are pressed. Although these task programs have different characteristics, their fundamental unit is the instruction unit. The basic steps of program execution in a processor are instruction fetching, decoding, execution, and storing, as illustrated in Figure 2.80(a). The most complicated part is the execution unit, shown in Figure 2.80(b), which can provide all types of hardware circuit to support different complex instructions. Each instruction has a corresponding critical data path to complete execution. Critical paths occupy only a small fraction of the total number of paths within a chip. However, the clock speed of a synchronous processor is determined by the worst delay of the critical paths. These critical paths usually map high-power-consuming and long-data-path instructions that are subject to single-instruction multiple-data (SIMD) instructions, such as divide (DIV), normalize (NORM), and multiply-and-accumulate (MAC). Long slack time occurs in non-critical path instructions.

(a)

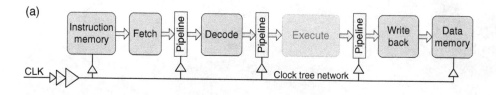

(b)

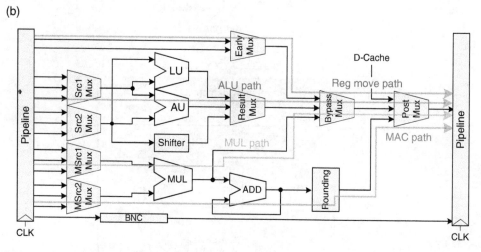

Figure 2.80 (a) Basic steps of program execution. (b) Critical paths during instruction execution

Different instructions present varying slack time, as shown in Figure 2.81(a). The measured slack time and its corresponding power consumption with different instructions are shown in Figure 2.81(b), with the supply voltage fixed at 1.8 V. In summary, a long slack time corresponds to a low power consumption because of the small supply voltage provided by the iDVS technique. Reducing the supply voltage in CMOS circuits affects the propagation delay T_d expressed in Eq. (2.80), which is inversely proportional to V_{DD} and the maximum operating frequency f_{max}:

$$T_d \propto \frac{V_{DD}}{(V_{DD} - V_t)^n} \propto \frac{1}{f_{max}} \tag{2.80}$$

where V_t is the threshold voltage determined by foundry process parameters.

For simplicity, instructions can be classified into different power groups, namely Groups 1–4, for the iDVS technique to provide the corresponding V_{DD}.

The test chip of a 23-stage ring oscillator in a 0.18 μm CMOS process uses 1.8 V core devices. Figure 2.82 shows the maximum operating frequency f_{max} and the total power consumption with respect to the supply voltage variation. According to the data measured at the supply voltage of 1.8 V, the unit of the y-axis is the normalized operating frequency and the normalized power consumption. The linear relationship between circuit operating frequency and supply voltage V_{DD} ranges from 1.2 to 1.8 V. If V_{DD} is scaled down from 1.8 to 1.4 V, then f_{max} is still larger than half of the maximum operating frequency at 1.8 V. Thus, the scaling range of V_{DD} for

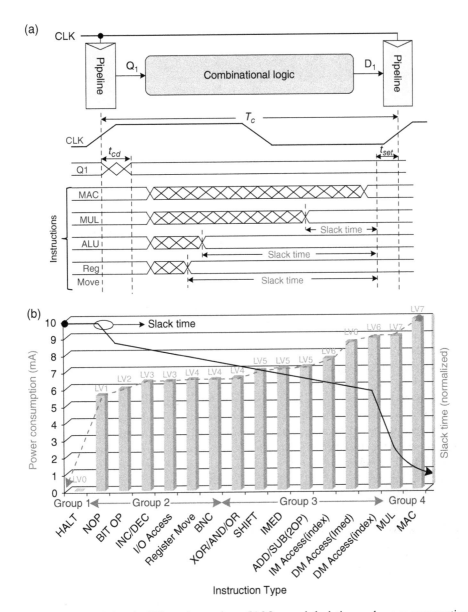

Figure 2.81 (a) Slack time in different instructions. (b) Measured slack time and power consumption in different instructions

normal operation in iDVS is from 1.2 to 1.8 V, excluding HALT and no operation (NOP) instructions. The minimum V_{DD} should be larger than 1 V; otherwise, the level-shift signal will experience serious delay when traveling from the level-shift circuit to peripheral I/O modules. Power reduction can be achieved if the iDVS technique reduces the V_{DD} of the instruction execution on the non-critical path while maintaining a high supply voltage on the critical paths to satisfy complex instruction timing requests. The iDVS dynamically adjusts V_{DD} based on the

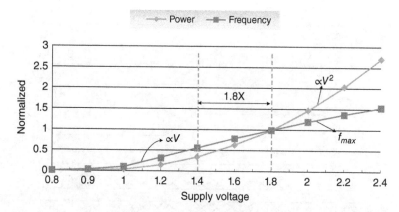

Figure 2.82 Normalized operating frequency and normalized power consumption vs. supply voltage in a 23-stage ring oscillator fabricated via the 0.18 μm CMOS process

instruction-cycle domain to guarantee sufficient power for the appropriate execution of instructions.

In conventional DVS systems, a voltage transition will stall the entire processor operation unless the required power for instructions is available, which results in serious degradation of the million instructions per second (MIPS) performance. By contrast, the iDVS technique uses an embedded all-digital LASC D-LDO regulator with low quiescent current to achieve high-speed voltage tracking capability and provide in-demand power for instruction execution. Moreover, the LASCD-LDO regulator receives assistance from the AIC circuit to avoid the aforementioned drawbacks. The AIC scheme can adaptively adjust the instruction execution cycle time to guarantee that each instruction is correctly executed during voltage tracking for high-performance iDVS operation. That is, the iDVS-based design processor does not change the processor clock frequency or stall the entire processor clock during DVS operation. Consequently, the clock frequency is kept constant, which is suitable for controlling peripheral I/O devices in the Soc.

2.6.3.3 Adaptive Instruction-Cycle Control

An instruction unit occupies one clock cycle in the reduced instruction set computing (RISC) design. A real-time adaptive instruction cycle should be performed in the AIC circuit to adapt to the scaling supply voltage level of the LASC D-LDO regulator.

Figure 2.83(a), (b) shows the topologies of the iDVS controller and the AIC circuit, respectively. The instruction critical path (ICP) shown in Figure 2.83(b) emulates the relative instruction group critical path delay, which is synthesized by the standard cell delay component after timing verification through the iDVS CAD design flow shown in Figure 2.84. Given that a processor has thousands of data paths and millions of logic gates, identifying the corresponding critical data path for each instruction and the relative operating voltage for the iDVS technique is an important issue. Manually analyzing the correlation between critical data paths and instructions is impractical. Circuit extraction tools obtain register-transfer-level (RTL) components and parasitic RC from the ICP to derive the parameters required by the ICP emulator. The extracted circuit netlist is used for simulation program with integrated circuit emphasis

Design of Low Dropout (LDO) Regulators 107

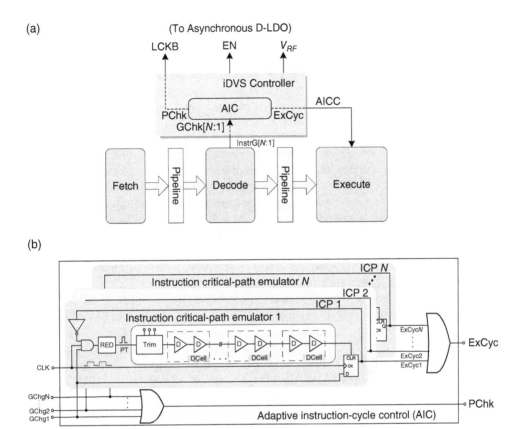

Figure 2.83 (a) Topology of the iDVS controller with the AIC circuit. (b) AIC circuit

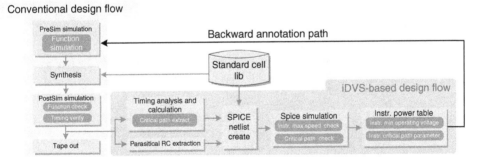

Figure 2.84 Standard cell library design flow with the iDVS design flow

(SPICE) simulations to obtain the minimum operating voltage for each instruction under PVT variations.

The design flow contains three steps, outlined as follows. First, hardware specifications are coded into the hardware description language (HDL) according to the traditional design flow to synthesize the cell-based circuit for post-simulation, which can check the function and verify timing. The critical path of each instruction at the post-stimulation stage can be extracted by the timing analysis tool. A SPICE simulator is used to establish the critical path table for correlating the minimum operating voltage to each corresponding instruction. The timing parameter of each ICP is also extracted to create the ICP in the AIC circuit. The final step is the backward annotation of each instruction power catalog and timing constraint to the iDVS controller in the HDL design. Considering the assistance of RC extraction and the timing analysis tools, the iDVS technique can fit any standard cell library provided by foundries.

Instructions with the same characteristics of data path or power consumption are grouped into one ICP emulator. Each ICP contains a RED, standard-cell delay components, a delay trimming module, and control logics. The delay trimming module is an option for minimizing mass-production deviation after a minor adjustment. Figure 2.85 shows the operating states of the iDVS controller with the timing diagrams, as presented in Figure 2.86. The DSP instruction cycle is synchronized with the edge-triggered clock signal CLK. In each cycle, different instructions as presented in Figure 2.81 are decoded to generate the instruction group signal InstrG[N:1]. When the DSP consecutively executes the instruction stream, the iDVS controller monitors the required power for each instruction according to the instruction power table generated by the CAD design flow.

Once iDVS detects that the required execution power of the next instruction is different from that of the current execution instruction group, the instruction group changes the signals GChg [N:1] to the AIC circuit.

In the next stage, the iDVS controller enters the tracking mode from its operating state to activate the LASC D-LDO regulator by setting the signal LCKB to high. As shown in Figure 2.86(a), given the characteristics of the DSP pipeline structure, the voltage transition command is issued before an instruction is executed prior to the one-clock cycle. Once RED detects the instruction group change signals, which are synchronized with CLK, RED will induce one pulse signal PT to the ICP emulator. The next operation of the AIC circuit is similar to the race condition, to test whether the instruction can finish execution within one instruction cycle at the present supply voltage V_{SUP}. If PT passes through the ICP emulator and simultaneously exceeds the rising edge of the CLK, then the AIC circuit will pull the signal ExCyc to low, which is synchronized by the iDVS controller to generate the signal AICC. Thus, the signal AICC is set to low to inform the DSP execution unit that an extra cycle is not required during the instruction cycle.

By contrast, the passing of PT through the ICP emulator, and the lag of the rising edge of the CLK, indicates that V_{SUP} provides insufficient power. DSP should insert an extra cycle to

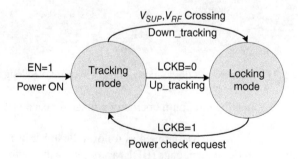

Figure 2.85 Operating states of the iDVS controller

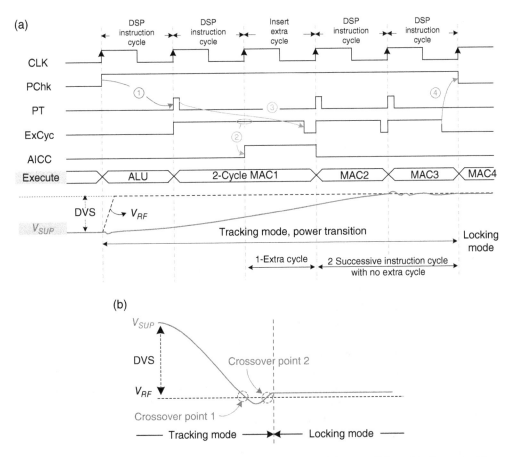

Figure 2.86 Timing diagram of iDVS operation. (a) Up-tracking condition. (b) Down-tracking condition

complete current instruction execution by setting AICC to high. According to Eq. (2.80) and Figure 2.82, no instruction is required to exceed two cycles for execution in the section of 1.4–1.8 V. Given the instruction for pre-decoding the pipeline structure and the fast transition response of the LASC D-LDO regulator, the iDVS-based DSP only requires one extra cycle during up-tracking voltage transition. If the iDVS controller detects a low level of AICC within two successive instruction cycles, then the supply voltage is well regulated and sufficient for instruction execution. The iDVS controller then withdraws the power check request signal PChk and returns to locking mode by setting LCKB to low.

During down-tracking voltage transition, the supply voltage V_{SUP} is sufficiently high to avoid blocking DSP execution. The control sequence of down-tracking voltage transition is shown as follows. First, the D-LDO regulator pulls the V_{SUP} to low. The iDVS controller then sends the group change signals GChg[N:1] to the AIC circuit and continuously monitors the comparison result of V_{SUP} with the reference voltage V_{RF}. Finally, the iDVS controller returns to locking mode by setting the LCKB to low until V_{SUP} and V_{RF} have two crossover points after signal AICC is maintained at low levels within two successive instruction cycles, as shown in

Figure 2.86(b). Simultaneously, the iDVS controller withdraws the power check request signal PChk. The supply voltage is adequate for executing instructions in locking mode. Therefore, the AIC mechanism achieves correct instruction execution during iDVS voltage transition without stopping the operation clock.

2.7 Switchable Digital/Analog-LDO (D/A-LDO) Regulator with Analog DVS Technique

2.7.1 ADVS Technique

In Soc applications, the sub-circuits can mainly be sorted into digital and analog sub-circuits. According to the analog/digital characteristics, different demands are made in supply voltage source qualities, such as driving capability, input/output voltage variation, transient response, noise immunity, and so on.

That is, power management is an essential design issue in Soc, which requires distributive voltage and current levels for distinct sub-circuits. Figure 2.87(a) shows the Soc with an analog sub-circuit and a digital sub-circuit, provided separately by supply voltages V_{OA} and V_{OB} from power management. Moreover, power reduction capability is another necessary feature in power management. DVS is the most efficient technique for power reduction. The DVS function is generally implemented in digital blocks, and thus the digital circuit power consumption can also be optimized by different operating instructions or tasks in Soc. To realize DVS-based power management, the power converter works with the system processor, which can send voltage indication signals to the power converter to adjust the output voltage. As shown in Figure 2.87(b), if Soc enters into data transmission operation, then the increase in V_{OB} satisfies the Soc speed requirement. The decrease in V_{OB} also conserves power when data transmission ends. Comparatively, analog circuits cannot use a similar DVS technique in digital circuits because the supply voltage for analog circuits should be kept constant to maintain some important performance. Moreover, when one SWR is used to provide V_{OA}, the switching voltage ripple at the V_{OA} will degrade accuracy, power supply rejection ratio (PSRR), and so on. By contrast, the LDO regulator can be utilized to suppress noise for high-quality supply voltage, sacrificing efficiency with a large dropout voltage when the LDO regulator directly converts V_{OA} from the input voltage V_{IN}. To reduce the power consumption on the dropout voltage of the LDO regulator, an LDO regulator is generally cascaded in series with a switching-based power converter. Consequently, complete conversion efficiency is expressed as:

$$\eta_{complete} = \eta_{switching} \cdot \eta_{LDO} \tag{2.81}$$

To maintain high efficiency and a high-quality supply voltage for analog sub-circuits, a switching-based power converter is responsible for a wide conversion ratio, whereas the LDO regulator is responsible for switching noise suppression with adequate dropout voltage. To prolong the battery life further using time for portable devices, we need to think whether a solution is available to improve the efficiency further. The dropout voltage of the LDO regulator remains a serious factor to limit complete conversion efficiency, although a switching-based power converter can benefit most complete conversion efficiencies because of the usage of the storage component, inductor, and capacitor. For example, the maximum $\eta_{complete}$ is

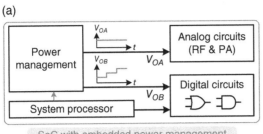

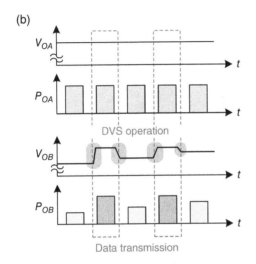

Figure 2.87 (a) Power management in Soc provides two distinct supply voltages, V_{OA} and V_{OB}, for digital and analog circuits, respectively. (b) DVS techniques at V_{OB} corresponding to the Soc operation compared with the fixed voltage at V_{OA}

approximately 81% when $\eta_{switching}$ is 90% and η_{LDO} is 90%. Consequently, the dropout voltage of the LDO regulator should be designed more carefully to improve efficiency.

Figure 2.88(a) illustrates the relationship between the dropout voltage V_{DPA} and the current of the power MOSFET in the analog LDO regulator. The performance of voltage ripple suppression depends on the dropout voltage across the pass transistor, which is regarded as a buffer between the input supply voltage and the output voltage. The power MOSFET in the triode region, which operates with a small dropout voltage, behaves as a 1 V controlled resistance. However, its PSR becomes worse than that in the deep saturation region. By contrast, a large dropout voltage in the deep saturation region improves the PSR but deteriorates the PCE, as shown in Figure 2.88(b). The definition of the ADVS technique for analog circuits is to scale the SWR output voltage dynamically, and simultaneously the dropout voltage V_{DPA} according to the output loading condition, and maintaining V_{OAR} constant for analog circuits. That is, the ADVS technique considers the trade-off between efficiency and output voltage ripple without affecting the performance of analog circuits in the Soc [23]. The cooperation between the SWR and the LDO becomes one design issue if the ADVS technique is applied in the Soc.

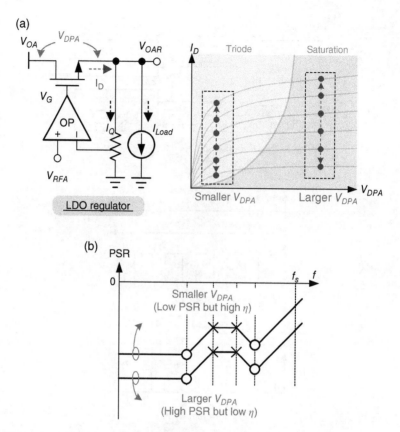

Figure 2.88 (a) Dropout voltage V_{DPA} of the LDO regulator. (b) Relationship between dropout voltage V_{DPA} and PSR

The drain current in the saturation region is shown in Eq. (2.82). The boundary condition between the triode region and the saturation region is defined as the V_{DS} that is equal to V_{OV} in Eq. (2.83). V_{DS} is proportional to the square of the drain current I_D if the channel length modulation parameter λ is neglected.

$$I_D = \frac{1}{2} k_n' \left(\frac{W}{L}\right) V_{OV}^2 (1 + \lambda V_{DS}) \text{ where } V_{DS} \geq V_{OV} \quad (2.82)$$

$$V_{DS} = V_{OV} \propto \sqrt{I_D} \quad (2.83)$$

As the loading current decreases, only a small dropout voltage is required to ensure proper functioning of the power MOSFET in the saturation region for good PSR. Therefore, a small dropout voltage can eliminate superfluous conduction power loss on the power MOSFET and enhance PCE.

Figure 2.89 compares the operations of both the DVS function in digital circuits and the ADVS function in analog circuits. When the Soc enters into operating mode, the DVS function that is indicated by the system processor is activated to optimize power consumption. The

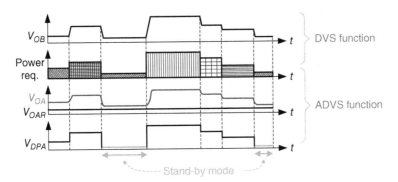

Figure 2.89 Operations of the DVS function in digital circuits and the ADVS function in analog circuits

ADVS function also enables dynamic dropout voltage adjustment of the LDO regulator to guarantee high PSR and PCE. This procedure can be realized by adjusting the output voltage level of V_{OA} corresponding to the loading current condition at V_{OAR}.

2.7.2 Switchable D/A-LDO Regulator

To realize versatile power management performance for Soc, reducing power loss in standby mode is a crucial consideration for prolonging battery life. As discussed earlier, the *C-free* LDO regulator encounters obstacles in stability under ultra-light load. In response to the disadvantages of the *C-free* LDO design, the output should dissipate a minimum load current to ensure stability but sacrifice power efficiency. Consequently, the switchable digital/analog low dropout (D/A LDO) regulator has been developed.

When the switchable D/A LDO regulator works in an analog operation, the power MOSFET is driven by an analog amplifier. Through the ADVS technique, the LDO regulator can exhibit good noise suppression. By contrast, the switchable D/A LDO regulator works in a digital operation if the Soc enters in silent or standby mode. The power MOSFET is driven by a digital controller, such as the D-LDO regulator. Although the capability of ripple suppression decreases, the sub-circuits in most applications can tolerate a large voltage variation in standby mode. Moreover, the dropout voltage is reduced to a relatively small value for an energy-efficient operation.

In prior studies, the current flowing through the resistor divider and the quiescent current are equal to tens or hundreds of microamperes. To reduce the loading current to a value lower than the minimum load current required by the conventional *C-free* LDO regulator, the D/A LDO regulator is switched from analog to digital operation. Thus, under ultra-light loads, the pole in the output moves toward the origin and stability decreases drastically. Without the assistance of the dummy load current, the power MOSFET is modulated by the digital controller rather than by the analog controller (the EA) to decrease the quiescent current significantly. Thus, the digital operation of the D/A LDO regulator breaks through the limitations of the minimum load current for the capacitor-free LDO regulator. The digitally operated LDO regulator confirms stability even under a no-load condition, while consuming only 50 nA quiescent current in the controller and 0.5 µA current in the resistor divider. One advantage of this regulator is that the digital controller features a low quiescent current and further enhances the PCE. In turn, a high current efficiency is achieved over a wide range of load current with a compact solution to Soc power management.

The concept of a switchable D/A LDO regulator is illustrated by the I–V characteristic curve in Figure 2.90. According to the loading current from the Soc power requirement, analog or digital operation is initiated to ensure voltage regulation and achieve high power efficiency. An analog operation uses the ADVS function to adjust the adaptive dropout voltage to maintain high efficiency.

The right section of Figure 2.90 illustrates the operating behavior of the power MOSFET M_Q in either analog or digital operation. In particular, the EA controls M_Q to determine an adequate gate voltage level that corresponds to the load current condition. Moreover, given the regulated output voltage, V_{OAR}, of the D/A LDO regulator, the loading current adjusts the dropout voltage to determine the adequate output voltage V_{OA} from the SWR. When the loading condition changes from heavy to light, V_{OA} changes from V_{OA_H} to V_{OA_L} and obtains a smaller dropout voltage while simultaneously decreasing the gate voltage level from V_{GH} to V_{GL} for an appropriate driving capability. If the loading current decreases under certain values, analog operation switches to digital operation to save power. During digital operation, the transistor M_Q is divided into several parallel SMUs. Portions of the SMUs are turned on and off by the digital controller according to the load current. The number of turned-on SMUs is denoted m. When the load current decreases continuously, the value of m shrinks. Thus, the switchable D/A LDO regulator selects suitable analog and digital operations to satisfy the demand from analog circuits in the Soc.

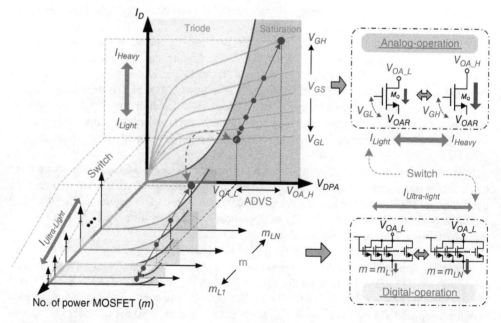

Figure 2.90 I–V characteristic curve of the switchable function in analog and digital operations

2.7.2.1 Switchable D/A LDO Regulator

A schematic of the switchable D/A LDO regulator is provided in Figure 2.91. Digital and analog operations indicate that SMUs are controlled by LASC [24] and the EA [25], respectively. The smooth switch technique (SST) decides whether digital or analog operation is used with respect to the loading current condition. The key point is a continuous and smooth switching procedure between digital and analog operations. The D/A selector is structured by an array of multiplexers and controlled by the load-dependent signals V_{AtoD} and V_{DtoA}. Pass transistors M_{Q1} to M_{QN} are controlled by the gate control signals V_{D1} to V_{DN} of the digital controller or the analog signal V_{GA} of the analog controller.

In analog operation, the EA generates the error signal V_{GA} to guarantee the output voltage V_{OAR} under different output loading current conditions, where each gate voltage, from V_{G1} to V_{GN}, of the pass transistors is connected to the error signal V_{GA}. By contrast, in digital operation, the LASC circuit generates thermometer control codes for each of the pass transistors when the Soc enters silent mode. The digital control method releases the minimum load limitations at only several microamperes for the system to conserve power effectively. Moreover, the frozen operation in the LASC circuit reduces the quiescent current to an ultra-low value for high efficiency. The switchable D/A LDO regulator achieves ripple suppression and energy-efficient operation in a distinct mode for high efficiency.

2.7.2.2 Smooth Switch Technique (SST)

The switchable D/A LDO regulator indicates that only one analog/digital operation can be enabled in case of a loading current change. The SST ensures a continuous and smooth takeover procedure between analog and digital operations to prevent undesirable oscillations that induce considerable output voltage ripples. The hysteresis window V_{hys} adjusts the reference voltages of the D/A LDO regulator during the switching procedure. The flowchart in Figure 2.92

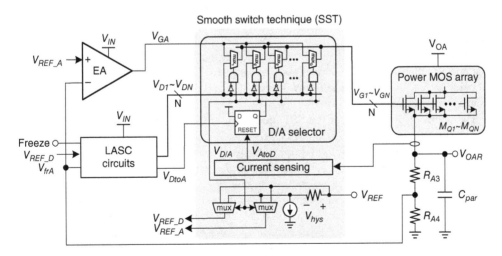

Figure 2.91 Schematic of the switchable D/A LDO regulator

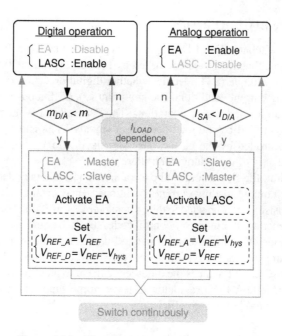

Figure 2.92 Flowchart of the switchable technique

describes the operation procedure during switching. The increasing value of m becomes larger than the predefined $m_{D/A}$, which triggers the operating procedure via the pulse signal V_{DtoA} when the LDO regulator switches from digital to analog operation as the loading current increases. Thus, the EA is activated and the reference voltage of LASC, that is V_{REF_D}, also changes from V_{REF} to the value of $V_{REF} - V_{hys}$. The EA and LASC temporarily operate simultaneously. Meanwhile, the EA gradually dominates control of the SMUs when the feedback voltage V_{frA} is larger than V_{REF_D}. The EA and LASC represent the master and the slave, respectively. Finally, the switching procedure is complete when the EA takes over the operation and disables LASC automatically. By contrast, when the LDO regulator switches from analog to digital operation during loading current decreases, the decreasing sensing current I_{SA} becomes smaller than the predefined $I_{D/A}$ and triggers the switching procedure by the pulse signal V_{AtoD} in Figure 2.91. The following switching procedure is complete via the aforementioned opposite procedure.

The operation waveforms of the SST are shown in Figure 2.93. The currents I_{LDO_D} and I_{LDO_A} flow through the SMUs and are controlled by LASC and the EA, respectively. In cases of increasing loading current, digital operation switches to analog operation, as illustrated in Figure 2.93(a). At $t = t_1$, I_{LDO_D} increases corresponding to the increasing number of turned-on SMUs. At $t = t_2$, if m is larger than $m_{D/A}$ then the EA is enabled and the reference voltage of LASC, V_{REF_D}, changes from V_{REF} to the value of $V_{REF} - V_{hys}$. Consequently, the EA and the LASC function have a master–slave relationship. During this period, the EA controls the remaining switched-off SMUs and I_{LDO_A} subsequently increases to regulate the output voltage V_{OAR}. Once the feedback voltage V_{frA} increases to a value higher than V_{REF_D} at $t = t_3$, I_{LDO_D} is reduced to zero and LASC is automatically disabled. Finally, the switching

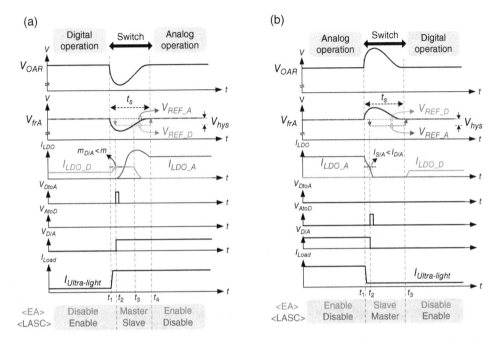

Figure 2.93 Control mechanism of the switchable technique between digital operation and analog operation. (a) Switching from digital operation to analog operation. (b) Switching from analog operation to digital operation

procedure is complete after V_{OAR} recovers at $t = t_4$. Analog operation takes over and enables only the EA. LASC is shut down to enable single analog loop control work.

Analog operation can be switched back to digital operation if the loading current decreases to a sufficiently low value, as shown in Figure 2.93(b). At $t = t_1$, the I_{LDO_A} controlled by the EA decreases continuously. At $t = t_2$, once the condition of I_{SA} is smaller than that of $I_{D/A}$, LASC is enabled and the reference voltage of the EA, V_{REF_A}, changes from V_{REF} to the value of $V_{REF} - V_{hys}$. Thus, I_{LDO_D} gradually increases while I_{LDO_A} decreases to zero. The switching procedure is complete when V_{OAR} is recovered at $t = t_3$.

According to the load condition, the SMUs are controlled by the analog or digital controller determined by the SST. Approximately 10% of the SMUs are driven by the digital controller under light loads. Meanwhile, 90% of the SMUs are turned off. During the D/A switching procedure under light-to-heavy load changing, the analog controller controls a portion of the power MOS units, while the digital controller controls other portions. The analog controller controls all power MOSFET units at the end of the load change. Finally, none of the SMUs is controlled by the digital controller. Consequently, the energy delivered to the output is continuous to retain a low output voltage variation. By reflecting the variation of the output voltage through the feedback network, the condition $m_{D/A} < m$ or $I_{SA} < I_{D/A}$ implies which controller is suitable for use. Thus, the SMUs are adequately controlled even if the loading condition changes faster than the switching time. Upon setting the reference voltage to $V_{REF} - V_{hys}$, stability is confirmed if a moderate or long period is used as the hysteresis window.

2.7.2.3 ADVS PMU with Combined SWR and Switchable D/A LDO Regulator for Soc

Among currently available advanced power management unit (PMU) designs, the single-inductor dual-output (SIDO) converter is used to further reduce PMU size, where the two output voltages are V_{OA} and V_{OB} for analog and digital sub-circuits in the Soc, respectively, as shown in Figure 2.94. The advantage of the SIDO converter is its compact size, which can be attributed to the usage of only one off-chip inductor compared with others that use multiple off-chip inductors [7,19,26–28]. Detailed design guidelines are provided in the next chapter.

For high efficiency, the ADVS technique can be implemented in the PMU design. That is, the combination of SWR and the switchable D/A LDO regulator can provide ADVS PMU for Soc. With regard to PMU design using the dual DVS techniques in Soc, the SIDO converter shown in Figure 2.95 has two outputs. The first output is for digital sub-circuits and the second, with the switchable D/A LDO regulator, is for analog sub-circuits.

With only one off-chip inductor, the power stage of the SIDO converter is composed of four power switches from M_1 to M_4. These switches transfer the energy from the input battery to both outputs, V_{OA} and V_{OB}. V_{OB} supplies digital circuits directly to the Soc. In DVS operation, the DVS indicator receives the control signal V_{Bit} from the system processor to generate the reference voltage V_{RFB} for V_{OB}. By contrast, the switchable D/A LDO regulator is cascaded in series with the V_{OA} of the SIDO converter to guarantee supply quality for analog circuits.

When the D/A LDO regulator is operated using the analog function, the load current condition at V_{OA} is reflected to the reference adjuster of ADVS modulation to adjust the dropout voltage dynamically, which subsequently leads to voltage ripple suppression or power conservation. If the output load condition at V_{OA} continuously decreases to an ultra-light condition, then the switchable D/A LDO regulator is switched to digital operation. Thus, the quiescent current decreases because the digital controller and the dropout voltage are minimized to enhance power efficiency further. The load-dependent technique indicates the switching procedure between analog and digital operations in the switchable D/A LDO regulator. Therefore, with the feedback control loop in the SIDO power module, the ADVS and DVS functions are achieved at V_{OA} and V_{OB}, respectively. The SIDO controller uses the current-programmed

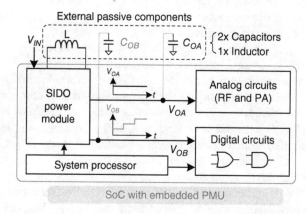

Figure 2.94 SIDO converter used in PMU design to deliver the advantage of a compact size solution

Design of Low Dropout (LDO) Regulators

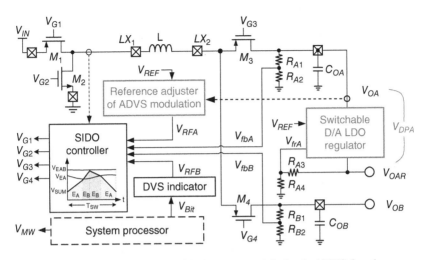

Figure 2.95 Structure of the SIDO power module for dual DVS functions

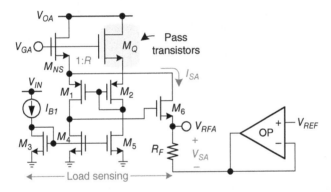

Figure 2.96 Reference adjuster in the ADVS technique

control scheme [28] to realize the energy delivery function at the power stage and ensure high power efficiency.

A schematic of the reference adjuster in ADVS is presented in Figure 2.96. The current flowing through the pass transistors must be monitored to adjust the dropout voltage dynamically in the analog LDO regulator. The load-sensing circuit obtains the supply current information. Then, the replica-sensed current I_{SA} generates the reference voltage V_{RFA} by resistor R_F to modulate the output voltage level of V_{OA} properly and achieve an optimal dropout voltage in the switchable D/A LDO regulator, which operates in the analog operation. V_{SA} reflects the dropout voltage corresponding to the loading current condition in the analog LDO regulator. That is, an adequate dropout voltage is maintained in the analog LDO regulator with regard to the loading current condition. High efficiency and PSR can be maintained in the analog LDO regulator. By contrast, the D/A switchable LDO regulator is switched to digital operation if the loading current condition is under light loads. The major reason for this procedure is to break through the limitation requested in the *C-free* LDO regulator. The digital operation of the D/A

LDO regulator reduces the minimum load current requirement to near zero [7]. Meanwhile, the dropout voltage is continuously minimized and the quiescent current can be reduced further to maintain high efficiency even under ultra-light loads. The switchable D/A LDO regulator ensures high efficiency over a wide load range and implements the ADVS technique in the PMU design.

References

[1] Fan, X., Mishra, C., and Sánchez-Sinencio, E. (2005) Single Miller capacitor frequency compensation technique for low-power multistage amplifiers. *IEEE Journal of Solid-State Circuits*, **3**, 584–592.

[2] Fan, X., Mishra, C., and Sanchez-Sinencio, E. (2004) Single Miller capacitor compensated multistage amplifiers for large capacitive load applications. *Proceedings of the IEEE International Symposium on Circuits and Systems*, May 2004, vol. **1**, pp. 23–26.

[3] Cannizzaro, S.O., Grasso, A.D., Mita, R., et al. (2007) Design procedures for three-stage CMOS OTAs with nested-Miller compensation. *IEEE Transactions on Circuits and Systems I: Regular Papers*, **54**(5), 933–940.

[4] Eschauzier, R.G.H., Kerklaan, L.P.T., and Huijsing, J.H. (1992) A 100-MHz 100-dB operational amplifier with multipath nested Miller compensation structure. *IEEE Journal of Solid-State Circuits*, **27**(12), 1709–1717.

[5] Pernici, S., Nicollini, G., and Castello, R. (1993) A CMOS low-distortion fully differential power amplifier with double nested Miller compensation. *IEEE Journal of Solid-State Circuits*, **28**, 758–763.

[6] Leung, K.-N. and Mok, P.-K.T. (2001) Analysis of multistage amplifier–frequency compensation. *IEEE Transactions on Circuits and Systems I (TCAS-I)* **48**(9), 1041–1056.

[7] Lee, Y.-H., Huang, T.-C., Yang, Y.-Y., et al. (2011) Minimized transient and steady-state cross regulation in 55-nm CMOS single-inductor d-output (SIDO) step-down DC–DC converter. *IEEE Journal of Solid-State Circuits*, **46**, 2488–2499.

[8] Hirairi, K., Okuma, Y., Fuketa, H., et al. (2012) 13% power reduction in 16b integer unit in 40nm CMOS by adaptive power supply voltage control with parity-based error prediction and detection (PEPD) and fully integrated digital LDO. *IEEE International Solid-State Circuits Conference Digest of Technical Papers*, February 2012, pp. 486–487.

[9] Okuma, Y., Ishida, K., Ryu, Y., et al. (2010) 0.5-V input digital LDO with 98.7% current efficiency and 2.7-μA quiescent current in 65nm CMOS. *Proceedings of the IEEE Custom Integrated Circuits Conference (CICC)*, September 2010, pp. 1–4.

[10] Onouchi, M., Otsuga, K., Igarashi, Y., et al. (2011) A 1.39-V input fast-transient-response digital LDO composed of low-voltage MOS transistors in 40-nm CMOS process. *Proceedings of the IEEE Asian Solid-State Circuits Conference (A-SSCC)*, November 2011, pp. 37–40.

[11] Peng, S.-Y., Huang, T.-C., Lee, Y.-H., et al. (2013) Instruction-cycle-based dynamic voltage scaling power management for low-power digital signal processor with 53% power savings. *IEEE Journal of Solid-State Circuits*, **48**(11), 2649–2661.

[12] Miyazaki, M., Ono, G., and Ishibashi, K. (2002) A 1.2-GIPS/W processor using speed-adaptive threshold voltage CMOS with forward bias. *IEEE Journal of Solid-State Circuits*, **37**, 210–217.

[13] Usami, K., Igarashi, M., Minami, F., et al. (1998) Automated low-power technique exploiting multiple supply voltages applied to a media processor. *IEEE Journal of Solid-State Circuits*, **33**, 463–472.

[14] Ickes, N., Gammie, G., Sinangil, M.E., et al. (2012) A 28 nm 0.6 V low power DSP for mobile applications. *IEEE Journal of Solid-State Circuits*, **47**, 35–46.

[15] Ashouei, M., Hulzink, J., Konijnenburg, M., et al. (2011) A voltage-scalable biomedical signal processor running ECG using 13 pJ/cycle at 1 MHz and 0.4 V. *IEEE International Solid-State Circuits Conference Digest of Technical Papers*, February **2011**, pp. 332–334.

[16] Sridhara, S.R., DiRenzo, M., Lingam S., et al. (2011) Microwatt embedded processor platform for medical system-on-chip applications. *IEEE Journal of Solid-State Circuits*, **46**(4), 721–730.

[17] Burd, T.D., Pering, T.A., Stratakos, A.J., and Brodersen, R.W. (2000) A dynamically voltage scaled processor system. *IEEE Journal of Solid-State Circuits*, **35**(11), 1571–1580.

[18] Lee, Y.-H., Chiu, C.-C., Peng, S.-Y., et al. (2012) A near-optimum dynamic voltage scaling (DVS) in 65nm energy-efficient power management with frequency-based control (FBC) for Soc system. *IEEE Journal of Solid-State Circuits*, **47**(11), 2563–2575.

[19] Lee, Y.-H., Peng, S.-Y., Wu, A.C.-H., *et al.* (2012) A 50nA quiescent current asynchronous digital-LDO with PLL-modulated fast-DVS power management in 40nm CMOS for 5.6 times MIPS performance. *Proceedings of the IEEE Symposium on VLSI Circuits*, pp. 178–179.

[20] Liu, Y. and Lin, M. (2009) On-line and off-line DVS for fixed priority with preemption threshold scheduling. IEEE Conference of Embedded Software and Systems, pp. 273–280.

[21] Wang, W. and Mishra, P. (2010) PreDVS: Preemptive dynamic voltage scaling for real-time systems using approximation scheme. *Proceedings of the 47th ACM/IEEE, Design Automation Conference (DAC)*, 2010, pp. 705–710.

[22] Peng, S.-Y., Lee, Y.-H., Wu, C.-H., *et al.* (2012) Real-time instruction-cycle-based dynamic voltage scaling (iDVS) power management for low-power digital signal processor (DSP) with 53% energy savings. *Proceedings of the IEEE Asian Solid-State Circuits Conference*, November 2012, pp. 377–380.

[23] Chen, W.-C., Su, Y.-P., Huang, T.-C., *et al.* (2014) Switchable digital/analog (D/A) low dropout regulator for analog dynamic voltage scaling (ADVS) technique. *IEEE Journal of Solid-State Circuits (JSSC)*, **49**(3), 740–750.

[24] Lee, Y.-H., Yang, Y.-Y., Chen, K.-H., *et al.* (2010) A DVS embedded power management for high efficiency integrated Soc in UWB system. *IEEE Journal of Solid-State Circuits*, **45**(11), 2227–2238.

[25] Das, S., Roberts, D., Lee, S., *et al.* (2006) A self-tuning DVS processor using delay-error detection and correction. *IEEE Journal of Solid-State Circuits*, **41**(4), 792–804.

[26] Lee, Y.-H. and Chen, K.-H. (2010) A 65nm sub-1V multi-stage low-dropout (LDO) regulator design for Soc systems. *Proceedings of the IEEE Midwest Symposium on Circuits and Systems (MWSCAS)*, August 2010, pp. 584–587.

[27] Le, H.-P., Chae, C.-S., Lee, K.-C., *et al.* (2007) A single-inductor switching DC–DC converter with five outputs and ordered power-distributive control. *IEEE Journal of Solid-State Circuits*, **42**(12), 2076–2714.

[28] Lee, Y.-H., Yang, Y.-Y., Wang, S.-J., *et al.* (2011) Interleaving energy-conservation mode (IECM) control in single-inductor dual-output (SIDO) step-down converters with 91% peak efficiency. *IEEE Journal of Solid-State Circuits*, **46**(4), 904–915.

3

Design of Switching Power Regulators

3.1 Basic Concept

DC/DC switching regulators (SWRs) are widely used in power management units because of their high efficiency, adjustable output voltage, and high driving capability [1–7]. Figure 3.1 shows the basic components of a DC/DC switching converter. As implied by the name "switching," switches S_1 and S_2 control the amount of energy delivered from the input DC voltage source v_{IN} to the output v_{OUT}. When S_1 and S_2 are on and off, respectively, energy is delivered from v_{IN} to v_{OUT}. When S_1 and S_2 are off and on, respectively, no energy is delivered from v_{IN} to v_{OUT}. By switching S_1 and S_2 alternatively, the energy delivered to v_{OUT} can be adequately controlled. Next, a low-pass filter is needed to filter out the high-frequency components produced in the switching operation. The DC component, which is equal to the average value in Fourier analysis, can be similarly derived. Finally, a DC voltage is presented at v_{OUT} to provide energy to the output load.

It should be mentioned here that S_1 and S_2 cannot be turned on concurrently, even for a very short time. Simultaneously turning on S_1 and S_2 drains the energy from input to ground and causes a large energy loss.

Allocating the turn-on time of S_1 and S_2 is an issue. When a certain output voltage for a certain load condition is required, the turn-on time ratio of S_1 and S_2 must remain the same to deliver constant energy to the output, no matter how many times S_1 and S_2 turn on alternately. First, a switching period (cycle) T_S is defined to easily achieve this. The relationship between switching frequency f_S and switching period T_S is:

$$f_S = \frac{1}{T_S} \tag{3.1}$$

Power Management Techniques for Integrated Circuit Design, First Edition. Ke-Horng Chen.
© 2016 John Wiley & Sons Singapore Pte Ltd. Published 2016 by John Wiley & Sons Singapore Pte Ltd.

Design of Switching Power Regulators

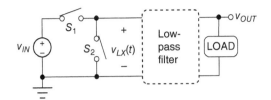

Figure 3.1 Basic components of a DC/DC switching converter

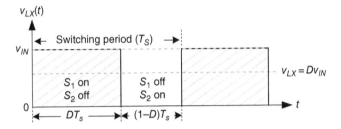

Figure 3.2 Periodic waveforms at the switching node

The time it takes to transfer energy from input to output in a switching cycle is the so-called duty cycle or duty ratio in Eq. (3.2) for DC/DC switching converters. The duty cycle is equal to the ratio of the turn-on time of S_1 to the switching period. Since the switching cycle repeats again and again, a fixed switching frequency operation occurs.

$$D = \frac{\text{Turn-on time of } S_1}{T_S} \tag{3.2}$$

The switching activity can easily be observed from the switching node LX that connects two switches S_1 and S_2, as depicted in Figure 3.2. Figure 3.2 also shows the on time of S_1 and S_2 in a switching cycle.

As depicted in Figure 3.2, when S_1 turns on, LX is connected to the supply voltage v_{IN} and $v_{LX}(t) = V_{IN}$, where V_{IN} is the DC component of $v_{IN}(t)$. When S_2 is turned on, LX is pulled down to the ground at zero voltage level. The low-pass filter helps filter out high-frequency harmonics to obtain the DC component of the signal $v_{LX}(t)$. That is, the average value of $v_{LX}(t)$ in a periodic duration is equal to the DC component derived from the Fourier transform analysis. The DC value of the switching node signal $v_{LX}(t)$ is:

$$\overline{v_{LX}}(t) = \frac{1}{T_S}\int_0^{T_S} v_{LX}(t)dt = DV_{IN} \tag{3.3}$$

Given that the output voltage is equal to the average value of $v_{LX}(t)$, we can conclude that the output voltage is equal to the input supply voltage multiplied by the duty cycle. Therefore, we can set the output voltage of the SWR with this topology using the duty cycle as indicated in Eq. (3.4). In other words, in a fixed switching frequency operation, modulating the duty cycle can effectively control the output voltage. This method is one of the most popular SWR control methods, called pulse width modulation (PWM).

$$D = \frac{v_{OUT}}{v_{IN}} \tag{3.4}$$

From Figure 3.2, D must take a value between 0 and 1 because $v_{LX}(t)$ is a periodic signal. In this type of SWR, the output voltage is equal to or smaller than the supply voltage. Only the "buck" or "step-down" voltage can be obtained (compared with the supply voltage). Therefore, this topology is a buck converter. In addition, other basic topologies of the SWR include the boost and buck-boost converters. A boost converter can convert the supply voltage to a higher level, while the buck-boost converter can realize either a step-up or step-down voltage level.

Similar to a buck converter, boost and buck-boost converters are composed of switches and a filter. The filter is realized by a second-order LC filter, which contains an off-chip inductor and an off-chip capacitor. The continuity of the inductor current decides the current to $LOAD$, and the capacitor stabilizes the output voltage. A summary of the three basic inductor-based SWR topologies is shown in Table 3.1. These topologies, which contain switches and a filter, are named the power stage because the topologies can provide large amounts of power to $LOAD$. These converters can guarantee both good energy-driving capability and high power-conversion efficiency.

Table 3.1 Summary of the three basic inductor-based SWR topologies

Name	Topology	Conversion ratio
Buck (step-down)		$M(D) = D$
Boost (step-up)		$M(D) = \dfrac{1}{1-D}$
Boost/boost (step-up and step-down)		$M(D) = \dfrac{-D}{1-D}$

Table 3.2 Comparison of distinct basic power management modules

	Linear regulators	Charge pumps	Switching regulators
Regulation type	Buck only	Buck/boost	Buck/boost
Noise	Low	Medium	High
Efficiency	Low	Medium	High
Power capability	Medium	Medium	High
Footprint area	Compact	Moderate	Large
Cost	Low	Medium	High
Complexity	Low	Medium	High
EMI	Low	Medium	High

The conversion ratio represents the ability to convert the output voltage from the supply voltage. Given a certain value of the duty cycle, the output voltage can be determined by a conversion ratio, which is represented by $M(D)$. When $M(D)$ is equal to one, the output voltage is equal to the supply voltage. As mentioned before, in a buck converter, the output voltage is equal to the input supply voltage multiplied by the duty cycle. Thus, the conversion ratio of the buck converter can be expressed as $M(D) = D$. The boost converter provides a step-up operation, in which the output voltage level is higher than the input voltage. The conversion ratio of the boost converter $M(D)$ can be expressed as $M(D) = 1/(1-D)$. The buck-boost converter carries out either a step-up or a step-down operation with conversion ratio $M(D) = -D/(1-D)$. As a result, different power management topologies realize distinct supply functions in different applications.

A comparison of distinct basic power management modules is listed in Table 3.2. A simple low-dropout (LDO) regulator generates a ripple-free output voltage but sacrifices power-conversion efficiency. By contrast, a switching-type power management results in a switching output voltage ripple and therefore degrades the quality of the voltage supply. The output voltage ripple may result in improper operations in noise-sensitive sub-circuits. Electromagnetic interference (EMI) problems caused by switching operations need to be solved to enhance the quality of the power management module. However, its high power-conversion efficiency is the main design consideration in battery-operated electronics.

3.2 Overview of the Control Method and Operation Principle

As discussed in Section 3.1, the output voltages of SWRs can be controlled by a duty cycle. We discuss how to obtain the signal of the duty cycle and how SWRs generate well-regulated voltages at different load conditions.

A straightforward method is to keep detecting the output voltage and adjust the duty cycle accordingly. If the output voltage is lower than the target value, increasing the duty cycle can transfer more energy to the output. If the output voltage is higher than the target value, decreasing the duty cycle can decrease the energy transferred to the output. Therefore, the output voltage can be controlled under all load conditions or input voltage variations. Such a control method is illustrated in Figure 3.3.

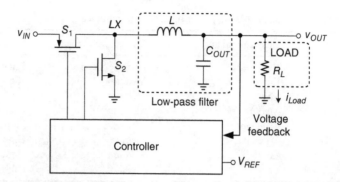

Figure 3.3 Closed-loop control method for switching buck converter

Without the controller, the output voltage is determined by a fixed duty cycle (which is preset according to the input and target output values). The energy is transferred to the output by the power stage without detecting the output voltage. This is an "open-loop control." Load variation can result in an output voltage apart from the target voltage. By contrast, a "closed-loop control" is constructed by voltage feedback. The output voltage is regulated at the target value by adjusting the duty cycle. The closed-loop control can be viewed as a feedback system, which modifies the output impedance of the buck converter for voltage regulation.

As shown in Figure 3.3, the target output voltage is set by a reference voltage V_{REF}, which is usually generated by a bandgap circuit. The output voltage is sent to the controller, after which it forms a voltage feedback loop. The controller can adjust the duty cycle according to the amount of output voltage deviation from V_{REF}.

The next step is to design the controller to achieve a closed-loop control. The classification of the SWR control methods is shown in Figure 3.4. The control methods can be classified into two main categories: fixed-frequency control and ripple-based control. The fixed-frequency control is the well-known PWM, which is mentioned in Figure 3.2. A clock signal is required to define the switching frequency. The fixed-frequency control contains a voltage-mode control, which has a basic voltage feedback as illustrated in Figure 3.3 and a current-mode control, which requires additional current-sensing information. By contrast, the ripple-based control does not require a clock signal to indicate the start of each switching cycle. Instead, the switching frequency varies with the requirement of output voltage or inductor current. The ripple-based control contains constant on time, constant off time, hysteresis, and V-square (V^2) control.

Figure 3.5 illustrates the voltage-mode controller in Figure 3.3. First, the difference between the voltage feedback V_{fb} and the reference voltage V_{REF} is required to indicate the trend of the duty cycle (i.e., increase or decrease). A single-ended error amplifier (EA), which is usually used, can generate the error signal v_C by amplifying the difference between v_{fb} and V_{REF}. Next, the duty cycle, which controls S_1 and S_2, must be generated according to v_C. Through the closed-loop control, v_{fb} becomes equal to V_{REF} to achieve voltage regulation. Here, R_{f1} and R_{f2} form a voltage divider to obtain the voltage feedback signal proportional to v_{OUT}. By adjusting the ratio of R_{f1} and R_{f2}, the target output voltage can be adjusted according to V_{REF}.

To achieve fixed-frequency control, the clock signal $CLK(t)$ defines the switching frequency and indicates the start of each switching cycle. The ramp signal v_{RAMP}, which is synchronized to $CLK(t)$, is required to generate the duty cycle. By comparing $v_C(t)$ with $v_{RAMP}(t)$, the duty cycle

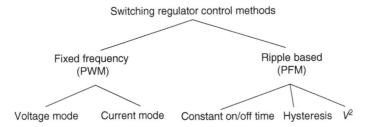

Figure 3.4 Classification of SWR control methods

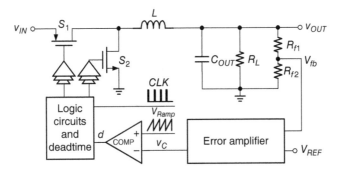

Figure 3.5 Architecture of the voltage-mode controlled buck converter

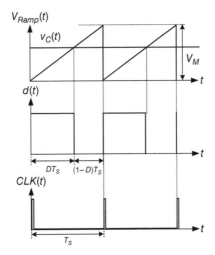

Figure 3.6 Operation waveforms of the voltage-mode buck converter

is generated as shown in Figure 3.6. By comparison, $v_C(t)$ is converted to the duty cycle, which indicates the DC-value deviation of $v_{OUT}(t)$ in the time domain. When $v_{RAMP}(t)$ is smaller than $v_C(t)$, $d(t)$ is high, S_1 is on, and S_2 is off. When $v_{RAMP}(t)$ is higher than $v_C(t)$, $d(t)$ is low, S_1 is off, and S_2 is on. This control method, which contains the output voltage feedback, is the voltage-mode control.

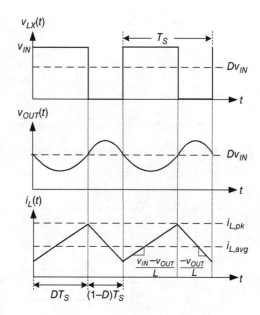

Figure 3.7 Steady-state waveforms of the buck converter

The power-stage waveforms of the buck converter in a steady state are illustrated in Figure 3.7. In a switching cycle T_S, the switching node LX is connected to v_{IN} and ground in DT_S and $(1-D)T_S$, respectively. Accordingly, the voltage across the inductor L is $v_{IN} - v_{OUT}$ and $-v_{OUT}$, respectively. Thus, the slope of the inductor current is $(v_{IN}-v_{OUT})/L$ and $-v_{OUT}/L$, respectively. With the constant $v_{IN}(t)$ and load condition, the system operates in a steady state. $v_{OUT}(t)$ and $i_L(t)$ return to the same values at the end of each switching cycle. The voltage second balance principle of L and the charge balance principle of C_{OUT} are met. Stable energy is delivered to the output, and $v_{OUT}(t)$ is well regulated at the value of Dv_{IN}. The average inductor current $i_{L,avg}$ is equal to the load current in the steady state.

When the load current increases, the load transient response waveforms are illustrated in Figure 3.8. In region (1) the buck converter operates in a steady state, and $i_{L,avg}(t)$ is equal to $i_{LOAD}(t)$. If the load current suddenly increases in region (2), the buck converter cannot provide sufficient energy to the output, that is, $i_{L,avg}(t)$ is less than $i_{LOAD}(t)$. Therefore, an undershoot voltage occurs at $v_{OUT}(t)$. Through the voltage feedback loop, the duty cycle starts to increase until $v_{OUT}(t)$ recovers to its original regulated value. At this time, the buck converter operates in a steady state in region (3). The inductor current is equal to the new load current level.

When the load condition varies (i.e., load transient), the voltage-mode controlled converter can detect the perturbation at v_{OUT} through the voltage feedback loop and modulate the duty cycle to regulate v_{OUT}. When v_{IN} varies (i.e., line transient), the duty cycle must be adjusted to retain v_{OUT} at the regulated value. However, the controller can modulate the duty cycle only through the voltage feedback loop. Therefore, v_{OUT} must deviate from its original regulated

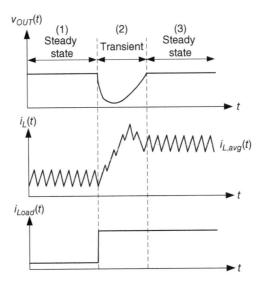

Figure 3.8 Load transient waveforms of the buck converter

value to help adjust the error signal and duty cycle. This induces large overshoot/undershoot voltages at v_{OUT}. For instance, when v_{IN} increases, v_{OUT} also increases without adjusting the duty cycle. Next, the voltage feedback decreases v_C, thereby decreasing the duty cycle. Finally, v_{OUT} recovers to its original regulated value. Is obtaining information on v_{IN} before affecting v_{OUT} possible? The answer is obvious if another path exists to reflect the perturbation from the line voltage. That is, a feedforward technique can help improve the line transient response.

It can be noticed that the inductor current not only directly indicates different loading current conditions but also includes v_{IN} information. By sensing the inductor current, current-sensing information is added to the controller as shown in Figure 3.9. Current-sensing information can provide v_{IN} information directly before affecting v_{OUT}. In other words, current sensing provides a feedforward path to improve the performance of the line transient. With current-sensing information, this control method is called current-mode control. By contrast, voltage feedback is still required for voltage regulation because current feedback cannot obtain v_{OUT} information directly.

The architecture of the current-mode controlled buck converter is illustrated in Figure 3.10. Excluding the original circuit blocks in voltage-mode control, the sensed inductor current v_S is added to v_{RAMP} to generate the summation signal, v_{SUM}. Operation waveforms are illustrated in Figure 3.11. In contrast to the voltage-mode control, which compares $v_{RAMP}(t)$ with $v_C(t)$ to determine $d(t)$, the current-mode control compares the sensed inductor current $v_S(t)$ with $v_C(t)$ (i.e., ignoring $v_{RAMP}(t)$ here for simplification). When $d(t)$ is high, S_1 is turned on to simultaneously increase the inductor current and $v_S(t)$. The duty cycle is determined when $v_S(t)$ is equal to $v_C(t)$. That is, when $v_S(t)$ reaches $v_C(t)$, $d(t)$ turns to low, S_1 turns off, and S_2 turns on to decrease $v_S(t)$. Since the duty cycle is determined by the peak value of $v_S(t)$, this control method is also called peak-current-mode control in conventional power electronics.

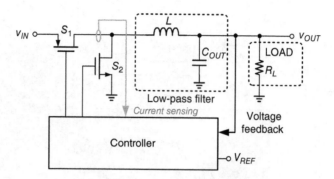

Figure 3.9 Adding current-sensing information to the buck converter

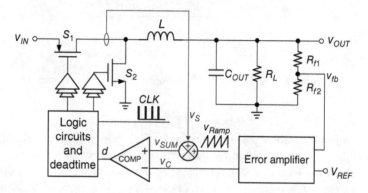

Figure 3.10 Architecture of the peak-current-mode controlled buck converter

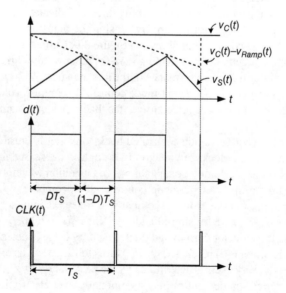

Figure 3.11 Operation waveforms of the peak-current-mode buck converter

Design of Switching Power Regulators

Table 3.3 Comparison of voltage- and current-mode controls

	Voltage-mode control	Current-mode control
Voltage feedback	Yes	Yes
Current feedback	No	Yes
Feedforward	No	Yes
Typical compensation method	Type III (PID)	Type II (PI)

$v_{RAMP}(t)$ is required for output voltage stabilization because sub-harmonic oscillation occurs when D is larger than 0.5 in the current-mode control [8]. The condition for preventing sub-harmonic oscillation is expressed as:

$$m_a \geq \frac{v_{OUT}}{2L} \quad (3.5)$$

where m_a is the slope of $v_{RAMP}(t)$.

A comparison of the voltage-mode control and current-mode control is listed in Table 3.3. The corresponding compensation method is also introduced to distinguish between the advantages and disadvantages of the two control methods.

3.3 Small Signal Modeling and Compensation Techniques in SWR

As discussed in Section 3.2, the power stage of the buck converter can be controlled using several control methods to achieve regulation performance. These control methods include the voltage feedback and current feedforward, which form outer and inner loops, respectively. System stability is considered through modeling and small signal analysis of the power stage, controller, and whole system.

3.3.1 Small Signal Modeling of Voltage-Mode SWR

As discussed above, the duty cycle is an ideal DC value according to v_{IN} and v_{OUT}. In practice, perturbation may occur in the input DC source, output loading, and circuit devices in the controller. Therefore, negative feedback control helps suppress the perturbation to present a stable and well-regulated output voltage by automatically adjusting the duty cycle. In the frequency domain, the duty cycle $d(s)$ is expressed as in Eq. (3.6), where D represents the DC component of the duty cycle to define the DC bias point (quiescent operating point), and the AC component $\hat{d}(s)$ indicates the small signal carried on the DC bias point. From the perspective of the converter, D indicates the duty cycle in a steady state if v_{OUT} is well regulated, while $\hat{d}(s)$ represents a small signal variation, which contains line/load variation and noise components.

$$d(s) = D + \hat{d}(s) \quad (3.6)$$

where $D \gg d(t)$.

As mentioned in Section 3.1, the average value of $v_{LX}(s)$ is equal to $d(s)$ multiplied by v_{IN}. From the average perspective, the output voltage of the buck converter can be viewed as a fraction of $v_{LX}(s)$ using L, C_{OUT}, and R_L:

$$v_{OUT}(s) = d(s)v_{IN} \frac{\frac{1}{sC_{OUT}} \| R_L}{sL + \frac{1}{sC_{OUT}} \| R_L} \tag{3.7}$$

where $v_{OUT}(s) = V_{OUT} + \hat{v}_{OUT}(s)$ and $d(s) = D + \hat{d}(s)$.

A small signal model can be obtained by removing the DC components. The small signal model and the transfer function from duty to output are shown in Figure 3.12 and Eq. (3.8), respectively. In other words, for a given operating point (i.e., input voltage V_{IN}, output voltage V_{OUT}, load resistor R_L, and duty cycle $D = V_{OUT}/V_{IN}$), the small signal model of the buck-converter power stage can be expressed as in Figure 3.12. This transfer function, which represents the small signal from duty to output, is defined as the duty-to-output transfer function $G_{vd}(s)$:

$$G_{vd}(s) = \frac{\hat{v}_{OUT}(s)}{\hat{d}(s)} = V_{IN} \cdot \frac{1}{1 + s\frac{L}{R_L} + s^2 L C_{OUT}} \tag{3.8}$$

From Eq. (3.8), $G_{vd}(s)$ of the buck converter contains a second-order polynomial in the denominator, meaning that the filter provides two poles in the buck-converter system. Considering such a characteristic, we can rewrite Eq. (3.8) using the second-order polynomial form:

$$G_{vd}(s) = \frac{1}{1 + \frac{2\zeta s}{\omega_0} + \left(\frac{s}{\omega_0}\right)^2} \tag{3.9}$$

where ζ is the damping ratio and ω_0 is the undamped natural frequency. The damping ratio describes how oscillations decay in a system after a disturbance. The undamped natural frequency is the frequency at which a system tends to oscillate in the absence of a damping force. In DC/DC converter systems and filter engineering, the standard second-order polynomial form is usually used:

$$G_{vd}(s) = \frac{1}{1 + \frac{s}{Q\omega_0} + \left(\frac{s}{\omega_0}\right)^2} \tag{3.10}$$

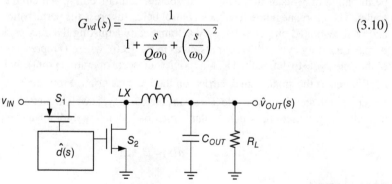

Figure 3.12 Duty-to-output small signal model of the buck converter power stage

where Q is the quality factor that represents the dissipation in a system, that is, the maximum stored energy divided by the dissipated energy per cycle. Moreover, the Q factor indicates the form of two poles. If $Q < 0.5$, then two separate real poles exist. If $Q > 0.5$, then complex conjugate pairs exist. The relationship of ζ and Q is expressed as:

$$Q = \frac{1}{2\zeta} \tag{3.11}$$

By comparing Eq. (3.8) with Eq. (3.10), Q and ω_0 can be expressed as:

$$Q = \frac{R_L}{\sqrt{\frac{L}{C_{OUT}}}} \quad \text{and} \quad \omega_0 = \frac{1}{\sqrt{LC_{OUT}}} \tag{3.12}$$

The natural frequency is determined by taking the square root of the product of the inductance and capacitance. The Q factor is related to R_L, L, and C_{OUT}. The normalized Bode plot of Eq. (3.10) with different Q is illustrated in Figure 3.13. The natural frequencies with different Q are normalized to 1. The phase decreases from 0° to 180° because two poles exist. However, the phase shapes near 1 rad/s are different. Under a low Q condition, the two poles are separate, the phase slowly decreases, and the magnitude does not peak. Under a high Q condition, complex

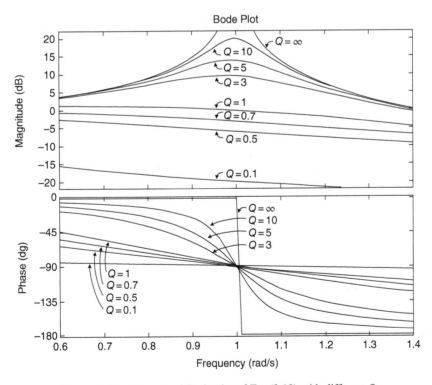

Figure 3.13 Normalized Bode plot of Eq. (3.10) with different Q

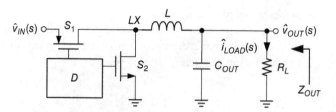

Figure 3.14 Disturbances in the power stage

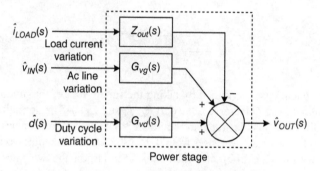

Figure 3.15 Complete power-stage model of a DC/DC converter with an open loop

conjugate poles induce peaking in magnitude, and the phase decreases dramatically at 1 rad/s. Different Q conditions greatly affect the frequency compensation for stabilizing the system.

Considering all the small signal sources in the power stage, except for the duty cycle variation, the input line voltage and loading current condition variation also disturb the output voltage. The duty cycle, line, and load variation can be viewed as three small signals to the system. Therefore, the line-to-output and output impedance transfer functions are derived in Eqs. (3.13) and (3.14), respectively, to describe the disturbance as shown in Figure 3.14 (similar to the way that the duty-to-output transfer function is derived).

$$G_{vg}(s) = \frac{\hat{v}_{OUT}(s)}{\hat{v}_{IN}(s)} = D \cdot \frac{1}{1 + s\dfrac{L}{R_L} + s^2 L C_{OUT}} \quad (3.13)$$

$$Z_{OUT}(s) = \frac{\hat{v}_{OUT}(s)}{\hat{i}_{LOAD}(s)} = \frac{sL}{1 + s\dfrac{L}{R_L} + s^2 L C_{OUT}} \quad (3.14)$$

The complete model of the power stage with an open loop is illustrated in Figure 3.15. Three independent disturbance sources – duty cycle, AC line voltage, and load current variations – are present. The output voltage of the power stage can be expressed by the superposition of these three independent disturbance sources:

$$\hat{v}_{OUT}(s) = G_{vd}(s)\hat{d}(s) + G_{vg}(s)\hat{v}_{IN} - Z_{OUT}(s)\hat{i}_{LOAD} \quad (3.15)$$

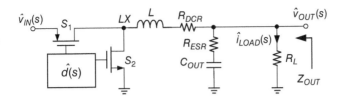

Figure 3.16 Disturbances in the power stage with parasitic resistances

where the control-to-output, line-to-output, and output impedance are shown in Eqs. (3.16), (3.17), and (3.18), respectively:

$$G_{vd}(s) = \frac{\hat{v}_{OUT}(s)}{\hat{d}(s)}\bigg|_{\substack{\hat{v}_{IN}=0 \\ \hat{i}_{LOAD}=0}} \quad (3.16)$$

$$G_{vg}(s) = \frac{\hat{v}_{OUT}(s)}{\hat{v}_{IN}(s)}\bigg|_{\substack{\hat{d}=0 \\ \hat{i}_{LOAD}=0}} \quad (3.17)$$

$$Z_{OUT}(s) = \frac{\hat{v}_{OUT}(s)}{\hat{i}_{LOAD}(s)}\bigg|_{\substack{\hat{v}_{IN}=0 \\ \hat{d}=0}} \quad (3.18)$$

In reality, the inductor and the capacitor have their individual parasitic resistances: direct current resistance R_{DCR} and equivalent series resistance R_{ESR}, respectively. As illustrated in Figure 3.16, R_{DCR} is especially important to the efficiency evaluation and current balance in multi-phase converter applications. In addition, R_{ESR} introduces a zero and affects the frequency response. Therefore, in such applications, R_{DCR} and R_{ESR} are required to be included in $G_{vd}(s)$.

The parasitic R_{ESR} in C_{OUT} introduces an LHP zero, which must be considered in stability analysis. Through the derivation method of $G_{vd}(s)$, ω_0 is independent of R_{DCR} and R_{ESR}. However, R_{DCR} and R_{ESR} affect the value of Q, which changes the shape of the phase near ω_0 and influences system stability. For simplicity, we usually ignore the variation in Q in frequency response analysis. Considering R_{ESR}, $G_{vd}(s)$ is approximately shown as:

$$G_{vd}(s) \approx V_{IN} \frac{1 + \dfrac{s}{C_{OUT}R_{ESR}}}{1 + s\dfrac{L}{R_L} + s^2 L C_{OUT}} \quad (3.19)$$

3.3.2 Small Signal Modeling of the Closed-Loop Voltage-Mode SWR

A basic voltage-mode controlled DC/DC buck converter with all circuit blocks is shown in Figure 3.17(a). The corresponding model of the whole system is illustrated in Figure 3.17(b). The feedback network $H(s)$, which is composed of two feedback resistors R_{f1} and R_{f2},

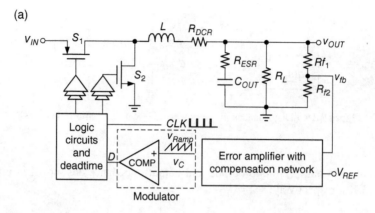

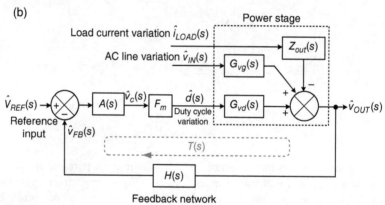

Figure 3.17 Voltage-mode controlled DC/DC buck converter: (a) schematic of the whole converter; (b) corresponding system model

is derived in Eq. (3.20). Using the feedback network, the output voltage information v_{FB} is fed back as $H(s)v_{OUT}$:

$$H(s) = \frac{R_{f2}}{R_{f1} + R_{f2}} \qquad (3.20)$$

The error signal between the feedback voltage $\hat{v}_{FB}$ and the reference voltage $\hat{v}_{REF}$ is amplified by the EA, which contains the compensation network and is represented by $A(s)$. The modulator compares the error signal $\hat{v}_C$ with v_{RAMP} to generate the duty cycle, and its modulation gain is represented by F_m. According to Figure 3.6, the transfer function from $\hat{v}_C$ to $\hat{d}$ can be expressed as:

$$F_m = \frac{\hat{d}(s)}{\hat{v}_C(s)} = \frac{1}{V_M} \qquad (3.21)$$

where V_M is the amplitude of the signal v_{RAMP}.

A complete voltage-mode buck converter can be modeled as shown in Figure 3.17. Three transfer functions – G_{vd}, G_{vg}, and Z_{out} – are defined to represent the relationships between the control input, output, and disturbances. G_{vd} is the open-loop control-to-output transfer function, G_{vg} is the open-loop line-to-output transfer function, and Z_{out} is the equivalent output impedance. The negative feedback control is used through the feedback network $H(s)$, compensator network $A(s)$, and modulator gain F_m to alleviate the variation caused by the disturbances.

The complete output voltage closed-loop expression, which considers the perturbations from the reference voltage, input source, and loading current variations, is shown in Eq. (3.22). Negative feedback can suppress the perturbation, stabilize the system, and maintain the voltage regulation. The loop gain $T(s)$, which is the product of the forward and feedback gains in the voltage feedback loop, is expressed as in Eq. (3.23).

$$\hat{v}_{OUT}(s) = \hat{v}_{REF}(s)\frac{1}{H(s)} \cdot \frac{T(s)}{1+T(s)} + \hat{v}_{IN}\frac{G_{vg}(s)}{1+T(s)} - \hat{i}_{LOAD}\frac{Z_{OUT}(s)}{1+T(s)} \qquad (3.22)$$

$$T(s) = G_{vd} \cdot A(s) \cdot F_m \cdot H(s) = G_{vd} \cdot A(s) \cdot F_m \text{ if } H(s) = 1 \qquad (3.23)$$

The loop gain $T(s)$ should be analyzed to design a stable buck converter. Generally, achieving high DC gain at low frequencies, a 45° phase margin, and at least a 10 dB gain margin is demanded. However, $G_{vd}(s)$ contains two poles that result in a 180° phase delay and instability as derived in Eq. (3.19). $A(s)$, which has not yet been determined in the previous discussion, can adjust the DC gain and generate poles and zeros to compensate for the insufficient phase margin. The compensation methods are introduced in Section 3.3.3.

The phase margin determines the amount of ringing when the load changes. The transient responses of different phase margins are shown in Figure 3.18. For a fast transient response, a 60° phase margin is suitable. However, sustaining 60° for all input, output, and load conditions is not easy. Thus, at least 45° is sufficient. Note that the ringing frequency in Figure 3.18 is related to the crossover frequency f_c of $T(s)$.

Aside from the gain and phase margin, a high bandwidth is required to obtain a fast transient response. Most importantly, the design of the highest crossover frequency is limited within 1/5 or 1/10 of the converter switching frequency f_S to ensure that the switching noise does not increase.

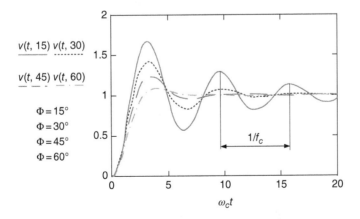

Figure 3.18 Transient response of different phase margin

3.3.2.1 Frequency Compensation Design in Voltage-Mode SWR

In frequency compensation design, the crucial issue is to design the transfer from the output to the control. The path from the output to the control contains the sensor gain $H(s)$, compensated EA $A(s)$, and modulation gain F_m. Thus, compensation techniques can be applied on any one of the transfer functions. For a voltage-mode buck converter, the loop transfer function $T(s)$ can be expressed as in Eq. (3.24) if $H(s) = 1$. One zero and two poles perform at the power stage. ω_O is the frequency of the double poles, which are composed of the output filter L and C_{OUT}. The equivalent series resistance zero $\omega_{Z(ESR)} = 1/(C_{OUT}R_{ESR})$ is derived from the ESR and output filter capacitor. The Bode plot of the control-to-output transfer function $G_{vd}(s)$ is illustrated in Figure 3.19.

$$T(s) = G_{vd}(s) \cdot F_m \cdot A(s) \tag{3.24}$$

where

$$G_{vd}(s) = V_{IN} \frac{1 + \dfrac{s}{\omega_{Z(ESR)}}}{1 + \dfrac{s}{Q\omega_O} + \dfrac{s^2}{\omega_O^2}}, \quad \omega_O \approx \frac{1}{\sqrt{LC_{OUT}}}, \quad \omega_{Z(ESR)} = \frac{1}{R_{ESR}C_{OUT}}, \quad Q \approx \frac{R_L}{\sqrt{\dfrac{L}{C_{OUT}}}}$$

Several design cases of $A(s)$ are analyzed and discussed as follows. The frequency response and its corresponding transient response are also shown.

First, we use a proportional compensation (P-compensation). That is, the compensation network contains a low DC gain K_L, as shown in Eq. (3.25). Thus, $T(s)$ can be expressed as in Eq. (3.26).

$$A(s) = -K_L \tag{3.25}$$

$$T(s) = F_m \cdot V_{IN} \frac{1 + \dfrac{s}{\omega_{Z(ESR)}}}{1 + \dfrac{s}{Q\omega_O} + \dfrac{s^2}{\omega_O^2}} \cdot (-K_L) \tag{3.26}$$

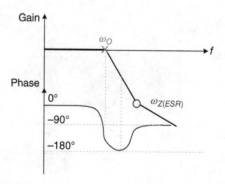

Figure 3.19 Bode plot of the power stage

The corresponding Bode plot is illustrated in Figure 3.20(a). With the low DC gain of K_L and LC double poles ω_O, the phase margin deteriorates by 180° before the unit gain frequency. Although the LHP ESR zero can increase the phase margin by 90°, the phase margin is not improved because the ESR zero is at frequencies that are significantly higher than the unit gain frequency. The 0° phase margin induces an unstable operation in the buck converter. Thus, the output voltage cannot be regulated. The Bode plot from the simulation results is shown in Figure 3.20(b). High-frequency poles that inherently exist in the EA cause the K_L and phase to drop as the frequency increases. Figure 3.20(c) indicates the corresponding simulation results. The voltage-mode buck converter is not stable and oscillates between 1.32 and 1.42 V.

A simple method that ensures the stable operation of the voltage-mode buck converter without modifying $A(s)$ is the use of an output capacitor with a high ESR value. Placing the ESR zero below the unit gain frequency can help improve the phase margin. As shown in Figure 3.21 (a), if $\omega_{Z(ESR)}$ is placed at lower frequencies, a larger phase margin can be obtained to increase the stability. As depicted in Figure 3.21(b), $\omega_{Z(ESR)}$ is placed at three times ω_O. The lowest phase still has 57° within unit gain frequency. The corresponding transient response from light to heavy load and vice versa is shown in Figure 3.21(c). The output voltage V_{OUT} is regulated near the predefined value with stable operation.

However, if any additional zeros do not improve the phase margin, then the DC gain cannot be increased to a higher level because the phase margin decreases below zero. This occurs because of the inherent high-frequency poles in the EAs. A low DC gain indicates poor load regulation and low output voltage accuracy. In addition, if the ESR zero is considered as a compensation zero to cancel the effect of double poles, then $\omega_{Z(ESR)}$ moves toward the origin to increase the system stability. The design criterion of ESR zero is that the ESR zero must place below $3\omega_O$ to ensure enough phase margin, since a poor phase margin results in a large ringing in case of load change. As shown in Figure 3.22, the lowest phase degrades from 56° to 45° and 40° if the ESR zero $\omega_{Z(ESR)}$ changes from $3\omega_O$ to $4\omega_O$ and $5\omega_O$, respectively. If $\omega_{Z(ESR)}$ is smaller than $3\omega_O$, the phase margin can be increased again to 90°, as shown in Figure 3.21(a).

The output voltage can be well regulated by a high ESR. However, $\omega_{Z(ESR)}$ is composed of the ESR and the output capacitor. The current flowing in or out of the capacitor induces a voltage drop across the ESR. As indicated in Figure 3.21(c), a large voltage ripple (i.e., 15 mV in the simulation) and undershoot/overshoot voltage (i.e., 33 mV in the simulation) occur because of the IR drop voltage formed by the current variations across the ESR. Using the ESR zero can certainly simplify the compensation technique at the cost of regulation performance.

To improve the regulation performance, an integrator is used to increase the DC gain as expressed in Eq. (3.27). Thus, Eq. (3.26) can be modified to Eq. (3.28) with a high DC gain of K_H.

$$A(s) = -\frac{K_H}{s} \tag{3.27}$$

$$T(s) = F_m \cdot V_{IN} \frac{1 + \dfrac{s}{\omega_{Z(ESR)}}}{1 + \dfrac{s}{Q\omega_O} + \dfrac{s^2}{\omega_O^2}} \cdot \left(-\frac{K_H}{s}\right) \tag{3.28}$$

After using an integrator, the loop transfer function contains three poles according to Eq. (3.28). One pole at the origin is formed by the ideal integrator, thereby contributing a

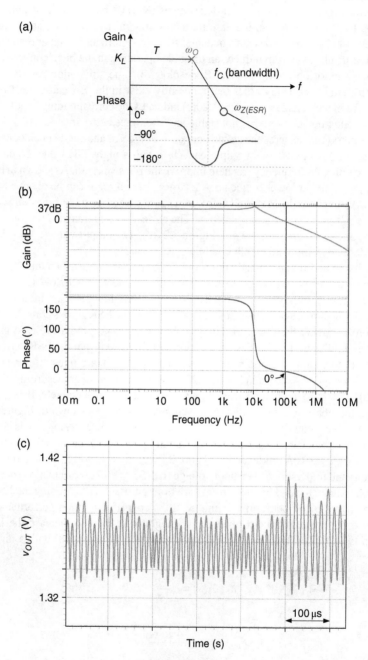

Figure 3.20 (a) Bode plot, (b) Bode plot by simulation results, and (c) unstable V_{OUT} in the simulation results when $A(s) = -K_L$

Design of Switching Power Regulators

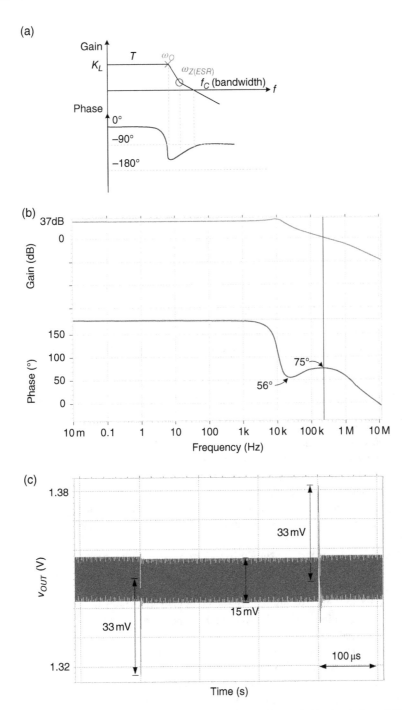

Figure 3.21 (a) Bode plot, (b) Bode plot by simulation results, and (c) simulation of V_{OUT} when $A(s) = -K_L$ with a large ESR

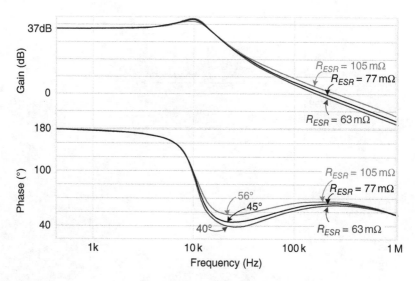

Figure 3.22 Comparison of the Bode plots when the ESR zeros are placed at $3\omega_O$ (R_{ESR} = 105 mΩ), $4\omega_O$ (R_{ESR} = 77 mΩ), and $5\omega_O$ (R_{ESR} = 63 mΩ).

90° delay. After the compensation with an ideal integrator, two possible scenarios may occur as shown in Figure 3.23.

In Figure 3.23(a), if the DC gain is increased to a higher level, the system becomes evidently unstable because the phase margin is significantly lower than zero. In Figure 3.23(b), the simulated phase margin becomes −40 dB/decade. In the simulation results, the dominant pole contributed by the integrator is not at the origin because a non-ideal EA has a limited low-frequency gain. The corresponding transient response in Figure 3.23(c) indicates an unstable output voltage. Lowering the DC gain can basically force the loop-transfer function to incur a lower crossover frequency f_C before seeing the LC double poles. Hence, if the f_C is designed to be substantially smaller than the LC double poles, the buck converter can be regarded as a one-pole system with an improved phase margin of 90°. Unfortunately, a lower DC gain degrades the output voltage accuracy, and a lower f_C slows down the transient response of the converter.

The other possible scenario is shown in Figure 3.24(a) without decreasing the DC gain; a large-compensation capacitor is placed at the output of the EA to further push the dominant pole toward the origin. As depicted in Figure 3.24(b), the phase margin becomes 88° with a relatively low bandwidth. The double poles can be viewed as high-frequency poles compared with the low bandwidth. Thus, the system becomes more stable. However, the dynamic response is sacrificed, as shown in Figure 3.24(c). For any load change, the transient recovery time is very long. This compensation method may be used to simplify the power module design in some applications where a fast transient response is not the major concern.

In summary, the two possible scenarios cannot have high DC gain and large bandwidth for regulation performance and fast transient response concurrently. To conquer these problems, two additional LHP zeros are needed below the crossover frequency f_C to increase the phase margin if a high DC gain is required at the same time. Type III compensation, which is commonly used in voltage-mode buck converters, is introduced to have two LHP zeros.

Design of Switching Power Regulators

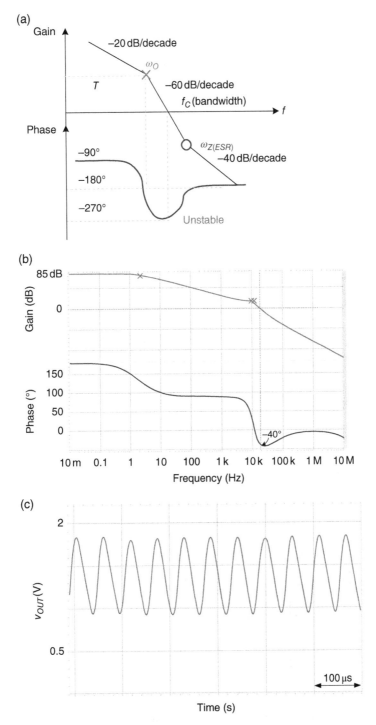

Figure 3.23 (a) Bode plot, (b) Bode plot by simulation results, and (c) simulation of V_{OUT} with high-DC-gain EA

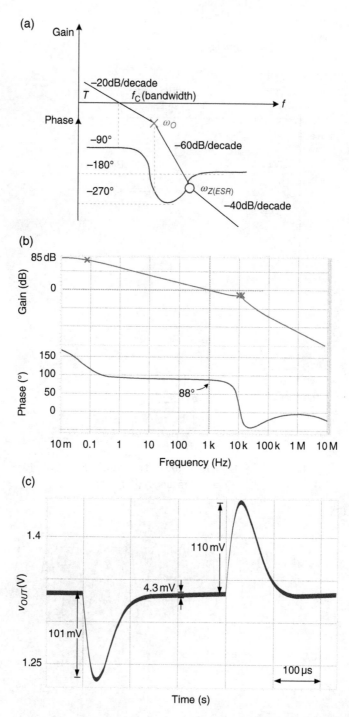

Figure 3.24 (a) Bode plot, (b) Bode plot by simulation results, and (c) simulation of V_{OUT} when a large-compensation capacitor is placed at the output of a high-gain EA

Design of Switching Power Regulators

Equation (3.29) depicts Type III compensation containing three LHP poles and two LHP zeros to obtain a high DC gain and a large bandwidth at the same time:

$$A(s) = -\frac{K_H}{s} \cdot \frac{\left(1+\dfrac{s}{\omega_{z1}}\right)\left(1+\dfrac{s}{\omega_{z2}}\right)}{\left(1+\dfrac{s}{\omega_{p1}}\right)\left(1+\dfrac{s}{\omega_{p2}}\right)} \qquad (3.29)$$

The corresponding frequency response is illustrated in Figure 3.25. With the help of two zeros ω_{z1} and ω_{z2}, the 180° phase delay caused by LC double poles can be compensated. The other two high-frequency compensation poles ω_{p1} and ω_{p2} are used to suppress high-frequency noise through the effect of −20 dB/decade contributed by each pole. Since ω_{z1} and ω_{z2} compensate the phase delay of ω_O, $T(s)$ can be viewed as a one-pole system within the bandwidth, which is unconditionally stable. In other words, high DC gain and adequate phase margin can be achieved simultaneously by Type III compensation. Usually, ω_{z1} is placed below ω_O to achieve the phase lead while ω_{z2} is placed above ω_O to decrease the phase drop. To ensure a suitable phase margin and a large bandwidth, the crossover frequency f_c is usually designed between ω_{z2} and ω_{p1}. To further enlarge the bandwidth, placing ω_{p1} at the location of ESR zero, $\omega_{Z(ESR)}$, can effectively extend the crossover frequency and speed up the transient response. After using Type III compensation, the loop transfer function can be modified as:

$$T(s) = F_m \cdot V_{IN} \frac{1+\dfrac{s}{\omega_{Z(ESR)}}}{1+\dfrac{s}{Q\omega_o}+\dfrac{s^2}{\omega_o^2}} \cdot \left[-\frac{K_H}{s} \cdot \frac{\left(1+\dfrac{s}{\omega_{z1}}\right)\left(1+\dfrac{s}{\omega_{z2}}\right)}{\left(1+\dfrac{s}{\omega_{p1}}\right)\left(1+\dfrac{s}{\omega_{p2}}\right)} \right] \qquad (3.30)$$

If using Type III compensation, the simulated Bode plot is shown in Figure 3.25(b). A high DC gain of 85 dB and 50° of phase margin can ensure good regulation and stable operation. As shown in Figure 3.25(c), compared with a large-compensation capacitor with high-gain EA, the transient response time of Type III compensation is much faster. Both undershoot and overshoot are much smaller. With a small ESR value, the voltage ripple is small. Zoomed-in transient responses with Type III compensation are shown in Figure 3.26.

The EA with Type III compensation is depicted in Figure 3.27, which contains input and feedback impedances Z_I and Z_F. With an ideal EA, the transfer function can be expressed by the ratio of Z_F to Z_I expressed as:

$$A(s) = \frac{\hat{v}_C}{\hat{v}_o} = -\frac{Z_F}{Z_I}$$

where $\qquad (3.31)$

$$Z_F = \frac{1}{sC_3} \bigg\| \left(R_2 + \frac{1}{sC_1}\right) \text{ and } Z_I = R_1 \bigg\| \left(R_3 + \frac{1}{sC_2}\right)$$

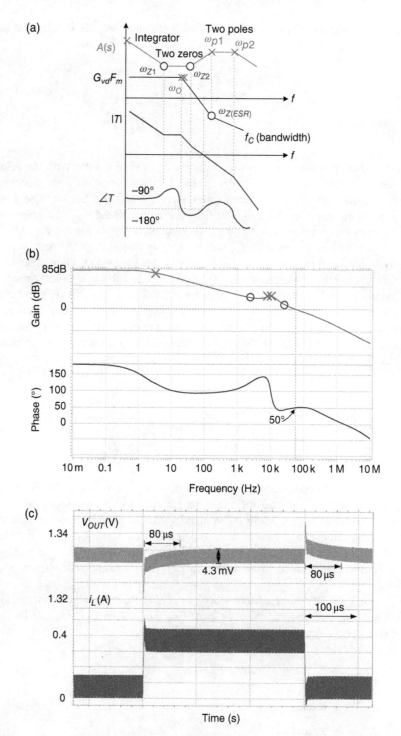

Figure 3.25 (a) Bode plot, (b) Bode plot by simulation results, and (c) simulation of V_{OUT} if Type III compensation is used

Design of Switching Power Regulators

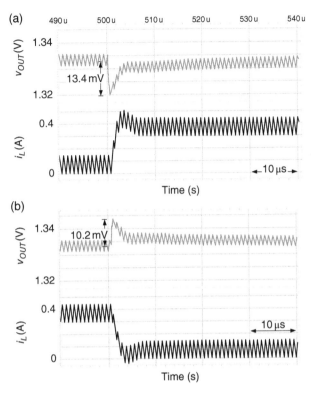

Figure 3.26 If Type III compensation is used, zoomed-in transient response is shown in case of (a) light-to-heavy load change and (b) heavy-to-light load change

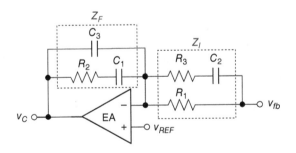

Figure 3.27 Type III compensation technique

By comparing Eq. (3.31) with Eq. (3.30), the locations of three poles and two zeros are listed as:

$$\omega_{z1} = \frac{1}{R_2 C_1}, \quad \omega_{z2} = \frac{1}{C_2(R_1 + R_3)}$$

$$\omega_{p0} = 0, \quad \omega_{p1} = \frac{1}{R_3 C_2}, \quad \text{and } \omega_{p2} = \frac{1}{R_2 \left(\frac{C_1 C_3}{C_1 + C_3} \right)} \approx \frac{1}{R_2 C_3} \text{ if } C_1 \gg C_3 \quad (3.32)$$

To generate phase compensation near ω_O from ω_{z1} and ω_{z2}, R_1, R_2, C_1, and C_2 inevitably take larger values. Since ω_{p1} and ω_{p2} are used to suppress the switching noise without affecting the stability, $R_3 \ll R_1$ and $C_3 \ll C_1$ are needed to prevent phase degradation.

The large resistor and capacitor values derived are not easily implemented with silicon. Hence, Type III includes numerous external compensation components. Cost, printed circuit board (PCB) area, and switching noise are known to significantly deteriorate the converter performance.

As mentioned earlier, a higher ESR value can provide a 90° phase lead within a crossover frequency. Here, using the ESR zero can simplify the compensation network without limiting the bandwidth. Thus, the compensation network requires only a compensation zero, a dominant pole at the origin, and a noise-suppressing high-frequency pole. In other words, Type II compensation can be used if the ESR zero is employed. The expression of Type II compensation and its corresponding $T(s)$ are shown in Eqs. (3.33) and (3.34), respectively:

$$A(s) = -\frac{K_H}{s} \cdot \frac{\left(1 + \dfrac{s}{\omega_{z1}}\right)}{\left(1 + \dfrac{s}{\omega_{p1}}\right)} \quad (3.33)$$

$$T(s) = F_m \cdot V_{IN} \frac{1 + \dfrac{s}{\omega_{Z(ESR)}}}{1 + \dfrac{s}{Q\omega_o} + \dfrac{s^2}{\omega_o^2}} \cdot \left[-\frac{K_H}{s} \cdot \frac{\left(1 + \dfrac{s}{\omega_{z1}}\right)}{\left(1 + \dfrac{s}{\omega_{p1}}\right)}\right] \quad (3.34)$$

The Bode plot is illustrated in Figure 3.28(a). Once more, the dominant pole is ideally at the origin. ω_{z1} is placed below ω_O to provide a 90° phase lead, and ω_{ESR} also compensates for the phase delay caused by ω_O. Here, ω_{z1} and ω_{ESR} replace the functions ω_{z1} and ω_{z2} in Type III compensation. An adequate phase margin can be achieved with fewer external components compared with Type III compensation. However, the output ripple is larger than that of Type III compensation because of the higher ESR, as shown in Figure 3.28(c).

Based on the discussion above, the compensation methods that stabilize the operation of buck converters can be classified into four types, as summarized in Table 3.4. A combined low DC gain and ESR zero results in the fastest transient response because of the large bandwidth. Integrator compensation requires a large-compensation capacitor, resulting in a slow transient response. Introducing two zeros by either Type III or Type II compensation with the ESR zero can increase the bandwidth with a high DC gain and shorten the transient response time. Type III compensation requires six external components that occupy a large PCB area. The corresponding detailed conditions in the simulation tool "Hspice" and the simulation results are listed and compared in Table 3.5. With a large ESR, the compensation networks are simplified at the cost of voltage ripple and undershoot/overshoot voltage performance. In terms of low voltage ripple, good transient, and regulation performances, Type III compensation is the most commonly used technique for voltage-mode buck converters. A complete architecture of the voltage-mode buck converter is illustrated in Figure 3.29.

Design of Switching Power Regulators 149

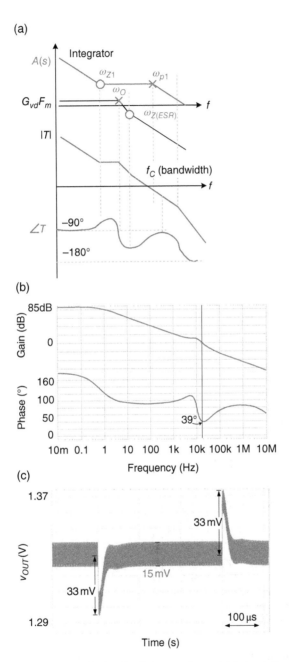

Figure 3.28 (a) Bode plot, (b) Bode plot by simulation results, and (c) simulation of V_{OUT} with Type II compensation and a large ESR

Table 3.4 Comparison of different compensation methods in voltage-mode buck converters

Method	$A(s)$	Topology	Ripple/transient/regulation
Low DC gain + ESR zero	$-K_L$		Larger/fastest/poor
Integrator	$-\dfrac{K_H}{s}$		Smaller/slowest/good
Type III	$-\dfrac{K_H}{s} \cdot \dfrac{\left(1+\frac{s}{\omega_{z1}}\right)\left(1+\frac{s}{\omega_{z2}}\right)}{\left(1+\frac{s}{\omega_{p1}}\right)\left(1+\frac{s}{\omega_{p2}}\right)}$		Smaller/faster/good
Type II + ESR zero	$-\dfrac{K_H}{s} \cdot \dfrac{\left(1+\frac{s}{\omega_{z1}}\right)}{\left(1+\frac{s}{\omega_{p1}}\right)}$		Larger/faster/good

Table 3.5 Comparison of Hspice simulation results in different compensation methods

Method	Hspice conditions	Voltage ripple (mV)	Recovery time (μs)	Overshoot/sundershoot (mV)
Low DC gain + ESR zero	K_L = 37 dB, R_{ESR} = 105 mΩ	15	7	33/33
Integrator	K_H = 85 dB, C_1 = 2000 pF, R_{ESR} = 30 mΩ	4.3	240	101/110
Type III (or PID)	K_H = 85 dB, C_1 = 22 pF, C_2 = 68 pF, C_3 = 0.2 pF, R_1 = 100 kΩ, R_2 = 2.2 MΩ, R_3 = 1 kΩ, R_{ESR} = 30 mΩ	4.3	80	13.4/10.2
Type II (or PI) + ESR zero	K_H = 85 dB, C_1 = 150 pF, C_2 = 0.2 pF, R_1 = 150 kΩ, R_{ESR} = 105 mΩ	15	88	33/33

L = 4.7 μH, C_{OUT} = 47 μF, f_s = 1 MHz, R_L = 18/3.6 Ω (light/heavy load).

3.3.3 Small Signal Modeling of Current-Mode SWR

The design of the compensation network for the voltage-mode buck converter is complex because of its complex poles at the power stage. The input voltage feedforward technique for improving the line transient response further increases the complexity of the design.

Design of Switching Power Regulators 151

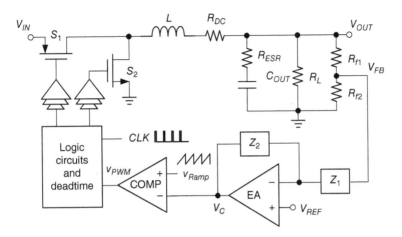

Figure 3.29 Complete architecture of the voltage-mode buck converter

In addition, another current sensor is required to achieve an over-current protection (OCP) function, which is necessary in the power management unit to increase reliability. Therefore, we consider what kind of control technique can address the problem met by the voltage-mode control technique. The current-mode control is one solution that is popular in present power management circuits. Current sensing inherently exists in current-mode controlled converters, as illustrated in Figure 3.10. With the sensed inductor current, the input voltage information is obtained simultaneously. The control system and compensation network are simpler in current-mode control techniques than in voltage-mode control techniques.

Before discussing frequency compensation in current-mode control buck converters, small signal modeling is again derived. The power-stage model of the buck converter, including G_{vd} and G_{vg}, has already been derived in Figure 3.15. However, the transfer functions represent only the duty-to-output and input-to-output characteristics. In the current-mode control, a transfer function and a model related to the inductor current are required. Here, duty-to-current and input-to-current transfer functions must be derived. Figure 3.30 illustrates the derivations of small signal models for duty-to-current and input-to-current transfer functions.

With the average concept, i_L can be expressed as V_{LX} divided by the impedance, as seen by V_{LX} to ground in Eq. (3.35). The DC quiescent operating point and perturbation for each signal are expressed in Eq. (3.36).

$$i_L(s) = d(s) \cdot v_{IN}(s) \cdot \frac{1}{sL + \frac{1}{sC_{OUT}} \| R_L} \tag{3.35}$$

$$i_L(s) = I_L + \hat{i}_L(s),\ v_{IN}(s) = V_{IN} + \hat{v}_{IN}(s),\ \text{and}\ d(s) = D + \hat{d}(s) \tag{3.36}$$

Following Eqs. (3.35) and (3.36), the duty-to-current and line-to-current transfer functions, defined as $G_{id}(s)$ and $G_{ig}(s)$, respectively, can be obtained in Eqs. (3.37) and (3.38).

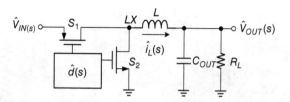

Figure 3.30 Small signal model of the buck converter power stage for the derivation of the duty-to-current and input-to-current transfer functions

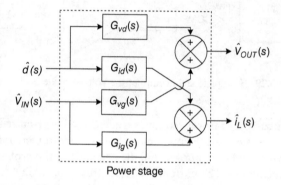

Figure 3.31 Complete power-stage model of a DC/DC converter (open loop)

$$G_{id}(s) = \frac{\hat{i}_L(s)}{\hat{d}(s)} = \frac{V_{IN}}{DR_L} \cdot \frac{1}{1 + s\dfrac{L}{R_L} + s^2 LC_{OUT}} \tag{3.37}$$

$$G_{ig}(s) = \frac{\hat{i}_L(s)}{\hat{v}_{IN}(s)} = \frac{D}{R_L} \cdot \frac{1}{1 + s\dfrac{L}{R_L} + s^2 LC_{OUT}} \tag{3.38}$$

By combining Figure 3.15 with Eqs. (3.37) and (3.38), the complete power-stage model of a DC/DC converter, including i_L, is illustrated in Figure 3.31.

The output voltage of a power stage can be expressed in Eq. (3.39) by the superposition of two independent disturbance sources. Similarly, the inductor current of the power stage can be expressed in Eq. (3.40) by the superposition of two independent disturbance sources.

$$\hat{v}_{OUT}(s) = G_{vd}(s) \cdot \hat{d}(s) + G_{vg}(s) \cdot \hat{v}_{IN} \tag{3.39}$$

$$\hat{i}_L(s) = G_{id}(s) \cdot \hat{d}(s) + G_{ig}(s) \cdot \hat{v}_{IN} \tag{3.40}$$

where control-to-current and line-to-current are shown in Eqs. (3.41) and (3.42), respectively.

$$G_{id}(s) = \left.\frac{\hat{i}_L(s)}{\hat{d}(s)}\right|_{\hat{v}_{IN}=0} \tag{3.41}$$

Design of Switching Power Regulators 153

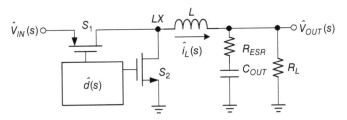

Figure 3.32 Disturbances in the power stage with parasitic resistances

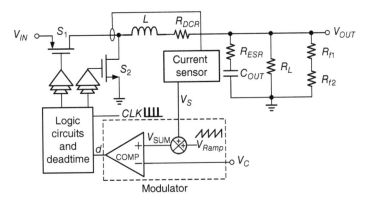

Figure 3.33 Current-mode controlled DC/DC buck converter with the current-sensing loop closed

$$G_{ig}(s) = \left.\frac{\hat{i}_L(s)}{\hat{v}_{IN}(s)}\right|_{\hat{d}=0} \tag{3.42}$$

Furthermore, in Figure 3.32, if the ESR value of the output capacitor is considered and the R_{DCR} of the inductor is neglected, then Eqs. (3.37) and (3.38) are modified to Eqs. (3.43) and (3.44), respectively, with an ESR zero.

$$G_{id}(s) \approx \frac{V_{IN}}{DR_L} \cdot \frac{1 + \dfrac{s}{C_{OUT}R_{ESR}}}{1 + s\dfrac{L}{R_L} + s^2 L C_{OUT}} \tag{3.43}$$

$$G_{ig}(s) \approx \frac{D}{R_L} \cdot \frac{1 + \dfrac{s}{C_{OUT}R_{ESR}}}{1 + s\dfrac{L}{R_L} + s^2 L C_{OUT}} \tag{3.44}$$

Figure 3.33 shows a current-mode controlled DC/DC buck converter with the current-sensing loop closed. The inductor current is sensed by the current sensor and added to V_{RAMP} to generate the summation signal V_{SUM}. By comparing V_{SUM} with v_C, the duty cycle is generated. In contrast to voltage-mode control, the slope of V_{RAMP} should follow the inequality of

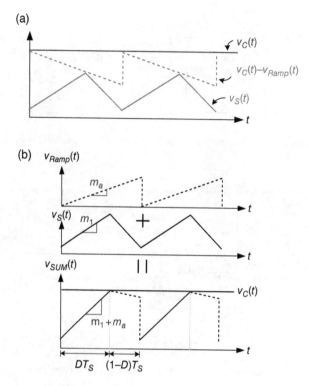

Figure 3.34 Operation waveforms of the peak current-mode buck converter: (a) v_S compared with the subtraction result of v_C and V_{RAMP} to determine the duty and (b) v_{SUM}, which is the summation of v_S and V_{RAMP}, compared with v_C to determine the duty

Eq. (3.5) to prevent sub-harmonic oscillation. Here, the control signal v_C can be temperately viewed as a current reference voltage that controls the peak current of the buck converter. v_C can sometimes be used to stand for the approximate output loading current.

The operation waveform is usually illustrated as in Figure 3.34(a). The duty cycle is intuitively determined when the peak of the sensed inductor current v_S reaches the control signal v_C. To avoid sub-harmonic oscillation, V_{RAMP} is introduced for slope compensation. Instead, the duty cycle is determined when the peak of v_S reaches $v_C - V_{RAMP}$. To simplify the derivation of the current-mode control model, the three signals v_C, V_{RAMP}, and v_S are combined and illustrated in a different manner, as shown in Figure 3.34(b). The slope of V_{RAMP}, which suppresses sub-harmonic oscillation, is m_a, while the slope of v_S is m_1, which is proportional to $v_{IN} - v_{OUT}$. By combining V_{RAMP} and v_S, the slope v_{SUM} is equal to $m_1 + m_a$ before v_{SUM} reaches v_C. $m_1 + m_a$ helps determine the duty cycle in the current-mode control.

Using the same derivation concept as F_M in voltage-mode control, the modulation gain is the reciprocal of the ramp amplitude V_M. Here, given that the slope during $(1 - D)T_S$ does not involve duty cycle determination, the effective amplitude of V_M can be viewed as $(m_1 + m_a)T_S$. The modulation gain is then derived in Eq. (3.45), where M is used to indicate the ratio of the compensation slope m_a and the inductor current charging slope m_1, as shown in

Eq. (3.46). The characteristic of the current-mode control will be completely different, if different M values are applied in Eq. (3.45), as discussed later.

$$F_m = \frac{\hat{d}(s)}{\hat{v}_C(s)} = \frac{1}{V_M} = \frac{1}{(m_1+m_a)T_S} = \frac{1}{Mm_1T_S} \quad (3.45)$$

$$M = 1 + \frac{m_a}{m_1} \quad (3.46)$$

Figure 3.35 shows the current-mode control equivalent model with the current loop closed. The current sensor, which senses the inductor current value I_L and converts it to the voltage signal V_S, has a transfer function of resistance R_i, as shown in Eq. (3.47). If the switching frequency is constant, the current-mode control in PWM can be considered a sample-and-hold system with the sampling instant occurring at DT_s [9]. Thus, the current sampling effect is modeled and converted into a continuous-time representation as shown in Eq. (3.48), where $H_e(s)$ contains two RHP zeros. In this analysis, $H_e(s)$ is accurate from 0 to half of the switching frequency.

$$R_i = \frac{v_S}{i_L} \quad (3.47)$$

$$H_e(s) \approx \frac{s^2}{\omega_n^2} + \frac{s}{\omega_n Q_n} + 1 \quad (3.48)$$

where

$$Q_n = -\frac{2}{\pi} \text{ and } \omega_n = \frac{\pi}{T_s}$$

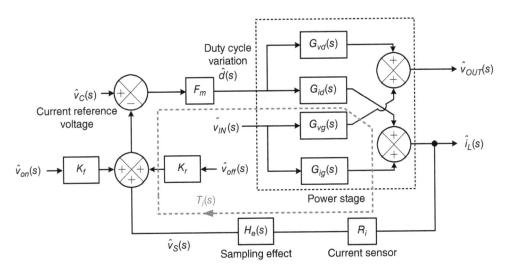

Figure 3.35 Current-mode control equivalent model with current loop closed

The K_f and K_r listed in Eqs. (3.49) and (3.50), respectively, provide the feedforward voltage across the inductor during on-time and off-time periods, respectively [9]. During on-time and off-time periods, the voltage across the inductor is $v_{IN} - v_{OUT}$ and v_{OUT}, respectively.

$$K_f = -\frac{DT_s R_i}{L} \cdot \left(1 - \frac{D}{2}\right) \tag{3.49}$$

$$K_r = \frac{(1-D)^2 T_s R_i}{2L} \tag{3.50}$$

The perturbations on the on- and off-time periods are $\hat{v}_{on}$ and $\hat{v}_{off}$, respectively, as shown in Eqs. (3.51) and (3.52), respectively, and affect the duty cycle.

$$\hat{v}_{on} = \hat{v}_{IN} - \hat{v}_{OUT} \tag{3.51}$$

$$\hat{v}_{off} = \hat{v}_{OUT} \tag{3.52}$$

The current-mode control in Figure 3.35 contains a current loop with the transfer function $T_i(s)$. By comparing the sensed inductor current with the current reference voltage, the duty cycle can be obtained by the modulator. At the power stage, the inductor current is generated according to the duty cycle. The current loop gain $T_i(s)$ is derived as:

$$T_i(s) = \frac{LR_i}{R_L(1-D)T_s M} \cdot \frac{1 + \dfrac{s}{C_{OUT} R_{ESR}}}{1 + s\dfrac{L}{R_L} + s^2 L C_{OUT}} \cdot H_e(s) \tag{3.53}$$

The control-to-output transfer function is desirable in demonstrating the characteristic of the current-mode control. With the closed current loop $T_i(s)$, the current-mode control-to-output transfer function is derived in Eq. (3.54). $F_p(s)$ and $F_h(s)$ are shown in Eqs. (3.55) and (3.56), respectively.

$$\frac{\hat{v}_{OUT}}{\hat{v}_C} \approx \frac{D \cdot \left[M(1-D) - \left(1 - \dfrac{D}{2}\right)\right]}{\dfrac{L}{R_L T_S} + M(1-D) - 0.5} \cdot F_p(s) \cdot F_h(s) \tag{3.54}$$

$$F_p(s) = \frac{1 + s C_{OUT} R_{ESR}}{1 + \dfrac{s}{\omega_p}} \tag{3.55}$$

where

$$\omega_p = \frac{1}{R_L C_{OUT}} + \frac{T_S}{LC} \cdot [M(1-D) - 0.5]$$

$$F_h(s) = \frac{1}{1 + \dfrac{s}{\omega_n Q} + \dfrac{s^2}{\omega_n^2}} \tag{3.56}$$

Design of Switching Power Regulators

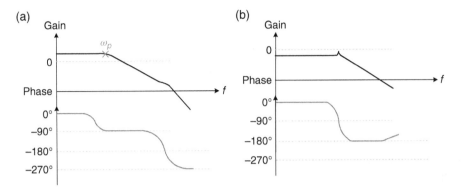

Figure 3.36 Bode plot of the current-mode control-to-current (a) with an appropriate small value of M and (b) with a large value of M

where

$$Q = \frac{1}{\pi[M(1-D)-0.5]}$$

Figure 3.36 illustrates the Bode plots of the current-mode control-to-current in two different cases. The typical characteristic of current-mode control is shown in Figure 3.36(a) with an appropriate small value of M. That is, m_a is equal to m_2, which follows Eq. (3.5). The DC loop gain is larger than zero to regulate the inductor current value. The control-to-current transfer function contains one dominant pole contributed by ω_p. It is worth mentioning that ω_p is approximately equal to $1/R_L C_{OUT}$ with a small value of M. In other words, the current-mode control buck converter can be viewed as a one-pole system whose dominant pole is located at the output and is contributed by R_L and C_{OUT}. With only one pole, the compensation design is much simpler than that in voltage-mode control and is a useful approximation when designing and simulating.

However, a large value of M indicates a large compensation slope and the possible formation of complex poles, because ω_p moves slightly toward higher frequencies and one of the high-frequency poles moves toward low frequencies. Since a large M represents a ratio of compensation slope much larger than the sensed inductor current, the current information is largely suppressed. Thus, the DC gain drops below zero. The characteristic of typical current-mode control disappears. With a large compensation slope, the current sensed signal is ignored. Therefore, the duty cycle is merely determined by comparing v_C with V_{RAMP}, which is similar to the voltage-mode control method and coincides with complex poles existing in the voltage-mode controlled buck converter. With an excessively large M indicating a large compensation slope, the current-mode control degenerates to the voltage-mode control.

By contrast, the line-to-output transfer function can be used to demonstrate the characteristic of the line transient response. The line-to-output transfer function in the current-mode control is derived as:

$$\frac{\hat{v}_{OUT}}{\hat{v}_{IN}} = \frac{D \cdot \left[M(1-D) - \left(1-\frac{D}{2}\right)\right]}{\frac{L}{R_L T_S} + M(1-D) - 0.5} \cdot F_p(s) \cdot F_h(S) \tag{3.57}$$

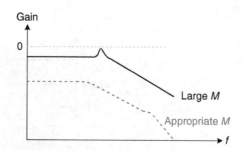

Figure 3.37 Comparison of the line-to-output transfer function in the current-mode control with appropriate small and large values of M

Figure 3.37 illustrates a comparison of the current-mode line-to-output transfer function with an appropriate small value and large value of M. The line-to-output transfer function can indicate the input noise suppression ability of converters. With an appropriate value of M, low DC gain can largely suppress input-voltage perturbation. In contrast, with a larger value of M, high DC gain deteriorates the ability of input-voltage perturbation suppression. Furthermore, it coincides with the advantage of the current-mode control in terms of good line regulation performance because it has inherent feedforward characteristics. On the contrary, poor input-voltage perturbation suppression occurs in the voltage-mode control buck converters because input information does not directly feed into the voltage-mode controller.

3.3.3.1 Frequency Compensation Design in Current-Mode SWR

In deriving a small signal model of the current loop, the control signal v_C can be moderately considered a current reference voltage that controls the peak current of a buck converter. In reality, the current reference needs to be adjusted according to different output loading conditions. To generate a dynamic current reference voltage, the output voltage is fed back for comparison with the reference voltage V_{REF}, as shown in Figure 3.38. When v_{OUT} is lower than V_{REF}, the inductor current I_L is smaller than the output loading current i_{LOAD}. A lower v_{OUT} also causes the output voltage to drop. Therefore, the dynamic increase in the current reference can increase i_L to meet the requirement of i_{LOAD}. Through the voltage feedback loop, current reference generation and voltage regulation can be achieved simultaneously.

With voltage feedback, the current-mode model in Figure 3.35 can be modified to that in Figure 3.39. The control-to-output transfer function has previously been derived in Eq. (3.54). Hence, the equivalent model of the complete current-mode model with voltage feedback loop can be simplified as shown in Figure 3.40.

Using the simplified model, the approximate $T(s)$ in Figure 3.40 can easily be obtained:

$$T(s) \approx \frac{\hat{v}_{OUT}(s)}{\hat{v}_C(s)} \cdot A(s) = K \cdot \frac{1 + sC_{OUT}R_{esr}}{1 + \frac{s}{\omega_p}} \cdot A(s) \qquad (3.58)$$

where $\omega_p \approx \frac{1}{R_L C_{OUT}}$ and K represents the DC gain derived in Eq. (3.54).

Design of Switching Power Regulators 159

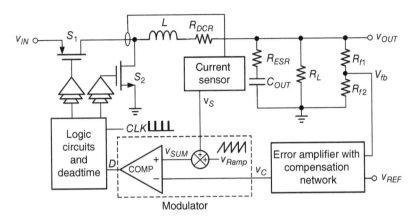

Figure 3.38 Current-mode controlled DC/DC buck converter with the current-sensing closed loop and outer voltage feedback loop

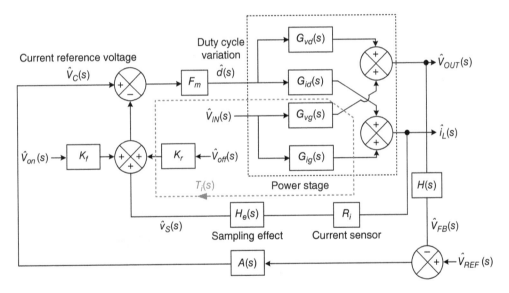

Figure 3.39 Complete current-mode model with voltage feedback loop

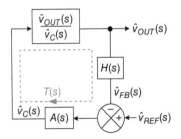

Figure 3.40 Equivalent model of the complete current-mode model with voltage feedback loop

The current-mode control contains one pole and one zero. Similar to the voltage-mode control, the output capacitor and its ESR value form a zero. The pole is supplied by the output capacitor and load resistance. The Bode plot of the simplified current-mode control-to-output model is illustrated in Figure 3.41. The load resistance R_L varies with different output loading conditions. Thus, the location of ω_p varies accordingly. At a light load, ω_p moves toward the origin. At a heavy load, ω_p moves toward higher frequencies.

The frequency response characteristic of the current mode is less complex than that of the voltage mode. Thus, the compensation network design can be simplified. Furthermore, the EA with compensation network is expressed by $A(s)$. An adequate $A(s)$ must be adapted to different load conditions in the following analysis. By introducing the high-DC gain K_H for voltage regulation, the Bode plot can be illustrated, as shown in Figure 3.42. The DC gain is approximately equal to $K_H + K$ ($\approx K_H$ because $K_H \gg K$). Here, $T(s)$ is a one-pole system with dominant pole ω_p. At both light and heavy loads, the phase margin is larger than 90°.

The simulation results of the current-mode control with high DC gain K_H are shown in Figure 3.43. As depicted in Figure 3.43(a), the phase margin at light and heavy loads is 31°

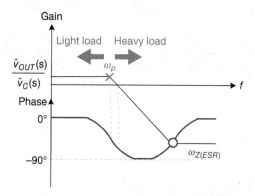

Figure 3.41 Bode plot of the simplified current-mode control-to-output model

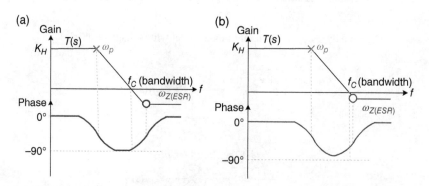

Figure 3.42 Bode plot of the simplified current-mode model with a high DC gain of K_H at (a) light and (b) heavy loads

and 24°, respectively, indicating that the converter is stable. Interestingly, v_{OUT} is unstable in Figure 3.43(b). Instability results because the designed bandwidth is too large. As a rule of thumb, the crossover frequency should be lower than 1/10 or 1/5 of the switching frequency. A crossover frequency of 6 MHz is inappropriate in this case because of the switching frequency of 1 MHz. Moreover, the transfer function in Eq. (3.58) is an approximate model that does not include high-frequency poles and zeros. Additionally, a low-frequency dominant pole must be introduced.

An integrator with a transfer function, shown in Eq. (3.59), is ideally used to insert a dominant pole at the origin. In circuit implementation, a large-compensation capacitor is placed

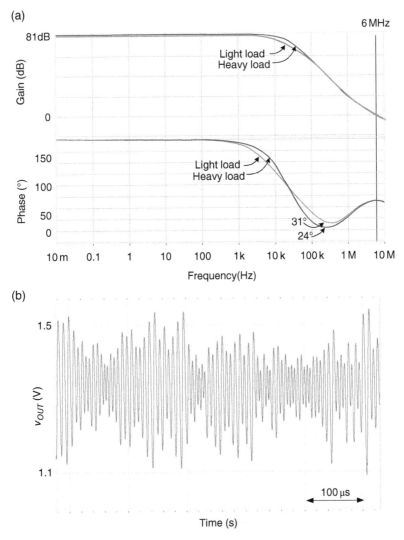

Figure 3.43 (a) Bode plot from the simulation results and (b) simulation results of V_{OUT} with a high DC gain of K_H

at the output of the EA to further direct the dominant pole toward the origin. The corresponding $T(s)$ is expressed in Eq. (3.60).

$$A(s) = -\frac{K_H}{s} \quad (3.59)$$

$$T(s) \approx K \cdot \frac{1 + sC_{OUT}R_{esr}}{1 + \frac{s}{\omega_p}} \cdot \left(-\frac{K_H}{s}\right) \quad (3.60)$$

Figure 3.44 illustrates the Bode plot of the simplified current-mode model with an integrator. The crossover frequency is lower than ω_p and 1/10 of the switching frequency at light or heavy loads because of the pole at the origin.

The simulated Bode plot in Figure 3.45(a) indicates a large sufficient phase margin and a low crossover frequency. In reality, the dominant pole is located at low frequencies if a finite EA gain is obtained. As depicted in Figure 3.45(b), stable operation and good regulation can be obtained. However, a large undershoot/overshoot and a slow transient response occur as a result of the small bandwidth. To achieve stable operation and faster transient response, Type II compensation or PI compensation is commonly adopted in the current-mode control.

Type II compensation contains two poles and one zero. As shown in Figure 3.46, ω_{p1} serves as the dominant pole and ω_{z1} is introduced to compensate the phase delay caused by the output pole ω_p. ω_{p2} at higher frequencies is used to suppress high-frequency switching noise. As illustrated in Figure 3.46(a), (b), load variations result in different phase margins – although stable operation is ensured with an appropriate ω_{z1} design.

The circuit implementation of Type II compensation is illustrated in Figure 3.47. Assuming that the ideal EA has an infinite gain, Eq. (3.61) can be derived. Two poles and one zero are expressed in Eqs. (3.62)–(3.65), respectively.

$$A(s) = -K \cdot \frac{\left(1 + \frac{s}{\omega_{z1}}\right)}{\left(1 + \frac{s}{\omega_{p1}}\right)\left(1 + \frac{s}{\omega_{p2}}\right)} \quad (3.61)$$

$$K = \frac{1}{R_{f1}C_2} \quad (3.62)$$

$$\omega_{p1} = 0 \quad (3.63)$$

$$\omega_{z2} = \frac{C_2 + C_1}{C_2 C_1 R_1} \approx \frac{1}{C_2 R_1} \text{ if } C_1 \gg C_2 \quad (3.64)$$

$$\omega_{p2} = \frac{1}{C_1 R_1} \quad (3.65)$$

The Bode plot of the simulation results is shown in Figure 3.48. The dominant pole is not located at the origin but at low frequencies, because of the finite gain of the EA. A stable and

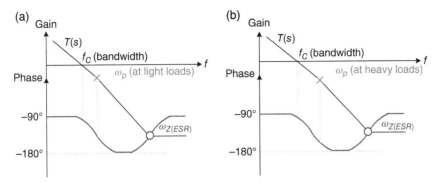

Figure 3.44 Bode plot of the simplified current-mode model with an integrator at (a) light and (b) heavy loads

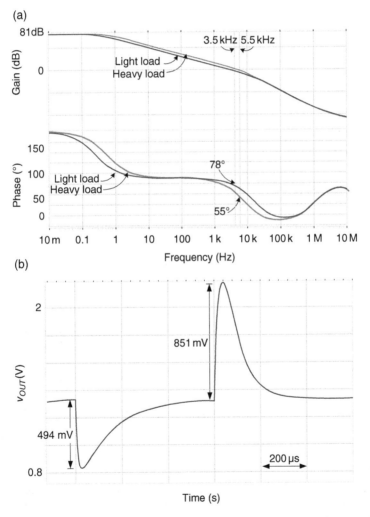

Figure 3.45 (a) Bode plot of simulation results and (b) simulation results of V_{OUT} when a large-compensation capacitor is placed at the output of a high-gain EA

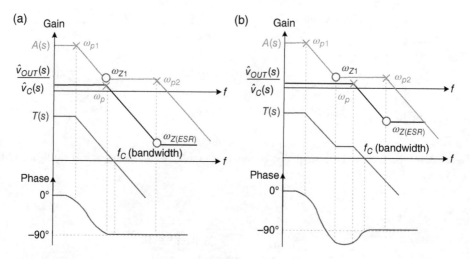

Figure 3.46 Bode plot of the simplified current-mode model with Type II compensation at (a) light and (b) heavy loads

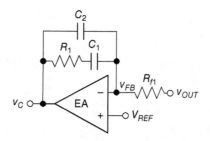

Figure 3.47 Topology of Type II (PI) compensation

well-regulated output voltage is shown in Figure 3.48(b), even when the phase margins are slightly different at light and heavy loads. Furthermore, the undershoot/overshoot and transient response are significantly improved because the bandwidth is extended by introducing a zero in Type II compensation.

The detailed simulation waveforms of the current-mode control buck converter are illustrated in Figure 3.49. The current sensor determines the inductor current as signal v_S when the high-side MOSFET is turned on. As shown in Figure 3.49(b), the DC value of v_S increases with an increase in inductor current. v_{SUM} represents the total waveform of v_S and V_{RAMP}. When v_{SUM} reaches v_C, the duty cycle is determined. In Figure 3.49(a), (b), v_C is higher at heavy loads to increase the inductor current level. This process verifies that v_C can generate the current-control voltage through the voltage feedback loop.

An alternative method of implementing Type II compensation is illustrated in Figure 3.50. Instead of v_{FB}, C_1 and C_2 are connected to the ground. The transfer function can be derived using Eq. (3.66). The poles and zero are expressed in Eqs. (3.67)–(3.69), respectively. Similar to the original Type II compensation, the alternative Type II compensation contains two poles

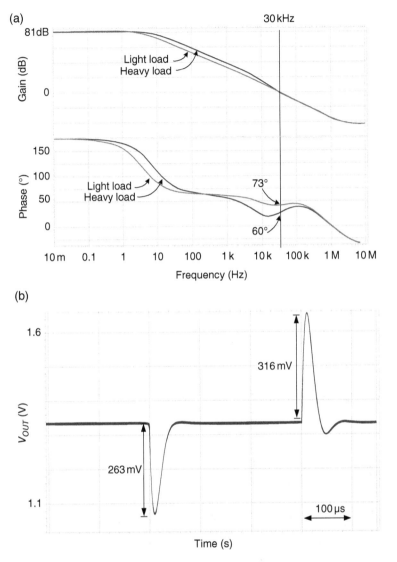

Figure 3.48 (a) Bode plot of the simulation results and (b) simulation results of V_{OUT} with Type II compensation

and one zero. The dominant pole is contributed by the output resistance R_O and the capacitance C_1 of the EA. The corresponding $T(s)$ is expressed in Eq. (3.70).

$$A(s) = -K_H \cdot \frac{\left(1 + \dfrac{s}{\omega_{z1}}\right)}{\left(1 + \dfrac{s}{\omega_{p1}}\right)\left(1 + \dfrac{s}{\omega_{p2}}\right)} \qquad (3.66)$$

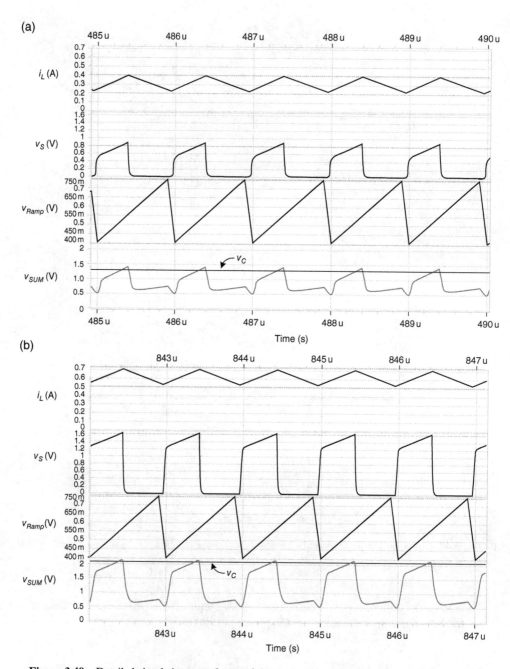

Figure 3.49 Detailed simulation waveforms of the current mode at (a) light and (b) heavy loads

Design of Switching Power Regulators 167

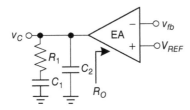

Figure 3.50 Alternative topology of Type II compensation

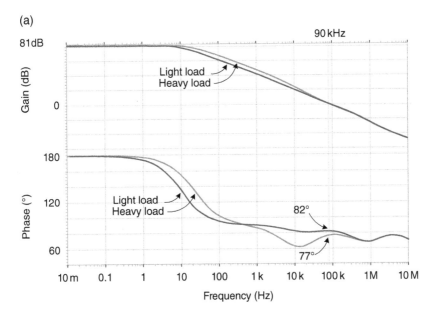

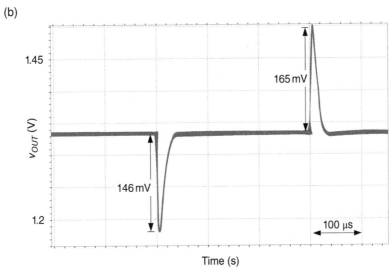

Figure 3.51 (a) Bode plot of the simulation results and (b) simulation results of V_{OUT} with the alternative Type II compensation

$$\omega_{p1} \approx \frac{1}{C_1 R_O} \qquad (3.67)$$

$$\omega_{z1} = \frac{1}{C_1 R_1} \qquad (3.68)$$

$$\omega_{p2} \approx \frac{1}{C_2 R_1} \qquad (3.69)$$

$$T(s) \approx K \cdot \frac{1 + sC_{OUT}R_{esr}}{1 + \dfrac{s}{\omega_p}} \cdot \left(-K_H \cdot \frac{\left(1 + \dfrac{s}{\omega_{z1}}\right)}{\left(1 + \dfrac{s}{\omega_{p1}}\right)\left(1 + \dfrac{s}{\omega_{p2}}\right)} \right) \qquad (3.70)$$

Table 3.6 Comparison of different compensation methods in current-mode buck converters

Method	A(s)	Topology	Ripple/transient/regulation
High DC gain	$-K_H$		Unstable
Integrator	$-\dfrac{K_H}{s}$		Small/slowest/good
Type II (PI)	$A(s) = -K \cdot \dfrac{\left(1 + \dfrac{s}{\omega_{z1}}\right)}{\left(1 + \dfrac{s}{\omega_{p1}}\right)\left(1 + \dfrac{s}{\omega_{p2}}\right)}$ where $K = \dfrac{1}{R_{f1}C_2}, \omega_{p1} = 0,$ $\omega_{z2} \approx \dfrac{1}{C_2 R_1}, \omega_{p2} = \dfrac{1}{C_1 R_C}$		Small/fast/good
Alternative Type II (PI)	$A(s) = -K_H \cdot \dfrac{\left(1 + \dfrac{s}{\omega_{z1}}\right)}{\left(1 + \dfrac{s}{\omega_{p1}}\right)\left(1 + \dfrac{s}{\omega_{p2}}\right)}$ where $\omega_{p1} \approx \dfrac{1}{C_1 R_O}, \omega_{z2} = \dfrac{1}{C_1 R_1},$ $\omega_{p2} \approx \dfrac{1}{C_2 R_1}$		Small/fastest/good

Table 3.7 Comparison of Hspice simulation results with different compensation methods

Method	Hspice condition	Ripple (mV)	Recovery time (μs)/f_C (kHz)	Overshoot/undershoot (mV)
High DC gain	K_H = 81 dB	NA	NA	NA
Integrator	K_H = 81 dB, C_1 = 2000 pF	6	400/3.5	494/851
Type II (PI)	K_H = 81 dB, C_1 = 50 pF, C_2 = 2 pF, R_1 = 150 kΩ	6	110/30	263/316
Alternative Type II (PI)	K_H = 81 dB, C_1 = 50 pF, C_2 = 2 pF, R_1 = 150 kΩ	6	54/90	13.4/10.2

L = 4.7 μH, C_{OUT} = 4.7 μF, f_s = 1 MHz, R_L = 4.3/2.15 Ω (light/heavy load), R_{ESR} = 30 mΩ.

The simulation results are shown in Figure 3.51. If C_1, C_2, and R_2 have the same values, the resultant crossover frequency is extended from 30 to 90 kHz. The transient response in Figure 3.51(b) indicates a smaller overshoot/undershoot and a shorter recovery time than the original Type II compensation. Thus, we can conclude that a larger bandwidth and a better transient response can be achieved by the alternative Type II compensation with the same component values as the original Type II compensation.

A comparison of the different compensation methods for current-mode buck converters is summarized in Table 3.6. Without inserting any pole and zero, the current-mode buck converter becomes unstable because the switching noise involves its bandwidth. By adding a large capacitor at the output of the EA, the integrator provides stable operation and slow transient response. To accelerate the transient response, Type II compensation introduces two poles and one zero. An alternative Type II compensation topology is adopted to further extend the bandwidth of the current-mode buck converter. A faster transient response can be achieved (Table 3.7).

References

[1] Huang, H.-W., Chen, K.-H., and Kuo, S.-Y. (2007) Dithering skip modulation, width and dead time controllers in highly efficient DC–DC converters for system-on-chip applications. *IEEE Journal of Solid-State Circuits*, **42** (11), 2451–2465.

[2] Chen, K.-H., Chang, C.-J., and Liu, T.-H. (2008) Bidirectional current-mode capacitor multipliers for on-chip compensation. *IEEE Transactions on Power Electronics*, **23**(1), 180–188.

[3] Patounakis, G., Li, Y.W., and Shepard, K.L. (2004) A fully integrated on-chip DC–DC conversion and power management system. *IEEE Journal of Solid-State Circuits*, **39**(3), 443–451.

[4] Alimadadi, M., Sheikhaei, S., Lemieux, G., et al. (2007) A 3GHz switching DC–DC converter using clock-tree charge-recycling in 90nm CMOS with integrated output filter. *IEEE International Solid-State Circuits Conference Digest of Technical Papers (ISSCC)*, February 2007, pp. 532–533.

[5] Hsieh, C.-Y. and Chen, K.-H. (2008) Adaptive pole–zero position (APZP) technique of regulated power supply for improving SNR. *IEEE Transactions on Power Electronics*, **23**(6), 2949–2963.

[6] Ma, F.-F., Chen, W.-Z., and Wu, J.-C. (2007) A monolithic current-mode buck converter with advanced control and protection circuit. *IEEE Transactions on Power Electronics*, **22**(5), 1836–1846.

[7] Mulligan, M.D., Broach, B., and Lee, T.H. (2007) A 3MHz low-voltage buck converter with improved light load efficiency. *IEEE International Solid-State Circuits Conference Digest of Technical Papers (ISSCC)*, February 2007, pp. 528–529.

[8] Erickson, R.W. and Maksimovic, D. (2001) *Fundamentals of Power Electronics*, 2nd edn. Kluwer Academic Publishers, Norwell, MA.

[9] Ridley, R.B. (1991) A new, continuous-time model for current-mode control. *IEEE Transactions on Power Electronics*, **6**(2), 271–280.

4

Ripple-Based Control Technique Part I

Control methods for DC/DC converters can be simply classified into three types: current-mode control, voltage-mode control, and ripple-based control. The transient response of current-mode or voltage-mode control is constrained by the bandwidth, which is mainly determined by the operating mode (continuous conduction mode (CCM) or discontinuous conduction mode (DCM)) and its compensation technique. Generally speaking, the compensation technique determines both converter stability and regulation performance. By contrast, ripple-based control has the advantage of fast transient response while maintaining system stability without any complicated compensation network. Another advantage of ripple-based control is low quiescence because of its simple structure, which can extend the battery usage time in portable electronics.

In recent years, portable electronics powered by batteries have needed their heavy and light load efficiencies extended to increase the battery usage time, which is a major concern in the design of wearable electronics. Ripple-based control can be used for such applications because of its fast transient response and high efficiency to meet the specifications of high slew rate and extensive load current ranges, respectively. For example, on-time control, which is one of the ripple-based control techniques, features fast transient response and high efficiency in case of loading current variations. As the term "constant on-time" implies, the constant value is composed of the product of the on-time period and the input voltage to obtain the feedforward function. By contrast, the off-time period can be automatically extended for high efficiency when the loading current decreases continuously at light loads. In a steady state, the off-time period is kept approximately constant. Thus, the converter behaves as a pseudo-constant frequency operation. In case of any loading current variation, the off-time period inversely varies with the load current to achieve a fast transient response. Constant on-time control has become a well-known control technique because of its advantages of feedforward function, fast transient response, and high efficiency at light loads to meet the demand of high-quality portable electronics.

Power Management Techniques for Integrated Circuit Design, First Edition. Ke-Horng Chen.
© 2016 John Wiley & Sons Singapore Pte Ltd. Published 2016 by John Wiley & Sons Singapore Pte Ltd.

In subsequent sections, a review of basic ripple-based control topologies reveals the characteristics of each technique, including their advantages and disadvantages. The preliminary design presents the operation principle and stability analysis. In turn, several techniques are introduced to improve the regulation performance, including electromagnetic interference reduction, output voltage ripple reduction, and line/load regulation improvement. Finally, the bootstrap architecture, which is more adequate for high input voltage, is presented for high-power applications.

4.1 Basic Topology of Ripple-Based Control

Ripple-based control commonly refers to DC/DC converters that use output voltage ripple information to compare with the reference voltage and control power switches on/off to determine adequate power delivery paths. Figure 4.1 shows a generic block diagram of a DC/DC buck converter with ripple-based control technique. A power stage with power switches and inductance–capacitance (LC) filter stores and releases energy. Thus, the power stage converts a high input voltage (V_{IN}) to a low output voltage (V_{OUT}). The output voltage is fed directly or through a feedback divider to the ripple-based controller, where the feedback signal is denoted as V_{FB}. The ripple-based controller basically includes a pulse signal modulator and a comparator (CMP). It can be used to determine the output V_{CMP} of the CMP by comparing V_{FB} with V_{REF}. Determining the output through the pulse signal modulator depends on the ratio of V_{OUT} and V_{IN}. As such, the output signal V_{DRI} can be used to determine the PWM signal of the power stage to regulate V_{OUT}.

Ripple-based control monitors the output voltage level and directly determines whether the delivered energy is sufficient. Given the fast transient response of the comparator, the power stage can be immediately controlled to deliver energy from V_{IN} to V_{OUT} corresponding to any sudden load transient variations. Thus, the output voltage is well regulated because periodic charging and discharging of adequate energy to the output capacitor occur through the use of an output voltage ripple as the PWM signal.

Ripple-based control techniques can be further classified according to on/off-time determination using the pulse signal modulator, as shown in Figure 4.2. The control parameters typically

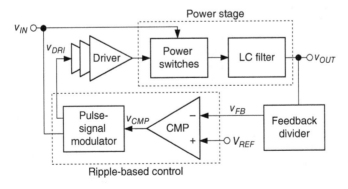

Figure 4.1 Generic block diagram of the ripple-based control technique

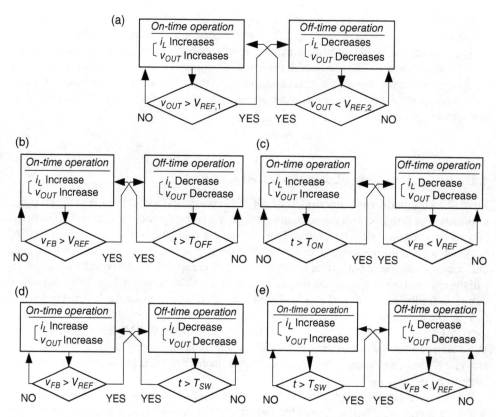

Figure 4.2 Types of ripple-based control: (a) hysteretic-mode control; (b) constant off-time with peak voltage control; (c) constant on-time with valley voltage control; (d) constant frequency with peak voltage control; (e) constant frequency with valley voltage control

used to determine power delivery paths include peak/valley voltage, on/off-time of power switches, and clock-controlled or clock-free operation.

On-time and off-time periods are defined as the time periods when the inductor current increases and decreases, respectively. Within a switching cycle, inductor voltage–second balance in steady state is established if the inductor current waveform increases and decreases portions equally. Considering energy conservation and complementary theory, the capacitor charge–second balance also ensures that no extra net charge is retained at the output capacitor in steady state.

The subsequent discussion shows how each control method determines the pulse width, including its starting and ending times. We can call the determination of starting and ending times as the setting (S) and resetting (R) points of the pulse width, respectively. For example, the hysteretic control shown in Figure 4.2(a) utilizes peak and valley voltages to compare two predefined reference voltages, namely the upper bound $V_{REF,1}$ and the lower bound $V_{REF,2}$, and determine the R and S points of one PWM signal, respectively. In this implementation, two comparators are needed.

However, the off-time control shown in Figure 4.2(b) utilizes a constant off-time (COT) period to determine the next pulse starting time (S point) and one upper bond reference voltage, which is the peak voltage detection, to determine the ending time (R point) of the pulse. By contrast, the on-time control shown in Figure 4.2(c) utilizes a constant on-time period to determine the ending time (R point) of the pulse width and one lower bond reference voltage, which is the valley voltage detection, to determine the next pulse starting time (S point). In this implementation, only one comparator is needed, as shown in Figure 4.2(b) or (c).

The methods shown in Figure 4.2(a)–(c) are called clock-free control because they lack the constant frequency generated by the clock generator. Filtering out the switching noise is sometimes difficult, because the switching frequency is variable and unpredictable. Therefore, a constant frequency with peak voltage control, as shown in Figure 4.2(d), utilizes one constant switching frequency and peak voltage detection to regulate the output. Similarly, Figure 4.2(e) uses constant frequency with valley voltage control to regulate the output voltage. Given that the constant switching period (T_{SW}) is known, the output filter is easily designed to remove any switching noise.

The characteristics of a boost converter, which is different from a ripple-based control buck converter, induce several design challenges if ripple-based control is used. For example, the use of a discontinuous inductor current for the output node results in difficult current sensing through the equivalent series resistance of the output capacitor because it lacks one part of the inductor current information if the low-side power switch is turned on. Moreover, the inherent RHP zero causes difficulty during stability analysis. Generally speaking, the RHP zero should not be smaller than the UGF because the RHP zero increases the gain but degrades the phase. The compensated system loop is designed to push the RHP zero beyond the UGF.

As a result, the maximum UGF is constrained not only by its inherent switching frequency but also by the existing RHP zero. The transient response performance degrades because of the limited UGF. After describing the ripple-based control buck converter, the ripple-based control boost converter will be discussed by introducing specific techniques used to obtain fast transient response without being affected by the RHP zero.

Detailed analyses and design principles for different types of ripple-based control buck converters are also discussed in subsequent sections. The asynchronous power stage of the buck converter, which contains a high-side switch (M_S), asynchronous diode (D), inductor (L), and capacitor (C), is used to simplify the analysis. Moreover, before stability is considered and analyzed in detail later, an ESR in the output capacitor is assumed to be sufficiently large so that the output voltage ripple is proportional to the inductor current ripple.

4.1.1 Hysteretic Control

Figure 4.3 shows the architecture of a buck converter with hysteretic control. The hysteretic window (V_H) in the comparator can determine the DC voltage level and the peak-to-peak voltage ripple at the output. In the steady state, the output voltage is limited within the predefined hysteresis window.

Figure 4.4 illustrates the mechanism of the waveforms of feedback voltage (v_{FB}), driving signal (v_G), inductor current (i_L), and output loading condition (i_{Load}). Comparing v_{FB} with the upper and lower boundaries V_{REF} and $V_{REF} + V_H$, respectively, determines the on-time and off-time periods, where V_H is the hysteresis window of the comparator. When the converter

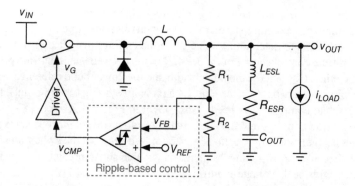

Figure 4.3 DC/DC buck converter with hysteretic control

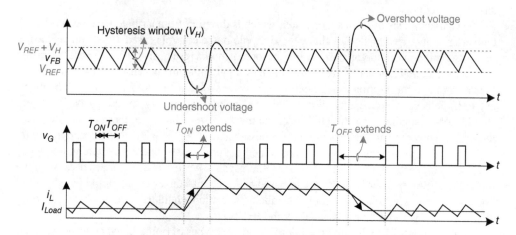

Figure 4.4 Waveforms of the buck converter with hysteretic-mode control

operates in the on-time period, M_S turns on to store energy in the inductor L. Thus, i_L increases and causes an increase in v_{OUT} and v_{FB}. When v_{FB} exceeds the value $V_{REF} + V_H$, the on-time period expires. The converter switches to the off-time period. In other words, M_S turns off to release the energy stored in the inductor L to the output. Thus, i_L decreases and causes a decrease in v_{OUT} and v_{FB}. When v_{FB} is less than V_{REF}, the off-time period expires to repeat another on-time period in the next switching cycle.

When the load current changes from light to heavy, the undershoot voltage at v_{OUT} triggers an extension of the on-time period until v_{FB} crosses the value $V_{REF} + V_H$. By contrast, in case of heavy-to-light load change, the overshoot voltage at v_{OUT} extends the off-time period and prevents extra energy from being delivered to the output. When v_{FB} is less than V_{REF}, the off-time ends and the on-time period starts again. The extension of the on/off-time period can speed up the transient response because of effective energy control to the output. Compared with PWM control, the hysteretic mode of ripple-based control can instantly adjust the duty cycle equivalent to 100% or 0% even during light-to-heavy or heavy-to-light load transient response, respectively. As for PWM, the changing speed of the duty cycle depends on the UGF of the system loop, and the duty cycle can only increase or decrease gradually according to the slow change of output voltage of the error amplifier. Specifically, hysteretic mode control achieves fast transient response.

The switching frequency (f_{SW}) of hysteretic mode control is expressed in Eq. (4.1). A major disadvantage of hysteretic mode control is the large variation of f_{SW}, a variation that is caused by the variable ESR of the output capacitor, output inductor, and input and output voltages. The effective hysteresis window $V_{H(eff)}$ can be expressed in Eq. (4.2) if non-ideal effects are considered.

$$f_{SW} = \frac{R_{ESR}}{V_{H(eff)}L} \cdot \frac{(v_{IN} - v_{OUT}) \times v_{OUT}}{v_{IN}} \quad (4.1)$$

where

$$V_{H(eff)} = V_H \cdot \left(1 + \frac{R_2}{R_1}\right) - \frac{L_{ESL}}{L} \cdot v_{IN} + T_{OFF} \cdot \frac{v_{IN} - v_{OUT}}{L} \cdot R_{ESR} + T_{ON} \cdot \frac{v_{OUT}}{L} \cdot R_{ESR} \quad (4.2)$$

Consequently, defining the operating frequency in comparison with other clock-based control techniques is difficult. Hysteretic mode control is unsuitable for several circuits that are sensitive to EMI because of the varying switching frequency, which apparently decreases when the converter operates in the DCM. As shown in Figure 4.5, the value of f_{SW} and the output voltage waveform in the CCM are different from those in the DCM. In DCM operation, the inductor current is reset to zero before the beginning of the next on-time period. Assuming that R_{ESR} of the output capacitor is sufficiently large to have an in-phase relationship between v_{OUT} and i_L, the constant value of the hysteresis window indicates that constant power is delivered to the inductor because of the constant on-time period, $T_{ON,1} = T_{ON,2} = T_{ON,3}$. By contrast, the

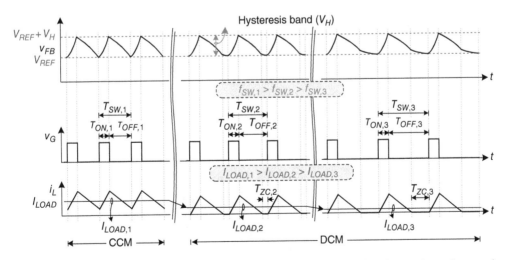

Figure 4.5 Characteristics of frequency variation in the buck converter with hysteretic-mode control when working in CCM and DCM under different loading conditions

energy dissipation rate slows down if a longer off-time period exists at light loads compared with that at heavy loads. As a result, the off-time period is extended if a constant on-time period is set; in other words, $T_{OFF,1} < T_{OFF,2} < T_{OFF,3}$. Thus, the switching frequency decreases; in other words, $f_{SW,1} > f_{SW,2} > f_{SW,3}$. Consequently, the switching loss is reduced to improve the power-conversion efficiency.

Hysteretic-mode control exhibits a simple structure and fast transient response. However, determining the hysteretic window V_H to achieve the expected specifications and identify the trade-off between output voltage ripple and regulation performance is complex because the output voltage ripple and switching frequency depend heavily on V_H, v_{IN}, v_{OUT}, L, and several parasitic components.

4.1.2 On-Time Control

Figure 4.6(a) shows the architecture of the buck converter with on-time control. Figure 4.6(b) illustrates simplified block diagrams for easy analysis in this book. The characteristics of the

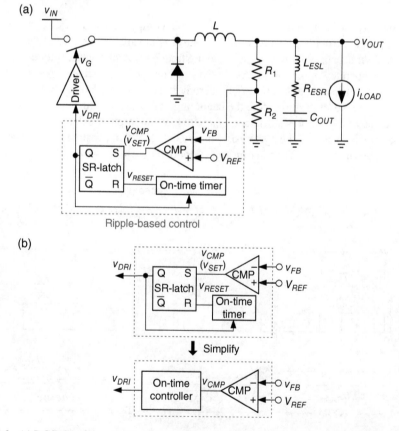

Figure 4.6 (a) DC/DC buck converter with constant on-time control technique. (b) Simplification of the on-time controller

on-time controller with slew rate (SR) latch include one constant on-time timer and one comparator, which are used to determine the on-time and off-time periods, respectively. During the on-time period, the increasing inductor current causes a simultaneous increase in v_{OUT} and v_{FB}. After a time period predefined by the on-time timer, the resetting pulse can be determined and the on-time period is ended to start the off-time period. Similarly, the decreasing inductor current causes a decrease in v_{OUT} and v_{FB} during the off-time period. When v_{FB} crosses the lower boundary V_{REF}, the comparator determines the setting pulse and outputs a low-to-high signal to end the off-time period and initiate the next on-time period.

Figure 4.7 shows the steady-state and transient waveforms. Undershoot voltage at v_{OUT} and v_{FB} occurs when the load current changes from light to heavy. The on-time period starts instantly once v_{FB} is less than V_{REF}. Several on-time periods are triggered continuously with the addition of the minimum off-time period between each on-time period, until v_{FB} is higher than V_{REF} again. Although the maximum duty cycle is constrained by the minimum off-time, the duty cycle can change instantly from the steady-state value to the maximum value in case of load transient response. Thus, on-time control can achieve a fast transient response compared with the gradual adjustment in PWM control by an error-voltage amplifier. By contrast, the off-time period is extended to ensure that surplus energy can be adequately discharged to output loading because of the overshoot voltage at v_{OUT} and v_{FB} in case of a heavy-to-light load change. Consequently, determining adequate power delivery paths during load transient response can benefit fast charging or discharging at the output capacitor by prolonging the on-time or off-time periods, respectively.

The worst case in transient response and overshoot voltage is any sudden load release during the on-time period if a constant on-time period is used, because the on-time period cannot be instantly terminated until the predefined constant on-time period expires. Unexpected power delivery paths result in a large overshoot voltage with surplus energy at the output. In particular, the overshoot voltage may cause permanent damage to the back-end circuits in the nanometer process, which implies low voltage stress ability.

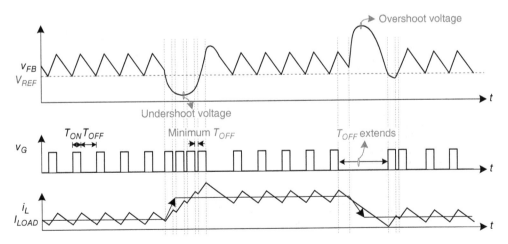

Figure 4.7 Waveforms of the DC/DC buck converter with constant on-time control in the CCM operation

The switching frequencies in CCM and DCM are expressed in Eqs. (4.3) and (4.4), respectively:

$$f_{SW} = \frac{v_{OUT}}{v_{IN} \cdot T_{ON}} \qquad (4.3)$$

$$f_{SW} = \frac{2Li_{LOAD}v_{OUT}}{T_{ON}^2(v_{IN}-v_{OUT})v_{IN}} \qquad (4.4)$$

The switching frequency f_{SW} in the CCM operation without any internal clock generators can easily be defined by v_{IN}, v_{OUT}, and T_{ON}. If a predefined on-time period is determined, then f_{SW} in the CCM can be kept constant. By contrast, f_{SW} in the DCM is proportional to the output loading condition i_{Load}, which is similar to the pulse frequency modulation (PFM) control [1].

Figure 4.8 shows the reduction in switching power loss because of the decrease in switching frequency at light loads. During the predefined constant on-time period, the inductor current increases to the same peak value and induces the same energy, which is delivered to the output. In the beginning of the off-time period, the inductor current first decreases to zero. At this moment, the output voltage is still higher than the rated voltage, indicating that a switching operation to deliver energy to the output is unnecessary. Before v_{FB} reaches V_{REF}, the output capacitor mainly provides energy to the output loading. Figure 4.8 shows that the light loading condition extends the off-time period ($T_{OFF,1} < T_{OFF,2}$) because the power dissipation rate becomes slow at light loads ($T_{ZC,1} < T_{ZC,2}$). As a result, the switching period should be extended ($T_{SW,1} < T_{SW,2}$) if the same energy is stored in the inductor during one constant on-time phase ($T_{ON,1} = T_{ON,2}$). When the loading current decreases continuously, the extended switching period implies a continuously reduced switching frequency ($f_{SW,2} < f_{SW,1}$) and reduces the switching power loss for high power efficiency. The constant on-time control has the advantage of an approximately constant switching frequency design in the CCM operation. Meanwhile, the converter can automatically transfer to variable frequency control in the DCM operation to save power without adding any mode decision control circuits. A seamless transition from

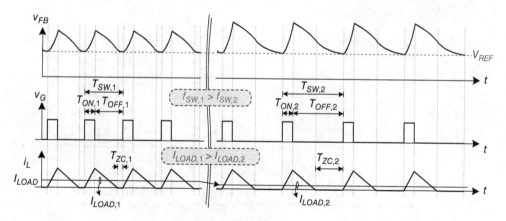

Figure 4.8 Waveforms of the DC/DC buck converter with constant on-time control in the DCM operation

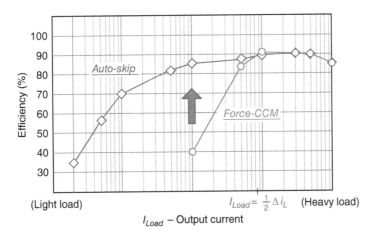

Figure 4.9 Power-conversion efficiency of the DC/DC converter with constant on-time control

CCM to DCM, and vice versa, occurs in this type of control method. As such, the constant on-time control has become well known in recent portable electronics because of its power-saving ability at light loads.

Figure 4.9 shows an example of efficiency improvement by DCM operation in the constant on-time control buck converter. When the operating mode of the converter is set as Force-CCM mode, the inductor current exhibits a triangular waveform without any zero current detection regardless of whether heavy or light loading conditions occur. When the average load current is smaller than half the inductor current ripple ($I_{LOAD} < 1/2\Delta i_L$), the inductor current crosses the zero current and is negative, where Δi_L is the inductor current ripple. This operating mode can result in an approximately constant switching frequency. However, the light-load power-conversion efficiency will degrade drastically because the switching loss dominates the power loss and the negative inductor current consumes extra conduction loss. By contrast, when the operating mode of the converter is set as Auto-Skip and when the load current is larger than half the inductor current ripple ($I_{LOAD} > 1/2\Delta i_L$), the inductor current is operated in the CCM. When the load current is smaller than half the inductor current ripple ($I_{LOAD} < 1/2\Delta i_L$), the converter automatically changes the operating mode to DCM and the switching frequency can decrease dynamically, corresponding to the decreasing loading conditions. The switching power loss can thus be reduced.

4.1.3 Off-Time Control

Figure 4.10 displays the architecture of a buck converter with the constant off-time control. In contrast to the COT, the constant off-time control uses one constant off-time timer instead of a constant on-time timer. Thus, the comparator compares v_{FB} and the upper boundary V_{REF} to determine the on-time period. During the on-time period, the increasing inductor current causes an increase in v_{OUT} and v_{FB}. When v_{FB} exceeds V_{REF}, the resetting pulse can be determined and the on-time period expires to start the off-time period. Similarly, the decreasing inductor current causes a decrease in v_{OUT} and v_{FB}. The constant off-time timer provides a predefined off-time

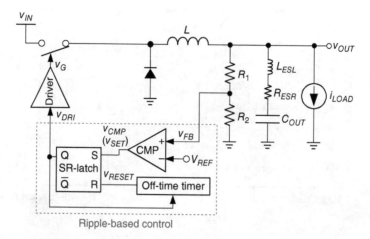

Figure 4.10 DC/DC buck converter with constant off-time control

period to determine the setting pulse when the off-time period is over, so that the converter can initiate the next on-time period. Given that the off-time period is constant, the switching frequency can vary within a small range in steady state. With the constant off-time control, the on-time period also varies with the corresponding adequate value to meet the requirements of conversion ratio and loading conditions. It indicates the switching frequency cannot be decreased to improve efficiency and to extend battery usage time because the on-time period is reduced in the DCM operation. As such, the constant off-time control seems unsuitable for portable electronics because this control has a constant off-time period and fails to improve efficiency.

Figure 4.11 shows the steady-state and transient response waveforms. When the load current changes from light to heavy, the undershoot voltage at v_{OUT} and v_{FB} triggers the on-time period until v_{FB} is higher than V_{REF}. By contrast, in case of a load change from heavy to light, the converter keeps operating in numbers of predefined off-time-period fragments with the addition of the minimum on-time period. In other words, compared with hysteretic mode control, off-time control exhibits the same performance of the load transient when a light-to-heavy load transient occurs, but exhibits slow performance when a heavy-to-light load transient occurs. However, compared with PWM control with an error-voltage amplifier, instantly adjusting the adequate power delivery paths during a load transient can still result in fast charging and discharging at the output capacitor with adjustable on-time and off-time periods, respectively. A disadvantage of constant off-time control is that the switching frequency increases at light loads because of the decrease in on-time period. Thus, constant off-time control is unsuitable for portable electronics. Furthermore, in case of a light-to-heavy load change during constant off-time period, another disadvantage is the unexpected power delivery paths that result in a large undershoot voltage and insufficient energy delivered at the output.

In steady state, the switching frequency in CCM operation is derived as:

$$f_{SW} = \frac{v_{IN} - v_{OUT}}{v_{IN} T_{OFF}} \tag{4.5}$$

Ripple-Based Control Technique Part I   181

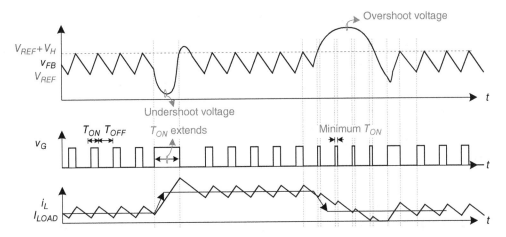

Figure 4.11 Waveforms of the DC/DC buck converter with constant off-time control in the CCM operation

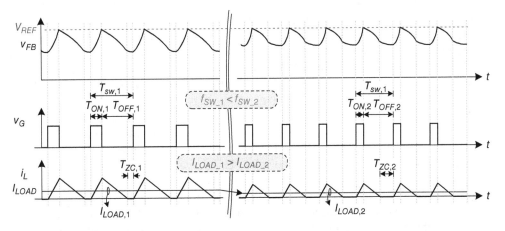

Figure 4.12 Waveforms of the DC/DC buck converter with constant off-time control in the DCM operation

Equation (4.5) shows that the switching frequency can easily be designed and kept constant without the need for any internal clock generators if v_{IN}, v_{OUT}, and T_{OFF} are well known. However, light loading conditions cause high switching frequencies when the converter operates in DCM. The switching frequency in DCM is expressed in Eq. (4.6), based on the principle of inductor voltage–second balance:

$$f_{SW} = \left(T_{OFF} + \frac{L \cdot i_{LOAD} \cdot v_{OUT}}{(v_{IN} - v_{OUT})v_{IN}} + \sqrt{\frac{(L \cdot i_{LOAD} \cdot v_{OUT})^2}{((v_{IN} - v_{OUT})v_{IN})^2} + \frac{2T_{OFF}}{(v_{IN} - v_{OUT})v_{IN}}} \right)^{-1} \quad (4.6)$$

Figure 4.12 shows that the reduced on-time period ($T_{ON,1} > T_{ON,2}$) causes the lower peak inductor current corresponding to the decrease in loading current because of the constant off-time period ($T_{OFF,1} = T_{OFF,2}$) in each switching cycle. Energy storage and release at the output capacitor can be in a state of equilibrium. Thus, the output voltage can be regulated. However, the increase in switching frequency ($f_{SW,1} < f_{SW,2}$) at light loads results in a high switching power loss and low power-conversion efficiency.

Constant off-time control exhibits a good transient response and a simple structure that can be used to design the switching frequency in the CCM. However, constant off-time control exhibits poor power-conversion efficiency in the DCM. By contrast, peak voltage control with constant off-time control can have a constant current-driving capability in power converters for commercial light-emitting diode (LED) drivers, because LEDs need constant current-driving control.

4.1.4 Constant Frequency with Peak Voltage Control and Constant Frequency with Valley Voltage Control

Figure 4.13(a), (b) shows the constant frequency with peak voltage control and the constant frequency with valley voltage control, respectively. To our knowledge, constant on-time and off-time controls determine the on-time and off-time periods and the beginning times of the off-time and on-time, respectively. Similarly, the constant frequency with peak voltage control determines the beginning of the on-time and the off-time by simply using the clock and peak voltage detection, respectively. By contrast, the constant frequency with valley voltage control determines the beginning of the on-time and the off-time by using valley voltage detection and the clock, respectively.

Moreover, the switching frequency is constant so that it induces a reduction in EMI, because the EMI filter can be well designed if the switching frequency is known. However, a constant switching frequency results in a slightly slow transient response and high power consumption compared with the previous control topologies. This outcome is attributed to the operation maintaining a constant switching frequency in the CCM and DCM without the ability to increase or decrease the switching frequency for fast transient response or power saving, respectively. The switching power loss cannot be reduced by the decrease in switching frequency, so that the power-conversion efficiency with the DCM operation is worse than that with the COT control. An additional load-dependent controller can be used to dynamically adjust the switching frequency at light loads at the cost of circuit complexity, thus improving the power-conversion efficiency.

The inductor current ripple is derived through inductor charging or discharging of the output capacitor if the ESR is sufficiently large. Without using any error amplifiers, a fast transient response can be obtained compared with current-mode/voltage-mode PWM control. An extra sawtooth signal can be added to the feedback path to improve the noise margin between the two input terminals of the comparator. Double-pulse phenomena will not occur after enlarging the noise margin because the difference between the two inputs of the comparator is sufficiently large to have high noise immunity. The reference voltage V_{REF} can also be replaced by an integrating error amplifier with two inputs, namely V_{REF} and v_{OUT}, to improve accuracy. Two feedback signals from the output contribute to the design of the V-square (V^2) control, which can ensure fast transient response and high accuracy at the same time.

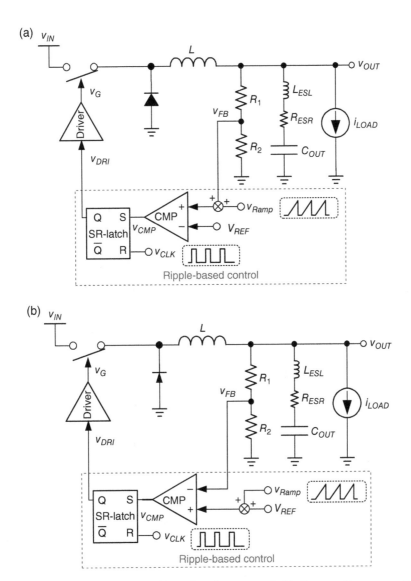

Figure 4.13 (a) Constant frequency with peak voltage control. (b) Constant frequency with valley voltage control

4.1.5 Summary of Topology of Ripple-Based Control

The common advantages of ripple-based control methods are listed as follows:

1. The modulation signal is derived directly from the comparator without using any error amplifiers.
2. The comparator has a wide bandwidth for fast transient response.

3. A simple structure can be achieved because of the absence of error amplifiers and an extra compensation network.
4. Clock-free operation, excluding constant frequency control, does not need any clock generators.
5. High power-conversion efficiency is obtained because of the variable switching frequency at light loads in the hysteretic mode and constant on-time controls.

Simple ripple-based control without any complex compensation network has the advantage of low cost. In particular, a wide bandwidth guarantees a fast transient response because of the utilization of a comparator. Furthermore, when operating at light loads, the switching frequency is automatically adjusted with the implementation of a zero current detector according to the output loading condition without additional complex frequency tuners. In other words, the converter is controlled by variable frequency modulation (VFM) to save significant switching power loss for high power-conversion efficiency at light loads for portable power electronics. Consequently, ripple-based control has become a suitable choice in many power management designs.

However, ripple-based control has several practical problems and limitations, as follows:

1. Trade-off between sub-harmonic instability and output voltage ripple owing to the selection constraint of the output capacitor.
2. EMI caused by the poorly defined switching frequency of the clock-free characteristic (except for constant frequency control).
3. Jitter behavior caused by low noise immunity.
4. Inaccurate DC regulation caused by low gain loop and direct peak/valley control.

Some of these problems can be improved by recently developed techniques. Table 4.1 shows a comparison of the different structures of ripple-based control in DC/DC buck converters. The superior performance of on-time control has led to its popularity in portable electronics.

Table 4.1 Comparison of different structures of ripple-based control in buck converter

	Transient response	Frequency in CCM	Frequency in DCM	Efficiency in DCM
Hysteretic control	Excellent	Difficult to determine and keep constant	Decreased frequency with decreased load	Very good
On-time control	Very good	Easy to determine and keep constant	Decreased frequency with decreased load	Excellent
Off-time control	Very good	Easy to determine but difficult to keep constant	Increased frequency with decreased load	Very poor
Constant frequency with peak voltage control	Good	Constant	Constant	Poor
Constant frequency with valley voltage control	Good	Constant	Constant	Poor

4.2 Stability Criterion of On-Time Controlled Buck Converter

Among the above topologies, on-time control is the most popular nowadays because the converter with on-time control features a fast transient and high efficiency. This kind of power converter is the most likely solution for Soc power supply, with its wide load range and minimized variation at the supply voltage in transient response and steady state. Thus, we need to derive the stability criterion of on-time control. This will be discussed later. Other topologies can use the same criterion to guarantee converter stability. However, similar to PWM control, only constant frequency control suffers from sub-harmonic oscillation. In order to prevent sub-harmonic oscillation, slope compensation can also be taken into consideration.

4.2.1 Derivation of the Stability Criterion

A power stage, with power switches and an LC filter, has energy storage and power transfer functions. Figure 4.14 shows such an LC filter. The inductor is modeled as one ideal inductor, L, and one parasitic DC resistance, R_{DCR}. The output capacitor is modeled as three components, including an equivalent series inductance, L_{ESL}; an ESR, R_{ESR}; and an ideal C_{OUT}. The output voltage waveform is generated by the voltage variation across each component due to the inductor current variation. The two parasitic components L_{ESL} and R_{ESR} in the output capacitor should be considered during ripple-based control because the output voltage waveform is significantly affected by these parasitic components. Thus, stability should be increased even under the influence of parasitic effects.

Figure 4.15 shows the waveforms of the inductor current and the voltages across each component in the output capacitor. The AC component of the inductor current flows into the output capacitor and results in v_{ESL}, v_{ESR}, and v_{COUT}, which are contributed by L_{ESL}, R_{ESR}, and C_{OUT}, respectively. The corresponding equations for these voltage ripples are derived as Eqs. (4.7)–(4.9):

$$\text{On-time phase: } v_{ESR}(t) = R_{ESR}\left(\frac{\Delta I \cdot t}{D \cdot T_{SW}} - \frac{\Delta I}{2}\right)$$

$$\text{Off-time phase: } v_{ESR}(t) = R_{ESR}\left(\frac{\Delta I}{2} - \frac{\Delta I \cdot t}{(1-D) \cdot T_{SW}}\right) \tag{4.7}$$

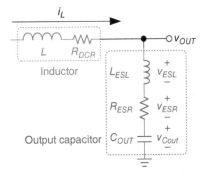

Figure 4.14 Modeling the output filter including the ideal inductor, the capacitor, and their corresponding parasitic characteristics

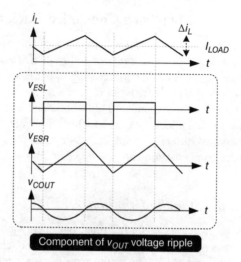

Figure 4.15 Inductor current waveform and voltage waveform of each component in the complete output capacitor modeling

$$\text{On-time phase:} \quad v_{ESL}(t) = \frac{L_{ESL} \cdot \Delta I_L}{D \cdot T_{SW}}$$
$$\text{Off-time phase:} \quad v_{ESL}(t) = \frac{L_{ESL} \cdot \Delta I_L}{(1-D) \cdot T_{SW}} \quad (4.8)$$

$$\text{On-time phase:} \quad v_{COUT}(t) = \frac{\Delta I_L \cdot t^2}{2C_{OUT} \cdot D \cdot T_{sw}} - \frac{\Delta I_L \cdot t}{C_{OUT}}$$
$$\text{Off-time phase:} \quad v_{COUT}(t) = \frac{\Delta I_L \cdot t}{C_{OUT}} - \frac{\Delta I_L \cdot t^2}{2C_{OUT} \cdot (1-D) \cdot T_{SW}} \quad (4.9)$$

$$v_{COUT}(t) = \frac{\Delta I_L \cdot t}{C_{OUT}} - \frac{\Delta I_L \cdot t^2}{2C_{OUT} \cdot (1-D) \cdot T_{SW}}$$

As a result, different output waveforms are generated by different quantities contributed by v_{ESL}, v_{ESR}, and v_{COUT}. Figure 4.16 shows four examples to illustrate the waveforms with different values of L_{ESL}, R_{ESR}, and C_{OUT}.

To compare the difference, L_{ESL} and R_{ESR} are varied, whereas C_{OUT} is set as constant. In case (a) with a small L_{ESL} and a small R_{ESR}, the voltage ripple caused by the inductor current charging/discharging C_{OUT} dominates v_{OUT}. The inductor current information is obtained by differentiating the output voltage ripple. Although a large C_{OUT} can be used to filter out the voltage ripple of the SWR and thus obtain a better quality of supply voltage, the small voltage ripple indicates that the differentiated voltage ripple will be small. In other words, the inductor current derived by the differentiating function has low noise immunity if a large C_{OUT} is used. The large inductor current information derived by the differentiating function and the low output ripple obtained by using a large C_{OUT} have conflicting requirements. Furthermore, in case (b) with a

Ripple-Based Control Technique Part I

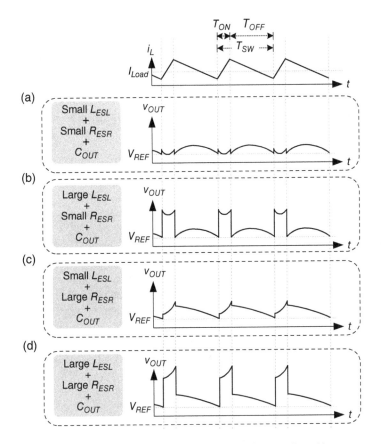

Figure 4.16 Output voltage ripple composed of different quantities contributed by v_{ESL}, v_{ESR}, and v_{COUT} because of the different values of ESL, ESR, and C_{OUT}, respectively

large L_{ESL} and a small R_{ESR}, the voltage ripple caused by the ESL dominates v_{OUT}. The voltage ripple caused by a large L_{ESL} also deteriorates all the output ripples. Specifically, the use of the differentiating function to derive the inductor current will be more difficult because the differentiated signal is influenced by the ESL effect. In case (c) with a small L_{ESL} and a large R_{ESR}, the voltage ripple caused by R_{ESR} dominates v_{OUT}. The output ripple is in phase with the inductor current; that is, a large R_{ESR} can be used to generate the inductor current. In case of any switch turning on/off, the voltage ripple caused by the ESL will not affect the current-sensing performance. As such, ripple-based control always uses a large R_{ESR} to derive the inductor current information. Considering one possible scenario in which a large L_{ESL} and a large R_{ESR} occur simultaneously, as shown in case (d), the inductor current generated by R_{ESR} will initially deteriorate when the switch is turned on/off. Although the DC level of the sensing signal will change, the inductor current information can be derived by ripple-based control. Thus, we can conclude that the inductor current information can be derived through the use of a large R_{ESR}, which is the method used in conventional ripple-based control, or through the differentiating function that will be introduced in subsequent sections. Importantly, two methods are possibly

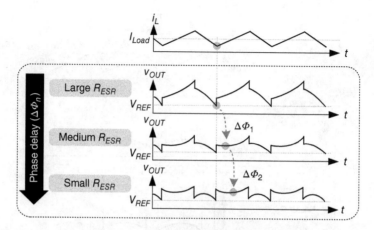

Figure 4.17 Relationship between i_L and output voltage ripple under different R_{ESR} values

influenced by the large ESL. This finding indicates that several techniques that can inhibit the ESL effect should be proposed to ensure the robust operation of ripple-based control.

Considerable attention should be focused on selecting the appropriate output capacitor with certain quantities of parasitic components to guarantee the stability of ripple-based control. As shown in Figure 4.17, the output capacitor with different R_{ESR} values results in different output ripple characteristics. This phenomenon can be translated to the phase delay ($\Delta\Phi$) related to the inductor current i_L caused by different parasitic R_{ESR} values. A small R_{ESR} value results in a long phase delay observed at the lowest value of v_{OUT}. Given a small R_{ESR} value, the system stability decreases because of an insufficient linear relationship between the inductor current ripple and the output voltage ripple. A small R_{ESR} value will decrease the system stability.

The stability criterion derived by using aspects of the S-domain and the time domain will be reviewed in detail to establish design guidelines.

4.2.1.1 S-Domain Derivation (RHP Pole Consideration)

If the DC operating points are determined, then an AC test signal can be fed into the node of V_{REF} in Figure 4.18 to analyze the stability in the S-domain.

Based on the Padé approximant, the transfer function from the reference voltage to the output is approximated to a fourth-order equation, as expressed in Eqs. (4.10) and (4.11):

$$\frac{\hat{v}_{OUT}(s)}{\hat{v}_{REF}(s)} \approx \frac{(R_{ESR}C_{COUT} \cdot s + 1)}{\left(1 + \dfrac{s}{Q_1\omega_1} + \dfrac{s^2}{\omega_1^2}\right)\left(1 + \dfrac{s}{Q_2\omega_2} + \dfrac{s^2}{\omega_2^2}\right)} \tag{4.10}$$

where

$$\omega_1 = \frac{\pi}{T_{ON}}, \quad Q_1 = \frac{2}{\pi}, \quad \omega_2 = \frac{\pi}{T_{SW}}, \quad Q_2 = \frac{T_{SW}}{\left(R_{ESR}C_{OUT} - \dfrac{T_{ON}}{2}\right) \cdot \pi} \tag{4.11}$$

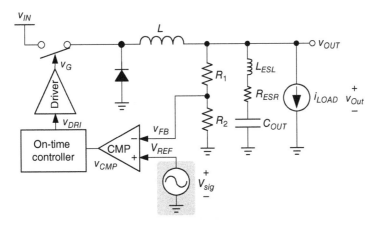

Figure 4.18 Setup for deriving the transfer function

When the on-time value T_{ON} is sufficiently small and ω_1 locates at high frequencies, the transfer function can be simplified to a second-order equation:

$$\frac{\hat{v}_{OUT}(s)}{\hat{v}_{REF}(s)} \approx \frac{(R_{ESR}C_{COUT} \cdot s + 1)}{\left(1 + \dfrac{s}{Q_2\omega_2} + \dfrac{s^2}{\omega_2^2}\right)} \tag{4.12}$$

Given that Q_2 must be larger than zero, the inequality in Eq. (4.13) can be derived to eliminate the unexpected RHP poles:

$$Q_2 = \frac{T_{SW}}{\left(R_{ESR}C_{OUT} - \dfrac{T_{ON}}{2}\right) \cdot \pi} > 0 \tag{4.13}$$

Then, the stability criterion is obtained:

$$R_{ESR} \cdot C_{OUT} > \frac{T_{ON}}{2} \tag{4.14}$$

The time constant and the selection of R_{ESR} should meet the following requirement:

$$R_{ESR} > \frac{T_{ON}}{2C_{OUT}} \tag{4.15}$$

4.2.1.2 Time Domain Derivation (Convergence Consideration)

When the stability criterion is derived based on the constraint set by the output capacitor, only C_{OUT} and R_{ESR} should be considered. The L_{ESL} effect can be ignored because the voltage ripple caused by the ESL is constant if v_{IN}, v_{OUT}, L, and L_{ESL} are kept constant. Figure 4.19 shows the

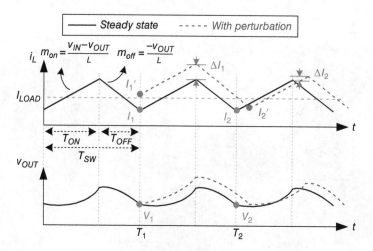

Figure 4.19 Steady-state and perturbed waveforms caused by perturbation

steady-state and perturbed waveforms caused by perturbation in the time domain. The net change in output voltage should be zero after one switching period, as expressed in Eq. (4.16). In the CCM, the switching period T_{SW} is the summation of T_{ON} and T_{OFF}, which denote the on-time and off-time periods, respectively. The on-time and off-time inductor current slopes are expressed in Eq. (4.17).

$$V_2 - V_1 = \frac{1}{C_{OUT}} \int_0^{T_{SW}} [i_L(t) - i_{LOAD}]dt + (I_2 - I_1)R_{ESR} = 0 \qquad (4.16)$$

where

$$T_{SW} = T_{ON} + T_{OFF} \text{ and } \begin{cases} \dfrac{di_L(t)}{dt} = \dfrac{v_{IN} - v_{OUT}}{L}, & \text{during the on-time period} \\ \dfrac{di_L(t)}{dt} = \dfrac{-v_{OUT}}{L}, & \text{during the off-time period} \end{cases} \qquad (4.17)$$

In this study, T_{OFF} can be expressed as follows:

$$T_{OFF} = \left(\frac{v_{IN}}{v_{OUT}} - 1\right) T_{ON} \qquad (4.18)$$

By substituting Eq. (4.18) into Eq. (4.16), we derive

$$\frac{v_{IN}}{v_{OUT}} \cdot \frac{T_{ON}}{C_{OUT}} (I_1 - I_{OUT}) + \frac{T_{ON}^2}{2C_{OUT}L} \cdot \frac{v_{IN}}{v_{OUT}} (v_{IN} - v_{OUT}) + R_{ESR}(I_2 - I_1) = 0 \qquad (4.19)$$

after evaluating the integration and simplification.

To conclude the stability criterion and linearization (Eq. (4.19)), the small perturbation signals ΔI_1 and ΔI_2 are considered in the steady-state inductor currents I_1 and I_2. These currents, as shown in Eq. (4.20), are at T_1 and T_2, respectively, according to the inductor current waveform in Figure 4.19.

$$\begin{cases} I_1 = i_{LOAD} - \dfrac{T_{ON}}{2L}(v_{IN} - v_{OUT}) \\ I_2 = i_{LOAD} - \dfrac{T_{ON}}{2L}(v_{IN} - v_{OUT}) \end{cases} \quad (4.20)$$

Consequently, the perturbed inductor currents I'_1 and I'_2 caused by perturbation at T_1 and T_2, respectively, are expressed as:

$$\begin{cases} I'_1 = \Delta I_1 + i_{LOAD} - \dfrac{T_{ON}}{2L}(v_{IN} - v_{OUT}) \\ I'_2 = \Delta I_2 + i_{LOAD} - \dfrac{T_{on}}{2L}(v_{IN} - v_{OUT}) \end{cases} \text{ and } I'_2 - I'_1 = \Delta I_2 - \Delta I_1 \quad (4.21)$$

The substitution of Eq. (4.21) into Eq. (4.19) derives

$$\Delta I_1 \left(R_{ESR} - \dfrac{v_{IN}}{v_{OUT}} \dfrac{T_{ON}}{C_{OUT}} \right) - \Delta I_2 R_{ESR} = 0 \quad (4.22)$$

$\Delta I_2 / \Delta I_1$ must be gradually converged to zero in steady state to increase the stability. Thus, the inequality is derived as Eq. (4.23) and simplified as Eq. (4.24):

$$\left| \dfrac{\Delta I_2}{\Delta I_1} \right| = \left| \dfrac{R_{ESR} - \dfrac{v_{IN}}{v_{OUT}} \dfrac{T_{ON}}{C_{OUT}}}{R_{ESR}} \right| < 1 \quad (4.23)$$

$$R_{ESR} \cdot C_{OUT} > \dfrac{T_{ON}}{2} \cdot \dfrac{v_{IN}}{v_{OUT}} \quad (4.24)$$

Finally, based on the definitions expressed in Eqs. (4.25) and (4.26), Eq. (4.27) is obtained to validate that the stability is not only related to the relationship of R_{ESR} and C_{OUT}, but also affected by the switching frequency.

$$f_{SW} = \dfrac{v_{OUT}}{v_{IN}} \cdot \dfrac{1}{T_{ON}} \quad (4.25)$$

$$f_{ESR} = \dfrac{1}{2\pi \cdot R_{ESR} C_{OUT}} \quad (4.26)$$

$$f_{SW} > \pi f_{ESR} \quad (4.27)$$

Capacitor choices are clearly limited to guarantee system stability. The stability criterion expressed in Eq. (4.27) also corresponds to the frequency response shown in Figure 4.20.

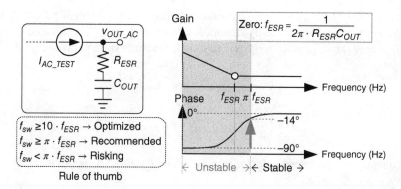

Figure 4.20 Test model of the output capacitor and its frequency response with the stability criterion

The obtained AC response of the output (v_{OUT_AC}) indicates the design guideline. The combination of R_{ESR} and C_{OUT} contributes one zero to reduce the phase delay between inductor current ripple and output voltage ripple. System stability can be guaranteed because the phase is boosted from $-90°$ to $-14°$ at the preferred operation switching frequency. Consequently, we need to ensure that the switching frequency is sufficiently high to derive an adequate in-phase relationship between the two signals.

Based on the stability criterion, the switching frequency should be sufficiently high when a specific capacitor with a certain ESR value is used. When operating at high switching frequencies, the problem of phase delay can be solved. However, infinitely increasing the switching frequency to solve the stability issue is impractical because unreasonable switching power loss degrades the power-conversion efficiency. Moreover, the delay of the comparator influences stability. As such, any type of delay should be considered when the switching frequency is beyond megahertz values. By contrast, a large ESR value can increase the system stability.

However, a large ESR significantly increases the undershoot/overshoot voltage during transient response and results in a large output voltage ripple. The ESR value also varies with temperature and influences the average level of the output voltage. That is to say, the variation of output voltage and switching noise easily influences the supplied circuits. As a result, a suitable selection of capacitors with an adequate ESR is important. A robust converter cannot be influenced by the variation in ESR. The system stability and output performance of ripple-based control depend on the careful selection of output capacitor. The following examples show the design with the specialty polymer capacitor (SP-CAP). A stable operation can possibly be ensured if the switching frequency is increased to high.

Example 4.1: The basic on-time controlled DC/DC buck converter is affected by the R_{ESR} value

Figure 4.21 shows the basic structure of the on-time controlled DC/DC buck converter. The specifications are listed in Table 4.2. Moreover, the other parameters are set as follows: $R_1 = 200$ kΩ and $R_2 = 400$ kΩ.

With a feedback loop, the expected v_{OUT} is 900 mV, f_{SW} is 1 MHz, and the duty ratio is 18% when the system is stable. Based on the stability criterion derived using Eq. (3.8) or Eq. (3.21), the stability of on-time control relies strongly on R_{ESR}.

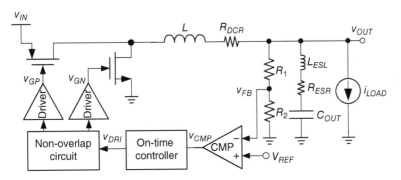

Figure 4.21 Basic on-time controlled DC/DC buck converter

Table 4.2 Basic specifications used in Example 4.1

V_{IN}	V_{OUT}	V_{REF}	L	R_{DCR}	C_{OUT}	L_{ESL}	T_{ON}	f_{SW}
5 V	900 mV	600 mV	4.7 µH	0 mΩ	4.7 µF	0 nH	180 ns	1 MHz

The simulation results with different values of R_{ESR} (i.e., 100, 28, and 10 mΩ) are observed and shown in Figure 4.22(a)–(c), respectively (Table 4.3).

The simulation tool used is SIMPLIS. With an adequate R_{ESR} value, Figure 4.22(a), (b) shows stable operating waveforms. Moreover, v_{OUT} is regulated. A comparison of Figure 4.22(a), (b) reveals that the large output ripple, $v_{OUT,pp}$, is caused by a large R_{ESR}. Figure 4.22(a) shows that the R_{ESR} value of 100 mΩ results in a $v_{OUT,pp}$ of approximately 16 mV. Figure 4.22(b) shows that the R_{ESR} value of 28 mΩ results in a $v_{OUT,pp}$ of approximately 6 mV. The valley voltage of v_{OUT} shown in Figure 4.22(a), (b) is 900 mV in both cases, because v_{OUT} is monitored using valley voltage control. However, the average of v_{OUT}, $v_{OUT,avg}$, can be calculated as:

$$v_{OUT,avg} = V_{REF} \cdot \left(\frac{R_1 + R_2}{R_2}\right) + \frac{1}{2}v_{OUT,pp} \qquad (4.28)$$

As a result, the $v_{OUT,avg}$ of 908 mV shown in Figure 4.22(a) has the worse offset voltage compared with the $v_{OUT,avg}$ of 903 mV shown in Figure 4.22(b). When the voltage accuracy requirement is strict, the inductor, output capacitor, and switching frequency should be carefully designed because $v_{OUT,pp}$ depends strongly on the ripple of i_L, C_{OUT}, and R_{ESR}.

In Figure 4.22(c), a small R_{ESR} of 10 mΩ does not result in the smallest $v_{OUT,pp}$ because R_{ESR} is relatively small to guarantee stable operation. Unstable operation increases the $v_{OUT,pp}$ and Δi_L. In other words, an unstable v_{OUT} may result in an unexpected overvoltage to induce overstress on back-end circuits. In addition, the peak inductor current may be considerably large and damage the components at the output stage.

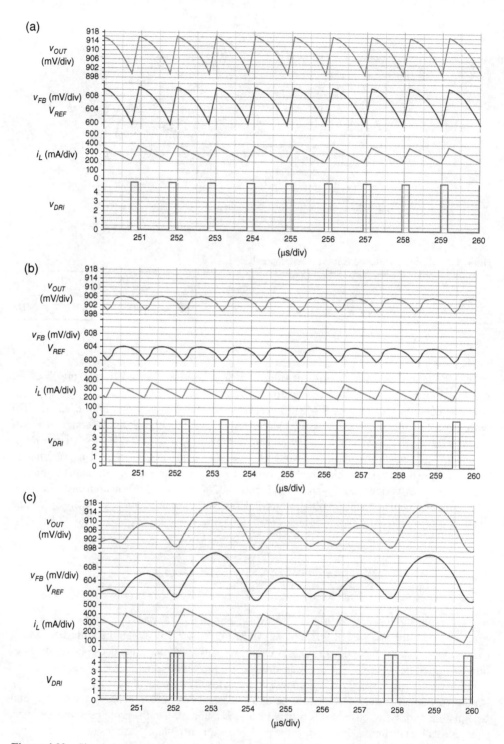

Figure 4.22 Simulation results of the on-time controlled DC/DC buck converter in SIMPLIS if R_{ESR} is (a) 100 mΩ, (b) 28 mΩ, and (c) 10 mΩ

Table 4.3 Value of R_{ESR} for three cases

Case	R_{ESR} (mΩ)
(a)	100
(b)	28
(c)	10

SIMPLIS Setting

Figure 4.23 shows the circuit setup in the circuit simulation software SIMPLIS, which corresponds to that shown in Figure 4.21. The high-side or low-side switch has parallel diodes that form a practical MOSFET.

Figures 4.24 and 4.25 provide zoom-in partial drawings of Figure 4.23. There are four critical devices to be set. Figure 4.26 illustrates the windows for setting parameters.

Figure 4.26(a) is a window to edit the device parameter of the output capacitor, whose value is 4.7 μF. "Level" is set as "2" to activate the value setting of ESR. In different cases, the values of ESR are set as 100, 28, and 10 mΩ.

Figure 4.26(b) is a window to edit the device parameter of the inductor, whose value is 4.7 μH. "Series Resistance" is the R_{DCR}, set as 0 mΩ.

Figure 4.26(c) is a window to edit the device parameter of the change in loading conditions. First "Wave shape" is selected as "One pulse," then "Time/Frequency" can be set for the time of load change. In this setting, the loading condition remains at light loads until the time is 500 μs, and then it changes from light to heavy load in 2 μs. In turn, the loading condition remains at heavy loads until the time is 550 μs, and then it changes from heavy to light load in 2 μs. Besides, the value of the load change is set by adjusting the value of "Resistance load" in Figure 4.24.

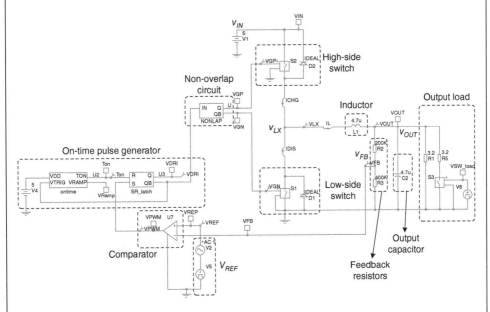

Figure 4.23 Setup circuit in SIMPLIS

Figure 4.26(d) is a window to edit the device parameter of the comparator in the feedback path. Care should be taken in setting "Input Resistance," because the value of this resistance has a load effect on the voltage value of V_{FB} fed from V_{OUT} through feedback resistance.

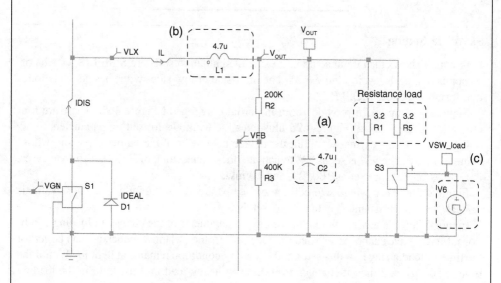

Figure 4.24 Zoom-in drawing of output filter and loading

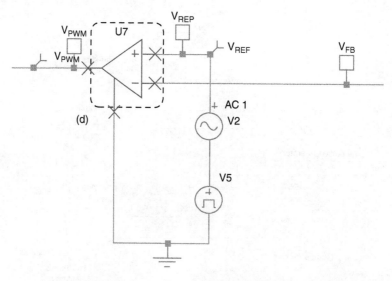

Figure 4.25 Zoom-in drawing of feedback path.

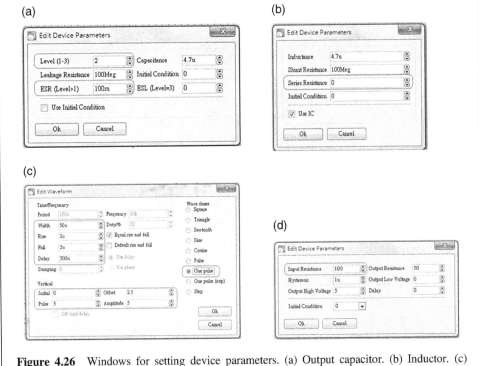

Figure 4.26 Windows for setting device parameters. (a) Output capacitor. (b) Inductor. (c) Output load. (d) Comparator

Example 4.2: Increasing the switching frequency to meet the stability criteria if low R_{ESR} is used

SP-CAP with a value of 100 μF has R_{ESR} value of 10 mΩ. This finding indicates that f_{ESR} is approximately 78 kHz. Specifically, f_{SW} should be greater than 240 kHz to obtain the maximum phase delay of 14°, as shown in Figure 4.27(a). Selecting the appropriate SP-CAPs to increase the stability of the on-time controlled system is easy, but the cost is high and the magnitude of the output ripple is large. Meanwhile, the R_{ESR} of the multi-layer ceramic capacitor (MLCC) is only several milliohms when considering the same output capacitor of 100 μF. In Figure 4.27(b), the value of R_{ESR} is 2 mΩ and the value of f_{ESR} is approximately 800 kHz, resulting in the lowest allowable switching frequency that is close to or even higher than 2.4 MHz. However, a high switching operation induces a large switching power loss and a worse jitter problem.

Generally, the MLCC is unsuitable for conventional adaptive on-time controlled methods. Thus, the appropriate output capacitor must be selected to increase system stability.

4.2.2 Selection of Output Capacitor

The switching regulator design must deal with a multi-dimensional trade-off among the input voltage, output voltage, output filter including a capacitor and an inductor, and switching frequency to obtain a regulated output voltage with improved performance. The capacitor has energy storage and voltage ripple filter functions. Thus, the characteristic of the output

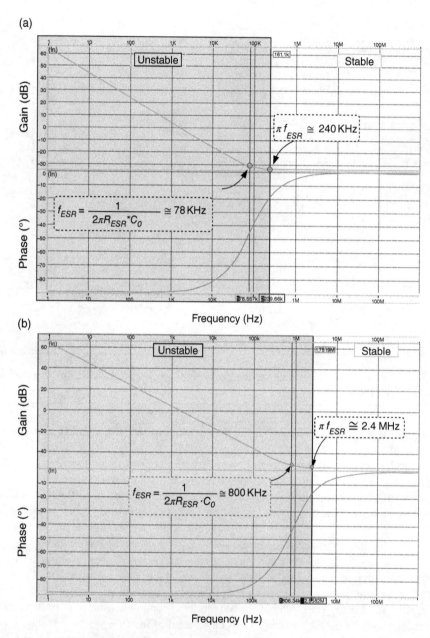

Figure 4.27 Frequency response of the output capacitor. (a) SP-CAP of 100 μF and R_{ESR} of 10 mΩ. (b) MLCC of 100 μF and R_{ESR} of 2 mΩ

capacitor is one of the factors used to determine the regularity and constancy of the output voltage. For ripple-based control in particular, the characteristics of the output capacitor, including the ideal capacitance, ESR, and ESL values, also determine whether the system is stable. Consequently, correctly selecting a capacitor with suitable properties has a direct influence on converter performance.

Table 4.4 Characteristics of different capacitor types

	Material	Electrolytic	Voltage stress (V)	Capacity (μF)
Al capacitor	Aluminum oxide	Polarity	4–400	470–10 000
TA capacitor	Tantalum oxide	Polarity	2.5–50	0.470–1000
Film capacitor	Plastic film	Non-polarity	50–1600	0.001–10
MLCC	Ceramic material	Non-polarity	6.3–250	0.001–100

Table 4.5 Advantages and disadvantages of different capacitor types

	Advantage	Disadvantage
Al capacitor	• High-voltage stress • Robust capacity over time	• Shortens lifetime when electrolytes are lost • Large volume
TA capacitor	• Large capacitance • Robust capacity over time	• Short circuit in failure
Film capacitor	• High-voltage stress	• Small capacitance • Few package options
MLCC	• Compact size • Good characteristic at high frequencies	• Capacity change over time • Easily broken and collapsed

Tables 4.4–4.6 summarize the characteristics of various types of capacitors. Common technology options include an aluminum (Al) capacitor, tantalum (Ta) capacitor, film capacitor, and MLCC. The tables also rank the performance of the technology options according to each characteristic. In the tables, F-C and T-C represent the variation in quality at different frequencies and temperatures, respectively. High-voltage and High-temperature represent the limitation of the highest voltage and temperature, respectively.

Al electrolytic capacitors are devices with anode and cathode polarities. These capacitors are composed of Al plates, aluminum oxide electrolyte, and dielectric. Al electrolytic capacitors can easily achieve large capacitance. Conventional Al electrolytic capacitors use liquid electrolyte, and the quality of the electrolyte can determine the capacity and lifetime. As a result, a high temperature induces the capacitors to lose electrolytes. Capacitance decreases and ESR increases, resulting in the failure of circuits to function. Poor F-C and T-C also make the Al electrolytic capacitors impractical for high-frequency applications.

SP-CAP with solid polymer electrolyte as the specialty polymer electrolyte has been created to resolve this problem. Based on Al electrolytic capacitor technology, the solid type has the advantage of a long lifetime. OS-CAPs use organic semiconductor electrolytes to obtain a similar performance.

Table 4.6 Different capacitor types

		Capacity	F-C	T-C	High-voltage	High-temperature	Size	Life	Cost	R_{ESR}
Al capacitor	Electrolyte	✓✓	✗	✗	✓	O	✓	✗	✓✓	Large
	OS-CAP	✓✓	✓✓	✓✓	✗	✗	O	✓✓	O	Large
	SP-CAP	✓	✓✓	✓✓	O	✓✓	O	✓✓	O	Large
TA capacitor (POS-CAP)		✓	O	✓✓	O	✓✓	✓	O	✓	Medium
Film capacitor		✗	✓✓	✓✓	✓✓	O	✗	✓✓	O	Small
MLCC		✓	✓✓	O	✓✓	✓✓	✓	✓✓	✓	Small

✗, Bad; O, Fair; ✓, Good; ✓✓, Excellent.

Solid Al electrolytic capacitors have good properties but a high price, and are used extensively in consumer applications. These capacitors are suitable for bypassing low-frequency noise and storing large amounts of energy. When SP-CAPs are applied in SWRs as the output capacitor, a large ESR value will result in a large output ripple and degrade the output voltage quality. Considering this disadvantage, SP-CAPs are rarely used as output capacitor when a power supply of high quality is required. By contrast, for conventional ripple-based control converters, SP-CAPs are a good choice because of their accurate and stable ESR properties.

Ta electrolytic capacitors use tantalum oxide as the electrolyte and have polarity. These capacitors exhibit improved performance at high temperature. As such, the lifetime of Ta electrolytic capacitors is longer than that of liquid Al electrolytic capacitors. Compared with Al electrolytic capacitors, Ta electrolytic capacitors exhibit a more stable capacitance, lower DC leakage, and lower impedance at high frequencies. The ESR value of Ta electrolytic capacitors is lower than that of Al electrolytic capacitors but higher than that of MLCC. Moreover, a small ESR value will result in a short circuit if the capacitance decreases. This short circuit will result in failure of the circuits to function.

Film capacitors are composed of thin plastic films as the dielectric. They have good stability, low inductance, and low ESR value, and are not polarized. As such, film capacitors are suitable for an AC signal. Compared with MLCCs, film capacitors have a significant capacitance within ±5% of its rated value. These capacitors can withstand voltages in the kilovolt range and high temperature. However, low capacitance and scarce package types are disadvantages that limit the flexibility of utilizing film capacitors.

MLCC is a non-polarity device composed of numerous thin ceramic layers. These capacitors have a compact size and are utilized extensively for switching power converters because of their small ESR and ESL compared with other types. Small parasitic values also imply a reduction in power loss and high performance.

As previously mentioned, the SP-CAP is the most common choice in conventional ripple-based control buck converters because system stability can be guaranteed simply by a large ESR value based on the derived stability criterion. However, a large ESR value results in a large output voltage ripple in steady state and degrades the overshoot/undershoot transient voltage variation during transient response. By contrast, the MLCC has become more attractive because of its low price, compact size, and low ESR value, as listed in Table 4.6. However, a low ESR value cannot provide a substantial time constant value to ensure stability for ripple-based control converters. The stability problem is a significant challenge for the designer in maintaining the superior characteristics of the ripple-based control technique while using MLCC as the output capacitor. In the subsequent section, we will introduce several techniques used to overcome this problem.

4.3 Design Techniques When Using *MLCC* with a Small Value of R_{ESR}

The ESR is a critical factor in determining the stability criterion. In the conventional ripple-based control technique, the value of R_{ESR} determined by the selection of the output capacitor should be sufficiently large to increase stability. Notable disadvantages are the increase in output voltage ripple in the steady state and the inevitable large undershoot/overshoot voltage that occurs in case of a load transient response. Owing to their inherent large output voltage ripples, conventional ripple-based control DC/DC converters are unsuitable for use in directly supplying power to sensitive analog circuits. In recent years, facilitating this technique and avoiding the use of large ESRs have become the trend in commercial products. These developed techniques will be discussed in the following.

4.3.1 Use of Additional Ramp Signal

If an additional ramp signal v_{Ramp} is used to enlarge the noise margin, then the converter with ripple-based control can have the advantage of low output voltage ripple through the use of a cheap ceramic capacitor that has a small R_{ESR}. The implementation of on-time control shown in Figure 4.28 uses two methods to enlarge the noise margin by the insertion of an external ramp signal. v_{OUT} is unable to linearly reflect the inductor current ripple if the output capacitor has a low ESR. By contrast, the small output voltage ripple is derived by charging/discharging C_{OUT}. The difference of the two terminals of the comparator can be enlarged by either adding an extra ramp to V_{REF}, as shown in Figure 4.28(a), or subtracting an extra ramp from v_{FB}, as shown in Figure 4.28(b). In other words, the noise margin is enlarged to ensure stability.

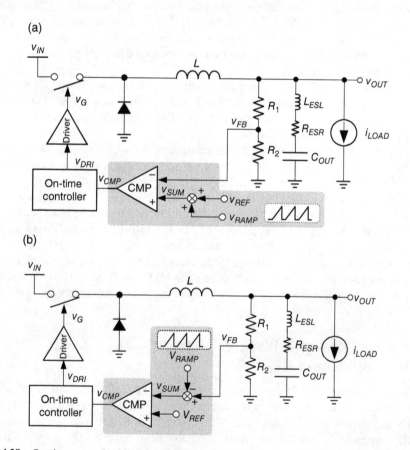

Figure 4.28 On-time control with an extra ramp signal for the use of a capacitor with a small R_{ESR}: (a) adding V_{Ramp} to V_{REF} and (b) subtracting V_{Ramp} from V_{FB}

Similar to the analysis presented in Section 4.2, Eq. (4.30) is an approximation of the transfer function from the reference voltage to the output expressed as follows:

$$\frac{\hat{v}_{OUT}(s)}{\hat{v}_{REF}(s)} \approx \frac{1}{\left(1 + \frac{s}{Q_1 \omega_1} + \frac{s^2}{\omega_1^2}\right)} \cdot \frac{\left(1 + \frac{s}{Q_1 \omega_2} + \frac{s^2}{\omega_2^2}\right)(R_{ESR} C_{COUT} \cdot s + 1)}{\left(1 + \frac{s}{Q_2 \omega_2} + \frac{s^2}{\omega_2^2}\right)\left(1 + \frac{s}{Q_1 \omega_2} + \frac{s^2}{\omega_2^2}\right) + \frac{m_{Slope,r}}{m_{ESR,f}} R_{ESR} C_{COUT} T_{SW} \cdot s^2} \quad (4.29)$$

where

$$\omega_1 = \frac{\pi}{T_{ON}}, \quad Q_1 = \frac{2}{\pi}, \quad \omega_2 = \frac{\pi}{T_{SW}}, \quad Q_2 = \frac{T_{SW}}{\left(R_{ESR} C_{OUT} - \frac{T_{ON}}{2}\right) \cdot \pi} \quad (4.30)$$

In this study, $m_{Slope,r}$, the slope of the additional ramp signal, and $m_{ESR,f}$, the off-time slope of the output voltage ripple expressed in Eq. (4.31) derived by using the ESR, have the same positive sign:

$$m_{ESR,f} = R_{ESR} \cdot V_{OUT}/L \quad (4.31)$$

When T_{ON} is sufficiently small, the transfer function in Eq. (4.29) can be simplified as follows:

$$\frac{\hat{v}_{OUT}(s)}{\hat{v}_{REF}(s)} \approx \frac{(R_{ESR} C_{COUT} \cdot s + 1)}{\left(1 + \frac{s}{Q_{2r} \omega_2} + \frac{s^2}{\omega_2^2}\right)} \quad (4.32)$$

where

$$Q_{2r} = \frac{T_{SW}}{\left[\left(2 \frac{m_{Slope,r}}{m_{ESR,f}} + 1\right) R_{ESR} C_{OUT} - \frac{T_{ON}}{2}\right] \cdot \pi} \quad (4.33)$$

Then the stability criterion is obtained:

$$\left(2 \frac{m_{Slope,r}}{m_{ESR,f}} + 1\right) R_{ESR} C_{OUT} > \frac{T_{ON}}{2} \quad (4.34)$$

In other words, the required R_{ESR} can be obtained:

$$R_{ESR} > \frac{T_{ON}}{2 C_{OUT}} \cdot \frac{1}{\left(2 \frac{m_{Slope,r}}{m_{ESR,f}} + 1\right)} \quad (4.35)$$

A comparison of Eqs. (4.15) and (4.35) reveals that the additional ramp contributes to the factor of $m_{Slope,r}$ so that we obtain an equivalent large R_{ESR} without the need for a physically large ESR. Consequently the MLCC, even with an ultra-small R_{ESR}, can be used in on-time control so that the output voltage ripple and the transient voltage variation can be small in the steady state and the transient response simultaneously.

4.3.2 Use of Additional Current Feedback Path

The use of an additional current feedback (CF) path is another practical technique if the ripple-based control uses a ceramic capacitor with a small R_{ESR}. Figure 4.29 shows the implementation of the on-time control with an additional CF path. R_{SEN} in Eq. (4.36) represents the converting transfer function from the current signal to the voltage signal. Thus, v_{SEN} is linearly proportional to the inductor current i_{SEN}. Then, v_{SEN} is added to v_{FB} so that v_{SUM} with a modified slope can adequately determine the switching operation point from the off-time to the on-time:

$$R_{SEN} = \frac{v_{SEN}}{i_{SEN}} \tag{4.36}$$

With an additional CF path, the new transfer function from the reference voltage to the output is derived as follows:

$$\frac{\hat{v}_{OUT}(s)}{\hat{v}_{REF}(s)} \approx \frac{(R_{ESR} C_{COUT} \cdot s + 1)}{\left(1 + \dfrac{s}{Q_1 \omega_1} + \dfrac{s^2}{\omega_1^2}\right)\left(1 + \dfrac{s}{Q_{2s} \omega_2} + \dfrac{s^2}{\omega_2^2}\right)} \tag{4.37}$$

where

$$\omega_1 = \frac{\pi}{T_{ON}}, \; Q_1 = \frac{2}{\pi}, \; \omega_2 = \frac{\pi}{T_{SW}}, \; Q_{2s} = \frac{T_{SW}}{\left[\left(\dfrac{R_{SEN}}{R_{ESR}} + 1\right) R_{ESR} C_{OUT} - \dfrac{T_{ON}}{2}\right] \cdot \pi} \tag{4.38}$$

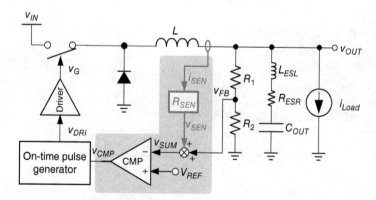

Figure 4.29 With an additional CF path, the ripple-based control can use a ceramic capacitor, even with a small R_{ESR}

Ripple-Based Control Technique Part I

If T_{ON} is sufficiently small, then the transfer function can be simplified to a second-order equation:

$$\frac{\hat{v}_{OUT}(s)}{\hat{v}_{REF}(s)} \approx \frac{(R_{ESR}C_{COUT} \cdot s + 1)}{\left(1 + \dfrac{s}{Q_{2s}\omega_2} + \dfrac{s^2}{\omega_2{}^2}\right)} \tag{4.39}$$

Equation (4.41) should be able to remove the unexpected RHP poles; that is, Q_{2s} is always larger than zero:

$$Q_{2s} = \frac{T_{SW}}{\left[\left(\dfrac{R_{SEN}}{R_{ESR}} + 1\right)R_{ESR}C_{OUT} - \dfrac{T_{ON}}{2}\right] \cdot \pi} > 0 \tag{4.40}$$

Therefore, the stability criterion is obtained:

$$\left(\frac{R_{SEN}}{R_{ESR}} + 1\right)R_{ESR}C_{OUT} > \frac{T_{ON}}{2} \tag{4.41}$$

Compared with Eq. (4.14), Q_{2s} shows that the current signal equivalently increases the value of R_{SER} by a factor of $(R_{SEN}/R_{ESR} + 1)$. Besides, the criterion in Eq. (4.42) can be extended from Eq. (4.24), and the stability is then easier to guarantee:

$$\left(\frac{R_{SEN}}{R_{ESR}} + 1\right)R_{ESR}C_{OUT} > \frac{T_{ON}}{2} \cdot \frac{V_{IN}}{V_{OUT}} = \frac{1}{2f_{SW}} \tag{4.42}$$

Without the need for a large R_{ESR}, the additional CF signal releases the stability constraint. Some general implementations of the CF path will be introduced next to discuss the individual characteristics of each implementation.

4.3.2.1 CF-type1 (Extra Sensing Resistor)

To sense the inductor current, the sensing resistor R_S should be directly in series with the inductor, as shown in Figure 4.30. The inductor current can easily be obtained because the sensing resistor is placed on the power delivery path. Through R_S, the feedback information v_{OUT0} can be used to obtain the extra inductor current information:

$$v_{OUT0}(t) = v_{OUT}(t) + [i_L(t) \cdot R_S] \tag{4.43}$$

The voltage v_S across R_S is composed of information from the output voltage and the inductor current. Thus, the stability criterion can be derived from Eq. (4.14) to obtain:

$$(R_{ESR} + R_S) \cdot C_{OUT} > \frac{T_{ON}}{2} \tag{4.44}$$

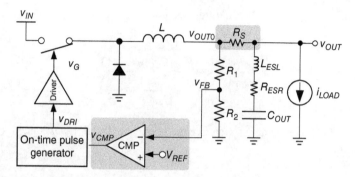

Figure 4.30 Insertion of the sensing resistor R_S in series can enhance system stability but can also cause extra power loss and voltage droop under different loading conditions

However, v_{OUT} is excluded from the feedback loop because R_S is used to obtain the inductor current information. In other words, directly monitoring v_{OUT} through the feedback loop is unnecessary. By contrast, the feedback loop can only modulate v_{OUT0} based on the valley voltage control and the reference voltage V_{REF}. The average value of v_{OUT0} is expressed in Eq. (4.45) as

$$v_{OUT0,avg} = V_{REF} \cdot \left(\frac{R_1 + R_2}{R_2}\right) + \frac{1}{2}v_{OUT0,pp} \tag{4.45}$$

When Eq. (4.43) is substituted in Eq. (4.45), Eq. (4.46) can reveal the relationship between $v_{OUT0,avg}$ and $v_{OUT,avg}$ to determine the deviation caused by different loading conditions. Moreover, the simplified Eq. (4.47) can show the deviation of $v_{OUT,avg}$ from the expected regulation value:

$$v_{OUT,avg} = v_{OUT0,avg} - \left(i_{L,avg} + \frac{1}{2}i_{L,pp}\right) \cdot R_S + \frac{1}{2}v_{OUT,pp} \tag{4.46}$$

$$v_{OUT,avg} = V_{REF} \cdot \left(\frac{R_1 + R_2}{R_2}\right) - \left(i_{L,avg} + \frac{1}{2}i_{L,pp}\right) \cdot R_S + \frac{1}{2}v_{OUT,pp} \tag{4.47}$$

Although easy to generate, the additional sensing resistor R_S in series causes the system to suffer from the disadvantages of extra power loss and voltage droop under different loading conditions. These disadvantages are undesired in recent portable electronics.

Example 4.3: Insertion of R_S to add more inductor current information

Figure 4.30 shows the basic structure of an on-time control DC/DC buck converter. Table 4.7 lists the basic specifications. The other parameters are set as follows: $R_1 = 200$ kΩ and $R_2 = 400$ kΩ.

Given the negative feedback loop, the system is sufficiently stable to obtain the expected V_{OUT} of 900 mV when f_{SW} is 1 MHz and the duty ratio is 0.18. Simulation results with different values of R_S, 60 and 20 mΩ, are observed, as shown in Figure 4.31(a), (b), respectively. The simulation results are also obtained by using the simulation tool SIMPLIS.

Table 4.7 Basic specifications used in Example 4.3

V_{IN}	V_{OUT}	V_{REF}	L	R_{DCR}	C_{OUT}	R_{ESR}	L_{ESL}	T_{ON}	f_{SW}
5 V	900 mV	600 mV	4.7 μH	0 mΩ	4.7 μF	0 mΩ	0 nH	180 ns	1 MHz

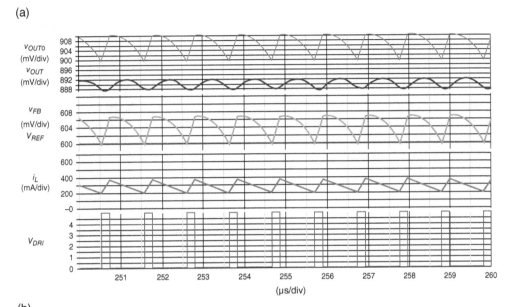

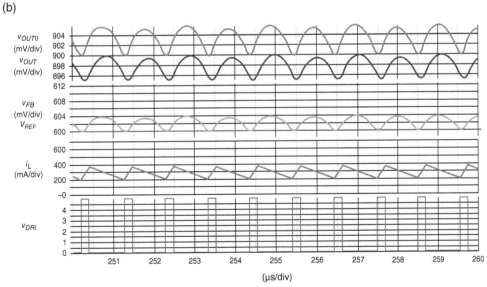

Figure 4.31 Simulation results obtained using SIMPLIS when the on-time controlled DC/DC buck converter uses different values of sensing resistor R_S: (a) 60 mΩ and (b) 20 mΩ

Figure 4.31(a) shows stable operating waveforms with an adequate R_{ESR} of 60 mΩ. Moreover, v_{OUT} is well regulated. The waveform of v_{OUT} with $R_{ESR} = 0$ mΩ clearly contributes little linear information to i_L. Through R_S, the waveform of v_{OUT0} contributes additional information to i_L. The linearity between v_{OUT0} and i_L is also increased, which validates the result derived using Eq. (4.43). Through R_1 and R_2, the feedback loop can regulate v_{OUT0} and provide a stable v_{OUT}. The waveforms also reveal that R_S results in a deviation voltage of −12 mV between v_{OUT0} and v_{OUT}, an outcome that proves the result derived using Eq. (4.48). In Figure 4.31(b), the small R_S of 20 mΩ is used. The deviation voltage between v_{OUT0} and v_{OUT} is reduced to −4 mV compared with the value of −12 mV in the case of $R_S = 60$ mΩ. Additionally, the waveforms reveal that $R_{ESR} = 20$ mΩ is relatively small to ensure system stability. Thus, v_{OUT} is slightly unstable (Table 4.8).

Figure 4.32 shows a transient response in case of $R_S = 60$ mΩ when the load changes from 280 to 560 mA at a time of 500 μs. Evidently, the deviation voltage between v_{OUT} and v_{OUT0} is different. By comparing the deviation voltage at $I_{Load} = 280$ and 560 mA, the deviation voltage increases because $i_{L,avg}$ increases. Moreover, v_{OUT0} is still well regulated and its valley is 900 mV even under different loading conditions. This phenomenon can also be expected when calculating Eq. (4.47). In other words, v_{OUT} exhibits a voltage droop that is strongly dependent on the value of R_S and i_{Load}. When high voltage accuracy is required, the structure shown in Figure 4.30 may not be a suitable method.

Table 4.8 Design parameters of two cases

Case	L	R_{DCR}	L_{ESL}	R_{ESR}	C_{OUT}	R_S (mΩ)
(a)	4.7 μH	0 mΩ	0 nH	0 mΩ	4.7 μF	60
(b)						20

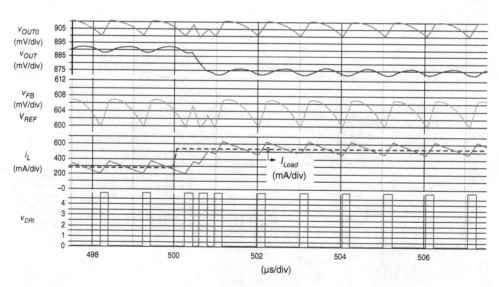

Figure 4.32 Transient response of the on-time controlled DC/DC buck converter if $R_S = 60$ mΩ when the load changes from 280 to 560 mA

SIMPLIS Setting

Figure 4.33 shows the setup circuit in the SIMPLIS simulation corresponding to Figure 4.30. Each high-side or low-side switch is placed parallel with a diode to form an equivalent MOSFET device. Besides, Figure 4.34 provides a zoomed-in view of the partial portion of Figure 4.33. The values of four critical devices should be decided to see the differentiation of the performance, where Figure 4.35 illustrates the setting windows for these parameters.

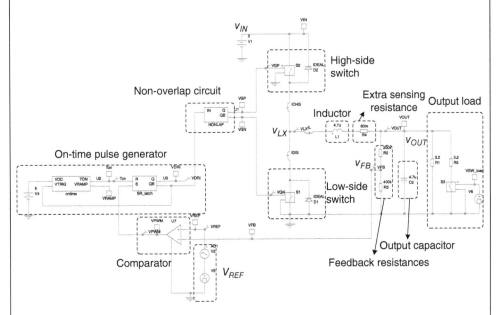

Figure 4.33 Setup circuit in SIMPLIS

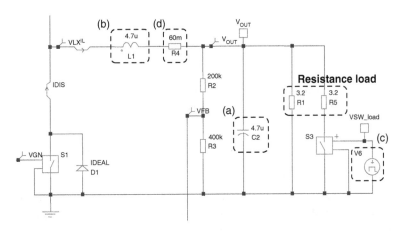

Figure 4.34 Zoom-in views of output filter and loading

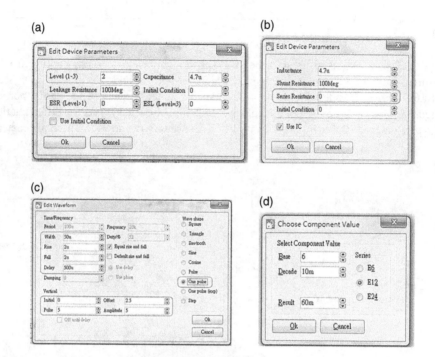

Figure 4.35 Windows for setting device parameters. (a) Output capacitor. (b) Inductor. (c) Output load. (d) Extra sensing resistance in series with the inductor

Figure 4.35(a) is a window to edit the device parameter of the output capacitor, whose value is 4.7 μF. "Level" is set as "2" to activate the value setting of ESR. However, to demonstrate the function of extra sensing resistance, the value of ESR is 0 mΩ now. Figure 4.35(b) is a window to edit the device parameter of the inductor, whose value is 4.7 μH. "Series Resistance" is the R_{DCR}, set as 0 mΩ. Figure 4.35(c) is a window to edit the device parameter of the change in loading conditions. "Wave shape" is first selected as "One pulse," then "Time/Frequency" can be set for the time of load change. In this setting, the loading condition remains at light loads until the time is 500 μs, and then changes from light to heavy load in 2 μs. In turn, the loading condition remains at heavy loads until the time is 550 μs, and then changes from heavy to light load in 2 μs. Besides, the value of the load change is set by adjusting the value of "Resistance load" in Figure 4.34. Figure 4.35(d) is a window to edit the device parameter of extra sensing resistance in series with the inductor. Its value is set at 60 and 20 mΩ in Example 4.3.

4.3.2.2 CF-type2 (Integral RC-Filter)

In Figure 4.36, R_{LPF} and C_{LPF} with a time constant of τ_{LPF} in Eq. (4.49) act as a low-pass filter (LPF) to integrate the signal variation at v_{LX}, which is the node between the high-side and low-side switches. Thus, on-time and off-time can denote the value of v_{IN} and ground, respectively, by monitoring v_{LX}:

$$\tau_{LPF} = R_{LPF} \cdot C_{LPF} \tag{4.48}$$

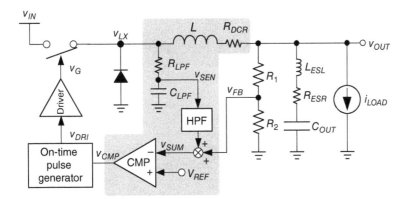

Figure 4.36 LPF composed of R_{LPF} and C_{LPF} with a time constant of τ_{LPF} integrating the signal variation at v_{LX}

The slope of v_{SEN} during the on-time and off-time periods, $m_{SEN,r}$ and $m_{SEN,f}$, can be expressed in Eqs. (4.50) and (4.51), respectively, when τ_{LPF} is larger than the switching period:

$$m_{SEN,r} = \frac{v_{IN} - v_{OUT}}{\tau_{LPF}} \qquad (4.49)$$

$$m_{SEN,f} = \frac{-v_{OUT}}{\tau_{LPF}} \qquad (4.50)$$

The average model is used to derive the DC value of v_{SEN}. In the steady state, the average current through R_{LPF} is zero. The average value of v_{LX} can be expressed in Eq. (4.51), where the complex value D is assumed constant to simplify the analysis:

$$v_{LX,avg} = v_{IN} \cdot D \qquad (4.51)$$

Considering the average model, the inductor can be viewed as shorted. The DC voltage (or average value) of v_{SEN} can be derived using Eq. (4.52), where $i_{L,avg}$ is the average inductor current:

$$v_{SEN,avg} = v_{OUT,avg} - i_{L,avg} \cdot R_L \qquad (4.52)$$

In Eqs. (4.49), (4.50), and (4.52), $v_{SEN}(t)$ represents information from the inductor current. In other words, $v_{SEN}(t)$ consists of the DC and AC information from the inductor current. When adding $v_{SEN}(t)$ to the feedback path to enhance stability, a high-pass filter (HPF) is needed to transmit the AC information from $v_{SEN}(t)$ so that the DC information from $v_{SEN}(t)$ will not worsen the accuracy of the v_{OUT} regulation. One approach that can provide a simple method to summarize v_{FB} and v_{SEN} is to short these two nodes directly. The design complexity is significantly reduced without the use of any mixing circuits. However, the phase delay characteristic between v_{FB} and v_{SEN} degrades the transient response. The accuracy of v_{OUT} regulation will also be deteriorated.

4.3.2.3 CF-type3 (Parallel RC-Filter)

In Figure 4.36, based on the integral function of the filters of C_{LPF} and R_{LPF}, the current-sensing information, v_{SEN}, has a phase delay corresponding to the actual inductor current information, i_L. Therefore, the load transient response is restricted because the output voltage variation cannot effectively be reflected back to the controller. In Figure 4.37, the filter of R_S and C_S, which is placed parallel to the inductor L, can have a similar inductor current-sensing effect. Transient results have a slight difference.

In this study, the parasitic DCR of the inductor, R_{DCR}, is utilized to sense the inductor current. No extra power loss occurs when no sensing resistor is used on the energy delivery path. The sensing pair of R_S and C_S is placed parallel with L and R_{DCR} to sense the current flowing through R_{DCR}. The voltage across C_S can be derived using

$$v_{SEN}(s) = \frac{i_L \cdot (R_{DCR} + sL)}{1 + sR_S C_S} = (i_L \cdot R_{DCR}) \cdot \left[\frac{1 + s(L/R_{DCR})}{1 + sR_S C_S} \right] \quad (4.53)$$

The DC term (i.e., $s = 0$) in Eq. (4.53) implies that the sensing result can be derived simply by using the R_{DCR} of the inductor. However, the frequency-dependent term, including one pole/zero pair, will influence the linearity of inductor current sensing. If the time constant values of L/R_{DCR} and $R_S C_S$ are equal, as shown in Eq. (4.54), then the voltage v_{SEN} across the capacitor C_S can be completely proportional to the inductor current i_L:

$$L/R_{DCR} = R_S C_S \quad (4.54)$$

The derived current-sensing signal can be independent of frequency without being affected by any time constant values from the inductor or sensing network. The values of L and R_{DCR}

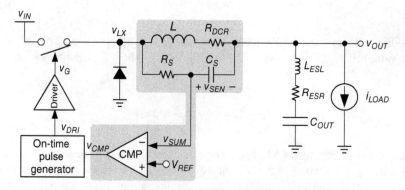

Figure 4.37 Parallel-sensing technique can enhance the sensing signal

should be known in advance to design the sensing network composed of R_S and C_S appropriately. A disadvantage of this method is the limited design flexibility. Another compromise method is to adjust the external resistor R_S value to ensure design flexibility.

During the on-time and off-time periods, v_{LX} is shortened to v_{IN} and ground, respectively. When charging and discharging the capacitor C_S during the on-time and off-time periods, respectively, the rising slope $m_{SEN,r}$ and the falling slope $m_{SEM,f}$ during the on-time and off-time periods can be expressed in Eqs. (4.55) and (4.56):

$$m_{SEN,r} = \frac{v_{IN} - v_{OUT}}{\tau_{SEN}} \tag{4.55}$$

$$m_{SEN,f} = \frac{-v_{OUT}}{\tau_{SEN}} \tag{4.56}$$

In this study, the switch on-resistance is assumed to be sufficiently small to be ignored, and the time constant τ_{SEN} is equal to the product of R_S and C_S:

$$\tau_{SEN} = R_S \cdot C_S \tag{4.57}$$

$v_{SUM}(t)$ can be derived using Eq. (4.58) to show that the amount of current ripple is effectively increased to improve system stability:

$$v_{SUM}(t) = v_{OUT}(t) + v_{SEN}(t) \tag{4.58}$$

However, the DC value of the sensed inductor current is also included. A disadvantage of the DC value of v_{SUM} is that it depends on the loading current conditions. Thus, this approach still suffers from voltage droop. Figure 4.38 shows the operating waveforms to explain the offset

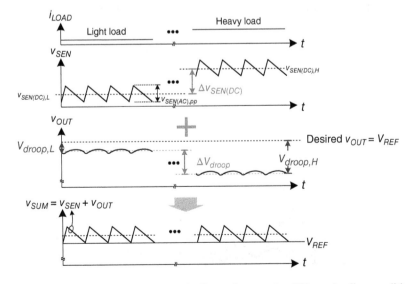

Figure 4.38 v_{OUT} of voltage droop and offset voltage under different loading conditions

voltage and the voltage droop at v_{OUT} under different loading conditions. Owing to the valley control, the valley of v_{SEN} is regulated at V_{REF}. Based on Eq. (4.58), the DC values of v_{OUT} and $v_{OUT(DC)}$ can be derived using Eq. (4.59). V_{droop} and v_{offset} in Eqs. (4.60) and (4.61) represent the voltage droop and the offset voltage, respectively. Notably, V_{droop} includes only $v_{SEN(DC)}$, whereas v_{offset} includes $v_{OUT(AC),pp}$ and $v_{SEN(AC),pp}$, which represent the peak-to-peak voltages of $v_{OUT(AC)}$ and $v_{SEN(AC)}$, respectively.

$$v_{OUT(DC)} = V_{REF} + V_{droop} + V_{offset} \tag{4.59}$$

$$V_{droop} = \frac{1}{2} v_{SEN(DC)} \tag{4.60}$$

$$v_{offset} = \frac{1}{2} v_{SEN(AC),pp} + \frac{1}{2} v_{OUT(AC),pp} \tag{4.61}$$

When MLCC is used for a small ripple, the small value of $v_{OUT(AC),pp}$ can be ignored. Thus, Eq. (4.61) is simplified as follows:

$$v_{offset} \approx \frac{1}{2} v_{SEN(AC),pp} \tag{4.62}$$

However, R_{SEN} should be sufficiently large to meet the inequality requirement in Eq. (4.41), which indicates that v_{SEN} is large. Moreover, the load-dependent voltage droop V_{droop} at v_{OUT} is caused by v_{SEN} under different loading conditions. By considering the regulation of $v_{OUT(DC)}$, Eq. (4.59) shows that $v_{SEN(AC),pp}$ results in a large offset voltage at v_{OUT}. Significantly, the change in $v_{SEN(DC)}$ under different loading conditions influences the value of V_{droop} at v_{OUT}.

In case of the load transient response shown in Figure 4.39, the larger $v_{SEN(AC)}$ is, the smaller Δv_{SUM} will be because the trend of v_{SEN} is opposite to that of v_{OUT}. Meanwhile, the on-time period becomes smaller than that in the steady state. Although the approach uses an additional CF path to stabilize the system, the performance of load regulation and transient response deteriorates.

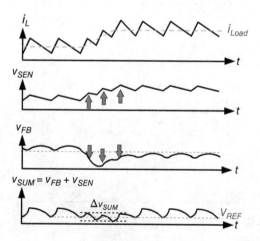

Figure 4.39 Deterioration of load transient response because of the additional CF path

Example 4.4: The DC value is seriously affected by the additional current-sensing value v_{SEN}; besides, the contribution of the additional sensing should be sufficiently large if we observe the change in C_S

Figure 4.40 shows the load transient response from 188 to 375 mA derived using the simulation tool SIMPLIS. The basic specifications are listed in Table 4.9. Based on Eq. (4.58), v_{OUT} is regulated at approximately 600 mV. The DC value of v_{SEN} increases when i_{Load} increases. Evidently, $v_{OUT(DC)}$ is influenced by v_{SEN}.

In this study, different characteristics are observed by monitoring different amplitudes of v_{SEN} if the size of C_S is changed but the same R_S value is kept. After using five values of C_S, Table 4.10 lists the critical parameters and Figure 4.41 shows the corresponding operating waveforms.

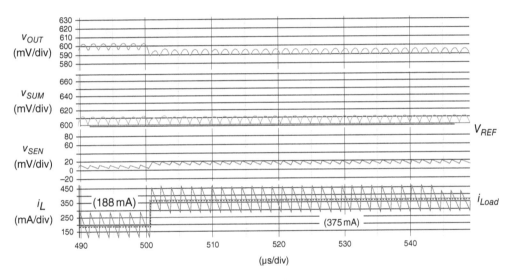

Figure 4.40 Load transient response

Table 4.9 Basic specifications used in Example 4.4

V_{IN}	V_{OUT}	V_{REF}	L	R_{DCR}	C_{OUT}	R_{ESR}	L_{ESL}	T_{ON}	f_{SW}
5 V	600 mV	600 mV	4.7 μH	40 mΩ	4.7 μF	0 mΩ	0 nH	180 ns	660 kHz

Table 4.10 Designing parameters for five cases

Case	L	R_{DCR}	L_{ESL}	R_{ESR}	C_{OUT}	R_S	C_S
(a)	4.7 μH	40 mΩ	0 nH	0 mΩ	4.7 μF	250 Ω	470 nF
(b)							100 nF
(c)							47 nF
(d)							1.0 μF
(e)							4.7 μF

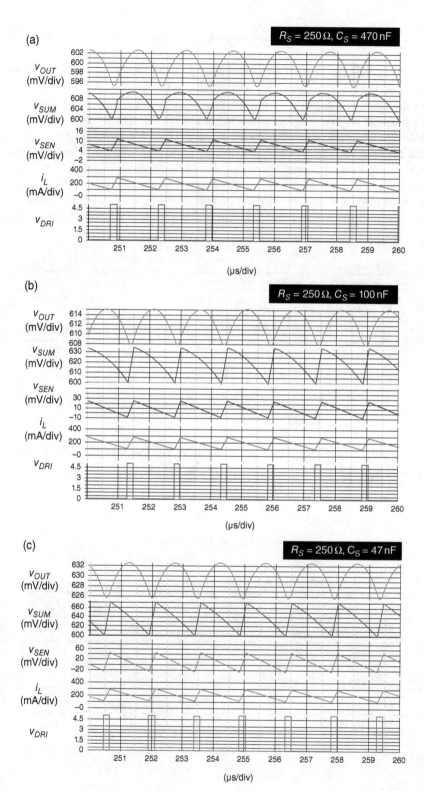

Figure 4.41 External components include $L = 4.7$ μH, $R_{DCR} = 40$ mΩ, $L_{ESL} = 0$ nH, $R_{ESR} = 0$ mΩ, and $C_{OUT} = 4.7$ μF. (a) $R_S = 250$ Ω and $C_S = 470$ nF. (b) $R_S = 250$ Ω and $C_S = 100$ nF. (c) $R_S = 250$ Ω and $C_S = 47$ nF. (d) $R_S = 250$ Ω and $C_S = 1.0$ μF. (e) $R_S = 250$ Ω and $C_S = 4.7$ μF

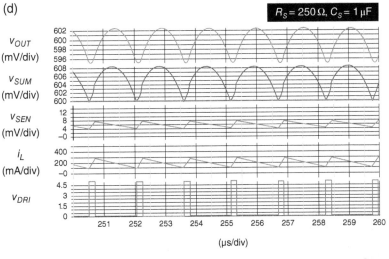

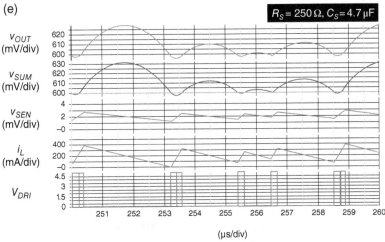

Figure 4.41 (*Continued*)

In Figure 4.41(a), R_S and C_S are 250 Ω and 470 nF, respectively, as determined using Eq. (4.54). Thus, v_{SEN} is proportional to i_L, and v_{FB} is composed of information from v_{OUT} and i_L. As a result, stability is guaranteed although R_{ESR} is zero. The valley of v_{SUM} is 600 mV because the two inputs of the comparator in the feedback loop are V_{REF} and v_{SUM}. However, the valley of v_{OUT} is equal to 600 mV − $v_{SEN(AC),pp}$ − $v_{SEN(DC)}$. In other words, the average value of v_{OUT} varies with different values of V_{REF} and $v_{SEN(AC),pp}$. Based on Eqs. (4.56) and (4.57), the time constant τ_{SEN} can determine $v_{SEN(AC),pp}$:

$$v_{SEN(AC),pp} = T_{ON} \cdot m_{SEN,r} = T_{ON} \cdot \left(\frac{v_{IN} - v_{OUT}}{\tau_{SEN}} \right) \tag{4.63}$$

Decreasing C_S causes a different value of $v_{SEN(AC),pp}$, which is the offset voltage at v_{OUT}. In Figure 4.41(b), τ_{SEN} is smaller than that in Figure 4.41(a) if the value of C_S decreases to 100 nF.

$v_{SEN(AC),pp}$ is 8 mV in Figure 4.41(a), whereas $v_{SEN(AC),pp}$ is 37.6 mV in Figure 4.41(b). Comparison of Figure 4.41(a) and (b) reveals that the value of $v_{SEN(AC),pp}$ is proportional to the value of C_S.

Furthermore, $v_{SEN(AC),pp}$ is 80 mV in Figure 4.41(c) because τ_{SEN} becomes smaller than its original value when R_S is 250 Ω and C_S is 47 nF. A larger $v_{SEN(AC),pp}$ results in a v_{FB} that is more proportional to i_L. Thus, the noise immunity of the feedback signal increases. However, a larger $v_{SEN(AC),pp}$ also causes a worse offset voltage at v_{OUT}. The value of $v_{OUT(DC)}$ in Figure 4.41(a)–(c) is 599, 611, and 630 mV, respectively. By contrast, the results reveal that τ_{SEN} increases with an increase in the value of C_S, as shown in Figure 4.41(d), (e). If R_S is 250 Ω and C_S is 1.0 μF, then the value of $v_{SEN(AC),pp}$ decreases to approximately 3.76 mV. v_{OUT} is still well regulated and its waveform is similar to that of v_{SUM}. However, the linearity between v_{SUM} and i_L is worse. As a result, the operation is unstable when a smaller τ_{SEN} with C_S = 4.7 μF is used, as shown in Figure 4.41(e).

In other words, stability is strongly determined by the value of τ_{SEN}. Compared with the conventional on-time control with a sufficiently large value of R_{ESR}, the ripple of v_{SEN} replaces the function ripple of V_{ESR}. Equation (4.64) defines $R_{eq,SEN}$ as the equivalent resistance for stability compensation:

$$R_{eq,SEN} = \frac{v_{SEN(AC),pp}}{i_{L,pp}} = \frac{L}{\tau_{SEN}} \tag{4.64}$$

The stability criterion expressed in Eq. (4.14) can be modified to obtain Eqs. (4.65) and (4.66) simultaneously, and thus reduce the practical ESR:

$$\left(R_{ESR} + R_{eq\_SEN}\right) \cdot C_{OUT} > \frac{T_{ON}}{2} \tag{4.65}$$

$$\left(R_{ESR} + \frac{L}{\tau_{SEN}}\right) \cdot C_{OUT} > \frac{T_{ON}}{2} \tag{4.66}$$

Example 4.5: Transient response is affected by different τ_{SEN} values because v_{SEN} and v_{OUT} exhibit opposite trends during transient response

According to the analysis results, another design consideration is the influence of different τ_{SEN} values on the transient response. Tables 4.11 and 4.12 list the basic specifications and several important parameters, respectively. Figure 4.42 shows the waveforms of the transient response if the same R_S of 250 Ω and different C_S of 470 nF, 100 nF, and 1.0 μF are used.

The time constant constituted by different R_S and C_S is defined as $\tau_{SEN,I}$, $\tau_{SEN,II}$, and $\tau_{SEN,III}$, corresponding to cases I, II, and III. Their relationship is shown as:

$$\tau_{SEN,II} < \tau_{SEN,I} < \tau_{SEN,III} \tag{4.67}$$

Table 4.11 Basic specifications used in Example 4.3

V_{IN}	V_{OUT}	V_{REF}	L	R_{DCR}	C_{OUT}	R_{ESR}	L_{ESL}	T_{ON}	f_{SW}
5 V	600 mV	600 mV	4.7 μH	40 mΩ	4.7 μF	0 mΩ	0 nH	180 ns	660 kHz

Table 4.12 Designing parameters and characteristics in three different cases

Case	L	R_{DCR}	R_S	C_S	Characteristic	Transient response
I	4.7 µH	40 mΩ	250 Ω	470 nF	Pole/zero cancellation	Optimum
II				100 nF	Phase lead	Insensitive
III				1.0 µF	Phase delay	Sensitive

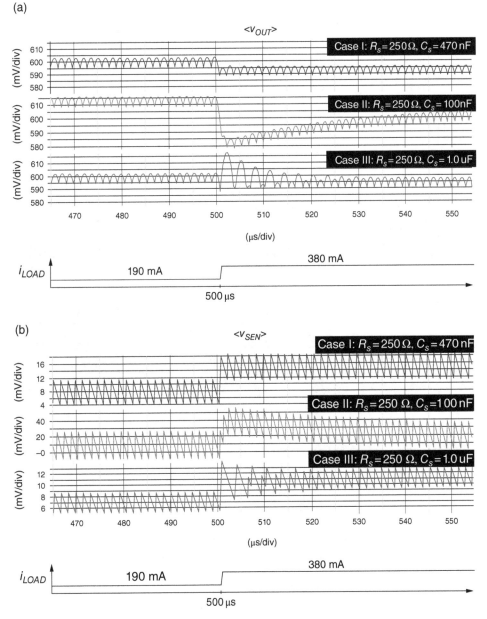

Figure 4.42 Waveforms of (a) v_{OUT}, (b) v_{SEN}, (c) v_{SUM}, and (d) i_L in case of three values of C_S (470 nF, 100 nF, and 1.0 µF) and the same R_S (250 Ω) during load transient response

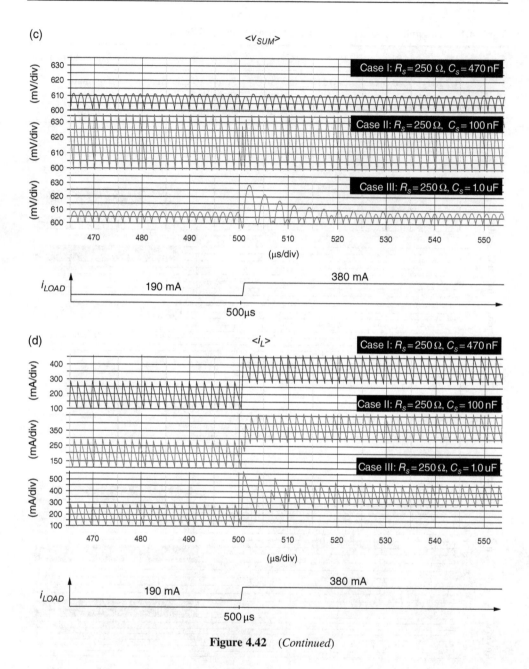

Figure 4.42 (*Continued*)

Figure 4.42(a)–(d) shows the waveforms of v_{OUT}, v_{SEN}, v_{FB}, and i_L, respectively. The waveforms of v_{SEN} reflect the similar waveforms of i_L, whereas the waveforms of v_{FB} reflect the information from v_{OUT} and v_{SEN}. The loading condition changes from 188 to 375 mA at 500 μs. In Figure 4.42(a), the waveforms of v_{OUT} in the three cases present different transient responses. The waveforms in case II exhibit a slow response, resulting in a significant undershoot voltage. The waveforms in case III exhibit a fast response, resulting in an overshoot voltage. We analyze this characteristic by observing the trends of v_{OUT} and v_{SEN} at the load transient response. As previously mentioned, v_{SEN} and v_{OUT} exhibit opposite trends during transient response. The variation of v_{SUM} becomes less evident. Therefore, the transient response is inhibited. The waveforms of v_{SEN} shown in Figure 4.42(b) reveal that different time constants reflect different responses of current sensing and determine the response of v_{OUT} during the transient period. In cases II and III, Eq. (4.54) does not hold. Different values of pole and zero in Eq. (4.53) result in the phase lead and phase delay functions corresponding to cases II and III, respectively. By contrast, the waveforms in case I exhibit an adequate response because the value of $\tau_{SEN,I}$ meets the requirement of Eq. (4.54). Considering the phase lead function, v_{SEN} in case II becomes more sensitive to i_L sensing, as shown in Figure 4.42(d). As a result, although voltage droop occurs at v_{OUT}, v_{SUM} shown in Figure 4.42(c) has a small variation, which slows down the transient response. Similarly, v_{SEN} in case III becomes less sensitive to i_L sensing because of the phase lead function.

Example 4.6: The influence on v_{OUT} ripples and power loss is not the same for different values of R_S and C_S with the same time constant τ_{SEN}

According to the analysis of Example 4.4, we discuss the performance in three cases which have different values of R_S and C_S but with time constant τ_{SEN} the same. Although the time constant is the same, the influence on v_{OUT} ripples and power loss is not the same in each case. Tables 4.13 and 4.14 list the basic specifications and the critical parameters, respectively.

The only different condition is that $R_{ESR} = 10$ mΩ compared with Example 4.4. Figure 4.43 gives three cases I, II, and III, with pairs of [R_S, C_S] equal to [250 Ω, 470 nF], [25 Ω, 4.7 μF], and [2.5 kΩ, 0.47 nF], respectively. Owing to the same τ_{SEN} in the three cases, v_{SEN} reflects the same waveforms in the steady-state period and the transient period. The peak-to-peak value and the DC value of v_{SEN} in the three cases are similar. As a result, v_{OUT} in the three cases has a similar droop and transient response. However, it's worth noticing that the i_{CS} are quite different in these cases, where i_{CS} is the current flow from C_S to v_{OUT}. With a smaller value of R_S and a larger value of C_S in Figure 4.43(b), the peak value of i_{CS}, $i_{CS,peak}$, is 180 mA. Basically, a higher peak value of i_{CS} is not expected because a larger current consumes more conduction loss. Besides, this current appearing during the on-time period can inject/sink the extra current into/from C_{OUT}. For this reason, in addition to the AC component of i_L, i_{CS} also flows through R_{ESR} and C_{OUT}. That's why the waveform of v_{OUT} shown in Figure 4.43(b) suffers from an obvious step at the switching time from on-time period to off-time period and vice versa, compared with Figure 4.43(a), (c).

Table 4.13 Basic specifications used in Example 4.5

v_{IN}	v_{OUT}	V_{REF}	L	R_{DCR}	C_{OUT}	R_{ESR}	L_{ESL}	T_{ON}	f_{SW}
5 V	600 mV	600 mV	4.7 μH	40 mΩ	4.7 μF	10 mΩ	0 nH	180 ns	660 kHz

Table 4.14 Designing parameters in three cases

Case	L	R_{DCR}	R_S	C_S
(a)	4.7 μH	40 mΩ	250 Ω	470 nF
(b)			25 Ω	4.7 μF
(c)			25 kΩ	47 nF

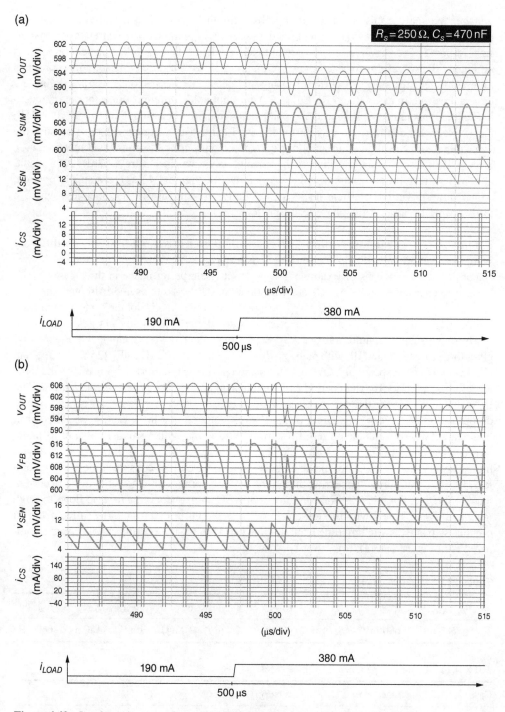

Figure 4.43 Load transient waveforms if (a) $R_S = 250\ \Omega$ and $C_S = 470$ nF, (b) $R_S = 25\ \Omega$ and $C_S = 4.7\ \mu$F, and (c) $R_S = 2.5$ kΩ and $C_S = 47$ nF where external components include $L = 4.7\ \mu$H, $R_{DCR} = 40$ mΩ, $L_{ESL} = 0$ nH, $R_{ESR} = 10$ mΩ, and $C_{OUT} = 4.7\ \mu$F.

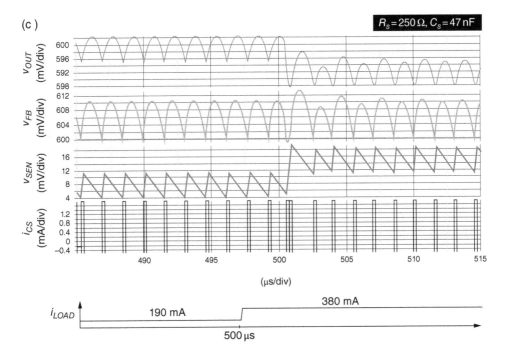

Figure 4.43 (*Continued*)

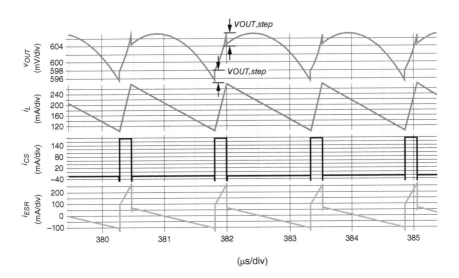

Figure 4.44 Operation waveforms for the case of $R_S = 25\ \Omega$ and $C_S = 4.7\ \mu F$

Figure 4.44 shows more detailed waveforms, including v_{OUT}, i_L, i_{CS}, and i_{ESR}. Here i_{ESR} is the current flowing through R_{ESR}, which can be derived from i_L and i_{CS}:

$$i_{ESR}(t) = i_L(t) - i_{L,avg} + i_{CS}(t) \tag{4.68}$$

The value of the extra step, $v_{OUT,step}$, in the waveform of v_{OUT} is determined by $i_{CS,peak}$ and the value of R_{ESR}, as expressed in Eq. (4.69). $v_{OUT,step}$ is only 1.8 mV, because R_{ESR} is 10 mΩ. However, the extra $v_{OUT,step}$ should be considered carefully because of the high-frequency noise and extra voltage ripple when R_{ESR} is large.

$$v_{OUT,step} = i_{CS,peak} \cdot R_{ESR} \tag{4.69}$$

As a result, we recommend that the designer chooses a larger R_S but a smaller C_S for small conduction power loss, less current density demand, and a small voltage ripple. Furthermore, a capacitance of several pico-farads also benefits the possibility of full integration.

> **Setting in SIMPLIS**
>
> Figure 4.45 illustrates the setup circuit via SIMPLIS corresponding to Figure 4.37. This circuit is composed of a comparator, on-time pulse generator, non-overlap circuit, high-side switch, low-side switch, inductor, parallel RC-filter, output capacitor, and output loading. High-side and low-side switches are parallel with a diode to form an equivalent MOSFET device.
>
> Figure 4.46 provides a zoomed-in view of the partial portion of Figure 4.45. Some critical devices should be defined, and Figure 4.47 illustrates the setting windows for those parameters.
>
> Figure 4.47(a) shows the window to edit the device parameter of the output capacitor, whose value is 4.7 μF. "Level" is set as "2" to activate the value setting of ESR. However, with the parallel RC-filter function of Example 4.4, the value of ESR is 0 mΩ. Figure 4.47(b) shows the window to edit the device parameter of the inductor, whose value is 4.7 μH. "Series Resistance" is R_{DCR}, set as 40 mΩ.

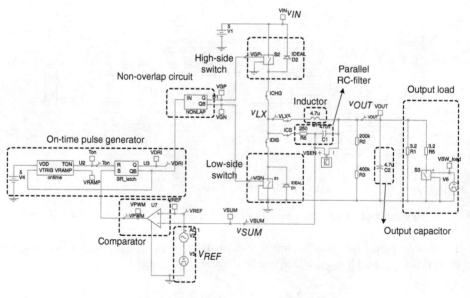

Figure 4.45 Setup circuit in SIMPLIS

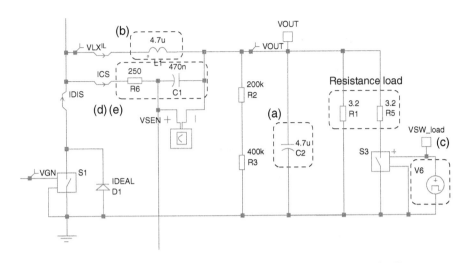

Figure 4.46 Zoom-in views of output filter, RC-filter, and output loading

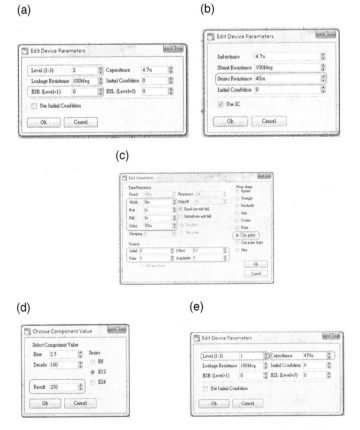

Figure 4.47 Windows for setting device parameters. (a) Output capacitor. (b) Inductor. (c) Output load. (d) Resistance of parallel RC-filter. (e) Capacitance of parallel RC-filter

Figure 4.47(c) shows the window to edit the device parameter of the change in loading conditions. "Wave shape" is selected first as "One pulse," then "Time/Frequency" can be set for the time of load change. In this setting, the loading condition remains at light load until a time of 500 μs, and then changes from light load to heavy load in 2 μs. In turn, the loading condition remains at heavy load until the time is 550 μs, and then changes from heavy load to light load in 2 μs. Besides, the value of the load change is adjusted by the value of "Resistance load," as labeled in Figure 4.24.

Figure 4.47(d), (e) shows the windows to edit the device parameters of resistance and capacitance of the parallel RC-filter, respectively. Different values are set in Examples 4.4 and 4.5.

4.3.2.4 CF-type4 (Parallel RC-Filter with AVP Function)

When Eq. (4.54) holds for pole/zero cancellation, the output exhibits resistive impedance rather than inductive or capacitive impedance. In other words, the converter features an adaptive voltage positioning (AVP) function. Voltage droop exists in different loading conditions. When the load transient changes from light to heavy load, as shown in Figure 4.40, v_{OUT} decreases because the instant load change extracts charges from the output capacitor. The output voltage is regulated at low voltage levels at heavy loading conditions because of the AVP. Then, the converter does not need to recharge the charge to the output capacitor, so the transient period is shortened. Similarly, the output voltage is regulated at high voltage levels at light loading conditions.

The voltage droop is dependent on the DC value of v_{SEN}, which is determined based on several parameters, as expressed in Eq. (4.53). In portable devices, the layout of the PCB and the size of the component are restricted. An inductor with a small size usually has a large DCR value. The bonding wires and conducting path on the PCB should also be considered in estimating the DCR value. The original current-sensing technique shown in Figure 4.37 is unsuitable to achieve adequate droop. Figure 4.49 below shows another RC-filter structure used to sense inductor current information. Using R_{S2} in parallel with C_S, as shown in Figure 4.48, $v_{SEN}(s)$ can be derived as in Eq. (4.70). Compared with Eq. (4.53), the DC gain can be adjusted

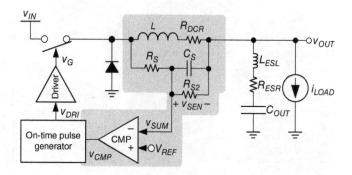

Figure 4.48 Sensing technique with additional parallel R_{S2} for the AVP function

using R_{S2}, as shown in Eq. (4.70). Thus, the DC value of v_{SEN} can be adjusted, as expressed in Eq. (4.71). Consequently, the extra parallel resistor R_{S2} can enhance the flexibility of the AVP function to achieve a fast transient response.

$$v_{SEN}(s) = \left[R_{DCR} \cdot i_L \cdot \left(\frac{R_{S2}}{R_S + R_{S2}} \right) \right] \cdot \frac{1 + s(L/R_{DCR})}{1 + s\left(\frac{R_S \cdot R_{S2}}{R_S + R_{S2}} \right) C_S} \quad (4.70)$$

$$v_{SEN(DC)} = R_{DCR} \cdot i_L \cdot \left(\frac{R_{S2}}{R_S + R_{S2}} \right) \quad (4.71)$$

4.3.2.5 CF-type5 (Parallel RC-Filter with Voltage Droop Cancellation)

In several applications, the v_{OUT} of the converter is required for good load regulation. Specifically, voltage droop is not expected at different loading conditions. In the aforementioned current-sensing technique with voltage droop, the voltage droop is caused by varying DC values of current-sensing information v_{SEN}, which is linear in i_L. Removing the DC component from the current-sensing information v_{SEN} is necessary to eliminate the effect of voltage droop.

As shown in Figure 4.49, a decoupling capacitor C_F is added directly to conduct the AC component of v_{SEN} but blocks the DC component of v_{SEN}. Moreover, the DC information from v_{OUT} is conducted using the feedback dividers R_1 and R_2. The output voltage can be regulated without voltage droop with the AC component of i_L and the DC component of v_{OUT}, respectively.

Although the extra capacitor C_F is added through the path of v_{SEN} to extract $v_{SEN(AC)}$, the voltage droop is removed but the offset voltage still remains, as expressed in Eq. (4.72) in comparison with Eq. (4.59), where v_{offset} in Eq. (4.73) represents the offset voltage:

$$v_{OUT(DC)} \approx \frac{1}{\beta} V_{REF} + v_{offset} \quad (4.72)$$

$$v_{offset} = \frac{1}{\beta} \cdot \left(\frac{1}{2} v_{SEN(AC),pp} \right) + \frac{1}{2} v_{OUT(AC),pp} \quad (4.73)$$

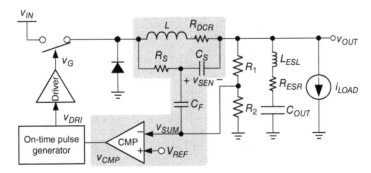

Figure 4.49 Current-sensing technique with decoupling capacitor C_F

When MLCC is used for a small ripple, the small value of $v_{OUT(AC),pp}$ can be ignored to simplify Eqs. (4.73) and (4.74):

$$v_{offset} = \frac{1}{\beta} \cdot \left(\frac{1}{2} v_{SEN(AC),pp}\right) \tag{4.74}$$

Figure 4.50 shows that the offset voltage is caused by $v_{SEN(AC),ripple}$. Figure 4.51 shows that the transient response is not improved, which is similar to the result shown in Figure 4.39 without extra C_F, because $v_{SEN(AC)}$ also exhibits an opposite trend to the variation of v_{FB} during transient response.

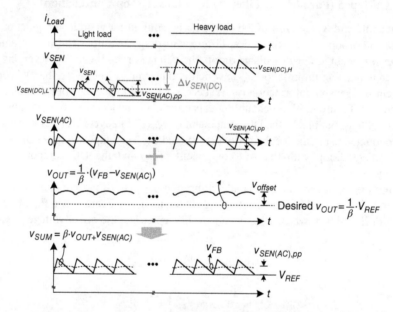

Figure 4.50 V_{OUT} has a voltage offset at different loading conditions

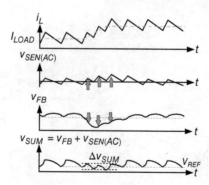

Figure 4.51 Deterioration of the load transient response because of the additional CF path

However, the direct use of C_F has poor capability to be a good HPF for conducting only the AC component of v_{SEN}. The other resistances and capacitances, such as R_1, R_2, R_S, C_S, and C_F, can influence the HPF bandwidth. In the aspect of a large signal, these resistances and capacitances also influence the driving capability of the AC component through C_F and the DC component through the feedback divider. The transfer function of this structure is complex. In other words, determining the weight of the AC component of i_L and the DC component of v_{OUT} for v_{FB} is difficult. Suitable stability and transient response are difficult to achieve at the same time.

Figure 4.52 shows the structure of another RC-filter used to achieve the same function. Two pairs of RC-filters are utilized. The first RC-filter is composed of R_S and C_S placed parallel with the inductor. The second RC-filter is composed of R_{S2} and C_{S2} placed parallel with C_S. Initially, these two RC-filters are assumed to be independent of each other to understand the function of this current-sensing technique easily. Based on Eq. (4.53), v_{S1} shown in Figure 4.52 includes information on the AC and DC components of i_L. v_{S1} is approximately expressed as follows:

$$v_{S1}(s) = \left[R_{DCR} \cdot i_L \cdot \frac{1 + s(L/R_{DCR})}{1 + sR_S C_S} \right] + v_{OUT} \qquad (4.75)$$

By intuition, the second RC-filter can be viewed as the LPF used to filter the AC component of v_{S1} so that v_{S2} includes only the DC component of v_{S1}. As a result, the difference between v_{S1} and v_{S2}, $v_{S,diff}$, represents the AC component of i_L, which is fed into the comparator. $v_{S,diff}$ is expressed as follows:

$$v_{S,diff} = v_{S1} - v_{S2} \qquad (4.76)$$

By contrast, v_{FB} and V_{REF} are also fed into the same comparator. Therefore, v_{SUM} is expressed as:

$$v_{SUM} = v_{FB} + v_{S,diff} \qquad (4.77)$$

Thus, the output signal of the comparator represents the response of the AC component of i_L and the DC component of v_{OUT}. Table 4.15 lists the signal and the corresponding components.

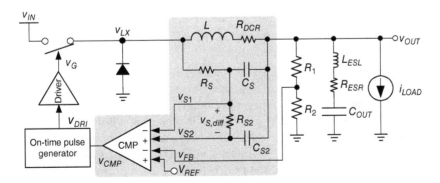

Figure 4.52 Current-sensing technique with two pairs of RC-filters

Table 4.15 Signal and corresponding components

Signal	Component
v_{S1}	DC + AC
v_{S2}	DC
$v_{S,diff}$	AC

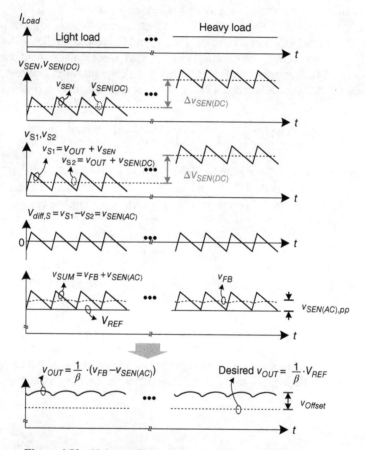

Figure 4.53 Voltage offset of v_{OUT} at different loading conditions

Similar to the result shown in Figure 4.50, the structure shown in Figure 4.52 can remove the voltage droop at the DC component of v_{SUM} at different loading conditions so that the voltage droop at V_{OUT} is removed. The operating waveform is shown in Figure 4.53.

The valley of v_{SUM} is regulated at v_{REF} because of the comparator in the feedback loop. However, the valley of v_{SUM} is not equal to the valley of v_{FB} because the average value of v_{SUM} is zero. The ripple of $v_{S,diff}$ generates a DC offset at v_{SUM}, ($V_{offset,vSUM}$), between the averages of v_{SUM} and v_{REF}, where v_{FB} is derived from Eq. (4.77) as:

$$v_{FB} = v_{SUM} + v_{S,diff} \tag{4.78}$$

$v_{offset,vSUM}$ is calculated as follows:

$$v_{offset,vSUM} = \frac{1}{2} v_{SUM,pp} \tag{4.79}$$

The DC offset of v_{OUT} can then be calculated as:

$$v_{offset,vOUT} = \frac{R_1 + R_2}{R_2} \cdot \frac{1}{2} v_{SUM,pp} \tag{4.80}$$

The DC offset of v_{OUT} is unexpected because the accuracy of v_{OUT} has deteriorated.

Furthermore, Figure 4.54 shows that an extra voltage-controlled voltage source (VCVS) linear amplifier is inserted between $v_{S,diff}$ and the comparator to control the weight of the AC component of i_L and the DC component of v_{OUT}. The gain of this VCVS linear amplifier is defined as G. Therefore, v_{SUM} is expressed as:

$$v_{SUM} = v_{FB} + G \cdot v_{S,diff} \tag{4.81}$$

Equation (4.82) can also be re-derived from Eq. (4.66):

$$\left(R_{ESR} + G \cdot \frac{L}{R_S C_S} \right) \cdot C_{OUT} > \frac{T_{ON}}{2} \tag{4.82}$$

However, these two RC-filters can influence their function in actual cases. The subsequent analyses determine the suitable parameters of these resistances and capacitances to avoid analyzing and using the complex transfer function from i_L to $v_{S,diff}$ of these two RC-filters.

First, we review the structure of the current-sensing technique with only one RC-filter, as shown in Figure 4.55. Based on the values of pole and zero in Eq. (4.75), Eq. (4.83) is utilized for pole/zero cancellation to achieve a fast transfer function:

$$L/R_{DCR} = R_S C_S \tag{4.83}$$

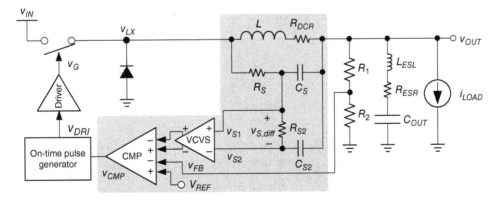

Figure 4.54 Current-sensing technique with two pairs of RC-filters and VCVS linear amplifiers

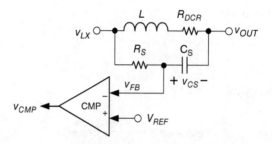

Figure 4.55 Structure of the current-sensing technique with one RC-filter

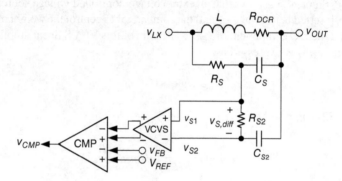

Figure 4.56 Structure of the current-sensing technique with two pairs of RC-filters

Figure 4.56 shows the structure of the current-sensing technique with two pairs of RC-filters. Given the transient response, R_S and C_S are also derived using Eq. (4.83) based on previous experience.

However, the values of R_{S2} and C_{S2} influence the aberration of pole/zero cancellation and the transient response of v_{S1} that generates the phase lead or phase delay function. Therefore, a significantly large aberration can inhibit the load transient response. Moreover, significantly large values of R_{S2} and C_{S2} further slow down the transient response of v_{S2} and inhibit the load transient response. As a result, the criteria expressed in Eqs. (4.84) and (4.85) are approximately set as follows:

$$R_{S2}C_{S2} < \frac{1}{10}R_S C_S \tag{4.84}$$

$$C_{S2} < \frac{1}{10}C_S \tag{4.85}$$

The second RC-filter, composed of R_{S2} and C_{S2}, is used to derive the function of the LPF. Considering the suitable bandwidth of the LPF, the inequality expressed in Eq. (4.86) is set as follows:

$$\frac{1}{2\pi R_{S2}C_{S2}} < \frac{f_{SW}}{4} \tag{4.86}$$

By rearranging, Eq. (4.87) can be derived from Eq. (4.86):

$$R_{S2}C_{S2} > \frac{2}{\pi} \cdot \frac{1}{f_{SW}} \quad (4.87)$$

The combination of Eqs. (4.82)–(4.86) results in Eqs. (4.88)–(4.90), which are employed to obtain the values of R_S, C_S, R_{S2}, and C_{S2}.

$$\begin{cases} (R_{ESR} + G \cdot \dfrac{L}{R_S C_S}) \cdot C_{OUT} > \dfrac{T_{ON}}{2} & (4.88) \\[6pt] \dfrac{2}{\pi f_{SW}} < R_{S2} C_{S2} < \dfrac{1}{10}(R_S C_S) = \dfrac{1}{10}(L/R_{DCR}) & (4.89) \\[6pt] C_{S2} < \dfrac{1}{10} C_S & (4.90) \end{cases}$$

Example 4.7: Simulation results provide four cases for comparison to validate Eq. (4.89) This simulation provides four cases for comparison to validate Eq. (4.89). Table 4.16 lists the basic specifications, and Table 4.17 lists the critical parameters. In this simulation, $R_1 = 0$ kΩ, $R_2 = 400$ kΩ, and v_{OUT} is equal to v_{FB}. In the four cases, different values of R_{S2} and C_{S2} are obtained, whereas R_S and C_S are constant and provide perfect pole/zero cancellation to L and R_{DCR}. The result reveals that the different time constants result in different capabilities of the LPF and different levels of distortion of $v_{S,diff}$.

Figure 4.57 shows the waveform with transient response for the four cases. In case (a), R_{S2} and C_{S2} are absent and the structure is similar to that of type4. The V_{OUT} waveform suffers from voltage droop at different loading conditions because v_{SEN} contains the AC and DC components of i_L. In case (b), R_{S2} is 2 kΩ and C_{S2} is 1 nF based on the criterion expressed in Eq. (4.89). Therefore, $v_{S,diff}$ exhibits a good result with only the AC component of i_L, and v_{OUT} exhibits good load regulation without extra voltage droop at different loading conditions. In

Table 4.16 Basic specifications used in Example 4.7

v_{IN}	v_{OUT}	V_{REF}	L	R_{DCR}	C_{OUT}	R_{ESR}	L_{ESL}	T_{ON}	f_{SW}
5 V	600 mV	600 mV	4.7 μH	100 mΩ	4.7 μF	0 mΩ	0 nH	180 ns	660 kHz

Table 4.17 Design parameters and characteristics of the four cases

Case	R_{DCR}	R_S	C_S	R_{S2} (kΩ)	C_{S2} (nF)	Characteristic
(a)	100 mΩ	100 Ω	470 nF	—	—	Has voltage droop
(b)				2	1	No voltage droop with good transient response
(c)				20	1	No voltage droop but poor transient response
(d)				0.2	1	Unstable because $v_{S,diff}$ has no AC component of i_L

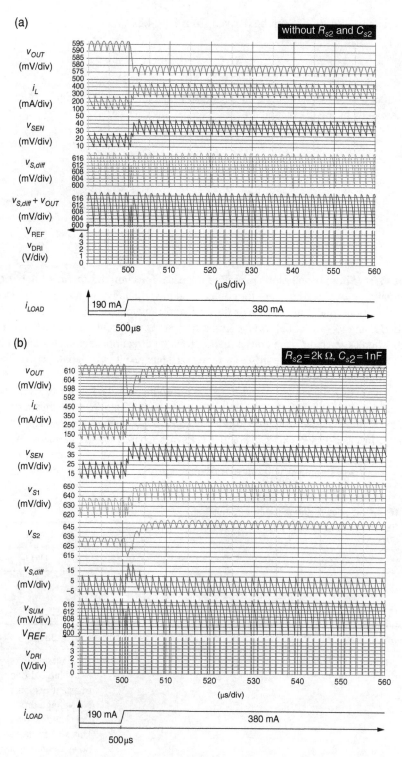

Figure 4.57 Load transient response (a) without R_{S2} and C_{S2}, (b) with $R_{S2} = 2$ kΩ and $C_{S2} = 1$ nF, (c) with $R_{S2} = 20$ kΩ and $C_{S2} = 1$ nF, and (d) with $R_{S2} = 0.2$ kΩ and $C_{S2} = 1$ nF

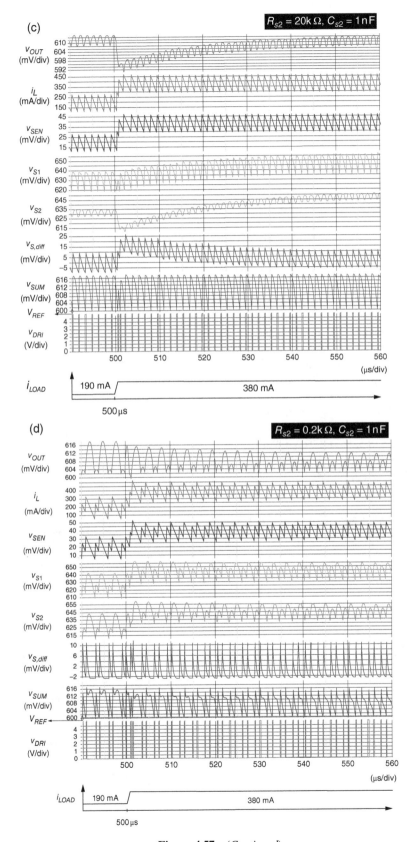

Figure 4.57 (*Continued*)

case (c), R_{S2} is 20 kΩ and C_{S2} is 1 nF. The condition of Eq. (4.89) holds, but the bandwidth of the values of R_{S2} and C_{S2} becomes lower than that in case (b). As a result, the transient response of v_{S2} becomes slower and still functions at steady state, although the voltage droop is removed. As a result, when the second inequality of Eq. (4.89) does not hold, the transient response becomes slower. In case (d), R_{S2} is 0.2 kΩ and C_{S2} is 1 nF. The first inequality of Eq. (4.89) does not hold, which is relative to the switching frequency. The RC-filter of R_{S2} and C_{S2} has no ability to filter out the AC component of i_L from v_{S2}. In other words, the AC component of i_L is almost removed from $v_{S,diff}$ because v_{S1} and v_{S2} remain the AC component of i_L. As a result, the lack of current information on feedback degrades the stability when the output capacitor with low ESR is used.

Example 4.8: This example demonstrates Eq. (4.90) to discuss the influence of different values of R_{S2} and C_{S2}

Table 4.18 lists the basic specifications. Here, $R_1 = 0$ kΩ and $R_2 = 400$ kΩ, v_{OUT} is equal to v_{FB}. Table 4.19 lists the parameters for seven cases, and Figure 4.58 shows the relative waveforms.

In case (a), R_{S2} and C_{S2} are absent, and the structure is similar to that of type4. The v_{OUT} waveform suffers from voltage droop at different loading conditions because v_{CS} contains both the AC and the DC component of i_L. In other cases, the time constants of R_{S2} and C_{S2} are the same and Eq. (4.89) holds. Cases (b), (c), and (e) perform well in terms of load regulation and transient response. However, the value of C_{S2} is 10 μF in case (d). Equation (4.90) does not hold, and the operation is unstable. In contrast, the values of R_{S2} and C_{S2} in case (f) satisfy Eqs. (4.89) and (4.90), but voltage droop occurs. In consideration of the too-large values of R_{S2} and the too-small values of C_{S2}, the leakage current of C_{S2} should be of concern. With a several mini-volt ripple of v_{SEN} across R_{S2} and C_{S2}, the current flowing through R_{S2} to v_{S2} would only be several nano-amperes. To emphasize the importance of the leakage current, the size of the leakage impedance is exaggerated to 100 MΩ in this case. Figure 4.59 shows the setting window of SIMPLIS to edit the device parameter of C_{S2}. Compared with R_{S2} of 20 MΩ and leakage

Table 4.18 Basic specifications used in Example 4.8

V_{IN}	V_{OUT}	V_{REF}	L	R_{DCR}	C_{OUT}	R_{ESR}	L_{ESL}	T_{ON}	f_{SW}
5 V	600 mV	600 mV	4.7 μH	100 mΩ	4.7 μF	0 mΩ	0 nH	180 ns	660 kHz

Table 4.19 The designing parameters and characteristics for different cases

Case	R_{DCR}	R_S	C_S	R_{S2}	C_{S2}	Characteristic
(a)	100 mΩ	100 Ω	470 nF	—	—	Voltage droop exists
(b)				2 kΩ	1 nF	Voltage droop is removed
(c)				2 Ω	1 μF	Voltage droop is removed
(d)				0.2 Ω	10 μF	Unstable operation because of violation of Eq. (4.90)
(e)				2 MΩ	1 pF	Voltage droop is removed
(f)				20 MΩ	0.1 pF	The function of R_{S2} and C_{S2} is degraded by leakage current

impedance of 100 MΩ, the leakage current is 1/5 of the current flowing through R_{S2}. In other words, the leakage current is 1/4 of the current flowing through C_{S2}. This ratio is so large that the filtering function of R_{S2} and C_{S2} is degraded. Therefore, v_{S2} extracts no DC component of v_{S1} and the droop of v_{OUT} is not removed. Last but not least, the droop of v_{S1} is caused by the droop of v_{OUT} through the conduction of C_{S2}. As a result, once the value of R_{S2} is increased for the decrease in value of C_{S2}, the leakage current should be carefully taken into consideration.

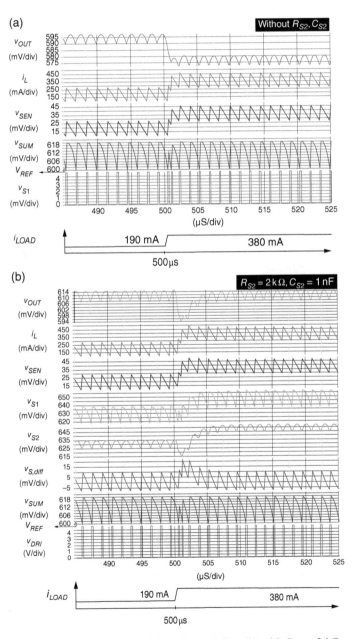

Figure 4.58 Load transient response (a) without R_{S2} and C_{S2}, (b) with $R_{S2} = 2$ kΩ and $C_{S2} = 1$ nF, (c) with $R_{S2} = 2$ Ω and $C_{S2} = 1$ μF, (d) with $R_{S2} = 0.2$ Ω and $C_{S2} = 10$ μF, (e) with $R_{S2} = 2$ MΩ and $C_{S2} = 1$ pF, (f) with $R_{S2} = 20$ MΩ and $C_{S2} = 0.1$ pF

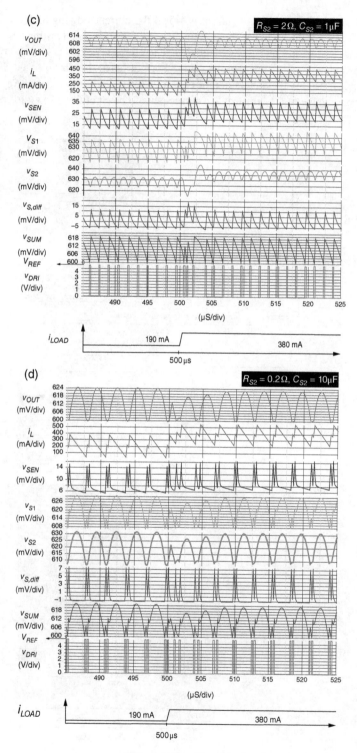

Figure 4.58 (*Continued*)

Ripple-Based Control Technique Part I

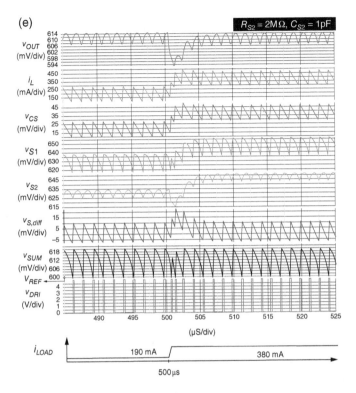

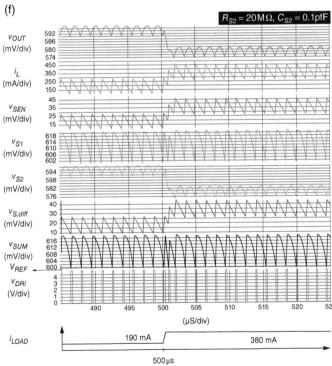

Figure 4.58 (*Continued*)

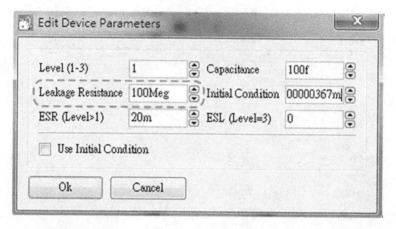

Figure 4.59 SIMPLIS window to edit the device parameters of C_{S2}

SIMPLIS Setting

Figure 4.60 illustrates the setup circuit via SIMPLIS corresponding to Figure 4.52.

This circuit is composed of a comparator, on-time pulse generator, non-overlap circuit, high-side switch, low-side switch, inductor, parallel RC-filter, output capacitor, and output load. The high-side switch and low-side switch are parallel with a diode to form an equivalent MOSFET device.

Figure 4.61 provides a zoomed-in view of the partial portion of Figure 4.60. Critical devices should be set, and Figure 4.62 illustrates the setting windows for these parameters.

Figure 4.62(a) is a window to edit the device parameter of the output capacitor, whose value is 4.7 µF. "Level" is set as "2" to activate the value setting of ESR. However, to demonstrate the parallel RC-filter function of Examples 4.7 and 4.8, the value of ESR is 0 mΩ.

Figure 4.62(b) is a window to edit the device parameter of the inductor, whose value is 4.7 µH. "Series Resistance" is R_{DCR}, set as 100 mΩ.

Figure 4.62(c) is a window to edit the device parameter of the change in loading conditions. "Wave shape" is selected first as "One pulse," then "Time/Frequency" can be set for the time of load change. In this setting, the loading condition remains at light load until the time is 500 µs, and then changes from light to heavy load in 2 µs. In turn, the loading condition remains at heavy load until the time is 550 µs, and then changes from heavy to light load in 2 µs. Besides, the value of the load change is set by adjusting the value of "Resistance load" in Figure 4.61.

Figure 4.62(d) is a window to edit the device parameters of three inputs summing the circuit. According to Eq. (4.77), the gain for v_{S1} and v_{S2} is 1 and −1, respectively, for extra $v_{S,diff}$. According to Eq. (4.77), the gain for v_{OUT} is 1. Besides, "Input Resistance" is set as 1G to avoid the load effect.

Block (e) in Figure 4.61 includes the resistances and capacitances of a parallel RC-filter. Different values are set in Examples 4.7 and 4.8.

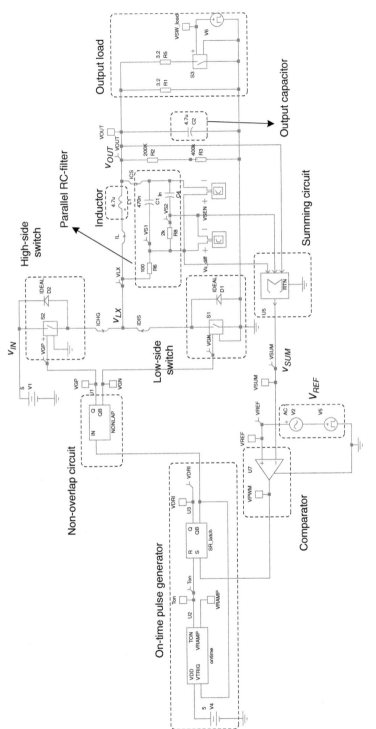

Figure 4.60 Setup circuit via SIMPLIS

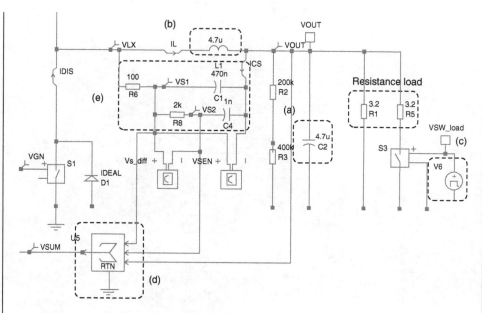

Figure 4.61 Zoom-in views of output filter, RC-filters, summing circuit, and loading

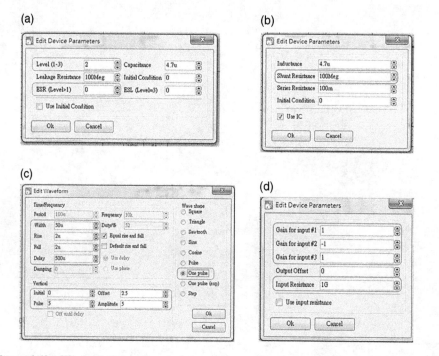

Figure 4.62 Windows for setting device parameters. (a) Output capacitor. (b) Inductor. (c) Output load. (d) Summing circuit

Example 4.9: This example shows the improvement of transient response in case of using VCVS if we change the values of R_S and C_S

This example discusses Eq. (4.82) for use of the VCVS in Figure 4.54. When different inductors are used, the values of R_S and C_S are designed according to the value of R_{DCR} as expressed in Eq. (4.89) for good transient response. In this example we have seven cases; the same value of C_S and different values of R_S to match the different values of R_{DCR}. According to the criterion of Eq. (4.88) for stability, the increase in G is designed for smaller values of R_S and R_{DCR}. Table 4.20 lists the basic specifications. Here, $R_1 = 0$ kΩ and $R_2 = 400$ kΩ, v_{OUT} is equal to v_{FB}. Table 4.21 lists the parameters for seven cases in which the values of R_{DCR}, R_S, and G are different. Based on the stable operation in case (a), the waveforms of the other cases are observed in Figure 4.63. By comparing cases (a) and (b), the smaller value of R_S leads to a lower $v_{SEN(AC),pp}$. As a result, v_{SUM} is composed of less information on $v_{SEN(AC)}$, although the transient response is improved. In case (c), the much smaller value of R_S (20 mΩ) is designed for R_{DCR} of 500 Ω. The $v_{SEN(AC),pp}$ in case (c) can be expected to become smaller than

Table 4.20 Basic specifications used in Example 4.9

v_{IN}	v_{OUT}	V_{REF}	L	R_{DCR}	C_{OUT}	R_{ESR}	L_{ESL}	T_{ON}	f_{SW}
5 V	600 mV	600 mV	4.7 μH	100 mΩ	4.7 μF	0 mΩ	0 nH	180 ns	660 kHz

Table 4.21 The designing parameters and characteristics for different cases

Case	R_{DCR} (mΩ)	R_S	C_S	R_{S2}	C_{S2}	G	$v_{SEN(AC),pp}$ (mV)	Characteristic
(a)	100	100 Ω	470 nF	2 kΩ	1 nF	1	17	
(b)	40	250 Ω				1	6.8	• Smaller $v_{SEN(AC)}$ • Faster transient response
(c)	20	500 Ω				1	3.4	• Smaller $v_{SEN(AC)}$ • Fast transient response • Worse aberration between v_{SUM} and i_L
(d)	20	500 Ω				5	3.4	• Similar to (a) because of same $G \cdot v_{S,diff}$
(e)	10	1 kΩ				1	1.7	• Serious aberration between v_{SUM} and i_L • Unstable because of too small value of $v_{SEN(AC)}$ relative to ripple of V_{FB}
(f)	10	1 kΩ				2	1.7	• Similar to (c) because of same $G \cdot v_{S,diff}$
(g)	10	1 kΩ				4	1.7	• Similar to (b) because of same $G \cdot v_{S,diff}$
(h)	10	1 kΩ				10	1.7	• Similar to (a) because of same $G \cdot v_{S,diff}$

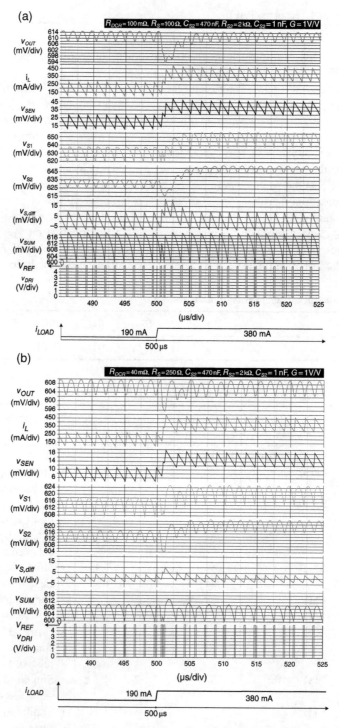

Figure 4.63 Load transient response with different $v_{SEN(AC),pp}$ and $G \cdot v_{S,diff}$ owing to different R_{DCR}, R_S, and G. (a) $G = 1$, $R_{DCR} = 100$ mΩ, and $R_S = 100$ Ω. (b) Faster transient response caused by smaller $v_{SEN(AC)}$. (c) Worse aberration between v_{SUM} and i_L. (d) Similar to (a) because of same $G \cdot v_{S,diff}$. (e) Unstable because of too small value of $v_{SEN(AC)}$ relative to ripple of V_{FB}. (f) Similar to (c) because of same $G \cdot v_{S,diff}$. (g) Similar to (b) because of same $G \cdot v_{S,diff}$. (h) Similar to (a) because of same $G \cdot v_{S,diff}$

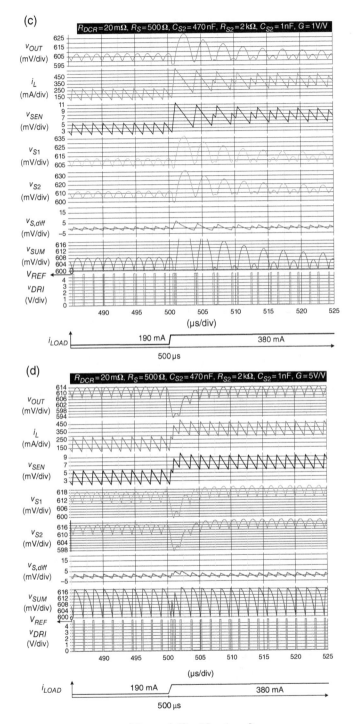

Figure 4.63 (*Continued*)

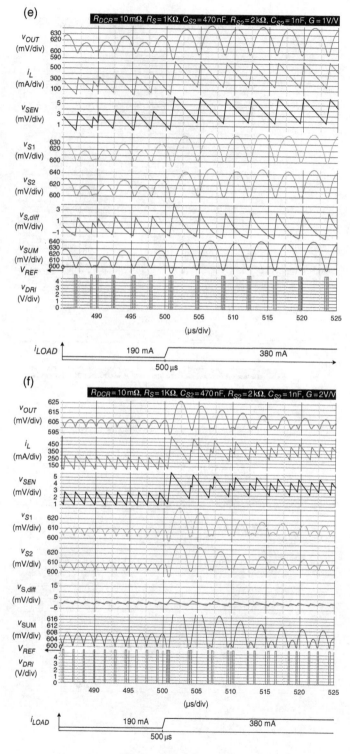

Figure 4.63 (*Continued*)

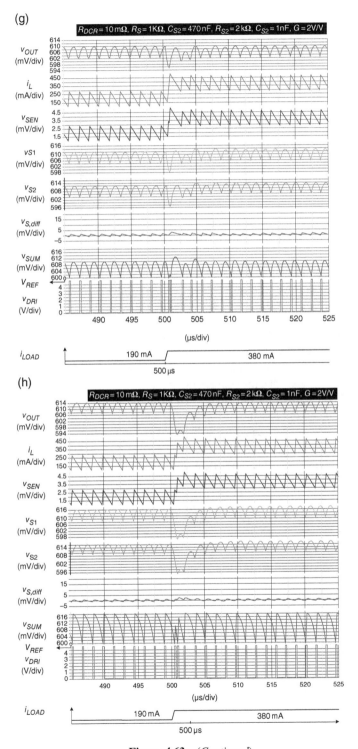

Figure 4.63 (*Continued*)

that in case (b). Figure 4.63(c) shows unstable operation because of an insufficient amount of information on $v_{SEN(AC),pp}$. In other words, the criterion in Eq. (4.88) does not hold. To stabilize the operation, the value of G is increased from 1 to 5 V/V to amplify the value of $v_{S,diff}$ in case (d). Comparing (a) and (d), $v_{SEN(AC),pp}$ shows a 1/5 change with a 1/5 change in the value of R_S. With the five-time change in G, the ripples of $v_{S,diff}$ in case (a) and (d) are almost similar. Consequently, both cases perform almost similarly in terms of stable operation and transient response. The remaining cases can be analyzed similarly. The operations in cases (e) and (f) are also unstable according to the criterion. Even though the values of $v_{SEN(AC),pp}$ are different in different cases, the final value of $v_{S,diff}$ determines whether the operation is stable or not. Consequently, cases (a), (d), and (h) perform similarly. Cases (b) and (g) also perform similarly, while cases (c) and (f) perform similarly.

Example 4.10: **The transient response is influenced by opposite trends among v_{OUT} and i_L**

As mentioned previously, the variations in v_{OUT} and i_L at different loading conditions exhibit opposite trends. Therefore, the transient response is inhibited. This simulation reveals that the transient response is influenced by the different values of G, resulting in a gain in amplifier $v_{S,diff}$ because $v_{S,diff}$ represents the AC component of i_L. Table 4.22 lists the basic specifications. In this simulation, $R_1 = 0$ kΩ, $R_2 = 400$ kΩ, and v_{OUT} is equal to v_{FB}.

Table 4.23 lists the parameters of three cases, and Figure 4.64 shows the relative waveforms.

Figure 4.64(a)–(c) shows that $v_{SEN(AC),pp}$ has the same value in all cases because of the same values of R_S and C_S. The values of G in all cases are 5, 10, and 5. The ripple of v_{SUM} has values of 12, 36, and 90 mV. The larger the ripple of $G \cdot v_{S,diff}$, the worse the transient response. The worst case (case (c)) shows that the converter responds to the changing load until the ratio of the variation in v_{OUT} over the ripple of $G \cdot v_{S,diff}$ is sufficiently large. Moreover, the on-time and off-time periods are not extended during light to heavy and heavy to light loads, respectively, because that ratio is not evident. In conclusion, although the extra gain in G can enhance the stability, the ratio of the tolerated variation in v_{OUT} of the ripple of $G \cdot v_{S,diff}$ should be carefully designed to obtain a suitable transient response.

Table 4.22 Basic specifications used in Example 4.10

v_{IN}	v_{OUT}	V_{REF}	L	R_{DCR}	C_{OUT}	R_{ESR}	L_{ESL}	T_{ON}	f_{SW}
5 V	600 mV	600 mV	4.7 μH	100 mΩ	4.7 μF	0 mΩ	0 nH	180 ns	660 kHz

Table 4.23 The designing parameters and characteristics

	R_{DCR}	R_S	C_S	R_{S2}	C_{S2}	G	$v_{SEN(AC),pp}$	$G \cdot v_{S,diff}$ (mV)
(a)	10 mΩ	1 kΩ	470 nF	2 kΩ	1 nF	5	1.7 mV	12
(b)						10		36
(c)						50		90

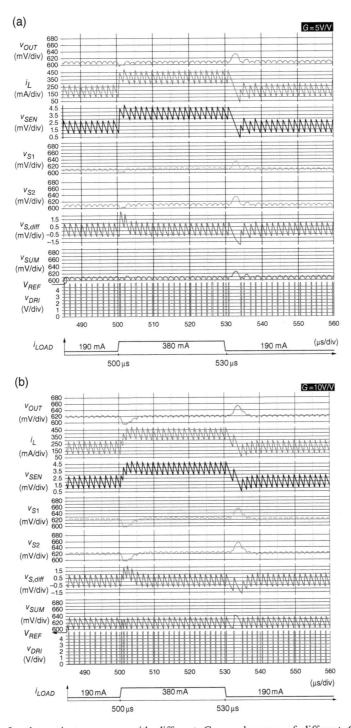

Figure 4.64 Load transient response with different $G \cdot v_{S,diff}$ because of different G. (a) $G = 5$. (b) $G = 10$. (c) $G = 50$

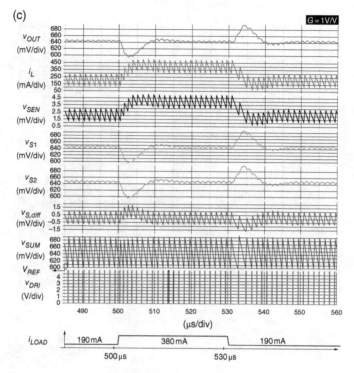

Figure 4.64 (*Continued*)

SIMPLIS Setting

Figure 4.65 illustrates the setup circuit via SIMPLIS corresponding to Figure 4.54. This circuit is composed of a comparator, on-time pulse generator, non-overlap circuit, high-side switch, low-side switch, inductor, parallel RC-filter, output capacitor, and output load. The high-side switch and low-side switch are parallel with a diode to form an equivalent MOSFET device.

Figure 4.66 provides a zoomed-in view of a partial portion of Figure 4.65. There are critical devices to be set, and Figure 4.67 illustrates the setting windows for these parameters.

Figure 4.67(a) is a window to edit the device parameter of the output capacitor, whose value is 4.7 μF. "Level" is set as "2" to activate the value setting of ESR. To demonstrate the parallel RC-filter function of Examples 4.9 and 4.10, the value of ESR is 0 mΩ.

Figure 4.67(b) is a window to edit the device parameter of the inductor, whose value is 4.7 μH. "Series Resistance" is R_{DCR}, set as 100 mΩ.

Figure 4.67(c) is a window to edit the device parameter of the change in loading conditions. "Wave shape" is first selected as "One pulse," then "Time/Frequency" can be set for the time of load change. In this setting, the loading condition remains at light load until the time is 500 μs, and then changes from light load to heavy load in 2 μs. In turn, the loading condition remains at heavy load until the time is 530 μs, and then changes from heavy load to light load in 2 μs. Besides, the value of the load change is set by adjusting the value of "Resistance load" as labeled in Figure 4.66.

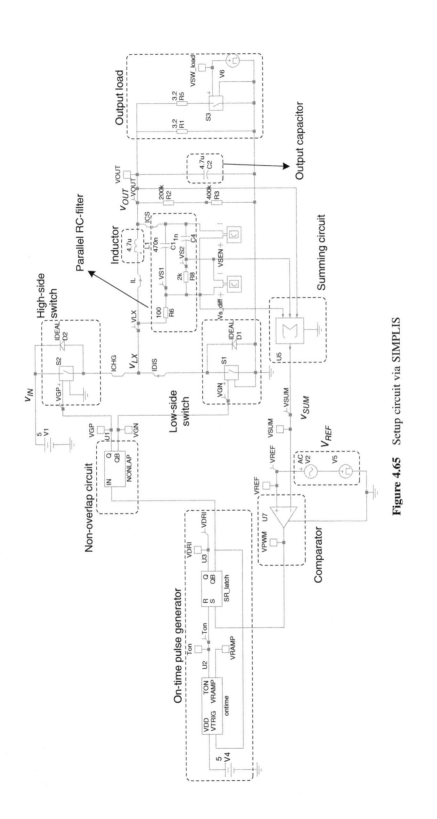

Figure 4.65 Setup circuit via SIMPLIS

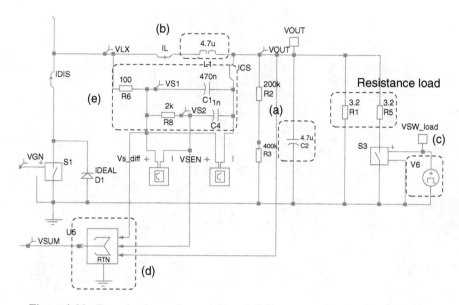

Figure 4.66 Zoom-in views of output filter, RC-filters, summing circuit, and loading

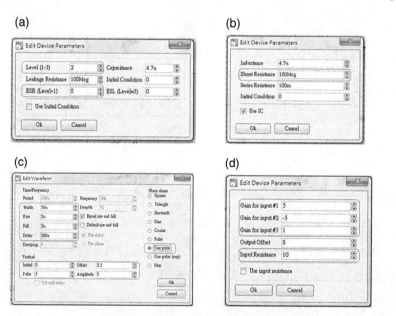

Figure 4.67 Windows for setting device parameters. (a) Output capacitor. (b) Inductor. (c) Output load. (d) Summing circuit

Figure 4.67(d) shows the window to edit the device parameter of a three-input summing circuit. According to Eqs. (4.76) and (4.81), the extra gain of G from VCVS indicates that the gains for v_{S1} and v_{S2} are set to have opposite sign and the same value. Taking $G = 5$, for example, the gains for v_{S1} and v_{S2} are 5 and -5, respectively. Different values are set in

Examples 4.9 and 4.10. According to Eq. (4.77), the gain for v_{OUT} is 1. Besides, "Input Resistance" is set as 1G to avoid the load effect.

Block (e) in Figure 4.61 includes the resistances and capacitances of a parallel RC-filter. Different values are set in Examples 4.9 and 4.10.

4.3.2.6 CF-type6 (Parallel RC-Filter with Accurate Regulation)

The CF-type6 is introduced in this subsection; it utilizes an extra CF path to ensure stability and achieve accurate regulation simultaneously. Figure 4.68 shows the structure of the CF-type6 modified from the structure shown in Figure 4.54. An additional valley detector is used. Figure 4.69 shows the operating waveform of the converter with CF-type6. The first RC-filter composed of R_S and C_S generates v_{S1}, which contains information from v_{OUT} and i_L. The valley detector and second RC-filter process V_{S1} to extract AC information from i_L, that is, remove the DC information from v_{OUT} and i_L. The valley sample-and-holder (S&H) uses v_{S1} to generate v_{VA}, which is equal to the valley value of v_{S1}. The second RC-filter, composed of R_{S2} and C_{S2}, processes v_{VA} to generate v_{S2}, which can also represent the valley value of v_{S1}. Afterward, the high-frequency noise can be suppressed. When v_{S1} and v_{S2} are fed into the VCVS, the differential output voltage of VCVS, that is, $v_{S,diff2}$, is determined, as expressed in Eq. (4.91). $v_{S,diff}$ can represent information from the AC component of i_L.

$$v_{S,diff2} = G \cdot v_{S,diff} = G \cdot (v_{S1} - v_{S2}) \tag{4.91}$$

In the comparison of $v_{S,diff}$ of CF-type5 shown in Figure 4.53 and $v_{S,diff}$ of CF-type6 shown in Figure 4.68, the average $v_{S,diff}$ of CF-type6 is not zero, whereas the valley of $v_{S,diff}$ of CF-type6 is zero. In other words, for CF-type6, Figure 4.68 shows that the valleys of v_{FB} and v_{SUM} are equal. No voltage offset between the valleys of v_{FB} and V_{REF} is observed because the valley of v_{SUM} is regulated at V_{REF} through the comparator in the feedback loop. In the comparison of

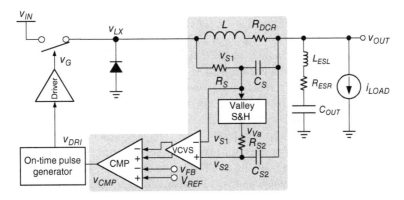

Figure 4.68 Current-sensing technique with two pairs of RC-filters, valley S&H, and VCVS linear amplifier

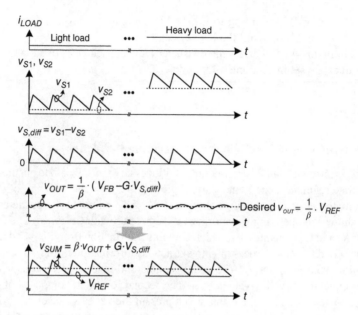

Figure 4.69 Operating waveform showing no voltage droop and offset voltage at v_{OUT}

v_{OUT} of CF-type5 shown in Figure 4.53 and v_{OUT} of CF-type6 shown in Figure 4.68, the voltage droop can be removed by the extra valley S&H in CF-type6.

4.3.3 Comparison of On-Time Control with an Additional Current Feedback Path

Table 4.24 lists the design issues of various types of ripple-based converters with additional CF. The voltage droop, offset voltage, and characteristics of CF-type1 to CF-type6 are compared. For the simplest method, CF-type1 can be used. However, R_S in series with the power delivery path aggravates power loss. CF-type2 is also easy to design, but the integral function of R_{LPF} and C_{LPF} slows down the transient response. CF-type3, which uses R_S and C_S in parallel with the inductor, can prevent aggravation of the transient response. The values of R_S and C_S are designed by considering pole/zero cancellation for optimum transient response. Based on the structure of CF-type3, an additional R_{S2} is added to CF-type4 to improve the flexible AVP function. The AVP features of voltage droop at different loading conditions can ensure a fast transient response. Based on the structure of CF-type3, an additional RC-filter composed of R_{S2} and C_{S2} is added to CF-type5 to remove the voltage droop of several applications that require good load regulation. However, the offset voltage still exists. Based on the structure of CF-type5, an extra valley S&H is added to CF-type6 to remove the offset voltage and ensure accurate regulation. In conclusion, a trade-off between achieving additional functions and sacrificing complex structures exists in an on-time control with additional CF path. Designers can select the most suitable structure based on the specifications listed in Table 4.25. This table lists

Table 4.24 Characteristics of different structures

Structure	Voltage droop	Offset voltage	Characteristics
CF-type1	●	●	• A series R_S • Extra power loss because of R_S • Easy to implement because of its simple structure
CF-type2	●	●	• An integral RC-filter • R_{LPF} and C_{LPF} • Slow transient response because of the integral function
CF-type3	●	●	• Basic structure of parallel RC-filters • R_{S2} and C_{S2}
CF-type4	●	●	• A parallel RC-filter and an additional R_{S2} • R_{S1}, R_{S2}, and C_{S1} • Voltage droop is adjustable for the AVP function
CF-type5	—	●	• Two sets of RC-filters • R_{S1}, R_{S2}, C_{S1}, and C_{S2} • Voltage droop is removed to ensure good load regulation
CF-type6	—	—	• Two sets of RC-filters and an additional valley S&H • R_{S1}, R_{S2}, C_{S1}, and C_{S2} • Voltage droop is removed to ensure good load regulation • Offset voltage is removed to ensure accurate regulation

Table 4.25 Issues and results of different structures

Table	Structure	Discussing issue	Result
Table 4.8	CF-type1	Different R_S	Different voltage droop
Table 4.10	CF-type3	Different C_S	Different amplitudes of v_{SEN} have an influence on stability
Table 4.12	CF-type3	Different C_S	Different pole locations lead to different transient responses
Table 4.14	CF-type3	Same time-constant value but different C_S and R_S	The influence on v_{OUT} and power loss of RC-filters
Table 4.17	CF-type4	Different R_{S2}	The capability of LPF and the different levels of distortion of $v_{S,diff}$
Table 4.19	CF-type4	Different C_{S2} and R_{S2}	Too large C_{S2} influences the time constant of C_S and R_S. The extra phase delay leads to unstable operation
		Same time-constant value	Too large R_{S2} with small C_{S2} suffers the problem of leakage current
Table 4.21	CF-type4	Different amplitude of v_{SEN} and $v_{S,diff}$	Stability
Table 4.23	CF-type4	Different amplitude of $v_{S,diff}$	Transient response

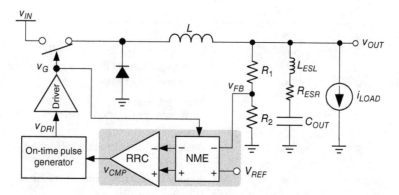

Figure 4.70 On-time control with RRC and NME to ensure stable operation even if a small R_{ESR} is used

the issues of the various techniques and simulation results of the relative simulation. These issues and results also provide important hints for design.

4.3.4 Ripple-Reshaping Technique to Compensate a Small Value of R_{ESR}

When R_{ESR} is small, extracting the inductor current ripple from the output voltage is difficult. The non-linear relationship between inductor current ripple and output voltage ripple causes sub-harmonic oscillation. Figure 4.70 shows the on-time control with ripple-reshaping technique, which includes ripple-recovered compensator (RRC) and noise margin enhancement (NME), to compensate for the small R_{ESR} effect.

The RRC technique with NME function is used to derive the ripple-reshaping function to compensate for the instability effect of a small R_{ESR}. The RRC provides a differentiation function by reversing the integration function of C_{OUT} and thus recovering the inductor current information. Moreover, the RRC determines the output DC level by comparing the feedback signal with the reference voltage. The NME is used to eliminate the ESL effect and increase the noise margin of the RRC technique.

4.3.4.1 RRC for Reversing the Integration Function by C_{OUT}

In Eq. (4.27), the conventional compensation technique utilizing R_{ESR} and C_{OUT} results in zero at the frequency of f_{ESR}. However, if R_{ESR} decreases because of the use of the MLCC, then the zero will move toward high frequencies. According to the analytical results shown in Figure 4.20, a high-frequency zero only has a slight effect on the boosting phase to increase stability. That is, the conventional compensation technique fails to stabilize the entire system. Therefore, the compensating zero generated by the RRC technique is used to boost the PM even without the ESR zero. Figure 4.71 shows that the compensating zero equivalently boosts the phase at approximately 90° so that the conventional stability criterion, as shown in Figure 4.20, can easily be achieved.

Figure 4.72 shows the circuit implementation of RRC. In the first stage of RRC, the input differential pair is composed of M_{P1} and M_{P2}, with the active load formed by N-MOSFETs

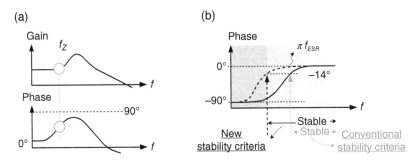

Figure 4.71 (a) Phase lead effect caused by a zero can improve the phase margin. (b) Phase lead effect caused by the RRC technique can increase system stability when a small ESR is used

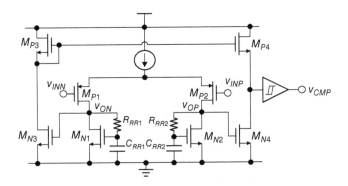

Figure 4.72 Circuit implementation of RRC

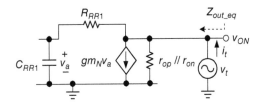

Figure 4.73 Small signal model of the first stage of RRC

(M_{N1} and M_{N2}), resistors (R_{RR1} and R_{RR2}), and capacitors (C_{RR1} and C_{RR2}). As a result, the RRC can generate a pole/zero pair to ensure phase lead compensation. By contrast, in the second stage of RRC, the analog signals (v_{OP} and v_{ON}) are converted to the digital control signal to obtain the full swing voltage v_{CMP}.

We use the common-mode small signal half-circuit model of the first stage of RRC, as shown in Figure 4.73, to derive the pole/zero pair. One test voltage v_t is applied to the v_{ON} node, and the corresponding current i_t flows into the circuit. As such, the equivalent output impedance Z_{out_eq} can be determined.

Kirchhoff's current law (KCL) theorem is used to express i_t as follows:

$$i_t = \frac{v_t}{r_{on}//r_{op}} + \frac{v_t}{R_{RR1} + \frac{1}{sC_{RR1}}} + gm_N \cdot v_a \qquad (4.92)$$

When r_o caused by the channel length modulation is neglected, the output impedance r_{out} is derived using Eq. (4.93) and the transfer function is derived using Eq. (4.94), where the pole and the zero locate at gm_n/C and $1/RC$, respectively:

$$r_{out} = \frac{v_t}{i_t} = \frac{1 + sR_{RR1}C_{RR1}}{gm_N + sC_{RR1}} \qquad (4.93)$$

$$A_{DM} = \left| \frac{v_{OP} - v_{ON}}{v_{INP} - v_{INN}} \right| = \frac{gm_P}{gm_N} \left(\frac{1 + sR_{RR1}C_{RR1}}{1 + s\frac{C_{RR1}}{gm_N}} \right) \qquad (4.94)$$

Consequently, if $R_{RR1} \gg 1/gm_N$, then the zero locates at lower frequencies compared with those of the pole. The differential input signal ($v_{INP} - v_{INN}$) will have a phase lead compared with the differential output signal ($v_{OP} - v_{ON}$). R_{D1} and C_{D1} are respectively set as 500 kΩ and 5 pF to ensure that the zero locates at approximately 63.6 kHz. The frequency response is shown in Figure 4.71. The compensating zero contributes a maximum phase delay of 14° located at approximately 200 kHz. Thus, the system operates at the lowest frequency of 200 kHz without the need for any large R_{ESR}. Meanwhile, the differential structure enhances the noise immunity and decreases the jitter and EMI effects. The RRC technique ensures stability, even though the 200 μF MLCC with only 1 mΩ ESR is used as the output capacitor.

4.3.4.2 NME for Tolerating L_{ESL}

The ESL in the output capacitor is another factor used to induce the distortion of the feedback voltage signal. The ESL effect on the output voltage ripple that influences the differential function of D-RRC expressed in Eq. (4.8) is rewritten as in Eqs. (4.95) and (4.96).

$$\text{On-time phase:} \quad v_{ESL\text{-}n} = v_{ESL}(t) = \frac{v_{IN} - v_{OUT}}{L} \cdot L_{ESL} \qquad (4.95)$$

$$\text{Off-time phase:} \quad v_{ESL\text{-}f} = v_{ESL}(t) = \frac{-v_{OUT}}{L} \cdot L_{ESL} \qquad (4.96)$$

The voltage ripple caused by the ESL, $v_{ESL}(t)$, is constant during the on-time and off-time phases. As a result, the ESL voltage ripple forms a pulse waveform, as shown in Figure 4.74.

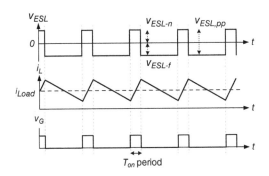

Figure 4.74 ESL voltage ripple at the output voltage

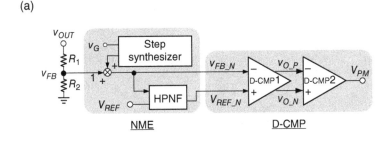

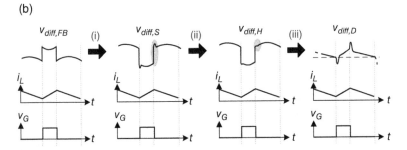

Figure 4.75 (a) Structure of the NME circuit. (b) Detailed waveforms of each output

The peak-to-peak value is the summation of Eqs. (4.93) and (4.94):

$$v_{ESL,pp} = \frac{v_{IN}}{L} \cdot L_{ESL} \qquad (4.97)$$

The voltage ripple value is proportional to v_{IN} and L_{ESL}. Generally, L_{ESL} is smaller than 1 nH. However, the ESL effect cannot be neglected if a high value of v_{IN} is commonly used to generate high power for commercial products. For example, $v_{ESL,pp}$ is as large as 42 mV when v_{IN} is 21 V, L_{ESL} is 1 nH, and L is 2 µH. The process of how the NME technique alleviates the ESL effect to improve D-RRC performance should be elucidated. Figure 4.75(a) shows the NME

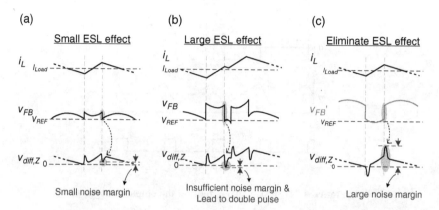

Figure 4.76 Function of the step generator. (a) Stable operation with a small ESL effect and without NME. (b) Unstable operation with a large ESL effect and without NME. (c) Stable operation with a large ESL effect and with NME

circuit, which consists of a step synthesizer and a high-frequency noise filter (HFNF). The differential voltage at different stages is defined as:

$$\begin{cases} v_{diff,FB} = v_{FB} - V_{REF} \\ v_{diff,S} = v_{INP} - V_{REF} \\ v_{diff,H} = v_{INP} - v_{INN} \\ v_{diff,D} = v_{OP} - v_{ON} \end{cases} \quad (4.98)$$

Figure 4.75(b) shows a simple waveform to illustrate the operation. Steps (i), (ii), and (iii) represent the corresponding functions of the (i) step synthesizer, (ii) HFNF, and (iii) RRC. Each function is described in detail as follows.

Figure 4.76(a), (b) indicates the necessity of NME through a comparison of the small L_{ESL} and the large L_{ESL}, respectively. In case of a small L_{ESL}, the down-stepping v_{ESL} at the beginning of the off-time period results in an undershoot of the differential signal $v_{diff,D}$ after the D-RRC process, but without the step generator circuit. The undershoot of the differential signal $v_{diff,D}$ may decrease to less than 0 V because of the ESL effect. Moreover, the incorrect triggering effect at the beginning of the off-time period instantly decreases the system stability. In other words, the double-pulse phenomenon will occur. Figure 4.76(c) shows the functions of NME for D-RRC when a small ESR is used. The feedback voltage signal can be modified from the ESL step to its opposite step. Thus, the feedback voltage signal can be depicted equivalently to v_{INP}. As a result, an overshoot of the differential signal $v_{diff,D}$ at the beginning of the off-time period significantly enhances the noise margin. $v_{diff,D}$ is reshaped and is in phase with the inductor current after the overshoot. By contrast, the undershoot of the differential signal $v_{diff,D}$, which is distorted drastically at the beginning of the on-time period, does not need to be considered because the on-time period is innately defined by the on-time pulse generator.

The opposite step must be generated and synchronized with the on-time period to eliminate the effect of v_{ESL}. However, the step generator has certain limitations in BW (bandwidth) and PM. If the PM is removed to obtain sufficient speed, then the differential output of the step generator, $v_{diff,S}$, as shown in Figure 4.77, will lead to the double-pulse phenomenon after the D-RRC. The HFNF circuit couples the high-frequency variation from one input terminal to another terminal of the D-RRC to alleviate the defect of the step generator. Consequently, v_{FB} is pre-regulated by the NME circuit to enhance noise immunity and regulated by the D-RRC technique. The differential output signal of the D-RRC, $v_{diff,D}$, is reshaped from the phase delay v_{FB} and is in phase with the inductor current during the off-time period.

Finally, Figure 4.78(a) shows the implementation of the step synthesizer, with one example of how to synthesize the step waveform into the feedback signal. Figure 4.78(b) shows that the summation function is derived using a voltage source v_{STEP} and a switch, which is controlled by the gate control signal v_{GP}. The voltage source can be generated by a capacitor. Figure 4.78(c) shows the implementation of another circuit with a buffer structure. By using the gate control signal v_{GP},

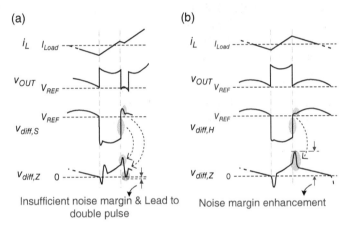

Figure 4.77 Waveforms illustrating the function of the NME and RRC techniques: (a) without HPNF and (b) with HPNF

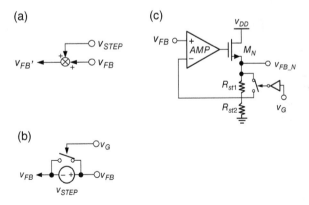

Figure 4.78 Implementation of the step synthesizer. (a) Concept of the step synthesizer. (b) Step synthesizer with a constant voltage source and a switch. (c) Step synthesizer with a buffer structure

the signals generated by the step synthesizer can be expressed in Eqs. (4.99) and (4.100). The magnitude of the step is equal to $v_{FB} \cdot (R_{st1}/R_{st2})$ and can be adjusted using R_{st1} and R_{st2}.

$$v'_{FB} = v_{FB} + v_{FB} \cdot \frac{R_{st1}}{R_{st2}}, \text{ when } v_{GP} \text{ is logic high} \quad (4.99)$$

$$v'_{FB} = v_{FB}, \text{ when } v_{GP} \text{ is logic low} \quad (4.100)$$

4.3.5 Experimental Result of Ripple-Reshaped Function

4.3.5.1 Chip Micrograph

The on-time controlled buck converter with RRC and NME is fabricated using the UMC 0.35 μm bipolar CMOS–DMOS (BCD) 40 V process. For applications with high conversion ratio and heavy driving current of 8 A, power MOSFETs are selected as discrete components. The high-side and low-side power MOSFETs are AOL1414 and AOL1412, respectively. The off-chip inductor and the capacitor are 1 μH and 220 μF (22 μF × 10), respectively. Table 4.26 lists the specifications of this converter.

The nominal switching frequency is near 300 kHz. The output voltage ranges from 0.75 to 3.3 V, with the input voltage defined by the laptop adapter or the desktop power supply. In other words, the highest input voltage is 21 V. Figure 4.79 shows the chip micrograph with active silicon area measuring approximately 3.61 mm² including the test circuits. Table 4.27 describes the function of the sub-circuits.

Figure 4.80 shows the prototype of the on-time controlled buck converter with D-CAP and NME. Here, MLCCs are used as the output capacitor. The equivalent of the R_{ESR} value is calculated as nearly 1 mΩ according to the estimated output ripple (v_{pp}) and Eq. (4.101):

$$\begin{aligned} v_{pp} &= v_{C_{OUT}} + v_{ESR} = \frac{v_{OUT}(1-D)}{8 f_{SW}^2 LC} + \frac{R_{ESR} v_{OUT}(1-D)}{f_{SW} L} \\ \Rightarrow R_{ESR} &= \left(v_{pp} - \frac{v_{OUT}(1-D)}{8 f_{SW}^2 LC} \right) \cdot \frac{f_{SW} L}{v_{OUT}(1-D)} \end{aligned} \quad (4.101)$$

Table 4.26 Performance of the on-time controlled converter with RRC and NME techniques

Process	UMC 0.35 μm BCD 40 V
Input voltage (v_{IN})	5–21 V
Output voltage (v_{OUT})	0.75–3.3 V
Supply voltage for chip (v_{DD})	5 V
Load range (i_{Load})	0.1–8 A
Inductor	1 μH
Output capacitor (MLCC)	220 μF (22 μF × 10)
R_{ESR}	1 mΩ
L_{ESL}	2.6 nH
Operation frequency	100–600 kHz
Output ripple	8–10 mV
Maximum efficiency	91%

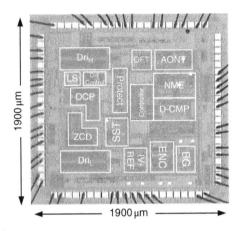

Figure 4.79 Chip micrograph

Table 4.27 Description of sub-circuits

RRC	Ripple-recovered compensator	ZCD	Zero-current detector
NME	Noise margin enhancement	OCP	Over-current protector
AONT	Adaptive on-time timer	PROTECTOR	Protect circuit
OFT	Minimum off-time timer	SST	Soft start
LS	Level shift	BG	Bandgap
D-T Control	Deadtime control	ENC	Enable controller
Dri_H	High-side driver	IV-REF	Biasing current/reference voltage generator
Dri_L	Low-side driver		

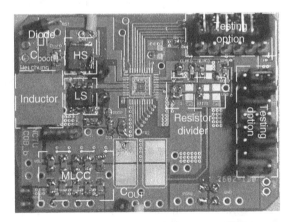

Figure 4.80 Prototype of the on-time controlled buck converter with RRC and NME

4.3.5.2 Steady-State and Load Transient Response

Figure 4.81 shows the steady state of the experimental results; v_{IN} is 5 V and the output voltage v_{OUT} is 1.5 V when i_{Load} is 1.5 A and the switching frequency is 300 kHz. i_L is the inductor current. v_{GH} is the driving signal for the high-side MOSFET. R_{ESR} is approximately 1 mΩ and C_{OUT} is 220 μF (22 μF × 10) because of the use of MLCC. According to the conventional

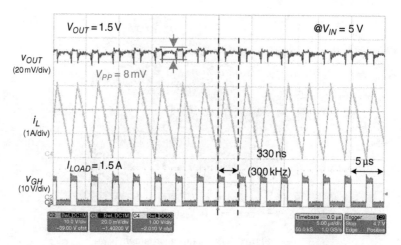

Figure 4.81 Waveforms in steady state with RRC and NME

stability criterion shown in Eq. (4.27), the switching frequency should be higher than 2 MHz. However, the experimental results show that the system stability is guaranteed even under a low switching frequency of 300 kHz because the RRC and NME techniques can reduce the limitation of the stability criterion. Thus, the switching loss can be significantly reduced. In particular, the voltage ripple at v_{OUT} can be smaller than 8 mV when v_{IN} is 5 V. Moreover, the output ripple and the inductor current are out of phase because of the MLCC.

Figure 4.82 demonstrates the functions of RRC and NME when v_{IN} is 5 V and v_{OUT} is 1.5 V. Figure 4.82(a) shows the stable operation attributed to the implementation of RRC and NME. In contrast, with the external option set by the testing circuit, the sub-harmonic oscillation waveform occurs when RRC and NME are disabled, as shown in Figure 4.82(b). RRC contributes the phase lead to the feedback signal, which results in a similar performance that utilizes the output capacitor with a large ESR.

Furthermore, if R_{ESR} is approximately 1 mΩ, the effect of ESL is considerable when v_{IN} is higher than 15 V. v_{ESL} is 40 mV, as shown in Figure 4.83. The figure shows the contribution of NME if Figure 4.83(a) and (b) are compared. The system with only the RRC circuit has inadequate noise margins, such that v_{OUT} is marginally stable. Figure 4.83(c) shows the seriously unstable waveforms without the aid of the RRC–NME technique.

Figure 4.84 shows the waveforms of v_{OUT} and the inductor current i_L operating in the CCM when the load current i_{Load} steps up from 1 to 8 A, or vice versa. Here, v_{IN} is 15 V and v_{OUT} is 1.5 V. Consequently, the switching frequency is 300 kHz. Undershoot and overshoot voltages are 20 and 38 mV, respectively. The transient recovery times are 20 and 25 μs, respectively. Figure 4.85 shows the waveforms operating in the DCM at light loads. The switching frequency f_{SW} is scaled down to 78 kHz to enhance efficiency, which is an advantage of the on-time control for high efficiency at light loads. Obviously, system stability can be guaranteed when the output operates under different loading conditions because of the implementation of RRC and NME.

4.3.5.3 Comparison with Other Techniques

Table 4.28 lists the comparison among prior literature in the design of constant on-time controlled DC/DC converters. The output ripple has been effectively reduced because of using the

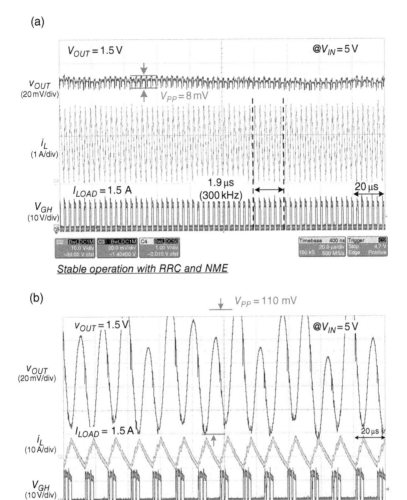

Figure 4.82 (a) Stable waveforms attributed to the implementation of RRC and NME at $v_{IN} = 5$ V, $v_{OUT} = 1.5$ V. (b) Unstable waveforms in the on-time controlled buck converter without RRC and NME at $v_{IN} = 5$ V, $v_{OUT} = 1.5$ V

MLCC. The system can also operate at a low switching frequency when a small ESR is used. It can demonstrate the high performance achieved by RRC and NME.

The RRC and NME conquer stability relative to the small ESR value and the large ESL effect in the COT buck converter. Even though the MLCC is used as the output capacitor, without conventional ESR compensation the RRC technique can still increase the system stability since the compensator contributes a phase lead similar to the phase delay (PD) controller. Besides, the differential structure can benefit the noise margin to decrease the jitter and the EMI effects. In contrast, the NME technique eliminates the effect of ESL to enhance the noise immunity. Furthermore, using a reliable on-time timer with an improved linear function, the near-constant switching frequency, which is adjusted to accommodate variable input voltage, can further

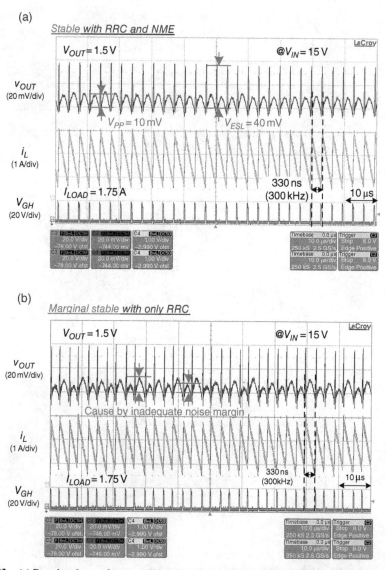

Figure 4.83 (a) Regulated waveforms attributed to the implementation of RRC and NME at v_{IN} = 15 V, v_{OUT} = 1.5 V. (b) Marginal stable waveforms with RRC only at v_{IN} = 15 V, v_{OUT} = 1.5 V. (c) Unstable waveforms in the on-time controlled buck converter without RRC and NME at v_{IN} = 15 V, v_{OUT} = 1.5 V

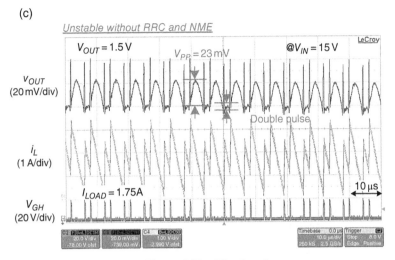

Figure 4.83 (*Continued*)

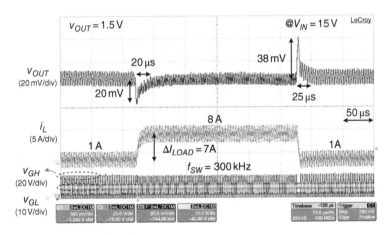

Figure 4.84 Load transient response at CCM operation with RRC and NME techniques

confirm the system stability. Owing to the MLCC with extremely small R_{ESR} value for general applications, the output ripple can be greatly reduced and thus the switching power loss can be decreased corresponding to the large R_{ESR} used to compensate conventional ripple-based control. Experiment results verify the correct and effective functions of the RRC and the NME in the strict case of small $R_{ESR} = 1$ mΩ and large $v_{ESL} = 40$ mV. Without scarifying the inherent advantages of the on-time control, RRC and NME for the MLCC applications can ensure a low ripple of 10 mV and a high efficiency of 91%.

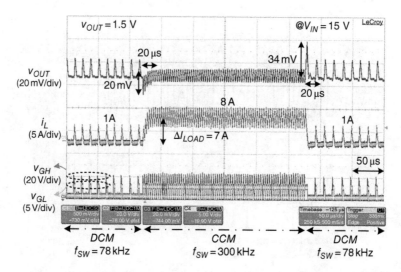

Figure 4.85 Load transient response at DCM operation at light loads with RRC and NME techniques

Table 4.28 Comparison table

	This work	[2] ISSCC	[3] T-PE	[4] T-CAS II	[5] JSSC	[6] T-PE	[7] T-PE
Control method	Ripple reshape	Quasi-V2 hysteretic	Virtual inductor current	Derivative-output ripple voltage (DOR)	Pseudo type III compensation	VIC ripple	QDI
Process (μm)	0.35	0.35	0.35	0.35	0.35	0.18	0.35
v_{IN} (V)	15	2.7–3.3	9	2.4	2.5–3.5	12	3.3
v_{OUT} (V)	1.5	0.9–2.1	4.5	1.8	0.8–2.4	3.3	2
Switching frequency, f_{SW} (MHz)	0.3	3	0.33	0.5	1	0.25	0.8
L (μH)	1	2.2	90	3.3	4.7	1	4.7
C_{CO} (μF)	220	4.4	220	4.7	4.7	286	4.7
Tolerance of minimum R_{ESR} (mΩ)	1	30	50	15	20	3	10
Minimum Δv_{ripple} (mV)	10	12	N/A	20	10	40	10
Δi_L (A)	7	0.45	N/A	0.45	0.5	N/A	0.49
Load range, i_{Load} (A)	1–8	0.05–0.5	N/A	0.25–0.7	0.1–0.6	N/A–18	0.01–0.5
Power efficiency	91%	93%	N/A	N/A	85% up	90%	93%

References

[1] Sahu, B. and Rincon-Mora, G.A. (2007) An accurate, low-voltage, CMOS switching power supply with adaptive on-time pulse-frequency modulation (PFM) control. *IEEE Transactions on Circuits and Systems I: Regular Papers*, **54**(2), 312–321.

[2] Su, F. and Ki, W.-H. (2009) Digitally assisted quasi-V^2 hysteretic buck converter with fixed frequency and without using large-ESR capacitor. *Proceedings of the IEEE ISSCC Digest of Technical Papers*, February 2009, pp. 446–447.

[3] Lin, Y.-C., Chen, C.-J., Chen, D., and Wang, B. (2012) A ripple-based constant on-time control with virtual inductor current and offset cancellation for DC power converters. *IEEE Transactions on Power Electronics*, **27**(10), 4301–4310.

[4] Mai, Y.Y. and Mok, P.K.T. (2008) A constant frequency output-ripple-voltage-based buck converter without using large ESR capacitor. *IEEE Transactions on Circuits and Systems II: Express Briefs*, **55**(8), 748–752.

[5] Wu, P.Y., Tsui, S.Y.S., and Mok, P.K.T. (2010) Area- and power-efficient monolithic buck converters with pseudo-type III compensation. *IEEE Journal of Solid-State Circuits*, **45**(8), 1446–1455.

[6] Chen, W.-W., Chen, J.-F., Liang, T.-J., Wei, L.-C., Huang, J.-R., and Ting, W.-Y. (2013) A novel quick response of RBCOT with VIC ripple for buck converter. *IEEE Transactions on Power Electronics*, **28**(9), 4299–4307.

[7] Lee, Y.-H., Wang, S.-J., and Chen, K.-H. (2010) Quadratic differential and integration technique in V^2 control buck converter with small ESR capacitor. *IEEE Transactions on Power Electronics*, **25**(4), 829–838.

5

Ripple-Based Control Technique Part II

5.1 Design Techniques for Enhancing Voltage Regulation Performance

5.1.1 Accuracy in DC Voltage Regulation

Conventional on-time control has the advantages of fast transient response and simple compensation compared with other control techniques, such as voltage-mode control (VMC) and current-mode control (CMC). However, the inaccuracy of output voltage regulation is an inherent drawback. The expected output voltage DC level v_{OUT} expressed in Eq. (5.1) is determined using the reference voltage V_{REF} and the voltage dividers R_1 and R_2:

$$v_{OUT} = V_{REF} \cdot \left(1 + \frac{R_1}{R_2}\right) \quad (5.1)$$

In on-time controlled converters, the output voltage is regulated but with an unexpected DC offset voltage deviation. According to Figure 4.6(a), the output voltage of the on-time controlled converter is regulated at the value calculated using Eq. (5.2). The offset voltage deviation is caused by the extra term v_{OUT_ripple}, which also represents the output voltage ripple:

$$v_{OUT} = V_{REF} \cdot \left(1 + \frac{R_1}{R_2}\right) + \frac{1}{2} v_{OUT\_ripple} \quad (5.2)$$

As previously mentioned in Section 4.2.1, the output voltage ripple is determined by many factors, such as the inductor current ripple, C_{OUT}, R_{ESR}, and L_{ESL}. Moreover, the output voltage

---

Power Management Techniques for Integrated Circuit Design, First Edition. Ke-Horng Chen.
© 2016 John Wiley & Sons Singapore Pte Ltd. Published 2016 by John Wiley & Sons Singapore Pte Ltd.

ripple is different under different operating modes, as derived using Eqs. (5.3) and (5.4) for CCM and DCM, respectively:

$$v_{pp(CCM)} \approx \frac{(1-D)}{8f_{SW}^2 LC} \cdot v_{OUT} + R_{ESR} \cdot \frac{(1-D)}{f_{SW} L} \cdot v_{OUT} + \frac{L_{ESL}}{L} \cdot v_{IN} \quad (5.3)$$

$$v_{pp(DCM)} \approx \frac{1}{RC} \left(\frac{1}{f_{SW}} + \frac{L}{2R_{Load}(1-D)} - \sqrt{\frac{2L}{R_{Load} \cdot f_{SW}(1-D)}} \right) \cdot v_{OUT}$$
$$+ R_{ESR} \cdot \frac{D(1-D)}{L \cdot f_{SW}} \cdot v_{OUT} + \frac{L_{ESL}}{L} \cdot v_{IN} \quad (5.4)$$

The accuracies of output voltage and load regulation are degraded because of the characteristics of the output voltage ripple when considering all parasitic effects. Therefore, the V^2 structure is implemented to enhance regulation performance. The basic design method of V^2 structure and other advanced techniques for high performance will be introduced in the next subsection.

5.1.2 V^2 Structure for Ripple-Based Control

5.1.2.1 V^2 On-Time Control

Figure 5.1 shows that the V^2 structure is implemented in on-time control by adding an extra output voltage feedback path. That is, the V^2 structure contains two voltage feedback paths traversing directly from the output to the comparator and is used to determine the switching operation. The V^2 structure is the feedback path generated by two parallel voltage paths. The first path is the original path that feeds the output voltage directly to the comparator. This path is also called the fast feedback path, because it reacts rapidly to the output voltage variation. As such, a good transient response can be derived using the V^2 structure. The other path from the output to the comparator consists of one error amplifier with a compensation network. The output of the error amplifier, $v_{EA}(t)$, is determined by the difference voltage of v_{OUT} and V_{REF}. This path is

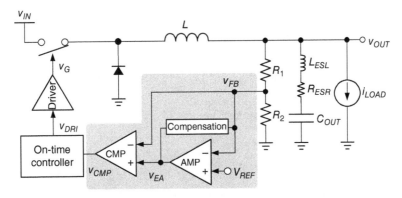

Figure 5.1 Schematic of V^2 on-time control

called the slow feedback path because the bandwidth is limited by the bandwidth of the error amplifier and the compensation network. Through the closed loop, $v_{EA}(t)$ in comparison with V_{REF} reflects the error in revising the DC offset between v_{OUT} and V_{REF}. As a result, the slow feedback path can improve the output DC regulation. Simultaneously, the fast feedback path can maintain the fast transient response.

As depicted in Figure 5.1, the compensation should be well designed because the V^2 structure uses the error amplifier on the feedback path. According to the superposition theory, the transfer function from the inverting terminal of the error amplifier to the output is the same as in Eq. (4.12).

Figure 5.2 shows the small single model of a buck converter with basic V^2 control. The system duty cycle d is composed of two factors, d_1 and d_2, which are the controls from the fast feedback path and the slow feedback path, respectively. The transfer function of the fast feedback path is from v_{OUT} to d_1, denoted as the transfer function F_{m1}, while the slow feedback path is from v_{OUT} to d_2 via the error amplifier, denoted as the transfer function F_{m2}. F_{m1} and F_{m2} stand for analog-to-digital conversion, where $-A_v$ is the DC gain of the EA.

Similarly, according to Eq. (4.12) and if R_{ESR} is large enough, a Type-II compensator can be utilized to guarantee stability. Figure 5.3(a), (b) illustrates the frequency response to reveal the distribution of poles and zeros. In Figure 5.3(a), the slow path with Type-II compensator provides one low-frequency pole/zero pair and another high-frequency pole. In contrast, the frequency response of the fast path with the comparator has low gain and high bandwidth. To simplify the stability analysis after compensation, the zero contributed by R_{ESR} and C_{OUT} is included in the feedback path. Consequently, the frequency response of the complete feedback path can be derived as the complete waveform at the bottom of Figure 5.3(a). Figure 5.3(b) illustrates the frequency response of the system loop gain composed of the complete feedback path and the power stage. The lowest-frequency pole is set as the dominant pole. The low-frequency zero contributed by the Type-II compensator and the ESR zero, generated by R_{ESR} and C_{OUT}, can be used to cancel the complex poles derived in Eq. (4.12). Finally, the high-frequency pole is used to decade high-frequency gain and to reduce high-frequency noise. As a result, system stability can be guaranteed.

On the contrary, the Type-I compensator is another choice for all possible compensation techniques. Figure 5.3(c), (d) illustrates the frequency response to reveal the distribution of poles and zeros. In Figure 5.3(c), the Type-I compensator provides one low-frequency pole

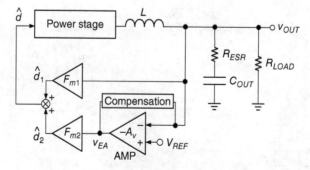

Figure 5.2 Small single model of buck converter with the basic V^2 control in a constant switching frequency

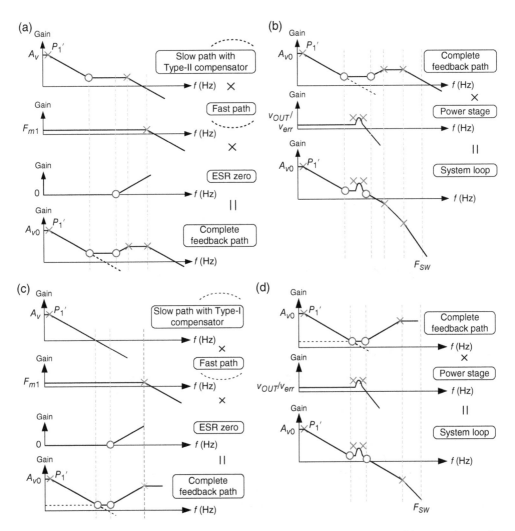

Figure 5.3 Frequency response of (a) feedback loop with Type-II compensator, (b) system loop gain compensated by Type-II compensator, (c) feedback loop with Type-I compensator, and (d) system loop gain compensated by Type-I compensator

as the dominant pole. The ESR zero, contributed by R_{ESR} and C_{OUT}, can cancel part of the effect from the complex poles but the system still needs another zero. Thus, another architectural zero, generated by the parallel structure composed of fast and slow feedback paths, locates at a frequency near the UGF of the voltage error amplifier. Similar to the case with Type-II compensator, the complete feedback path contains one low-frequency pole and two compensation zeros. Figure 5.3(d) shows the frequency response of the system loop gain. Within the UGF, the compensated system can guarantee stability with an adequate phase margin because the system apparently performs as a one-pole system due to the cancellation of complex poles by two compensation zeros.

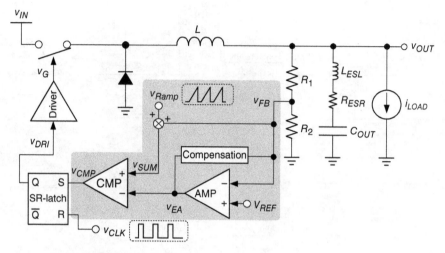

Figure 5.4 Schematic of V^2 constant-frequency valley-voltage control

5.1.2.2 V^2 Constant-Frequency Valley-Voltage Control

Figure 5.4 shows that the V^2 structure is implemented in the constant-frequency valley-voltage control technique introduced in Section 4.1.4. The inverting input of the comparator substitutes the reference voltage V_{REF} by the EA output voltage $v_{EA}(t)$ determined by comparing v_{OUT} and V_{REF}. $v_{EA}(t)$ is treated as one new V_{REF} because the voltage EA revises the DC offset by the comparison result of v_{OUT} and V_{REF}. Because the error amplifier is inserted in the control loop, the compensation network is needed to increase stability. The control-to-output transfer function of the power stage can be expressed as Eq. (5.5), where G_{d0}, the DC gain of the power stage, and the LC double poles at ω_0 are shown in Eq. (5.6). The ESR zero ω_{ESR}, R_{ESR} and C_{OUT} can be helpful in increasing the system stability if ω_{ESR} is less than $3\omega_0$ at the cost of a large output voltage ripple.

$$\frac{\hat{v}_{OUT}(s)}{\hat{d}} = G_{d0} \frac{1 + \dfrac{s}{\omega_{ESR}}}{1 + \dfrac{1}{Q\omega_0}s + \dfrac{1}{\omega_0^2}s^2} \tag{5.5}$$

where

$$\omega_{ESR} = \frac{1}{R_{ESR}C_{OUT}}, \quad \omega_0 = \frac{1}{\sqrt{LC_{OUT}}}, \quad Q = R_{Load}\sqrt{\frac{C_{OUT}}{L}} \tag{5.6}$$

To analyze the stability, Figure 5.3 is also utilized to understand the small single model of a basic V^2 control buck converter with constant switching frequency. Similar to the compensation skill for V^2 on-time control, a Type-II compensator can be used to stabilize the V^2 structure with the assistance of a large R_{ESR}. Equation (5.7) expresses the loop-gain transfer function of the buck converter with V^2 structure, where $F_{m1} = F_{m2} = F_m$:

$$T(s) = G_{d0}F_m A_v \cdot \frac{\left(1 + \dfrac{s}{\omega_{ESR}}\right)\left(1 + \dfrac{s}{\omega_z}\right)}{\left(1 + \dfrac{1}{Q\omega_0}s + \dfrac{1}{\omega_0^2}s^2\right)\left(1 + \dfrac{s}{\omega_p}\right)} \quad (5.7)$$

The Type-II compensator generates a low-frequency pole ω_p and a zero ω_z with gain improvement by the error amplifier. ω_p works as the system dominant pole, while ω_z and ω_{ESR} cancel the effect of the complex poles at ω_0. As a result, Type II with a large value of R_{ESR} can guarantee system stability. Alternatively, architectural parallel paths can generate a zero to replace the zero of the Type-II compensator. In other words, the compensator can be simplified to a Type-I compensator if an adequate UGF can be derived (Figure 5.5).

5.1.3 V^2 On-Time Control with an Additional Ramp or Current Feedback Path

An additional ramp signal or current feedback signal can increase the stability. However, the ripple of feedback voltage equivalently induces a larger DC voltage offset. Thus, the V^2 structure can be implemented to maintain the stability without scarifying the DC regulation. Figures 5.6 and 5.7 show the V^2 on-time control with an additional ramp signal and an additional current feedback path, respectively.

The current-sensing circuit can be determined by the approaches introduced in Section 4.3.2. With a voltage EA, v_{EA} adjusts the DC value to accurately regulate the output voltage. In other words, no matter whether the sensing voltage includes DC information or not, the DC value of v_{EA} can dynamically change corresponding to the DC value of v_{SUM} and the difference between v_{FB} and v_{REF}. If v_{SUM} includes DC information on the inductor current, the DC value of v_{EA} is almost determined by the loading conditions. One design issue that needs to be taken into consideration is that the DC value of v_{EA} would influence the operation of the amplifier. Too high or too low a value would cause the MOSFET to operate in a linear region and degrade the DC gain of the amplifier. Similar to current-mode PWM control, the loading current range is

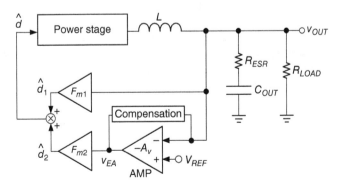

Figure 5.5 Small signal of basic V^2 control buck converter with a constant switching frequency

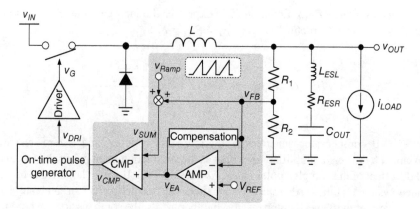

Figure 5.6 V^2 on-time control with an additional ramp signal

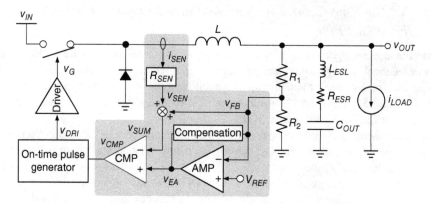

Figure 5.7 V^2 on-time control with an additional current feedback path

somehow constrained by the DC value of the current-sensing signal because of the limited headroom voltage. In contrast, Type-I or Type-II compensator can be used with similar analysis methods in V^2 on-time control to guarantee system stability.

Although the accuracy of DC regulation and stability can be guaranteed, the side effect of slow transient response caused by the additional signals is significant. According to the results of Eqs. (4.32) and (4.37), the quality factor Q in the denominator implies that the system can be more robust when the feedback signal accompanies an increasing amount of additional ramp signal or additional current ripple. However, the increasing amount of the additional signals equivalently reduces the gain to monitor the output voltage. As a result, the insensitivity to variation of output voltage degrades the transient response. To guarantee stability, the excess of extra signals scarifies the fast transient response, which is an inherent advantage of ripple-based control. Consequently, when these techniques are utilized, any over-designs should be avoided to keep the original advantages and to further achieve advanced performance.

5.1.4 Compensator for V^2 Structure with Small R_{ESR}

5.1.4.1 Type-III Compensator for Realizing V^2 Control with Small R_{ESR}

Based on the analysis in Section 5.1.2, the compensator should provide two zeros to cancel the complex poles. If the output capacitor C_{OUT} has large R_{ESR}, one of the zeros can be contributed by C_{OUT} and R_{ESR}. The other zero is contributed by the compensator on the feedback path. In case of large R_{ESR}, the Type-I and Type-II compensators are suitable candidates to guarantee stability. In contrast, if C_{OUT} contains a small R_{ESR}, the zero generated by R_{ESR} and C_{OUT} is located at high frequencies and is not useful for system stability. To cancel the complex poles at the power stage, two zeros must be provided by the compensator on the feedback path. Consequently, a Type-III compensator can be taken into consideration.

Figure 5.8(a), (b) shows that the Type-III compensator is applied to V^2 on-time control and V^2 constant-frequency peak-voltage control, respectively.

Figure 5.9(a) shows the structure of V^2 control with Type-III compensator. Figure 5.9(b) depicts the frequency response, and the transfer function is expressed in Eq. (5.8). An error

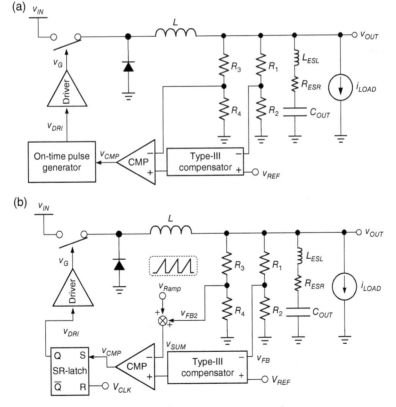

Figure 5.8 Type-III compensator for (a) V^2 on-time control, (b) V^2 constant-frequency peak-voltage control

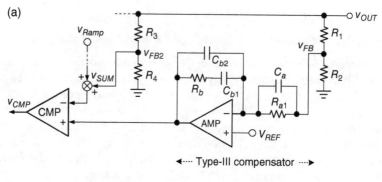

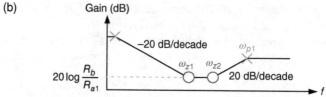

Figure 5.9 (a) Structure of V^2 control and Type-III compensator with one pole and two zeros. (b) Frequency response

amplifier with three capacitors and two resistors to implement the Type-III compensator can generate two zeros (ω_{z1} and ω_{z2}) and one pole (ω_{p1}), as in Eq. (5.9):

$$A(s) = -A_0 \cdot \frac{1}{s(C_{b1}+C_{b2})R_{a1}} \cdot \frac{(1+sR_{a1}C_a)(1+sR_bC_{b1})}{1+sR_b\left(\dfrac{C_{b1}\cdot C_{b2}}{C_{b1}+C_{b2}}\right)} \tag{5.8}$$

$$\omega_0 = \frac{1}{(C_{b1}+C_{b2})R_{a1}}, \quad \omega_{p1} = \frac{1}{R_b\left(\dfrac{C_{b1}\cdot C_{b2}}{C_{b1}+C_{b2}}\right)} \tag{5.9}$$

$$\omega_{z1} = \frac{1}{R_{a1}C_a}, \quad \omega_{z2} = \frac{1}{R_b C_{b1}}$$

Two zeros need to locate at frequencies near the complex poles, caused by the power stage. To ensure a negative feedback loop, the frequency of one of the zeros is required to be below the complex poles. Besides, according to the stability criterion of ripple-based control as in Eq. (4.27), two zeros, ω_{z1} and ω_{z2}, are suggested to be located at frequencies lower than 10% of the switching frequency. In contrast, the high-frequency pole, ω_{p1}, is suggested to be located at frequencies close to the high-frequency zero due to R_{ESR} for suppressing high-frequency noise.

Figure 5.10(a) shows the structure of V^2 control with Type-III compensator. Figure 5.10(b) depicts the frequency response whose transfer function is expressed in Eq. (5.10):

$$A(s) = -A_0 \cdot \frac{1}{s(C_{b1}+C_{b2})R_{a1}} \cdot \frac{(1+s(R_{a1}+R_{a2})C_a)(1+sR_bC_{b1})}{1+sR_b\left(\dfrac{C_{b1}\cdot C_{b2}}{C_{b1}+C_{b2}}\right)(1+sR_{a2}C_a)} \tag{5.10}$$

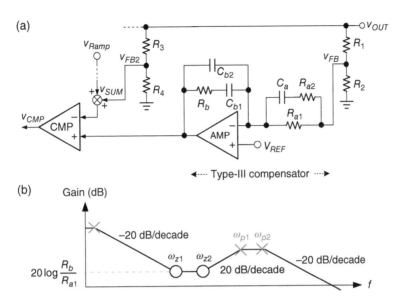

Figure 5.10 (a) Structure of V^2 control and Type-III compensator with two poles and two zeros. (b) Frequency response

An error amplifier with three capacitors and three resistors to implement a Type-III compensator can generate two zeros (ω_{z1} and ω_{z2}) and two poles (ω_{p1} and ω_{p2}), as shown in Eq. (5.11). Compared with Figure 5.9, this Type-III compensator provides an additional pole, ω_{p2}, which can benefit artifact noise suppression at high frequencies. Thus, this pole, ω_{p2}, is typically set to half of f_{SW}, or 10 times the UGF, f_{UGF}.

$$\omega_0 = \frac{1}{(C_{b1}+C_{b2})R_{a1}}, \ \omega_{p1} = \frac{1}{R_b \frac{C_{b1} \cdot C_{b2}}{C_{b1}+C_{b2}}}, \ \omega_{p2} = \frac{1}{R_{a2}C_{a1}}$$
$$\omega_{z1} = \frac{1}{(R_{a1}+R_{a2})C_{a1}}, \ \omega_{z2} = \frac{1}{R_b C_{b1}} \quad (5.11)$$

To comprehensively understand the similarities and differences compared with the previous V^2 structure, the Type-III compensator in Figure 5.9(a) can also be divided equally into two parallel paths, fast path$_1$ and slow path, as shown in Figure 5.11. Through capacitors C_a and C_{b2}, this fast path provides the pass of high-frequency signals. In other words, the inductor current ripple information can be equivalently recovered by the output voltage ripple so that the inductor current can be monitored to ensure stability. On the slow feedback path through the amplifier, the offset at the output can be modified. There is another fast path (fast path$_2$) through the voltage divider (R_3 and R_4) that can directly reflect the DC regulation of v_{OUT} so that the converter can instantly adjust the power delivery during the transient period. Consequently, the function of fast path$_1$ and fast path$_2$ is similar to the fast path in the previous V^2 structure, which can simultaneously monitor v_{OUT} for DC regulation and i_L for stability. The buck converter with

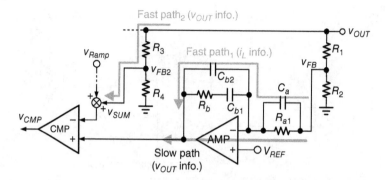

Figure 5.11 Different signal paths feeding back from the output

V^2 control and small ESR can be modulated with an ensured stability. And last but not least, if the constant-frequency clock signal is applied as shown in Figure 5.8(b), an extra ramp signal is necessary to avoid the sub-harmonic oscillation.

5.1.4.2 Type-III Compensator for Realizing Single Path V^2 Control with Small R_{ESR}

There are researches that provide simplification methods to further develop the structure of the Type-III compensator for V^2 control. The direct voltage path is removed so that the feedback path can be viewed as one single way through the compensator while the ripple-based control is still established. Figure 5.12(a), (b) shows the architecture of a single feedback path for V^2 on-time control and V^2 constant-frequency peak-voltage control, respectively.

To understand the concept for realizing ripple-based control by a single path, let's take Figure 5.12 as an example with the help of Figure 5.13 to depict the comparator including Type-III compensator in detail. The path through the Type-III compensator includes two paths, both a fast path and a slow path. The fast path provides a function of the high-pass filter to recover the inductor current ripple information from the output voltage ripple. The slow path provides DC information to regulate V_{OUT}. According to the above analysis, this approach does work. However, the limited bandwidth of the slow path cannot instantly reflect the output voltage level. As a result, the transient response is significantly deteriorated.

5.1.4.3 Differentiator for V^2 Structure with Small R_{ESR}

The above discussion provides a Type-III compensator to guarantee stability. However, the integration becomes a challenge because large capacitors and large resistors are necessary. For this reason, it is necessary to develop other approaches that provide the same function together with the possibility of integrated compensation. We introduce some ideas to make such an integrated design more feasible.

From the aspect of ripple-based control, the inductor current ripple is difficult to extract from the voltage feedback path because the output voltage ripple is not generally dominated by the

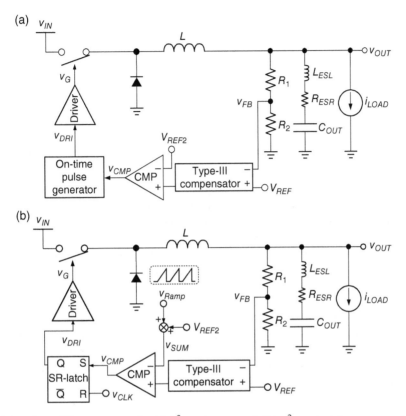

Figure 5.12 Type-III compensator for (a) V^2 on-time control, (b) V^2 constant-frequency peak-voltage control

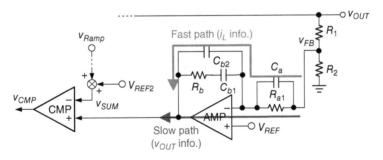

Figure 5.13 The comparator including Type-III compensator in detail

small R_{ESR}. The relationship between v_L and v_{OUT} is not linear. According to Eqs. (4.7)–(4.9), the output voltage can be derived as Eq. (5.12) including ESR and ESL effects:

$$v_{OUT}(t) = R_{ESR}i_L(t) + \frac{1}{C_{OUT}}\int i_L(t)dt + L_{ESL}\frac{d}{dt}i_L(t) \tag{5.12}$$

Obviously, only the first term of R_{ESR} contributes a linear function to reflect the inductor current ripple. When an output capacitor with low R_{ESR} is used, the ripple is mainly determined by the second and third terms. Besides, assuming the ESL effect is also ignored due to a small ESL value, the relationship between i_L and v_{OUT} is shown in Eq. (5.13), which expresses simply charging/discharging the output capacitor with the inductor current:

$$v_{OUT}(t) = \frac{1}{C_{OUT}} \int i_L(t) dt \tag{5.13}$$

This equation shows the non-linear relationship between $v_{OUT}(t)$ and $i_L(t)$. Feeding back the output voltage to the controller will cause unstable operation due to the out-of-phase relationship between $v_{OUT}(t)$ and $i_L(t)$. Therefore, the differentiator in Figure 5.14 is used in the V^2 structure to derive the inductor current information though the feedback output voltage. Because the slow feedback path through the amplifier is used to enhance the accuracy, the output voltage ripple has little impact on the operation. In contrast, the fast feedback loop needs the inductor current ripple information to determine correct switching timing. It's necessary to recover the inductor current ripple from v_{OUT}.

Equation (5.14) is derived to show the inductor current ripple recovered by the signal $v_{FB\_D}(t)$ after an ideal differentiator. In other words, $v_{FB\_D}(t)$ is linearly proportional to $i_L(t)$. With the assistance of a recovery function of the differentiator, the stability issue can be analyzed as discussed in Section 4.3. Consequently, even when R_{ESR} of the output capacitor is too small, the stability can be increased by the V^2 structure with an additional differentiator:

$$v_{FB\_D}(t) = v_{OUT}(t) \cdot \beta = \frac{d}{dt}\left(\frac{1}{C_{OUT}} \int i_L(t) dt\right) \cdot \beta = i_L(t) \cdot \frac{\beta}{C_{OUT}} \tag{5.14}$$

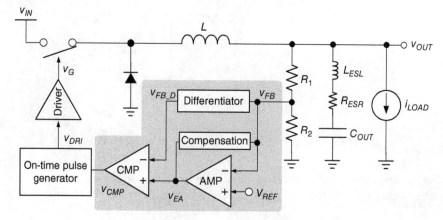

Figure 5.14 The differentiator can be used in the V^2 structure to derive the inductor current information though the feedback output voltage

where

$$\beta = \frac{R_2}{R_1 + R_2}$$

5.1.4.4 Equivalent Type-III Compensator for V^2 Control with Small R_{ESR}

If R_{ESR} is not large enough in the V^2 on-time control, the differentiator has the ability to extract the inductor current ripple through the fast feedback path. In other words, the zero is contributed by a differentiator to replace the equivalent ESR zero so that Eq. (4.27) holds to ensure system stability.

In contrast, because the complex poles exist at the power stage, two zeros should be provided on the feedback path, which is composed of slow and fast paths. Thus, a Type-I or Type-II compensator can be used to work on the slow feedback path. When a Type-I compensator is used, two zeros are contributed by the compensator and the architectural zero due to parallel paths. When a Type-II compensator is used, both zeros are contributed by the compensator.

The block circuit is illustrated in Figure 5.15. The function of the differentiator is equivalent to the function of a high-pass filter. In practice, the function should be seen as a band-pass filter, expressed as $B(s)$, because high-frequency poles exist inherently in the circuit. The critical design issue is to ensure the zero is adequate to contribute a phase lead function for differentiation, and meanwhile to ensure that the other pole is at a high enough frequency. The transfer function of the amplifier with compensation is expressed as $A(s)$. Whether for on-time control as in Figure 5.15(a) or for constant-frequency peak-voltage control as in Figure 5.15(b), the system contains complex poles.

Figure 5.16(a), (c) illustrates the frequency response for the system with Type-I compensator. Figure 5.16(b), (d) illustrates the frequency response for the system with Type-II compensator. The parallel structure of $A(s)$ and $B(s)$ can induce two zeros, as shown in Figure 5.16(a), (b). Except for the dominant pole, the rest of the poles locate at a frequency higher than the bandwidth (BW), so it is reasonable to ignore their effects. As a result, Figure 5.16(c), (d) shows that the complex poles are compensated by two zeros for on-time control and constant-frequency peak-voltage control, respectively.

It should be noticed that the additional differentiator induces some drawbacks. On the fast feedback path, the differentiator recovers the inductor current ripple from the output voltage ripple. However, some high-frequency poles exist practically in the differentiator circuit. The bandwidth of this fast feedback path then limits the signal conduction from $v_{OUT}(t)$ to $v_{CS}(t)$. Thus, the inherent fast transient response feature is drastically degraded, although the stability is increased.

5.1.5 Ripple-Based Control with Quadratic Differential and Integration Technique if Small R_{ESR} is Used

Through use of the differentiation function, the ripple of the feedback signal can be recovered from the distortion of v_{OUT} when a small R_{ESR} is used. In other words, the system stability can be guaranteed. Besides, the output voltage ripple and the transient dip can effectively be

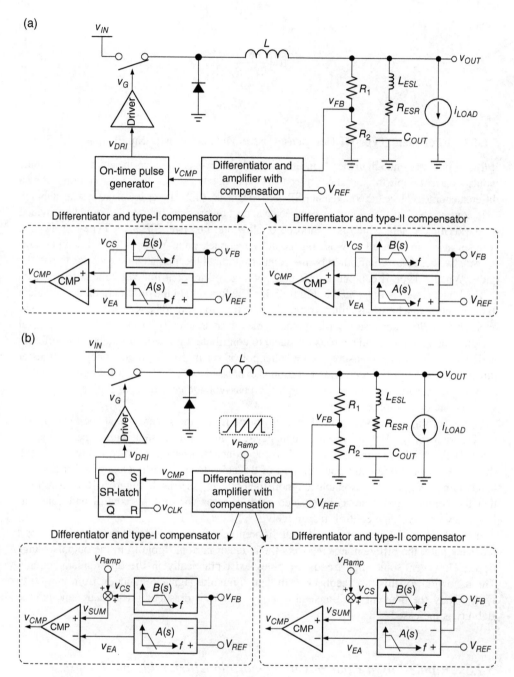

Figure 5.15 Equivalent Type-III compensator for (a) V^2 on-time control, (b) V^2 constant-frequency peak-voltage control

Ripple-Based Control Technique Part II

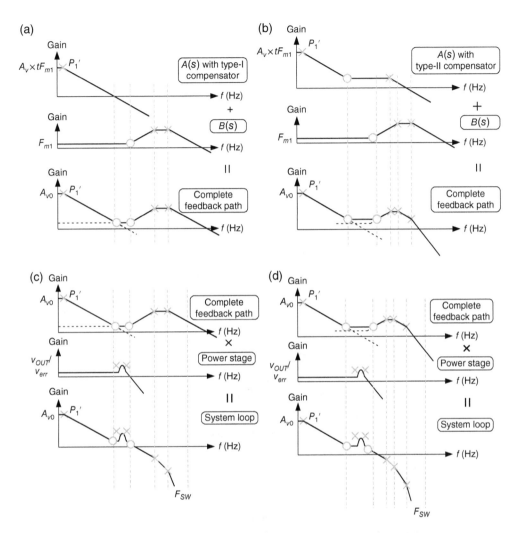

Figure 5.16 Frequency response of (a) feedback loop with Type-I compensator, (b) feedback loop with Type-II compensator, (c) system loop with Type-I compensator, (d) system loop with Type-II compensator

minimized. However, the existing R_{ESR} still affects the accuracy of the recovery signal. This unwilling distortion deteriorates the system stability. Furthermore, the inaccurate accuracy signal results in a worse load regulation.

The quadratic differential and integration (QDI) technique based on the V^2 control buck converter improves these performances. Figure 5.17 shows the V^2 control with the QDI technique. Based on the structure of V^2 control, the QDI technique is inserted on the fast feedback path.

Figure 5.18 shows a simple structure for comparison between the QDI technique and previous methods, based on V^2 control. Figure 5.18(a) shows the conventional V^2 control with large ESR. Figure 5.18(b), (c) shows the differentiator technique with small ESR and the

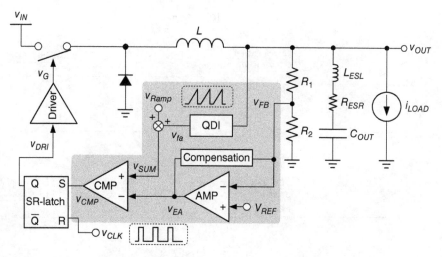

Figure 5.17 V^2 control with the QDI technique

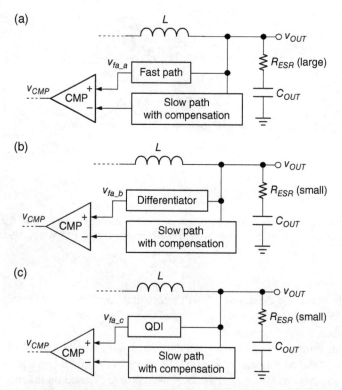

Figure 5.18 The structures with V^2 control: (a) conventional V^2 control for large ESR application; (b) V^2 control with differentiator technique for small ESR application; (c) V^2 control with QDI technique for small ESR application

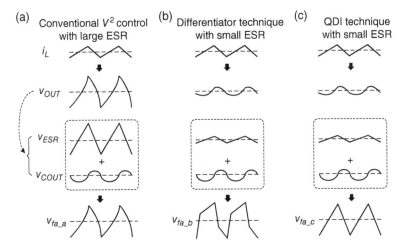

Figure 5.19 (a) Conventional V^2 control with large ESR. (b) Differentiator technique with small ESR. (c) QDI technique with small ESR

QDI technique with small ESR, respectively. The L_{ESL} on the capacitor is assumed small so that the ripple of L_{ESL} can be neglected. The signal v_{fa} (v_{fa_a}, v_{a_b}, and v_{fa_c} for each technique) represents a signal fed into the positive input node of the comparator. Corresponding to these structures, let's observe the characteristics of operating waveforms as shown in Figure 5.19.

Case (a) in Figure 5.19 shows conventional V^2 control with large R_{ESR}. The inductor current ripple flows into the output capacitor. v_{ESR} and v_{COUT} reflect linear and integral voltage ripples, respectively. With a large R_{ESR}, v_{OUT} ripple reflects the high linearity of i_L. The inductor current information is converted to v_{fa_a} through the fast feedback path. Because the ripple v_{fa_a} keeps high linearity to i_L, the system remains stable without using any extra techniques. Cases (b) and (c) in Figure 5.19 show the techniques with small R_{ESR}. With small R_{ESR}, v_{OUT} ripple reflects low linearity to that of i_L. To maintain stability, the differentiator is inserted on the fast path as shown in Figure 5.18, with the operating waveforms in Figure 5.19(b). By neglecting the ripple of ESL, Eq. (5.12) can be expressed as:

$$v_{OUT}(t) = R_{ESR}i_L(t) + \frac{1}{C_{OUT}}\int i_L(t)dt \qquad (5.15)$$

Using the conventional amplifier (AMP)-based differentiator as depicted in Figure 5.20, $v_{fa\_b}(t)$ can be expressed in the S-domain and the time domain, respectively:

$$\frac{v_b(s)}{v_a(s)} = (sC_1R_1 + 1) \qquad (5.16)$$

$$v_{fa\_b}(t) = C_1R_1v'_{OUT}(t) + v_{OUT}(t)$$
$$= R_{ESR}C_1R_1\frac{d}{dt}i_L(t) + \frac{C_1R_1}{C_L}i_L(t) + R_{ESR}i_L(t) + \frac{1}{C_L}\int i_L(t)dt \qquad (5.17)$$

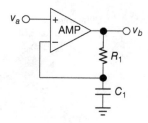

Figure 5.20 Basic amplifier-based differentiator

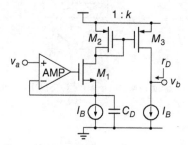

Figure 5.21 VCCS circuit with the differentiator function

Although the inductor current ripple factor is included in Eq. (5.17), the first and third terms still result in undesirable distortion in $v_{fa\_b}(t)$. In contrast, one voltage-control current-source (VCCS) circuit [1] as depicted in Figure 5.21 can perform a suitable differentiation function to eliminate the distortion and increase the accuracy. $v_{fa\_b}(t)$ after the VCCS circuit is expressed in the S-domain and the time domain, respectively:

$$v_b(s) = skC_D r_D v_a(s) \tag{5.18}$$

$$\begin{aligned}v_{fa\_b}(t) &= kC_D r_D \left(R_{ESR} \frac{d}{dt} i_L(t) + \frac{i_L(t)}{C_{OUT}} \right) \\ &= kC_D r_D R_{ESR} \frac{d}{dt} i_L(t) + \frac{kC_D r_D}{C_{OUT}} i_L(t)\end{aligned} \tag{5.19}$$

The parameter k is the current mirror ratio and r_D is the equivalent output resistance of the VCCS circuit. By comparing Eqs. (5.19) and (5.17), it's more accurate to recover the inductor current ripple by the VCCS circuit. Without a large ESR, the second term of Eq. (5.19) provides information on the inductor current ripple. However, the first term of Eq. (5.19), which is caused by parasitic ESR, is still undesirable. From the description in Figure 5.19(b), $v_{fa\_b}(t)$ is composed of four slopes during a switching period. The second rising stage and the fourth falling stage represent the inductor current ripple. The first rising stage and the third falling stage are caused by the ESR. In other words, a larger ESR leads to a larger distortion. As a result, $v_{fa\_b}(t)$ still suffers from the distortion problem owing to the existing ESR of the output capacitor.

In contrast, with the assistance of the QDI technique, the unwilling ESR-related distortion is completely removed. Consequently, a more accurate ripple of v_{fa} can achieve better stability. Thus, the QDI technique is expected to recover $v_{fa\_c}(t)$ as shown in Figure 5.19c. $v_{fa\_c}(t)$ is then proportional to the inductor current ripple. The QDI circuit not only gets an adequate function to modulate the duty cycle, but also purifies the recovery signal to enhance the signal-to-noise ratio (SNR). As shown in Figure 5.17, V^2 control with the QDI circuit contains two voltage feedback paths.

The first path, which is a slow feedback path, compares the output voltage with the reference voltage to decide the error signal. The second path, which is a fast feedback path, is handled by the QDI circuit to obtain the recovery signal and remove the dependence on the ESR. This second voltage path can still react rapidly to the output voltage variation to speed up the transient response. Therefore, a fast transient response can be achieved by the V^2 control scheme with the QDI circuit without adding other fast transient techniques.

5.1.5.1 Stability Analysis

Figure 5.22 shows the small signal model of the V^2 control with the QDI circuit, which is inserted between the output voltage node and the PWM modulator F_{m1}. The QDI circuit works as a differentiator and the transfer function is simply expressed as in Eq. (5.20), where C_Q and r_Q constitute an equivalent differential constant in the QDI circuit:

$$G_{VCCS}(s) = skC_Q r_Q \tag{5.20}$$

There are two parallel paths that contribute d_1 and d_2 to adjust the duty cycle d according to the output voltage. The first path for determining the factor d_1 is composed of $G_{VCCS}(s)$ and F_{m1}, which are the transfer functions of the QDI circuit and the comparator, respectively. The second path for determining d_2 is composed of $A_V(s)$ and F_{m1}, which are the transfer functions of the amplifier with Type-II compensation and the comparator, respectively. Type-II

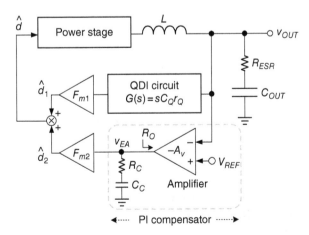

Figure 5.22 Small signal model of the V^2 control buck converter with the QDI technique

compensation is used because the zero generated by small R_{SER} and C_{OUT} locates at high frequencies and contributes no compensation. The transfer function of the Type-II compensator is expressed as in Eq. (5.21). Equation (5.22) shows the compensation pole/zero pair, where R_O is the equivalent output impedance of the AMP:

$$A_{AMP}(s) = A_{v0} \frac{1 + \frac{s}{\omega_z}}{1 + \frac{s}{\omega_p}} \quad (5.21)$$

$$\omega_z = \frac{1}{R_C C_C}, \, \omega_p = \frac{1}{R_O C_C} \quad (5.22)$$

Therefore, the output-to-duty transfer function can be derived as:

$$\frac{\hat{d}(s)}{v_{OUT}(s)} = \frac{\hat{d}_1(s) + \hat{d}_2(s)}{v_{OUT}(s)} = F_m \left(sC_Q R_Q + A_{v0} \frac{1 + \frac{s}{\omega_z}}{1 + \frac{s}{\omega_p}} \right) = A_{v0} F_m \left(\frac{\left(1 + \frac{s}{\omega_{zcom1}}\right)\left(1 + \frac{s}{\omega_{zcom2}}\right)}{1 + \frac{s}{\omega_p}} \right) \quad (5.23)$$

Not only can the zero be derived from Type-II compensation, but the structure of the parallel paths can also generate an additional zero. These two zeros, ω_{zcom1} and ω_{zcom2}, are shown as:

$$\omega_{zcom1}, \omega_{zcom2} = \frac{A_{v0} R_C}{2 C_Q R_Q R_O} \left(1 \pm \sqrt{1 - \frac{4 C_Q R_Q R_O}{A_v C_C R_C^2}} \right) \quad (5.24)$$

With the transfer function of the power stage in Eq. (5.25), the transfer function of the system loop with the QDI circuit is expressed as:

$$\frac{\hat{v}_{OUT}(s)}{\hat{d}(s)} = G_{d0} \frac{\left(1 + \frac{s}{\omega_{ESR}}\right)}{\left(1 + \frac{1}{\omega_0 Q} s + \frac{1}{\omega_0^2} s^2\right)} \quad (5.25)$$

$$T_{Loop,QDI}(s) = \frac{\hat{v}_{OUT}(s)}{\hat{d}(s)} \cdot \frac{\hat{d}(s)}{\hat{v}_{OUT}(s)} \approx A_v F_m G_{d0} \frac{\left(1 + \frac{s}{\omega_{ESR}}\right)\left(1 + \frac{s}{\omega_{zcom1}}\right)\left(1 + \frac{s}{\omega_{zcom2}}\right)}{\left(1 + \frac{1}{\omega_0 Q} s + \frac{1}{\omega_0^2} s^2\right)\left(1 + \frac{s}{\omega_p}\right)} \quad (5.26)$$

Figure 5.23 depicts the Bode plot of the V^2 control buck converter with the QDI technique. The system dominant pole ω_p is determined by the Type-II compensator. Two zeros, ω_{zcom1} and ω_{zcom2}, are used to cancel the effect of the complex pole from the LC filter at the output stage. With a small ESR, the zero ω_{ESR} typically appears at very high frequencies. As a result, system stability can be guaranteed even though a small ESR is used.

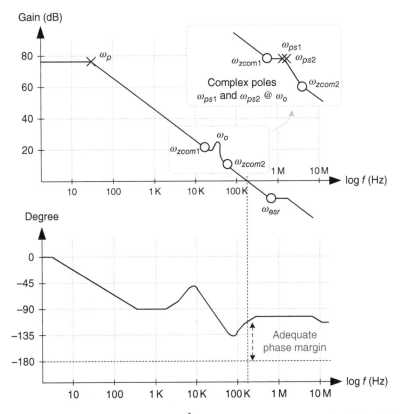

Figure 5.23 Frequency response of the V^2 control buck converter with the QDI technique

5.1.5.2 Circuit Implementation of QDI

Figure 5.24 shows a circuit implementation of the QDI technique including two stages, a quadratic differential circuit and a duty-based integral circuit. The quadratic differential circuit is composed of two cascaded VCCS structures. The VCCS structure can provide a differentiation function. After the first differentiation, the signal v_{df1} contains the undesirable first term in Eq. (5.27). After the second differentiation, the signal v_{df2} contains only information on the slope of the inductor current ripple, as shown in Eq. (5.28). In other words, the ESR effect is eliminated at this moment.

$$v_{df1}(t) = \frac{d}{dt} v_{OUT}(t) = \tau_1 \left(R_{ESR} \frac{d}{dt} i_L(t) + \frac{i_L(t)}{C_L} \right) \quad (5.27)$$

$$v_{df2}(t) = \frac{d}{dt} v_{df1}(t) = \tau_1 \tau_2 \left(\frac{1}{C_L} \frac{v_{IN} - v_{OUT}}{L} \right) \quad (5.28)$$

Through the integration stage, $v_{fa(t)}$ can be expressed as in Eq. (5.29). $v_{fa(t)}$ is purely proportional to the inductor current ripple i_L. The accuracy of v_s is then further enhanced compared with the previous results, which suffer from ESR-related distortion.

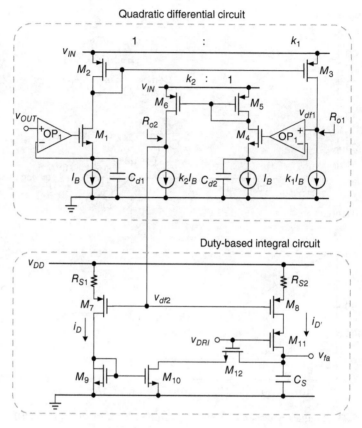

Figure 5.24 QDI implementation circuit

$$v_{fa}(t) = \int v_{df2}(t)dt = \frac{\tau_1\tau_2}{C_L}i_L(t) + \tau_3 \quad (5.29)$$

Here, τ_1, τ_2, and τ_3 are constants generated in the procedure of differentiation and integration. As mentioned in Eq. (5.18), the transfer function of the cascaded VCCS structure can be expressed as:

$$\frac{v_{df2}(s)}{v_{OUT}(s)} = (sk_1 C_{d1} R_{O1})(sk_2 C_{d2} R_{O2}) \quad (5.30)$$

In the integral circuit, according to the PWM signal, V_{DRI}, the currents i_D and $i_{D'}$ flow into the capacitor C_s during on-time and off-time periods, respectively, to achieve the integration function. The transfer function of the integral circuit can be derived in Eq. (5.31), where g_{m7} and g_{m8} are the transconductances of MOSFETs M_7 and M_8, respectively:

$$\begin{cases} \dfrac{v_{fa}(s)}{v_{df2}(s)} = \dfrac{1}{sC_s(1/g_{m8}+R_{s1})} & \text{during on-time period} \\ \dfrac{v_{fa}(s)}{v_{df2}(s)} = \dfrac{1}{sC_s(1/g_{m7}+R_{s2})} & \text{during off-time period} \end{cases} \quad (5.31)$$

Corresponding to the transfer function of QDI as shown in Figure 5.22, the exact transfer function of the QDI circuit is shown in Eq. (5.32) with the equivalent capacitance C_Q and resistance R_Q as in Eq. (5.33):

$$G_{QDI}(s) = \frac{v_s(s)}{v_{OUT}(s)} = \frac{(sk_1 C_{d1} R_{O1})(sk_2 C_{d2} R_{O2})}{sC_s(1/g_m + R_s)} = \frac{s}{C_s(1/g_m + R_s)/k_1 k_2 C_{d1} C_{d2} R_{O1} R_{O2}} = sC_Q R_Q \quad (5.32)$$

$$C_Q = \frac{k_1 k_2 C_{d1} C_{d2}}{C_s} \text{ and } R_Q = \frac{R_{O1} R_{O2}}{1/g_m + R_s} \quad (5.33)$$

The coefficients k_1 and k_2 are the current mirror ratios in the two-stage VCCS structure. According to Eq. (5.32), the equation in the S-domain between $v_{fa}(s)$ and $v_{OUT}(s)$ is derived in Eq. (5.34), and the equation in the time domain between $v_{fa}(t)$ and $v_{OUT}(t)$ is then derived in Eq. (5.35):

$$v_{fa}(s) = \frac{1}{s} \cdot \frac{k_1 k_2 C_{d1} C_{d2} R_{O1} R_{O2}}{C_s(1/g_m + R_s)} \cdot s^2 \cdot v_{OUT}(s) \quad (5.34)$$

$$v_{fa}(t) = \frac{k_1 k_2 C_{d1} C_{d2} R_{O1} R_{O2}}{C_s(1/g_m + R_s)} \int \left(\frac{d^2}{dt^2} \left(R_{ESR} i_L(t) + \frac{1}{C_L} \int i_L(t) dt \right) \right) dt$$

$$\Rightarrow \begin{cases} v_{fa}(t) = \dfrac{k_1 k_2 C_{d1} C_{d2} R_{O1} R_{O2}}{C_s C_L(1/g_{m8} + R_{s1})} i_L(t) \propto i_L(t) & \text{during on-time period} \\ v_{fa}(t) = \dfrac{k_1 k_2 C_{d1} C_{d2} R_{O1} R_{O2}}{C_s C_L(1/g_{m7} + R_{s2})} i_L(t) \propto i_L(t) & \text{during off-time period} \end{cases} \quad (5.35)$$

There are no ESR-dependent terms in Eq. (5.35), because the distortion caused by the ESR is eliminated by the QDI circuit. Consequently, the QDI function can recover a pure inductor current ripple from the output voltage ripple when a small ESR is used. Besides, compared with conventional current-sensing circuits, the QDI can also save much power consumption. Furthermore, the recovery function is not constrained by the load range. Therefore, large equivalent differential constants, C_Q and R_Q, can be generated without the implementation of large resistors and capacitors. Hence, the QDI circuit not only confirms the stability but also benefits the reduction in cost of the active silicon area. For example, the capacitors C_{d1} and C_{d2} have the same small value of 0.5 pF and the capacitor C_s is approximately equal to 2 pF, since the constants k_1 and k_2 are used to reduce the requirement of large on-chip capacitors.

Figure 5.25 depicts the operating waveform of the QDI circuit. According to Figure 5.24, the process of the QDI can be divided into three stages with the output voltage, v_{df1}, v_{df2}, and v_{fa}. The component of the voltage cross R_{ESR} deteriorates the recovering function, as explained in Figure 5.19. By this response the voltage components from R_{ESR} and C_{OUT} are depicted for v_{OUT}, v_{df1}, and v_{df2} to explain how to remove the influence from the voltage cross R_{ESR}. Through R_{ESR} and C_{OUT}, v_{OUT} information is derived in i_L. Through the differentiation in the first stage of QDI, $v_{out,ESR}$ with triangular waveform becomes $v_{df1,ESR}$ with rectangular waveform, and $v_{out,COUT}$ is recovered to $v_{df1,COUT}$ with triangular waveform. Through the differentiation in the second stage of QDI, $v_{df1,ESR}$ becomes $v_{df2,ESR}$ with spike, and $v_{df1,COUT}$

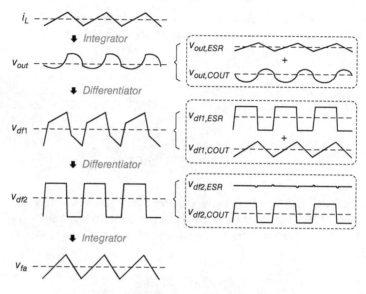

Figure 5.25 Operating waveform of the QDI circuit

becomes $v_{df2,COUT}$ with rectangular waveform. Through the integration function in the final stage, $v_{df2,COUT}$ is recovered to the triangular waveform while $v_{df2,ESR}$ contributes almost no result to v_{fa} because the spike of $v_{df2,ESR}$ is composed of high-frequency elements. As a result, the QDI can benefit the recovery function without being influenced by R_{ESR}.

5.1.6 Robust Ripple Regulator (R3)

Ripple-based control has an inherent fast transient advantage. As we mentioned before, the control of the power converter includes the set and reset operations to determine the starting of the on-time and the off-time period, respectively. The COT control has the ability to adjust the off-time period. Thus, the switching frequency can be changed in case of any load change. Once the load changes from light to heavy, the shortened off-time effectively increases the switching frequency for a fast transient response. On the contrary, in case of a heavy to light load change, both the high-side and the low-side are turned off to enter the diode emulation mode and further extend the off-time period. A load-dependent switching frequency can ensure the power-saving advantage. However, if considering in detail the scenario where the load changes during the on-time period, the COT control loses its ability to dynamically adjust the on-time period since the constant on-time is calculated by the input and output voltages. Here, we can say that the predefined on-time limits the flexibility of the on-time period if the load changes instantly and during the on-time period. To release the constraint set by the constant on-time, some ripple-based control methods can be thought of as possible control techniques to improve the overall performance.

Hysteresis control methods are one possible solution to get both flexibility in modulating on-time and off-time at the same time. First of all, observing voltage-ripple-based (VRB) hysteresis can have some disadvantages. The output voltage ripple limits the allowable voltage hysteresis window. That is to say, the smaller the hysteresis is, the smaller the noise margin is. This causes unstable operation if a small ESR value is used. Although the compensation offset voltage can be inserted in both the upper and the lower bonds to solve the unstable problem, the need for a large ESR sets the constraint of selecting the output capacitor. Furthermore, owing to the hysteresis operation, it is hard to retrieve the switching frequency and derive equally divided phases if a multiple-phase technique is needed. The advantage of the VRB hysteresis control is that the output voltage can be limited within the voltage hysteresis window and the accuracy can be ensured within a certain percentage.

In contrast, the current-ripple-based (CRB) hysteresis control can be a possible solution. The inductor current is regulated within the predefined hysteresis window, which is similar to the output voltage of the VRB regulated within the voltage hysteresis window. However, some obvious disadvantages of the CRB need to be addressed. The crucial problem is that the output voltage is floating without being regulated because there is no feedback from the output voltage. The CRB hysteresis control needs an additional voltage feedback loop to regulate the output voltage, as shown in Figure 5.26. The R_{SEN} block can sense the AC inductor current ripple. The feedback path needs to include one integrating error amplifier to improve the regulation performance of the output voltage. The output $v_{EA,L}$ of the error amplifier can be set as the lower bond of the CRB hysteresis window. With an additional and constant value of V_{HYS} as the hysteresis window, the output voltage ripple can be kept constant without being affected by any disturbance. As we know, the changing rate of $v_{EA,L}$ determines the transient response time of the CRB hysteresis control. Moreover, the changing rate of $v_{EA,L}$ is completely controlled by the compensation techniques used in the converter. There is a trade-off between the fast transient response, which benefits from CRB control and the regulation accuracy, which benefits from the error amplifier. Please refer to the compensation skills in Chapter 3. Type III is recommended to let the converter behave well with high accuracy and well-regulated output voltage.

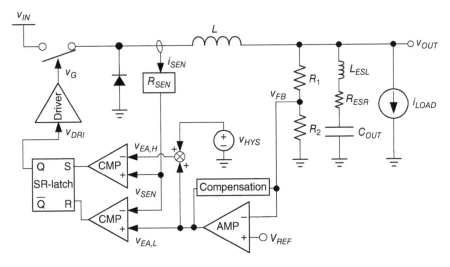

Figure 5.26 Conventional CRB hysteresis control with additional voltage feedback for improving output voltage regulation

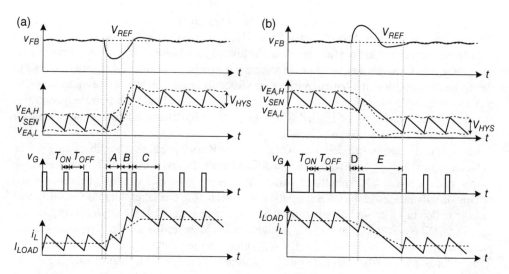

Figure 5.27 Operating waveform at (a) light-to-heavy load transient and (b) heavy-to-light load transient

Besides, the derivation of the switching frequency through the current ripple is still difficult and thus the interleaving technique is hard to apply. Furthermore, full inductor current sensing is a challenge if the sensing method is an on-chip technique. Any switching noise may deteriorate the sensing performance and thereby cause an undesired unstable operation with increased output voltage ripples.

Figure 5.27 shows the waveform in the transient process to illustrate the fast transient. The rising and falling slopes of v_{SEN} in the on-time and off-time periods are constant, respectively. In steady state, the on-time and off-time periods are fixed, modulated by the ripple of v_{SEN} and the fixed hysteretic window. The DC level of $v_{EA,L}$ is proportional to the loading conditions. During the transient from light to heavy load as shown in Figure 5.27(a), a dynamical increase in $v_{EA,L}$ can adjust the on-time and off-time periods. Periods A, B, and C illustrate that the on-time period extends and the off-time period shrinks. The duty cycle enlarges and the switching frequency becomes higher, so that sufficient power can be delivered instantly to the output to recover the voltage drop. During the transient from heavy to light load as shown in Figure 5.27 (b), periods D and E illustrate that the on-time period shrinks and the off-time period extends so that the transient performance is enhanced. Consequently, compared with other methods of ripple-based control, voltage-mode hysteresis control with an extra current loop can further improve the transient response because of the lack of constraints in the fixed on-time period, off-time period, and switching frequency.

In order to get rid of the switching noise, Figure 5.28 uses one voltage-to-current converter, which is a transconductance (gm) amplifier to generate a clean pseudo-inductor AC current ripple. One ripple capacitor C_{ripple} at the output of the gm amplifier can generate the ripple signal v_{SEN} to represent (reflect) the inductor current ripple. Owing to C_{ripple}, the switching noise can be reduced to get a noise-reduction sensing signal v_{CS}. In the frequency domain, the insertion of C_{ripple} can generate one low-frequency zero to alleviate the LC double pole effect. In other words, the whole system becomes a one-pole system. Type-I compensation

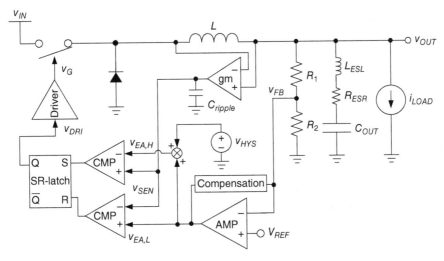

Figure 5.28 Robust ripple regulator using the regeneration of the inductor AC current ripple as PWM signal

can be used to achieve a fast transient response at the cost of the regulation performance, because a trade-off exists between system stability and regulation performance. The transient response can be as fast as that of COT control. For highly precise output supply requirements, Type II is recommended to compensate the whole system for high performance. High low-frequency gain and high bandwidth can be guaranteed at the same time after Type-II compensation. However, the BW is still limited within 10–20% of the switching frequency and thus the transient response is still slower than that of COT control. Consequently, the conclusion is that a good transient response and excellent regulation can be achieved with the implementation in Figure 5.28.

5.2 Analysis of Switching Frequency Variation to Reduce Electromagnetic Interference

Table 5.1 lists the characteristics and performance of three common control methods for the DC/DC buck converter. Using ripple-based on-time control (RBOTC) [2–21] for the buck converter meets the requirements of Soc because it exhibits the best performance compared with PWM CMC and PWM VMC [11, 22–29].

A comparison of the line transient responses of each control method shows that the PWM CMC exhibits better performance than the PWM VMC because the disturbance that occurs at the input voltage directly influences the inductor current, which can be sensed from the CF path. However, the response is constrained because the duty ratio is controlled by v_{EA}, the output of the voltage error amplifier with limited bandwidth. By contrast, RBOTC exhibits a faster transient response because the duty ratio is instantly adjusted by the variation that occurs at v_{OUT}. Similarly, RBOTC exhibits a fast load transient response. Considering the load range, PWM CMC exhibits poor performance because the current-sensing range is restricted by

Table 5.1 Characteristics of three common control methods for DC/DC buck converter

	PWM CMC	PWM VMC	RBOTC
Line transient response	Good	Poor	Excellent
Load transient response	Good	Good	Excellent
Load range	Poor	Good	Excellent
Conversion ratio range	Poor	Poor	Excellent
Conversion ratio	Poor	Good	Excellent
Quietest current	Poor	Good	Excellent
Output ripple	Excellent	Excellent	Poor (w/i SP-CAP) Excellent (w/i MLCC)
PFM at DCM for efficiency enhancement	Complex	Complex	Excellent (inherence)
Rejection of frequency variation at CCM	Excellent	Excellent	Poor

SP-CAP, specialty polymer capacitor. PFM, pulse frequency modulation.

the limited bandwidth of the current-sensing circuit. Moreover, the frequency response of PWM CMC and PWM VMC depends on the output loading conditions. As for the conversion ratio, PWM CMC exhibits poor performance because the current-sensing circuit has a limited bandwidth. Thus, an extreme duty ratio results in a short period that enables current information to be sensed. The current-sensing circuit also consumes extra power.

In contrast, the performance of the output ripple for RBOTC is constrained by the output capacitor for stability issue as discussed in Section 4.2. The constraint of selecting output capacitor with a large ESR is successfully overcome so that the output ripple can be further reduced if MLCCs are used [4–10, 29]. (This consideration will be introduced in detail later.)

Another constraint is that RBOTC is a clock-free architecture because of its ripple-based control architecture. As such, RBOTC suffers from severe switching frequency variation (Δf_{SW}) caused by disturbances from the change in input voltage (v_{IN}), output voltage (v_{OUT}), and loading current (i_{Load}).

In other words, a power converter with RBOTC becomes a frequency interference source in the range of $f_{SW} \pm 0.5\Delta f_{SW}$, which is a disadvantage. Analog circuits, such as radio frequency, audio system, analog-to-digital (ADC) converters, digital-to-analog (DAC) converters, and phase-locked loop (PLL) circuits, are sensitive to a certain level of Δf_{SW} because the frequency variation degrades their performance, as shown in Figure 5.29. Given the frequency interference with surrounding circuits through EMI, a constant f_{SW} in standing wave ratios is expected because the known noise spectrum with a small Δf_{SW} can be utilized to design the noise-filtering circuit for low EMI.

5.2.1 Improvement of Noise Immunity of Feedback Signal

Ripple-based control techniques inherently lead to an unexpected frequency variation because the modulation is determined by a small output voltage ripple and the reference voltage. Consequently, on the feedback path, the feedback signal v_{FB} with low noise immunity is easily disturbed by any perturbation, possibly from parasitic capacitive or inductive coupling paths.

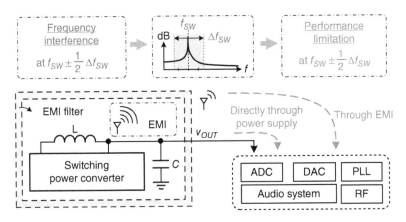

Figure 5.29 Frequency interference from SWRs limits the performance of analog circuits at $f_{SW} \pm 0.5\Delta f_{SW}$

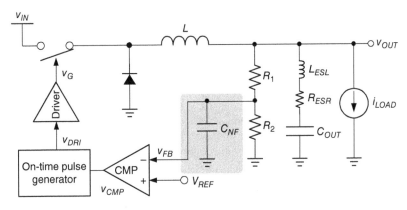

Figure 5.30 Noise filter on the feedback path

Increasing the quiescent current to suppress the voltage variation by coupling noise is one of the methods used to easily enhance the noise immunity. However, efficiency is sacrificed. The use of an additional ramp signal or additional inductor current information can also improve the noise immunity and enhance the stability. Several well-known techniques have recently been developed to solve these problems, and these will be introduced here.

5.2.2 Bypassing Path to Filter the High-Frequency Noise of the Feedback Signal

Figure 5.30 shows that a bypass capacitor C_{NF} can provide a noise-filtering path to filter out high-frequency interference at v_{FB}. However, a pole, which is constituted by $R_1 \| R_2$ and C_{NF}, causes phase delay on the feedback path. The phase margin is deteriorated and the system may suffer from sub-harmonic oscillation. Selection of the noise-filtering capacitor will influence converter stability and performance. Thus, the appropriate noise-filtering design is needed if noise immunity and system stability are considered simultaneously.

5.2.2.1 Feedforward Path to Enlarge the Feedback Signal

Figure 5.31 shows another approach to improve the noise immunity by adding one feedforward capacitor C_{FF} placed parallel with the feedback resistor R_1. C_{FF} provides the feedforward path for high-frequency components at v_{OUT} directly summed at v_{FB}. Specifically, the direct feedforward path equivalently increases the magnitude of the feedback voltage ripple of the feedback signal from the conventional feedback divider.

With the addition of C_{FF}, the transfer function of the output voltage to the feedback voltage is derived and expressed as follows:

$$\frac{\hat{v}_{FB}(s)}{\hat{v}_{OUT}(s)} = \frac{R_2}{R_1+R_2} \cdot \frac{(1+sR_1 C_{FF})}{(1+s(R_1\|R_2)C_{FF})} \quad (5.36)$$

Equivalently, the feedforward capacitor contributes one pole/zero pair. The feedforward capacitor can work correctly only if the position of the pole is higher than that of zero:

$$pole = \frac{1}{(R_1\|R_2)C_{FF}} >> zero = \frac{1}{R_1 C_{FF}} \quad (5.37)$$

Moreover, the designed zero is assumed to locate around the switching frequency to effectively conduct the inductor current ripple through the additional feedforward path, as expressed in Eq. (5.38). The transfer function of the inductor current to the feedback voltage can also be derived as in Eq. (5.40), if Eqs. (5.38) and (5.39) are substituted in Eq. (5.36):

$$(R_1\|R_2)\cdot C_{FF} \approx R_{ESR} C_{OUT} \quad (5.38)$$

$$\frac{\hat{v}_{OUT}(s)}{\hat{i}_L(s)} \approx R_{ESR} + \frac{1}{sC_{OUT}} \quad (5.39)$$

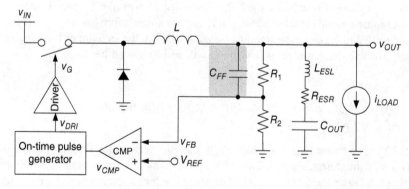

Figure 5.31 Additional feedforward capacitor C_{FF} improves noise immunity

$$\frac{\hat{v}_{FB}(s)}{\hat{i}_L(s)} \approx \frac{R_2}{R_1+R_2} \cdot \frac{1+s\left(\frac{R_1+R_2}{R_2}\right)R_{ESR}C_{OUT}}{sC_{OUT}} \quad (5.40)$$

The additional zero is effectively pushed toward the origin through increasing R_{ESR} by a factor of $(R_1+R_2)/R_2$. Consequently, C_{FF} not only increases the ripple of v_{FB}, but also releases the limitation of the large time constant constituted by R_{ESR} and C_{OUT}.

5.2.2.2 Active Controller to Enlarge the Feedback Signal (Patent-US 6958594)

Enlarging the difference voltage between two terminals, namely v_{FB} and V_{REF}, which are the two inputs of the comparator, is important to improve noise immunity. The definition of the noise margin indicates the difference voltage between the two inputs of the comparator. If the difference voltage is large, then the converter can be free from noise interference. In Figure 5.32, one current source I_f, which is controlled by the on-time controller, is used to increase the difference voltage between two terminals of the comparator corresponding to the on-time period.

The voltage on C_{NF} is charged by the on-time controlled current source. In this study, a low input voltage indicates a large on-time value and the need for a large difference voltage. By contrast, a high input voltage has a small on-time value and a small voltage difference enhancement in this technique. Considering the adaptive ramp signal generated by the constant current source, the difference voltage between v_{FB} and V_{REF} can be enlarged inversely proportional to the input voltage. In other words, distinguishing the voltage difference between v_{FB} and V_{REF} becomes easy. Thus, the noise immunity is improved. Figure 5.33(a) shows the waveforms without NME. Thus, any noise from the coupling effects will deteriorate the regulation performance. By contrast, Figure 5.33(b) shows the advantage of enlarging the difference voltage to obtain high noise immunity, where the parameter β is the ratio of R_2 over R_1+R_2. A large noise margin can tolerate a large noise from the coupling effects.

In Figure 5.33, the on-time period is relatively short to generate an adequate difference voltage under the high input voltage condition. One possible scenario may fail to ensure a sufficient difference voltage to inhibit the sub-harmonic effect. Thus, a robust and flexible method as

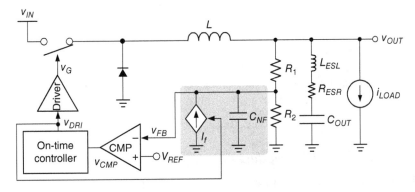

Figure 5.32 Noise margin is enhanced by a constant current source

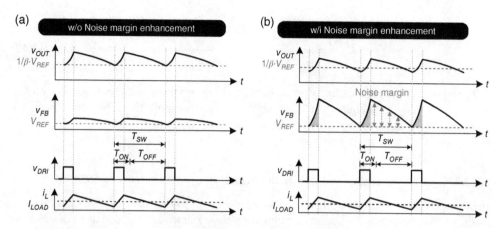

Figure 5.33 (a) Poor noise margin and low noise immunity. (b) Noise margin is enhanced by the addition of one current source corresponding to the adaptive on-time value

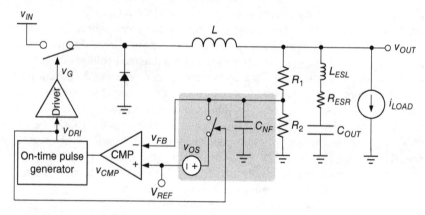

Figure 5.34 Noise margin is enhanced by a constant DC offset voltage

presented in Figure 5.34 can enhance the noise margin by using an additional constant offset voltage V_{OS} without being affected by the adaptive on-time period. Figure 5.35 shows the waveforms that illustrate the correct operation. Any noise will not cause large frequency variations. This phenomenon is assumed to be a pseudo-constant switching frequency control. The EMI problem can be alleviated by enhancing the noise margin. However, the additional offset voltage distorts the DC level of the feedback signal in case of the additional DC offset voltage. The offset voltage can be discharged through the resistors but can influence the DC regulation of the output voltage, which is an evident disadvantage.

5.2.3 Technique of PLL Modulator

The PLL modulator is applied to on-time control to maintain a constant operating frequency in steady state even under different v_{IN}, v_{OUT}, and i_{Load} [11, 12, 23–26]. The

Ripple-Based Control Technique Part II

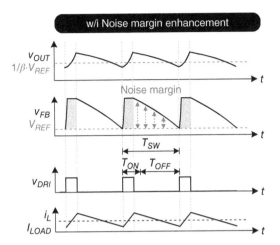

Figure 5.35 Noise margin is enhanced by the addition of one constant DC offset voltage without being affected by the adaptive on-time value

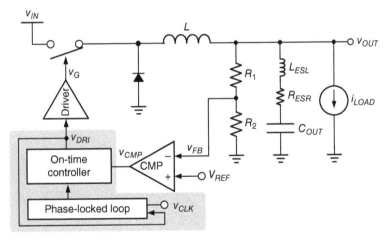

Figure 5.36 Switching frequency of the converter can be synchronized with v_{CLK} from the Soc through the PLL modulator

PLL modulator can control the on-time controller and adjust the corresponding on-time period based on the reference constant clock signal, which originates from the Soc, to synchronize the switching frequency without being influenced by any disturbances. By intuition, the performance can be improved significantly by the PLL modulator. However, the circuit architecture becomes complex. Thus, both the cost and the quiescent power loss increase (Figure 5.36).

5.2.4 Full Analysis of Frequency Variation under Different v_{IN}, v_{OUT}, and i_{Load}

5.2.4.1 Operation of On-Time Control Switching Converter

Figure 5.37 shows the conventional architecture of the DC/DC buck converter with on-time control. The pre-regulator converts the input voltage (v_{IN}) into the core voltage (v_{core}) to supply the controllers, which are designed by core devices. The feedback loop monitors the feedback voltage (v_{FB}) using the voltage divider to regulate v_{OUT}. The feedback loop mainly comprises the comparator (CMP), on-time timer, and slew rate (SR)-latch.

Figure 5.38 illustrates the timing diagram in the steady state, where the inductor current i_L conforms to the principle of voltage-second balance for stable regulation. The switching cycle

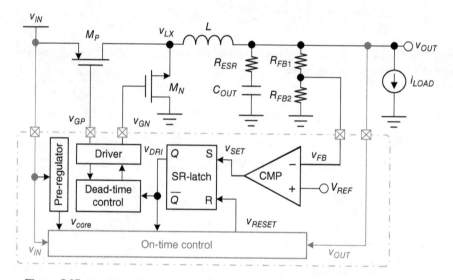

Figure 5.37 Architecture of conventional on-time controlled DC/DC buck converter

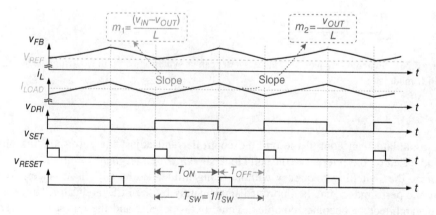

Figure 5.38 Timing diagram of the on-time controlled DC/DC buck converter

starts at the beginning of T_{ON}, which is triggered by the setting signal v_{SET} and terminated by the resetting signal v_{RESET}. v_{SET} is the output of the CMP and reflects the energy request from the output voltage. By contrast, v_{RESET} is the output of the on-time timer to switch the on-time period to the off-time period. The operation during T_{ON} refers to the charging path to increase the inductor current. T_{ON} is triggered when v_{FB} falls below the reference voltage V_{REF}. The duty ratio D is determined by the relationship between the input voltage v_{IN} and the output voltage v_{OUT}. Consequently, the switching frequency f_{SW} in the on-time control can be determined by the predefined T_{ON} and its corresponding D, where T_{ON} has some constraints to ensure system stability.

5.2.4.2 Analysis of Switching Frequency Variation

The switching frequency of the buck converter in the CCM is expressed as:

$$f_{SW} = D \cdot \frac{1}{T_{ON}} \tag{5.41}$$

Ideally, the slope of i_L is proportional to $v_{IN} - v_{OUT}$ and $-v_{OUT}$ during the on-time and off-time periods, respectively. The ideal duty ratio (D_{ideal}) is equal to the ratio of v_{OUT} to v_{IN} in a lossless ideal case, as shown in Eq. (5.42), if v_{OUT} is well regulated:

$$D_{ideal} = \frac{v_{OUT}}{v_{IN}} \tag{5.42}$$

Similarly, the ideal on-time period $T_{ON(ideal)}$ in the conventional architecture should be proportional to v_{OUT}/v_{IN} in Eq. (5.43) if the switching period T_{SW} ($=1/f_{SW}$) is constant:

$$T_{ON(ideal)} = \frac{v_{OUT}}{v_{IN}} \cdot T_{SW} = D_{ideal} \cdot T_{SW} = \frac{D_{ideal}}{f_{SW}} \tag{5.43}$$

The switching frequency f_{SW} can then be expressed as:

$$f_{SW} = D_{ideal} \cdot \frac{1}{T_{ON(ideal)}} \tag{5.44}$$

Moreover, Eq. (5.45) expresses the actual duty ratio (D_{actual}), which is defined as the ratio of the actual on-time period $T_{ON(actual)}$ to T_{SW} in Eq. (5.46):

$$D_{actual} = \frac{T_{ON(actual)}}{T_{SW}} = T_{ON(actual)} \cdot f_{SW} \tag{5.45}$$

$$\text{where } T_{SW} = T_{ON(actual)} + T_{OFF(actual)} \tag{5.46}$$

Therefore, the constant switching frequency f_{SW} is expressed as:

$$f_{SW} = \frac{1}{T_{SW}} = \frac{D_{actual}}{T_{ON(actual)}} \tag{5.47}$$

D_{actual} is larger than D_{ideal} because more energy should be derived from the power source if the possible power loss is considered when all parasitic effects are included. f_{SW} is also drastically perturbed by different v_{IN}, v_{OUT}, and loading current i_{Load}, because $T_{ON(ideal)}$ in Eq. (5.43) cannot compensate for the change in D_{actual}.

To completely analyze the influence of Δf_{SW}, Figure 5.39(a) is discussed with the addition of parasitic resistance, including the direct current resistance of the inductor (R_{DCR}), as well as the on-resistance $R_{on,P}$ and $R_{on,N}$ of the power MOSFETs, M_P and M_N, respectively. Figure 5.39(b) shows the actual and non-ideal power stage of the buck converter if parasitic resistances are considered. Figure 5.39(c), (d) illustrates the energy delivery paths, including paths I and II,

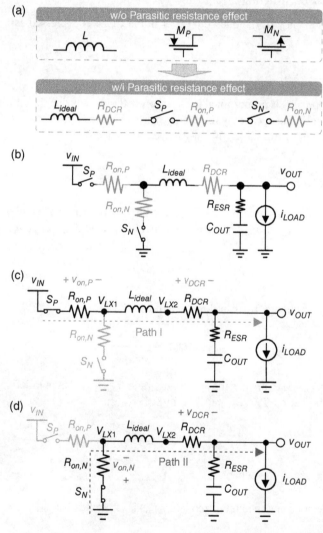

Figure 5.39 (a) Consideration of parasitic resistance. (b) Power stage of buck converter by considering the parasitic resistance. (c) Energy delivery path during T_{ON}. (d) Energy delivery path during T_{OFF}

during T_{ON} and T_{OFF}, respectively. Parasitic resistances result in voltage drops, $v_{on,P}$, $v_{on,N}$, and v_{DCR}. All voltage drops are dependent on i_{Load}, as shown:

$$\begin{cases} v_{on,P} = R_{on,P} \cdot i_{Load} \\ v_{on,N} = R_{on,N} \cdot i_{Load} \\ v_{DCR} = R_{DCR} \cdot i_{Load} \end{cases} \quad (5.48)$$

Figure 5.40 illustrates the frequency variation as a result of parasitic resistances. $v_{on,P}$, $v_{on,N}$, and v_{DCR} lead the voltage variation across the inductor L. The waveform shows that the voltage of nodes v_{LX1} and v_{LX2} affects the cross voltage of L_{ideal}, and subsequently results in the gradual steep rising and falling of slopes with extra term $-(v_{on,P} + v_{DCR})$ and $v_{on,N} + v_{DCR}$, respectively. In other words, the cross voltage of the inductor decreases and increases as i_L flows through paths I and II, respectively. The voltages of v_{LX1} and v_{LX2} with and without considering the parasitic resistance effect are listed in Table 5.2. Even if v_{IN} and v_{OUT} are not modified, the constant T_{ON} at different loading conditions results in frequency variation because different slopes of i_L occur at different loading conditions.

According to the voltage/second balance in steady state, Eqs. (5.49) and (5.50) are obtained by assuming that i_L rises and falls linearly during T_{ON} and T_{OFF}, respectively, where Δi_L is the peak-to-peak ripple current of i_L:

$$[(v_{IN} - v_{OUT}) - (v_{on,P} + v_{DCR})] = L_{ideal} \cdot \frac{\Delta i_L}{D_{actual} \cdot T_{SW}} \quad (5.49)$$

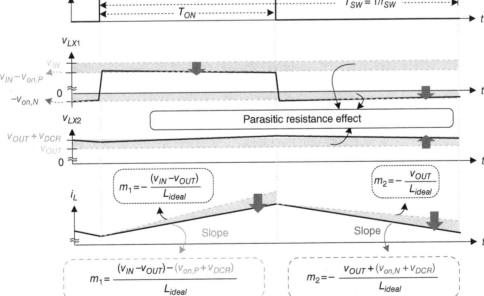

Figure 5.40 Influence of parasitic resistances on v_{LX1}, v_{LX2}, and i_L slope

Table 5.2 Voltage across the inductor w/i and w/o considering parasitic resistances

	Ideal case: w/o parasitic resistance effect		Actual case: w/i parasitic resistance effect	
	v_{LX1}	v_{LX2}	v_{LX1}	v_{LX2}
Path I	v_{IN}		$v_{IN} - v_{on,P}$	
Path II		v_{OUT}		$v_{OUT} + v_{DCR}$

$$[v_{OUT} + (v_{on,N} + v_{DCR})] = L_{ideal} \cdot \frac{\Delta i_L}{(1 - D_{actual}) \cdot T_{SW}} \quad (5.50)$$

D_{actual} in Eqs. (5.51) and (5.52) is derived from Eqs. (5.49) and (5.50), respectively, during T_{ON} and T_{OFF}:

$$D_{actual} = \frac{v_{OUT} + (R_{on,N} + R_{DCR}) \cdot i_{Load}}{v_{IN} - (R_{on,P} - R_{on,N}) \cdot i_{Load}} \quad (5.51)$$

$$D_{actual} = \frac{D_{ideal} + \dfrac{(R_{on,N} + R_{DCR}) \cdot i_{Load}}{v_{IN}}}{1 - \dfrac{(R_{on,P} - R_{on,N}) \cdot i_{Load}}{v_{IN}}} \quad (5.52)$$

By substituting Eq. (5.52) into Eq. (5.41), the switching frequency is derived in Eq. (5.53). Compared with Eq. (5.44), Eq. (5.53) reveals that the switching frequency is dependent on several parasitic factors:

$$f_{SW} = \frac{D_{ideal} + \dfrac{(R_{on,N} + R_{DCR}) \cdot i_{Load}}{v_{IN}}}{1 - \dfrac{(R_{on,P} - R_{on,N}) \cdot i_{Load}}{v_{IN}}} \cdot \frac{1}{T_{ON(actual)}} \quad (5.53)$$

Figure 5.41 presents four cases to depict Δf_{SW} at different v_{OUT}, v_{IN}, and i_{Load} according to Eq. (5.53). The desired f_{SW} is 800 kHz in cases (a) and (b), and the desired f_{SW} is 2.5 MHz in cases (c) and (d). In all these cases, the change in i_{Load} results in more significant Δf_{SW} when v_{OUT} is at low values. A comparison of cases (a) and (b) reveals that Δf_{SW} in case (b) is the worse because of the serious parasitic effect. Cases (c) and (d) also have similar results. Comparing the desired f_{SW} of 800 kHz with 2.5 MHz shows that the case with high desired f_{SW} has the worse Δf_{SW} at the load change from 0.3 to 1.7 A.

Figure 5.42 depicts Δf_{SW} with specific load changes versus the desired f_{SW} and v_{OUT}. This characteristic can also be obtained using Eqs. (5.52) and (5.53). Considering that v_{OUT}, v_{IN}, and the parasitic resistances remain constant, ΔD_{actual} with a certain load change remains constant. However, a high desired f_{SW} represents a short $T_{ON(actual)}$, and Δf_{SW} is amplified from D_{actual} by a large value of $1/T_{ON(actual)}$.

In conclusion, Δf_{SW} should not be neglected. A conventional $T_{ON(ideal)}$ as in Eq. (5.43) cannot compensate for the variation in D_{actual} as in Eq. (5.52). Furthermore, Eq. (5.53) reveals that

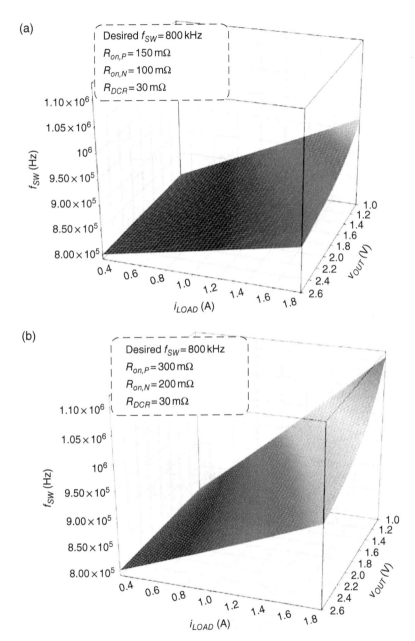

Figure 5.41 Relationship among f_{SW}, i_{Load}, and v_{OUT} when v_{IN} = 3.3 V: (a) desired f_{SW} = 800 kHz, $R_{on,p}$ = 150 mΩ, $R_{on,N}$ = 100 mΩ, R_{DCR} = 30 mΩ; (b) desired f_{SW} = 800 kHz, $R_{on,p}$ = 300 mΩ, $R_{on,N}$ = 200 mΩ, R_{DCR} = 30 mΩ; (c) desired f_{SW} = 2.5 MHz, $R_{on,p}$ = 150 mΩ, $R_{on,N}$ = 100 mΩ, R_{DCR} = 30 mΩ; (d) desired f_{SW} = 2.5 MHz, $R_{on,p}$ = 300 mΩ, $R_{on,N}$ = 200 mΩ, R_{DCR} = 30 mΩ

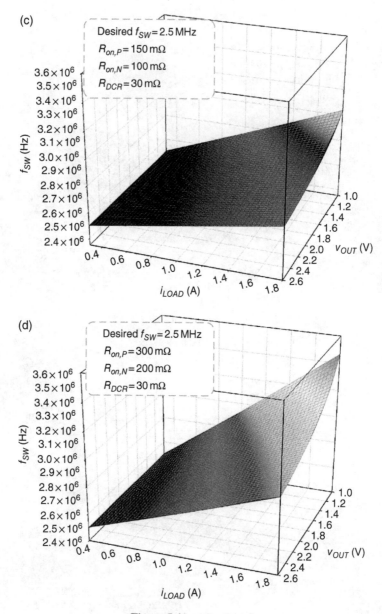

Figure 5.41 (*Continued*)

designing a proper T_{ON} to completely compensate for the parasitic effects is the challenge for on-time control to maintain a constant f_{SW}.

The designer should focus on reducing the frequency variation because the desired output voltage is not scaled down, and the switching does not increase significantly as in the past. However, the transistor density increases drastically for Soc applications because of the recent

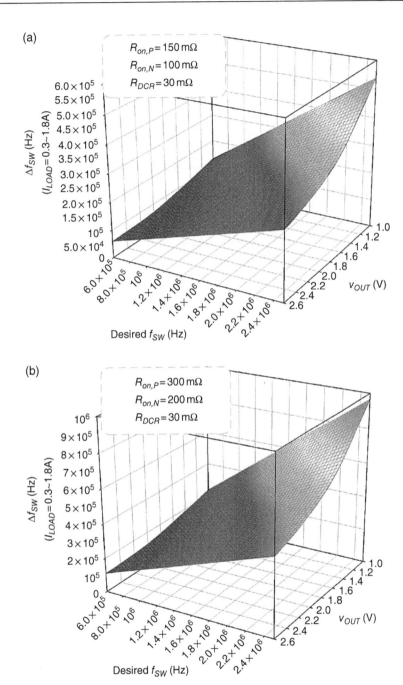

Figure 5.42 Variation of f_{SW} in load change from 0.3 to 1.8 A versus different desired f_{SW} and v_{OUT} when (a) $R_{on,p}$ = 150 mΩ, $R_{on,N}$ = 100 mΩ, R_{DCR} = 30 mΩ and (b) $R_{on,p}$ = 300 mΩ, $R_{on,N}$ = 200 mΩ, R_{DCR} = 30 mΩ

advances in nanometer-scale processes. Meanwhile, the power density increases significantly, and the supply voltage is scaled down, as shown in Figure 5.43. The power density increases, although the supply voltage is scaled down from 5 V in 0.5 μm to 0.85 V in 10 nm. The trend follows, and may even be beyond Moore's law. Consequently, power management systems or smart power converters, which feature high power and high performance, are urgently needed for today's consumer electronics. If the output voltage provided by power converters is scaled down below 1 V for circuits in the advanced process, f_{SW} will also increase to several megahertz for compact size and small output ripple. Predicting the variation of the switching frequency does not influence the Soc performance of some filters, and thus the reduction in switching frequency variation is expected.

Although Eq. (4.3) is a roughly derived result, understanding intuitively why the switching frequency varies in different v_{IN} and v_{OUT} is useful. However, D_{actual} is determined not only by v_{IN} and v_{OUT}, but also by i_{Load} and parasitic resistances, as shown in Eq. (5.51). Specifically, Eq. (5.51) manifests that the power MOSFET, inductor, and load range should be selected carefully and specified according to the tolerance of frequency variation. To release the frequency variation, a conventional T_{ON} as derived in Eq. (5.43) cannot compensate for D_{actual} in Eq. (5.51). Furthermore, this equation reveals the proper design of T_{ON} to compensate for completely parasitic facts, which is a challenge for on-time control with constant switching frequency. The following subsections will introduce useful techniques to maintain a constant switching frequency by considering the cases under different v_{IN}, v_{OUT}, and i_{Load}.

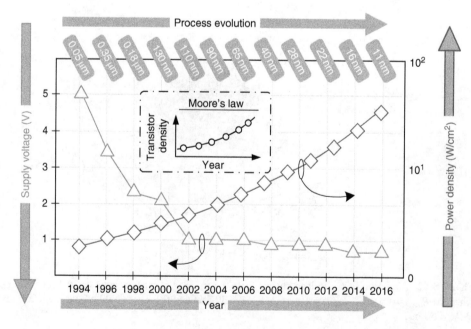

Figure 5.43 Trends of supply voltage and power density with the evolution of process technology

5.2.5 Adaptive On-Time Controller for Pseudo-Constant f_{SW}

5.2.5.1 Fixed On-Time Controller (Fixed T_{ON})

Figure 5.44 shows the fundamental structure of the basic fixed on-time controller to realize the fixed pulse of one on-time period. The current I_{REF} charges C_{ON} to increase v_{Ramp} from zero to V_{REF}. T_{ON} can be derived as in Eq. (5.44), with its characteristics proportional to V_{REF} and inversely proportional to I_{REF}. T_{ON} can be adjusted by different C_{ON}, I_{REF}, and V_{REF}:

$$T_{ON} = \frac{C_{ON}}{I_{REF}} \cdot V_{REF} \qquad (5.54)$$

Figure 5.45 illustrates the operating waveforms. The switching period T_{SW} is defined as a period that starts when v_{FB} falls below V_{REF} and terminates when the next switching starts again. At the beginning of T_{SW}, the on-time phase is triggered when v_{FB} falls below V_{REF}. v_{DRI} is set high from low. In the initial condition of the on-time pulse, v_{Ramp} is reset to zero by the switch S_{ON} controlled by the signal v_{DRI}. Thereafter, the current source starts to charge C_{ON}

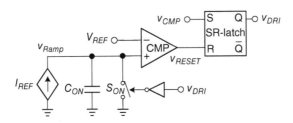

Figure 5.44 Model of the basic constant on-time controller

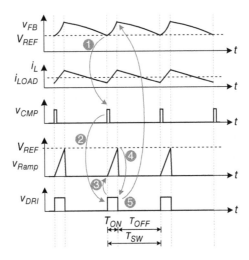

Figure 5.45 Waveforms to illustrate the operation of the on-time controller

with an increasing signal v_{Ramp} during the T_{ON} period. Third, when v_{Ramp} reaches the upper threshold, V_{REF}, the comparator output, v_{DRI}, is set from high to low and the T_{ON} period is terminated. Meanwhile, v_{Ramp} is reset to zero by v_{DRI} and prepared to trigger in the next period.

In this design, the switching frequency varies with different v_{IN} and v_{OUT}. The following topics examine the analyses and different methods to pursue the constant switching frequency independent of operating conditions, such as v_{IN}, v_{OUT}, i_{Load}, and the characteristics of components in the power stage.

5.2.5.2 Adaptive On-Time Controller Utilizing Information on v_{IN} and v_{OUT}

In Eq. (4.3), the switching frequency is determined by three factors, T_{ON}, v_{IN}, and v_{OUT}. A constant T_{ON} implies that the switching frequency varies at different v_{IN} and v_{OUT}. For example, this situation may occur if v_{IN} is provided by a battery in portable applications. When the electronic device continues to operate and the power from the battery is consumed for some time, the voltage level of the battery continuously decreases, and then v_{IN} decreases. This situation shows an approach to eliminate the switching frequency variation. The value of T_{ON} is adjusted according to the input voltage v_{IN} and the output voltage v_{OUT} in the adaptive on-time pulse generator. According to Eq. (4.3), T_{ON} is inversely proportional to v_{IN} while it is proportional to v_{OUT}, as expressed in Eq. (5.55) (Figure 5.46):

$$T_{ON} \propto \frac{v_{OUT}}{v_{IN}} \tag{5.55}$$

As a result, Eq. (5.56) shows that a pseudo-constant switching frequency can be achieved, such that the switching frequency is not affected by the variations of v_{IN} and v_{OUT}:

$$f_{SW(new)} \propto \frac{v_{OUT}}{v_{IN} \cdot \left(\dfrac{v_{OUT}}{v_{IN}}\right)} = \text{constant} \tag{5.56}$$

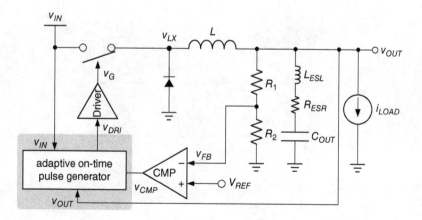

Figure 5.46 The adaptive on-time is adjusted by v_{IN} and v_{OUT}

Ripple-Based Control Technique Part II

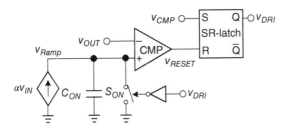

Figure 5.47 Model of the basic adaptive on-time control to obtain a pseudo-constant switching frequency

Figure 5.47 shows the basic structure of the adaptive on-time pulse generator to realize the pseudo-constant switching frequency in Eq. (5.55). The charging current source, which is equal to $\alpha \cdot v_{IN}$, is used to charge the capacitor C_{ON}, where α is a constant value.

Assuming that v_{Ramp} ramps up linearly, the increase or decrease in v_{OUT} results in the charging time as v_{Ramp} increases or decreases because the upper threshold voltage is v_{OUT}. By contrast, a different v_{IN} results in different charging currents of C_{ON} and in different rising speeds at v_{Ramp}. In other words, increasing/decreasing v_{IN} causes a faster or slower rising speed at v_{Ramp}, and then changes the charging time as v_{Ramp} decreases or increases. Consequently, T_{ON} is derived as Eq. (5.57) with its characteristics proportional to v_{OUT} and inversely proportional to v_{IN}:

$$T_{ON} = \frac{C_{ON}}{\alpha} \cdot \frac{v_{OUT}}{v_{IN}} \qquad (5.57)$$

The basic adaptive on-time control can maintain a pseudo-constant switching frequency f_{SW} by adjusting the on-time value corresponding to the varying v_{IN} and v_{OUT}.

Figure 5.48(a), (b) describes the different cases by comparing between constant and adaptive on-time. If we use case I as reference, when v_{IN} decreases, we observe different results in case II with constant on-time and case III with adaptive on-time. The duty cycle changes because the decrease in v_{IN} eases the decreasing slope of i_L. If T_{ON} remains constant, as shown in case II, then T_{OFF} will shrink accordingly and cause a short switching period T_{SW}. In other words, at heavy loads, f_{SW} can increase because of the constant T_{ON}. The f_{SW} variation is summarized in Table 5.3. By contrast, the on-time value can adjust dynamically according to v_{IN} and v_{OUT}, such that f_{SW} remains constant at different v_{IN} and v_{OUT} conditions. Case III shows that f_{SW} remains constant because of the dynamic T_{ON} when v_{IN} decreases. As shown in Figure 5.47, decreasing v_{OUT} creates a short time for v_{Ramp} to rise to the upper threshold, v_{OUT}, such that T_{ON} extends. Consequently, the desired T_{ON} for constant T_{SW} under different v_{IN} and v_{OUT} is summarized in Table 5.3.

In this work, the start-up operation should be considered carefully. When the converter is triggered from the shut-down situation, the initial value of v_{OUT} is zero. This zero value makes the on-time pulse extremely short, because v_{Ramp} rises quickly to v_{OUT} whenever C_{ON} is charged. As a result, C_{OUT} in the output stage is difficult to charge because the on-time period is extremely short, although v_{CMP} remains high to represent the need for energy replenishment. For this reason, the complete circuit design, as shown in Figure 5.49, will utilize the minimum

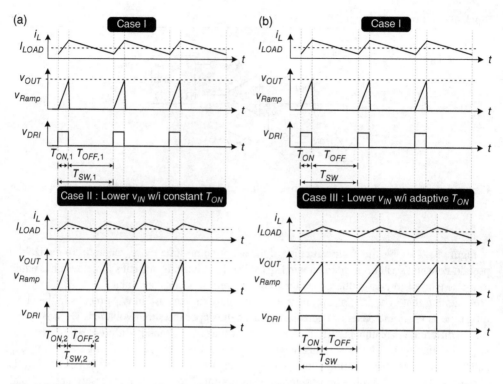

Figure 5.48 Characteristics of frequency change under different v_{IN} in comparison with (a) constant T_{ON} and (b) adaptive T_{ON}

Table 5.3 Characteristics of the summarized variation in on-time control

Condition	Duty variation (ΔD)	Switching frequency variation (Δf_{SW}) with constant T_{ON}	Desired T_{ON} for constant f_{SW}
$v_{IN} \Downarrow$	$\Uparrow$	$\Downarrow$	$\Uparrow$
$v_{IN} \Uparrow$	$\Downarrow$	$\Uparrow$	$\Downarrow$
$v_{OUT} \Downarrow$	$\Downarrow$	$\Uparrow$	$\Downarrow$
$v_{OUT} \Uparrow$	$\Uparrow$	$\Downarrow$	$\Uparrow$

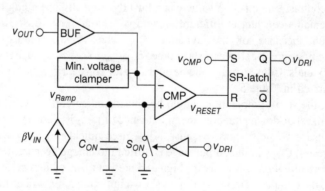

Figure 5.49 Model of the basic adaptive on-time control with buffer and minimum voltage clamper

voltage clamper to control the minimum voltage of the minus input of CMP, and thus ensure that the minimum on-time value during start-up is adequate. Moreover, v_{OUT} will be fed into CMP through a buffer (BUF) because v_{OUT} contains switching noise.

In contrast, it's also an important design issue to discuss how to implement the charging current circuit, which is proportional to v_{IN}. The following examples give the reader the circuit implementation for reference.

5.2.5.3 Circuit Implementation (1)

Figure 5.50 shows the circuit implementation of the Type-I charging current circuit for the model in Figure 5.47. Considering that the value of v_{IN} may be higher than the normal voltage of core devices in the process, v_{IN} is often fed first into voltage dividers to scale down its value. The voltage divider is composed of R_{ON} and R_{a1}. v_1 is proportional to v_{IN}:

$$v_1 = v_{IN} \cdot \frac{R_{a2}}{R_{a1} + R_{a2}} \qquad (5.58)$$

The amplifier with negative feedback then converts i_1 from v_1 and R_{a3}. Given the current mirror ratio of n between M_3 and M_4, i_{ON} can be derived as in Eq. (5.59). In other words, the charging current i_{ON} is proportional to v_{IN}:

$$i_{ON} = n \cdot \frac{v_1}{R_{a3}} = v_{IN} \cdot \left(n \cdot \frac{R_{a2}}{(R_{a1} + R_{a2}) \cdot R_3} \right) \qquad (5.59)$$

To adjust the on-time period for the user, the designer can set R_{a1}, R_{a2}, or R_{a3} as a discrete component for replacement with a specific value. The weakness of this circuit is that the voltage divider has a trade-off between silicon area and quiescent current. The amplifier with negative feedback also consumes extra silicon area for adequate compensation. Additionally, the operating range of v_{IN} is limited by the consideration of the input common-mode range of the amplifier.

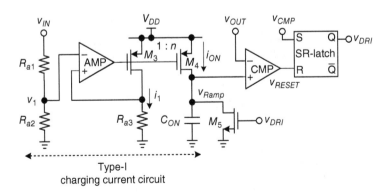

Figure 5.50 Implementation of Type-I charging current circuit

5.2.5.4 Circuit Implementation (2)

Figure 5.51 shows the implementation of the Type-II charging current circuit. This circuit uses a self-bias current structure composed of R_{ON} and M_1 to generate i_1. The current is mirrored by ratios n and m, and then generates i_{ON} as expressed in Eq. (5.60). The factor v_{GS} deteriorates the linearity between i_{ON} and v_{IN}:

$$i_{ON} = mn \cdot \frac{v_{IN} - v_{GS}}{R_{ON}} \quad (5.60)$$

However, when v_{IN} is much higher than v_{GS}, v_{GS} can be neglected and Eq. (5.60) can be simplified as:

$$i_{ON} \approx m \cdot n \cdot \frac{v_{IN}}{R_{ON}} \quad (\text{when } v_{IN} \gg v_{GS}) \quad (5.61)$$

In this design, R_{ON} can be set as a discrete component and replaced by an adequate value for the desired on-time period. The value of R_{ON} sacrifices no cost of the silicon area, such that the quiescent current is easily reduced and power loss is saved.

5.2.5.5 Adaptive On-Time Controller with Information on v_{LX} and v_{OUT}

Figure 5.52 shows the structure of the adaptive on-time pulse generator to alleviate the switching frequency variation caused by different loading conditions, as derived in Eq. (5.13).

Compared with the previous structure depicted in Figure 5.47, the voltage-controlled current source in Figure 5.52 depends on v_{LX}, which is the node connected to the high-side switch and the inductor. Through the on-resistance of the high-side switch, v_{LX}, as expressed in Eq. (5.62), changes with different loading conditions i_{Load} during the on-time period:

$$v_{LX} = v_{IN} - i_L(t) \cdot R_{ON} \quad (5.62)$$

To simplify the analysis, $i_L(t)$ is replaced by the average inductor current, which is equal to I_{Load}:

$$v_{LX} = v_{IN} - I_{Load} \cdot R_{ON} \quad (5.63)$$

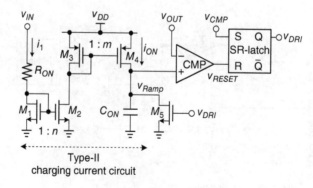

Figure 5.51 Implementation of Type-II charging current circuit

Ripple-Based Control Technique Part II

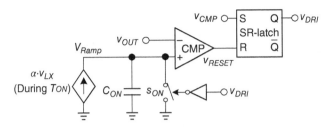

Figure 5.52 Structure of adaptive on-time controller with the addition of v_{LX} information

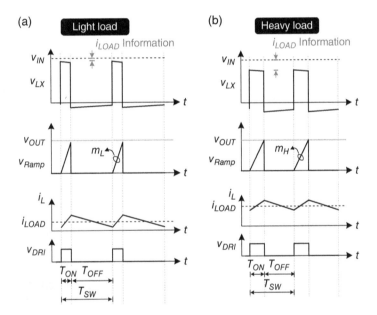

Figure 5.53 Waveforms illustrate the load current affecting the v_{Ramp} slope value (a) at light loads and (b) at heavy loads

Then, T_{ON} is derived in Eq. (5.64) and adjusted by v_{OUT}, v_{IN}, and I_{Load}:

$$T_{ON} = \frac{C_{ON}}{\alpha} \cdot \frac{v_{OUT}}{v_{IN} - I_{Load} \cdot R_{ON}} \quad (5.64)$$

As a result, if I_{Load} increases, then the T_{ON} period will increase according to the change in I_{Load}, such that the loading effect on duty can be eliminated. According to Eq. (5.13), although the factor of I_{Load} cannot be calibrated completely, the switching frequency variation can be more or less alleviated.

Figure 5.53 illustrates the operating waveforms for the adaptive on-time controller with the addition of v_{LX} information. As expressed in Eq. (5.62), I_{Load} information can be included via

v_{LX} as the derived equation in Eq. (5.63). Compared with different loading conditions, the voltage level of v_{LX} during the on-time period is lower at heavy loads, and then produces v_{Ramp} with a smaller slope, or $m_H < m_L$. As a result, v_{Ramp} spends a long time to ramp up to v_{OUT}, such that T_{ON} can be extended slightly at heavy loads. This approach is the simplest to apply to compensate for the frequency variation in I_{Load}.

5.2.5.6 Circuit Implementation (3)

Figure 5.54 shows an adaptive on-time controller. The charging current i_{ON} includes both i_1 and i_2, as expressed in Eq. (5.65). When the value of R_{a1} is equal to the value of R_{a2}/n, the second factor is eliminated, such that i_{ON} is completely proportional to v_{IN}:

$$I_{ON} = i_1 + i_2 = \left(\frac{v_{LX} - v_{Ramp}}{R_{a1}}\right) + \left(n \cdot \frac{v_{Ramp}}{R_{a2}}\right) = \frac{v_{LX}}{R_{a1}} + v_{Ramp} \cdot \left(-\frac{1}{R_{a1}} + \frac{n}{R_{a2}}\right) \quad (5.65)$$

As a result, the charging current i_{ON} flows into C_{ON}, and the on-time period is determined as expressed in Eq. (5.66), which is similar to Eq. (5.64):

$$T_{ON} = \frac{C_{ON}}{i_1 + i_2} \cdot v_{OUT} = C_{ON} \cdot R_{a1} \cdot \frac{v_{OUT}}{v_{LX}} \quad (5.66)$$

The value of R_{a1} equal to the value of R_{a2}/n can be achieved by good layout matching skills on the same chip without the influence of process variation. However, the on-time period is difficult to adjust to the desired switching frequency. Although R_{a1} and R_{a2} can be designed as discrete components, additional components require high costs and PCB area.

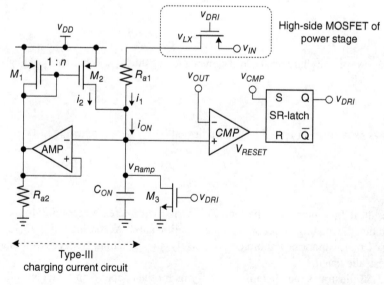

Figure 5.54 Adaptive on-time controller to obtain a reliable switching frequency without the influence of the variation in input voltage v_{IN}

According to Eq. (1.3), f_{SW} remains constant regardless of the variation from v_{OUT} and v_{IN} and i_{Load}. Compared with prior studies that utilize v_{IN} instead of v_{LX}, this work provides an accurate and linear solution to correctly minimize the variation of the switching frequency. Therefore, the approach can work as a PWM operation with constant switching frequency, although the internal clock is absent. Furthermore, by directly comparing the connection of R_{ON} to the nodes v_{IN} and v_{LX} in the previous design, which causes i_{ON} to flow continuously, a switch can be used to cut the current flowing path to prevent more power loss during the off-time period. However, this switch should be a high-voltage device if v_{IN} is higher than the normal voltage of core devices. In other words, the silicon area cost is higher. By contrast, when R_{ON} is connected to v_{LX}, the high-side switch can be utilized to generate i_{ON} only at the on-time period. v_{Ramp} is always lower than v_{OUT}. The advantages of this design include avoiding high-voltage damage caused by v_{IN} and wasteful power consumption. Moreover, the number of high-voltage devices can be reduced to minimize the cost.

5.3 Optimum On-Time Controller for Pseudo-Constant f_{SW}

v_{IN}, v_{OUT}, and the load current i_{Load} will influence the value of f_{SW}. Generally, f_{SW} increases if i_{Load} increases. For EMI consideration, the suppression of the frequency variation caused by i_{Load} is also expected. This section introduces the detailed analysis and importance of f_{SW}.

The previous literature proposed several techniques to alleviate variation because of the clock-free architecture RBOTC. An extra external clock or PLL is utilized at the expense of circuit complexity, silicon area, and cost [11, 12, 23–26]. By contrast, the techniques use v_{IN} and v_{OUT} information to generate adaptive on-time and maintain constant f_{SW} at different v_{IN} and v_{OUT} [7, 17, 29]. However, the performance remains constrained, and these works cannot ensure constant f_{SW} at different loading conditions because practical parasitic resistances are not considered. Although the load current information is applied by replacing v_{IN} with v_{LX}, the improvement in Δf_{SW} is not effective [7]. A previous technique also applies load information to revise the on-time period, but the compensation values are not analyzed. Deciding on the amount of compensation accurately is difficult [17] because once overcompensation occurs, Δf_{SW} deteriorates. Furthermore, deriving the quantitative Δf_{SW} analysis, the information of v_{OUT}, v_{IN}, and i_{Load} is insufficient because the assumption conditions are not suitable [17].

The literature in [29] utilized the RC network to sense i_L and adjust the adaptive on-time for pseudo-constant f_{SW}. However, the assumption of equal on-resistance value at both high- and low-side power MOSFETs is necessary but it is suitable because maintaining this assumption is difficult. Many resistors and the DCR of the inductor should also be matched exactly, which increases the design difficulty.

The reduction in Δf_{SW} needs to suitably control the on-time; alleviating Δf_{SW} also reduces EMI problems. Consequently, the quantitative analysis is complete and provides the predicting correction technique (PCT) to modulate an adaptive on-time for constant f_{SW} under different conditions, including v_{IN}, v_{OUT}, and wide-range i_{Load}. The complete parasitic resistances of each component are considered without any assumption or simplification. Only the driving signal of the high-side power MOSFET (v_{GP}) is necessary to achieve constant f_{SW} in the on-time circuit. This technique significantly reduces the complexity compared with those in previous works, which require additional v_{IN}, v_{OUT}, v_{LX}, current-sensing circuit, and parasitic resistance of components of the power stage.

5.3.1 Algorithm for Optimum On-Time Control

To completely compensate for Δf_{SW} using this approach, Eq. (5.67) is derived by rearranging Eqs. (5.49) and (5.50):

$$v_{IN} D_{actual} = v_{OUT} + D_{actual} \cdot (R_{on,P} \cdot i_{Load}) \\ + (1 - D_{actual}) \cdot (R_{on,N} \cdot i_{Load}) + R_{DCR} \cdot i_{Load} \quad (5.67)$$

An equivalent output voltage ($v_{OUT,eq}$) is defined by v_{OUT} and Δv_{par}, as expressed in Eq. (5.68), in which Δv_{par} in Eq. (5.69) represents the voltage variation when the parasitic effects are considered:

$$v_{OUT,eq} = v_{OUT} + \Delta v_{par} \quad (5.68)$$

$$\Delta v_{par} = R_{par} \cdot i_{Load} \quad (5.69)$$

The parasitic effects are a series of parasitic equivalent resistances (R_{par}), as expressed in Eq. (5.70):

$$R_{par} = D_{actual} \cdot R_{on,P} + (1 - D_{actual}) \cdot R_{on,N} + R_{DCR} \quad (5.70)$$

$v_{OUT,eq}$ is then expressed by D_{actual} in Eq. (5.71):

$$v_{OUT,eq} = D_{actual} \cdot v_{IN} \quad (5.71)$$

Figure 5.55 illustrates the average model for the power stage of the buck converter. All parasitic resistances are considered, and these parasitic effects are revealed by the equivalent ratio of the ideal transformer and the series of parasitic equivalent resistances R_{par} in Eq. (5.70). Instead of using v_{OUT} in the conventional design, $v_{OUT,eq}$ can be synthesized by the optimum on-time. This work utilizes v_{IN} and $v_{OUT,eq}$ to generate the optimum T_{ON} according to Eqs. (5.41) and (5.71). The solution to effectively alleviate Δf_{SW} can be realized at different v_{IN}, v_{OUT}, and i_{Load} conditions.

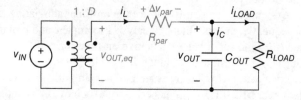

Figure 5.55 Average model for the power stage of the buck converter when all parasitic resistances are considered

5.3.2 Type-I Optimum On-Time Controller with Equivalent V_{IN} and $V_{OUT,eq}$

5.3.2.1 Architecture and Operation of Converter

The optimum on-time controller is implemented in the DC/DC buck converter as shown in Figure 5.56. This optimum on-time controller can achieve a thorough solution to predict an adequate T_{ON} and obtain a nearly constant T_{SW}. The calibration includes all the changes in v_{IN}, v_{OUT}, and i_{Load}. The optimum on-time controller technique includes the fully linear voltage-to-current generator (FLVCG), equivalent output-voltage synthesizer (EOVS), on-time modulator (OTM), and voltage clamper.

The FLVCG circuit converts v_{IN} into the current i_{ON} and obtains a linear relationship with v_{IN}. The EOVS circuit modulates $v_{OUT,eq}$ by v_{IN} and the driving signal v_{GP}. The OTM circuit outputs v_{RESET} to determine T_{ON} by i_{ON} and $v_{OUT,eq}$. According to Eq. (5.41), T_{ON} is designed as the value proportional to D_{actual}. Although numerous factors determine D_{actual} in Eq. (5.51), the PCT circuit uses v_{IN} and v_{GP} to generate an optimum T_{ON}, and thus compensate for D_{actual} without using any extra complex current-sensing circuits. Obtaining the on-resistance information of the power MOSFET and R_{DCR} of the inductor is also unnecessary.

Based on Eqs. (5.41) and (5.45), the optimum T_{ON} must be the value proportional to D_{actual}. Although various factors determine D_{actual} in Eq. (5.51), the optimum on-time controller can only use v_{IN} and v_{GP} to generate the optimum T_{ON} and compensate for D_{actual} to achieve the

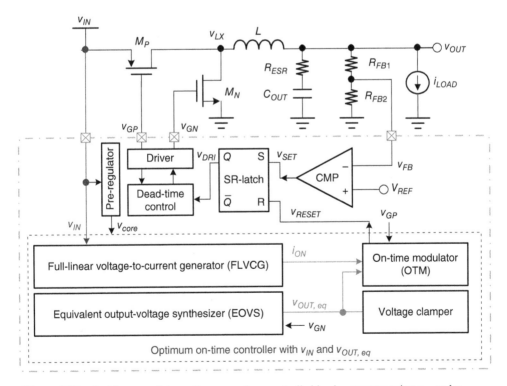

Figure 5.56 Architecture of the optimum on-time controlled buck converter using v_{IN} and $v_{OUT,eq}$

pseudo-constant f_{SW}. In other words, v_{OUT} and v_{LX} are unnecessary for v_{IN} and v_{GP}. According to Eq. (5.71), FLVCG converts v_{IN} into the current (i_{ON}), which is linear to v_{IN}. EOVS modulates $v_{OUT,eq}$ by v_{IN} and the driving signal v_{GP}. Then, OTM determines T_{ON} by i_{ON} and $v_{OUT,eq}$.

When $v_{OUT,eq}$ is directly realized by Eqs. (5.68) and (5.69), the complex current-sensing circuit is required because $v_{OUT,eq}$ includes many parasitic on-resistances. To reduce the complexity, $v_{OUT,eq}$ is synthesized by v_{IN} and v_{GP} according to Eq. (5.71). Consequently, T_{ON} is expressed as in Eq. (5.72) to achieve a pseudo-constant f_{SW}:

$$T_{ON} = \frac{v_{OUT\_eq}}{v_{IN}} \cdot T_{SW} = \frac{(D_{actual} \cdot v_{IN})}{v_{IN}} \cdot T_{SW} = \frac{D_{actual}}{f_{SW}} \tag{5.72}$$

Compared with the constant T_{ON} and the optimum T_{ON}, Figure 5.57 explains that the variable D_{actual} at different loading conditions has some effects on Δf_{SW}. Slopes $m_{1,L}$ and $m_{2,L}$ rise and fall at light loads, respectively. Slopes $m_{1,H}$ and $m_{2,H}$ rise and fall at heavy loads, respectively. Increasing i_{Load} results in a smaller $m_{1,H}$ than $m_{1,L}$ and a larger $m_{2,H}$ than $m_{2,L}$. As a result, T_{ON} is required to increase adequately at heavy loading conditions to maintain a constant f_{SW} at heavier loading conditions because of the expanding duty ratio, and vice versa.

The optimum on-time controller provides an adaptive T_{ON}, which is proportional to D_{actual} to compensate for the change in D_{actual} obtaining a nearly constant f_{SW}, as shown in Eq. (5.45). Without any complex sensing circuits, the optimum on-time controller generates $v_{OUT,eq}$ as an equivalent v_{OUT} proportional to D_{actual} and v_{IN}. By contrast, v_{Ramp} is a rising voltage with a slope value proportional to v_{IN}. T_{ON} is also determined by the period when v_{Ramp} starts to rise until its value is equal to $v_{OUT,eq}$. Consequently, T_{ON} can be expressed as in Eq. (5.73), in which the on-time capacitor C_{ON} is constant to obtain a nearly constant f_{SW}.

Figure 5.58 and Table 5.4 illustrate that the optimum on-time controller can modulate the adequate $v_{OUT,eq}$ and v_{Ramp}. Different loading conditions reflect the corresponding D_{actual} with

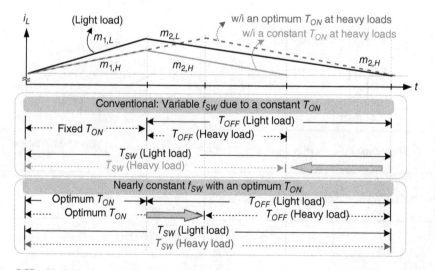

Figure 5.57 Variable f_{SW} with fixed T_{ON} and fixed f_{SW} with optimum T_{ON} at different load currents

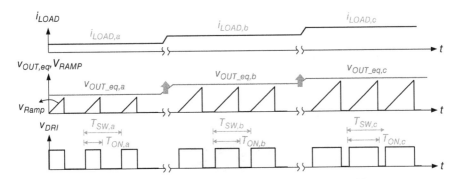

Figure 5.58 Operating waveforms of the optimum on-time controlled buck converter in case of increasing output loading current

Table 5.4 Relationship of the values in different loading conditions

i_{Load}	$i_{Load,a} < i_{Load,b} < i_{Load,c}$
D_{actual}	$D_a < D_b < D_c$
$v_{OUT,eq}$	$v_{OUT,eq,a} < v_{OUT,eq,b} < v_{OUT,eq,c}$
T_{ON}	$T_{ON,a} < T_{ON,b} < T_{ON,c}$
T_{SW}	$T_{SW,a} > T_{SW,b} > T_{SW,c}$
f_{SW}	$f_{SW,a} < f_{SW,b} < f_{SW,c}$

certain v_{OUT} and v_{IN} because of parasitic resistance effects. Figure 5.58 includes three cases with loading conditions $i_{Load,a}$, $i_{Load,b}$, and $i_{Load,c}$, where $i_{Load,a} < i_{Load,b} < i_{Load,c}$. A large i_{Load} expands D_{actual}, as shown in Eq. (5.51). Consequently, PCT predicts and generates a high $v_{OUT,eq}$ to generate a long T_{ON}. By contrast, a low $v_{OUT,eq}$ generates T_{ON} when D_{actual} shrinks at a small i_{Load}.

Figure 5.59 explains the method to recover the constant f_{SW} during the load transient. In light-load steady state, the values of D, f_{SW}, and T_{ON} are D_L, $f_{SW,L}$, and $T_{ON,L}$, respectively. In heavy-load steady state, the values of D, f_{SW}, and T_{ON} are D_H, $f_{SW,H}$, and $T_{ON,H}$, respectively. Comparing the light and heavy loads reveals that the expected result achieves an $f_{SW,L}$ equal to $f_{SW,H}$ to obtain a constant switching frequency. Now, we observe the change of D, f_{SW}, and T_{ON} if a light-to-heavy load transient is used as an example. When the load current changes from light to heavy, f_{SW} increases temporarily because D decreases and T_{ON} is not changed. In other words, $f_{SW,1}$ is larger than $f_{SW,L}$, and D_H is less than D_L. Meanwhile, the optimum on-time controller adjusts T_{ON} to obtain a large value according to the increasing duty, D_H. In other words, $T_{ON,H}$ is larger than $T_{ON,L}$. As a result, the shorter T_{ON} can modify f_{SW} to $f_{SW,H}$ from the increasing value, $f_{SW,1}$, in which $f_{SW,1}$ is larger than $f_{SW,H}$. Finally, the system enters into heavy-load steady state. According to Eqs. (5.47) and (5.71), T_{ON} adjusted by the optimum on-time controller can achieve an $f_{SW,L}$ equal to $f_{SW,H}$. By contrast, the procedure is similar, although the load current changes from heavy to light.

Figure 5.60 illustrates the timing diagram of the optimum on-time controller in any load current change. The EOVS circuit modulates $v_{OUT,eq}$ as $v_{OUT,eq,a}$ and T_{ON} as $T_{ON,a}$ in light

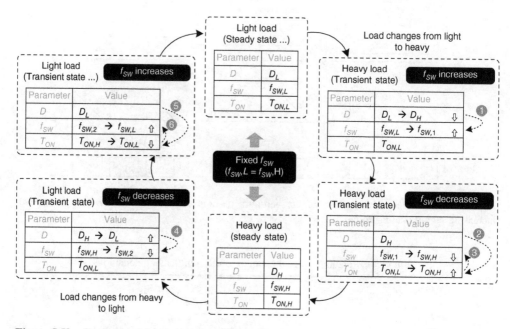

Figure 5.59 Operation of the optimum on-time controlled buck converter can keep f_{SW} constant when the output loading changes

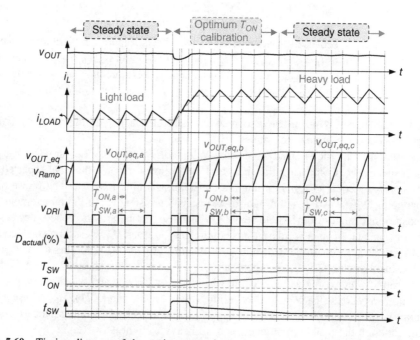

Figure 5.60 Timing diagram of the optimum on-time controlled buck converter to obtain a nearly constant switching frequency during light-to-heavy load transient

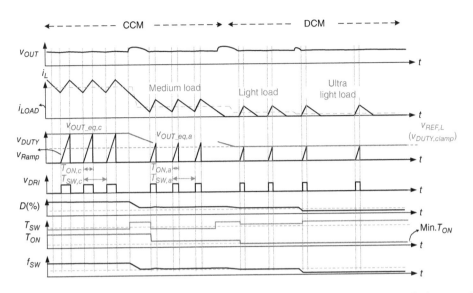

Figure 5.61 Timing diagram of the optimum on-time controlled buck converter to obtain a nearly constant switching frequency during heavy-to-light load transient

loads. If the load current changes from light to heavy, T_{SW} will decrease because T_{ON} is too short to provide adequate energy to manage the heavy loading conditions. In this period, the switching frequency is not constant temperately. Immediately, the optimum on-time controller processes T_{ON} calibration to adjust T_{ON}. According to v_{IN} and a large D_{actual} value, which reflect the information on parasitic effects and loading condition, the EOVS circuit obtains a large $v_{OUT,eq}$ corresponding to an optimum value $v_{OUT,eq,c}$. Subsequently, the OTM circuit expands T_{ON} to increase the effective value of T_{SW}. f_{SW} is regulated back to its value equivalent to that of light loads. Finally, f_{SW} is nearly constant without being affected by any disturbance.

The voltage level of $v_{OUT,eq}$ will be very high or low during the start-up period. An extreme load transient or DCM operation occurs because $v_{OUT,eq}$ is synthesized by v_{GP}. According to Eqs. (5.68) and (5.69), this window range is designed to tolerate $v_{OUT,eq}$. Consequently, the voltage clamper in Figure 5.56 is used to ensure the voltage level of $v_{OUT,eq}$ is within the adequate range. As shown in Figure 5.61, this technique thus achieves a pseudo-constant f_{SW} in the CCM. The on-time period is adjusted to a small value when the decreasing loading current results in a large D_{actual}. In the DCM, the voltage clamper limits the minimum on-time period, such that the inherent advantage to reduce f_{SW} is maintained for high efficiency.

5.3.2.2 Model

Figure 5.62 shows the structure to generate the adaptive on-time period with the calibration. The structure also alleviates the variation from the variations of v_{IN}, v_{OUT}, and i_{Load} to obtain a small f_{SW} variation. By multiplying D_{actual} and v_{IN}, $v_{OUT,eq}$ is obtained to represent the output information. D_{actual} represents the practical value when parasitic components are considered.

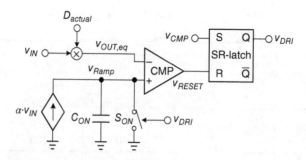

Figure 5.62 Structure of advanced adaptive on-time period with $v_{OUT,eq}$ for load calibration

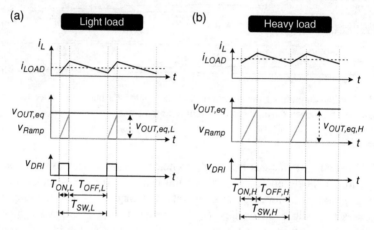

Figure 5.63 Operating waveforms illustrate the adaptive on-time period with the calibration: (a) at light loads; (b) at heavy loads

The comparator then determines T_{ON}, as shown in Eq. (5.73), which is proportional to D_{actual}:

$$T_{ON} = \frac{C_{ON}}{\alpha} \cdot \frac{v_{OUT\_eq}}{v_{IN}} = \frac{C_{ON}}{\alpha} \cdot \frac{D_{actual} \cdot v_{IN}}{v_{IN}} \propto D_{actual} \qquad (5.73)$$

Consequently, the pseudo-constant f_{SW} can be achieved by substituting Eq. (5.73) into Eq. (5.47). Figure 5.63 illustrates the operating waveforms for the on-time control with equivalent v_{OUT} information. According to Eq. (5.51), different loading conditions lead to different duty cycles. As defined in Eq. (5.45), the ratios of T_{ON} and T_{SW} reflect the duty cycle. In case of an increasing loading current, a large duty cycle leads to a high $v_{OUT,eq}$. In other words, $v_{OUT,eq,L}$ is less than $v_{OUT,eq,H}$, such that T_{ON} is extended.

Moreover, Eq. (5.73) shows that T_{ON} is proportional to D_{actual}, which contains v_{OUT} and v_{IN} information according to Eq. (5.51). This T_{ON} also compensates for the variation of v_{OUT} and v_{IN}, implying that the frequency variation caused by different v_{IN}, v_{OUT}, and i_{Load} conditions can be alleviated by the modified T_{ON}.

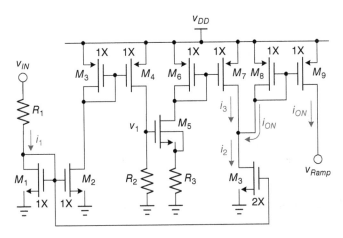

Figure 5.64 Implementation of the fully linear voltage-to-current generator

5.3.2.3 Circuit Implementation

Figure 5.64 shows the FLVCG circuit, which is implemented by core devices and low-voltage MOSFETs and is supplied by v_{DD}. Although v_{IN} is higher beyond the voltage level for core devices, the structure of the resistor R_1 and the diode-connected MOSFET M_1 benefits from not using a high-voltage MOSFET. i_1 and i_2 are generated by v_{IN}, as shown in Eq. (5.74), but i_1 is not completely linear to v_{IN} because of $v_{GS,M1}$, which represents the gate-to-source voltage of MOSFET M_1:

$$i_1 = \frac{1}{2}i_2 = \frac{v_{IN} - v_{GS,M1}}{R_1} \tag{5.74}$$

To compensate for the unexpected $v_{GS,M1}$, M_5 forms the source-generation structure to generate the compensated current i_3, as expressed in Eq. (5.75). $v_{GS,M1}$ and $v_{GS,M5}$ are nearly equal by setting M_1 and M_5 with the same aspect ratio and removing the body effect. Consequently, i_3 is composed of $v_{GS,M5}$ and derived in Eq. (5.76):

$$i_3 = \frac{v_1 - v_{GS,M5}}{R_3} = \frac{i_1 R_2 - v_{GS,M5}}{R_3} \tag{5.75}$$

$$i_3 = \frac{R_2 v_{IN} - (R_1 + R_2) v_{GS,M1}}{R_1 R_3} \tag{5.76}$$

Conducting i_2 and i_3 into the node by opposite directions generates i_{ON}, as derived in Eq. (5.77):

$$i_{ON} = i_2 - i_3 = \frac{(2R_3 - R_2) v_{IN} + (R_1 + R_2 - 2R_3) v_{GS,M1}}{R_1 R_3} \tag{5.77}$$

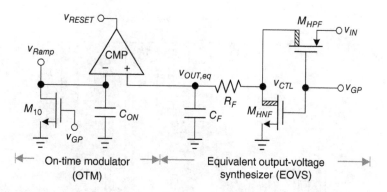

Figure 5.65 Implementation of an equivalent output-voltage synthesizer and an optimum on-time synthesizer

With equal values of $2R_1$, $2R_2$, and R_3, the current i_{ON} is obtained, as shown in Eq. (5.78). Consequently, i_{ON} is converted from v_{IN} by FLVCG and is proportional to v_{IN}. Then, the current i_{ON} is used to charge into C_{ON}, such that v_{Ramp} increases the voltage, and the increase in rate is proportional to v_{IN}:

$$i_{ON} = \frac{3}{2} \frac{v_{IN}}{R_1} \tag{5.78}$$

By contrast, Figure 5.65 shows the implementation of EOVS and OTM. All devices are implanted with low voltage, except for M_{HPF} and M_{HNF}, which are high-voltage MOSFETs because of v_{IN} and v_{GP}. The right portion is EOVS and structured by M_{HPF}, M_{HNF}, R_F, and C_F. The signals v_{IN} and v_{GP} control EOVS to produce $v_{OUT,eq}$. Signal $v_{OUT,eq}$ is expressed as Eq. (5.71) because v_{GP} drives the signal to control the high-side power MOSFET M_P, and the duty of v_{GP} is equal to the ratios of T_{ON} and T_{SW}.

When v_{Ramp} and $v_{OUT,eq}$ are compared, T_{ON} is determined by the OTM according to the duration when v_{Ramp} is below $v_{OUT,eq}$. Consequently, T_{ON} is proportional to D_{actual}, which is derived as Eq. (5.79) using $v_{OUT,eq}$, C_{ON}, and i_{ON}:

$$T_{ON} = \frac{C_{ON} \cdot v_{OUT,eq}}{I_{ON}} = \frac{2}{3} C_{ON} \cdot R_1 \cdot \frac{v_{IN} \cdot D_{actual}}{v_{IN}} \propto D_{actual} \tag{5.79}$$

Substituting Eq. (5.79) into Eq. (5.47) yields a thoroughly constant T_{SW} because of the independence of v_{IN}, v_{OUT}, D_{actual}, i_{Load}, and the parasitic resistances. The ripple of $v_{OUT,eq}$ is also derived approximately in Eq. (5.80). For example, to ensure the ripple is small enough in this work, C_F is 1 pF and R_F is 5 MΩ when v_{IN} is 3.3 V, v_{OUT} is 1.05 V, and f_{SW} is 2.5 MHz:

$$v_{OUT,eq,pp} = T_{ON} \cdot \frac{v_{IN}}{R_F \cdot C_F} \tag{5.80}$$

5.3.3 Type-II Optimum On-Time Controller with Equivalent V_{DUTY}

5.3.3.1 Architecture and Operation of Converter

In addition to the optimum on-time controller with $v_{OUT,eq}$ and v_{IN}, as depicted in Figure 5.56, the optimum on-time controller can be modified as shown in Figure 5.66. This modified optimum on-time controller comprises the constant-current generator (CCG), EDCS, OTM, and voltage clamper. This controller further removes the need for information on v_{IN}. In other words, v_{GP} is the only necessary information.

The CCG block generates the constant current i_{ON}, which is independent of variations of v_{IN}, v_{OUT}, and i_{Load}. The EDCS circuit modulates the equivalent duty-cycle voltage (v_{DUTY}) by driving the signal v_{GP}. The OTM circuit outputs v_{RESET} to determine the on-time period by i_{ON} and v_{DUTY}. Similar to the operation and performance of the optimum on-time controller in Figure 5.66, this optimum on-time controller achieves constant f_{SW} without the need for extra complex current-sensing circuits and the detection of the on-resistance of the power MOSFET and R_{DCR} of the inductor.

5.3.3.2 Model

According to Eq. (5.47), T_{ON} can be designed directly by D information instead of v_{IN} and v_{OUT} information, such that the frequency variation can be compensated. With a similar structure, the current, i_{CHA}, is used to charge the on-time capacitor C_{ON} to reach an upper bound and determine the value of T_{ON}. i_{CHA} should be constant and be independent of any perturbation from

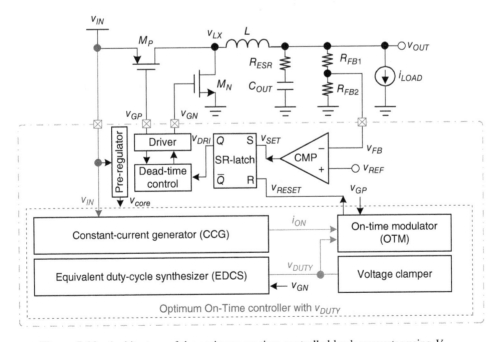

Figure 5.66 Architecture of the optimum on-time controlled buck converter using V_{DUTY}

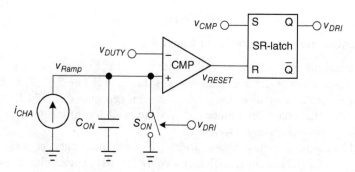

Figure 5.67 Structure of Type-II optimum on-time controller with v_{DUTY}

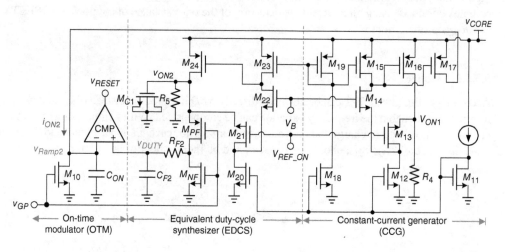

Figure 5.68 Implementation of Type-II optimum on-time controller including OTM, EDCS, and CCG

v_{IN}, v_{OUT}, i_{Load}, and so on. The upper bound is set by an equivalent duty, v_{DUTY}, whose value is proportional to D information (Figure 5.67):

$$T_{ON} = \frac{C_{ON} \cdot v_{DUTY}}{i_{ON}} = \frac{C_{ON}(k \cdot D_{actual})}{i_{ON}} \propto D_{actual} \qquad (5.81)$$

where k = constant

5.3.3.3 Circuit Implementation

Figure 5.68 provides another circuit implementation to realize the optimum T_{ON} with i_{CHA} and v_{DUTY}. The complexity of CCG is reduced compared with that of FLVCG, and the new EDCS removes the need for high-voltage MOSFETs. Furthermore, only one control signal v_{GP} is needed. V_{REF_ON} is a reference voltage with constant value, and v_B is a voltage to bias M_{22}

and M_{14}. In the CCG of the right portion, M_{13} is a voltage follower to determine v_{ON1} from V_{REF_ON}. Then, R_4 and M_{13}–M_{17}, which are structured by negative feedback, determine the current of M_{17} as the value

$$i_{ON} = \frac{V_{REF\_ON} + v_{GS13}}{R_4} \qquad (5.82)$$

In the EDCS of the middle portion, M_{21} is the voltage follower to determine v_{ON2} from V_{REF_ON}. Then, MOSFETs M_{21}–M_{24}, which are structured by negative feedback, provide the driving capability to regulate v_{ON2} as the value

$$v_{ON2} = V_{REF,ON} + v_{GS,M21} \qquad (5.83)$$

R_5 and M_{C1} are used to bias M_{24} and assist the regulation. v_{GP} then utilizes M_{PF}, M_{NF}, R_{F2}, and C_{F2} to generate v_{DUTY}:

$$v_{DUTY} = v_{ON2} \cdot D_{actual} \qquad (5.84)$$

Consequently, the OTM in the left portion uses v_{DUTY} and i_{ON2} to determine T_{ON}:

$$T_{ON} = i\frac{C_{ON} \cdot v_{DUTY}}{I_{ON2}} = C_{ON} \cdot \frac{(V_{REF\_ON} + v_{GS,M21}) \cdot D_{actual}}{(V_{REF\_ON} + v_{GS,M13}) \cdot \frac{1}{R_4}} \propto D_{actual} \qquad (5.85)$$

The optimum T_{ON} proportional to D_{actual} can be obtained because V_{RER_ON}, v_{GS21}, v_{GS13}, and C_{ON} are constant.

5.3.4 Frequency Clamper

Figure 5.69 shows the voltage clamper, including lower-bound and upper-bound clamps. When $v_{OUT,eq}$ is lower than the lower-bound voltage $V_{REF,L}$, M_{30} and M_{31} can drive M_{32} to clamp $v_{OUT,eq}$ at the voltage level of $V_{REF,L}$. By contrast, when $v_{OUT,eq}$ is higher than the upper-bound voltage $V_{REF,H}$, M_{35} and M_{36} can drive M_{37} to clamp $v_{OUT,eq}$ at the voltage level of $V_{REF,H}$. In other words, the swing range of $v_{OUT,eq}$ ensures the window between $V_{REF,L}$ and $V_{REF,H}$. Moreover, MOSFETs M_{31}, M_{32}, M_{36}, and M_{37} can be off and have no influence on $v_{OUT,eq}$ when $v_{OUT,eq}$ is within the window.

5.3.5 Comparison of Different On-Time Controllers

Table 5.5 compares the various on-time controllers. A comparison of the structures in Figures 5.62 and 5.67 demonstrates that the structure in Figure 5.62 is more complex because the charging current should be dependent on V_{IN} information to calibrate the variation of $v_{OUT,eq}$. By contrast, the structure in Figure 5.67 only requires the driving signal information, such that the structure becomes simpler. The circuit implementation in Figure 5.64 for the structure in Figure 5.62 should carefully consider the accuracy of resistors and the current mirror, as

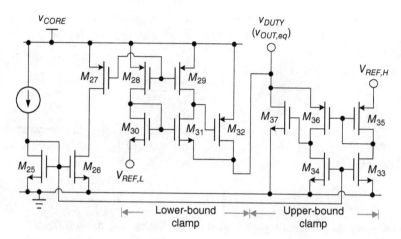

Figure 5.69 Voltage clamper

Table 5.5 Comparison of different on-time controllers

	Fixed T_{ON}	Basic adaptive T_{ON} with v_{IN} and v_{OUT}	Adaptive T_{ON} with v_{LX} and v_{IN}	Type-I optimum T_{ON} with $v_{OUT,eq}$ and v_{IN}	Type-II optimum T_{ON} with v_{DUTY}
Structure	Figure 5.44	Figure 5.47	Figure 5.52	Figure 5.62	Figure 5.67
Circuit implantation		Type I in Figure 5.50 Type II in Figure 5.51 Type III in Figure 5.54 FLVCG in Figure 5.64		Figure 5.65	Figure 5.68
Required information	v_{DRI} (v_{GP})	v_{IN} v_{OUT} v_{DRI} (v_{GP})	v_{LX} v_{OUT} v_{DRI} (v_{GP})	v_{IN} v_{OUT} v_{DRI} (v_{GP})	v_{DRI} (v_{GP})
Design issue		• Matching issue should be designed carefully • Charging current should be fully linear to v_{IN} • High-voltage devices are needed because of the use of v_{IN}			
Flexibility	Poor	Charging current dependent on v_{IN} is difficult to achieve			Excellent
Δf_{SW} by v_{IN} and v_{OUT}	Poor	Good	Good	Excellent	Excellent
Δf_{SW} by i_{Load}	Poor	Poor	Good	Excellent	Excellent

well as the issue of body effect. Owing to the mismatch value, Eq. (5.78) is difficult to achieve and causes a large variation in switching frequency. The desired switching frequency is also difficult to adjust. On the contrary, the circuit implemented in Figure 5.68 for the structure in Figure 5.67 provides more flexible and robust performance. The desired switching frequency can be adjusted by the value of R_4, which is a discrete component. The switching frequency variation is more independent of the process variation or mismatch problem. This structure reduces the required information on v_{OUT} and v_{IN}. It can release strict design considerations, including the matching design of the resistor and the current mirror, high linearity between v_{IN} and charging current, and use of high-voltage devices.

5.3.6 Simulation Result of Optimum On-Time Controller

The on-time control DC/DC converter is simulated in the UMC 28 nm CMOS process. In simulation results, Type-I PCT is implemented to demonstrate and observe the different fixed v_{OUT} in conventional constant on-time control and variable $v_{OUT,eq}$ in the optimum on-time control. Figure 5.70 demonstrates the function of the PCT for the desired f_{SW} = 800 kHz. The specifications are v_{IN} = 3.3 V, v_{OUT} = 1.05 V, L = 2.2 µH, and C_{OUT} = 4.7 µF. The parasitic effects are also considered, in which $R_{on,P}$ and $R_{on,N}$ are 150 and 100 mΩ, respectively, and the R_{DCR} of the inductor is 30 mΩ. The waveforms include v_{OUT}, i_L, v_{Ramp}, v_{OPT}, and v_{GP}. With conventional on-time control, Δf_{SW} is approximately 76 kHz when the i_{Load} change is 1.5 A. By contrast, the PCT ensures that the f_{SW} variation is lower than 3 kHz. Particularly, $v_{OUT,eq}$ increases as i_{Load} increases to achieve optimum T_{ON}.

Figure 5.71 demonstrates the function of the PCT if the desired f_{SW} is 2.5 MHz, where L is 1 µH. Under the same parameters of parasitic effects, the performance comparison is shown in Figure 5.70. In conventional on-time control, Δf_{SW} is approximately 410 kHz when the i_{Load} change is 1.5 A. By contrast, the PCT ensures that Δf_{SW} is lower than 4 kHz.

Table 5.6 shows the performance of the simulation result. By comparing Figures 5.70(a) and 5.71(a) with the conventional constant on time, the higher desired f_{SW} results in a larger Δf_{SW}, which is consistent with the analysis result. The PCT benefits ultra-low Δf_{SW}, as shown in both.

Figure 5.72 shows the performance of the simulation result with the desired f_{SW} = 800 kHz. f_{SW} is nearly constant, although the parasitic effect is more serious when $R_{on,P}$ is 300 mΩ and $R_{on,N}$ is 200 mΩ. The performance of the f_{SW} variation is defined by $\Delta f_{SW}/f_{SW}$ and $\Delta f_{SW}/\Delta i_{Load}$. The PCT has a performance of 0.375% $\Delta f_{SW}/f_{SW}$ and 2 kHz/A $\Delta f_{SW}/\Delta i_{Load}$. By contrast, conventional on-time has $\Delta f_{SW}/f_{SW}$ of more than 9.5% and $\Delta f_{SW}/\Delta i_{Load}$ of more than 50 kHz/A.

5.3.7 Experimental Result of Optimum On-Time Controller

The Type-II optimum on-time controlled buck converter is fabricated by UMC 28 nm CMOS technology. The specifications include v_{IN} = 3.3 V, v_{OUT} = 1.05 V, L = 1 µH, C_{OUT} = 4.7 µF, and f_{SW} = 2.5 MHz. The results show $R_{on,P}$ and $R_{on,N}$ are 300 and 200 mΩ, respectively. R_{DCR} of the inductor is 30 mΩ. Figure 5.73 shows the waveforms of the conventional on-time controller when i_{Load} changes from 1.7 to 0.3 A, and vice versa. The on-time period remains constant at different loading conditions in the conventional design. However, the change in slopes of

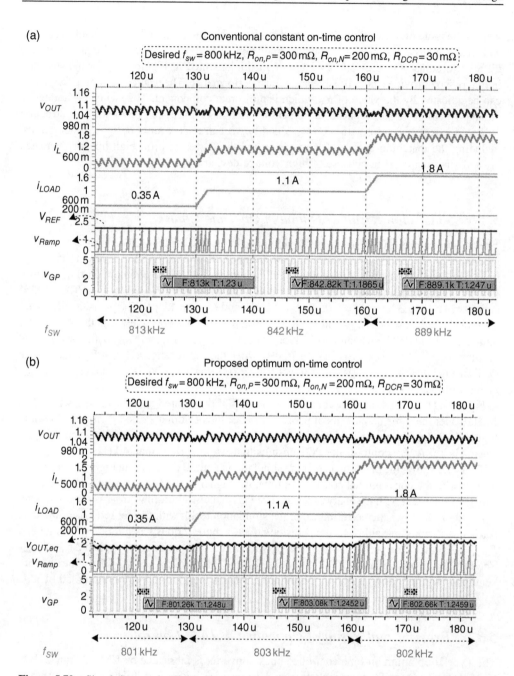

Figure 5.70 Simulation result of converter with desired f_{SW} = 800 kHz. (a) Constant on-time control. (b) Optimum on-time control with PCT

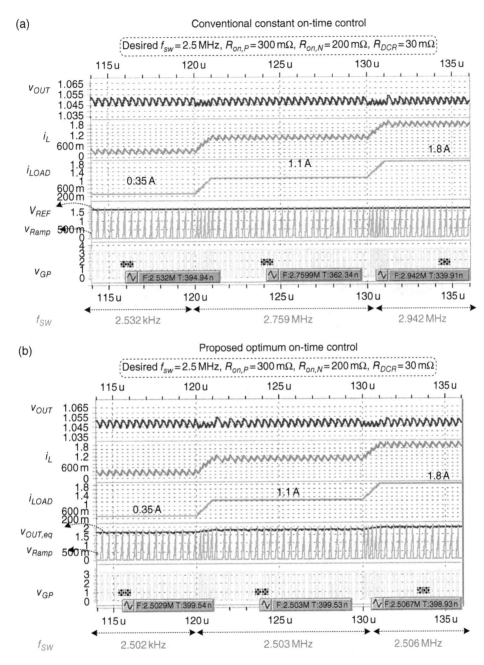

Figure 5.71 Simulation result of the converter with desired f_{SW} = 2.5 MHz. (a) Constant on-time control. (b) Optimum on-time control with PCT

Table 5.6 Comparison table of conventional COT converter and the COT converter with PCT technique

Control method	Constant on-time		Optimum on-time	
v_{IN} (V)	3.3		3.3	
v_{OUT} (V)	1.05		1.05	
L (µH)	1		2.2	
C_{OUT} (µF)	4.7		4.7	
Δi_{Load} (A)	1.5		1.5	
Desired f_{SW}	800 kHz	2.5 MHz	800 kHz	2.5 MHz
Δf_{SW} (kHz)	76	410	3	4
$\Delta f_{SW}/f_{SW}$ (%)	9.5	16.4	0.375	0.16
$\Delta f_{SW}/\Delta i_{Load}$ (kHz/A)	50.6	273.3	2	2.6

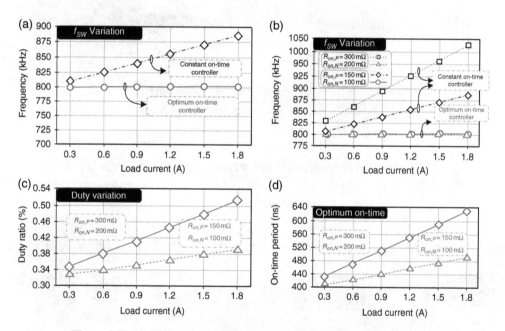

Figure 5.72 Switching frequency variation during different load conditions

i_L and D_{actual} is due to the parasitic effects. f_{SW} is 2.5 and 3.4 MHz when i_{Load} is 0.3 and 1.7 A, respectively. The f_{SW} variation is approximately 0.9 MHz for a 1.4 A change in i_{Load}.

By contrast, Figure 5.74 demonstrates the function of the optimum on-time controller. Figure 5.75 provides the zoomed-in waveforms in steady state. Although the slopes of i_L and D_{actual} continue to change at different loading conditions because of the parasitic effects,

Ripple-Based Control Technique Part II

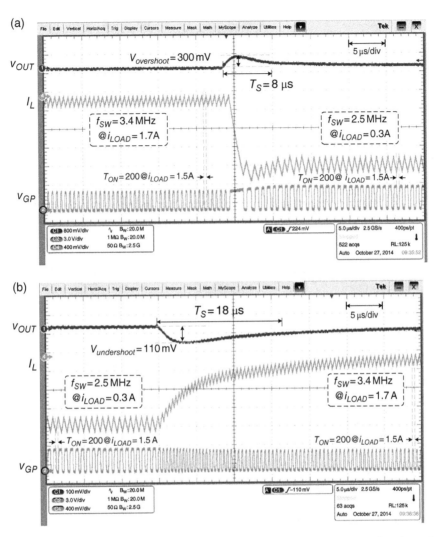

Figure 5.73 Frequency variation in constant on-time controlled buck converter when i_{Load} changes (a) from heavy to light load and (b) from light to heavy load

f_{SW} is nearly constant at 2.5 MHz during adjustable on-time periods. T_{ON} is 170 ns and the duty cycle is 0.35 at light loads, whereas T_{ON} is adjusted to 210 ns and its duty cycle is 0.52 at heavy loads. Δf_{SW} is approximately 8 kHz when the change in i_{Load} is 1.4 A.

Table 5.7 shows the comparison of prior methods. The performance of f_{SW} variation is indicated by $\Delta f_{SW}/f_{SW}$ and $\Delta f_{SW}/\Delta i_{Load}$. The designs in [11, 12, 25] achieve good performance at $\Delta f_{SW}/f_{SW}$ and $\Delta f_{SW}/\Delta i_{Load}$, but the extra clock signal and complex PLL-based loop should be employed. The design in [17, 29] also reduces the complexity of the controller, and these

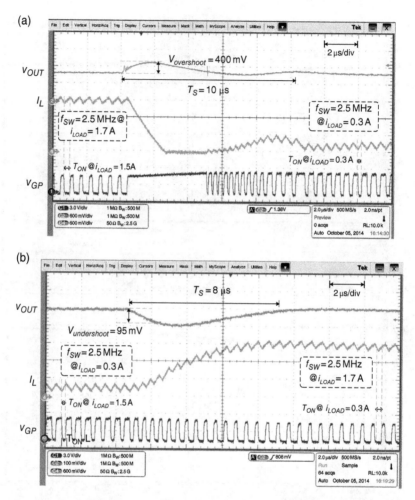

Figure 5.74 Pseudo-constant frequency in optimum on-time controlled buck converter when i_{Load} changes (a) from heavy to light load and (b) from light to heavy load

designs remove the need for an extra clock signal and achieve adaptive on-time, but the performance of Δf_{SW} in these designs is withdrawn.

Figure 5.76 shows the performance, which includes the constant on-time control, optimum on-time control, LCC [17], and SLCC [29]. The design in [17] using current sensing performs with $\Delta f_{SW}/f_{SW}$ of 9.5% and $\Delta f_{SW}/\Delta i_{Load}$ of 129 kHz/A. The design in [29] utilizes the RC network, but the f_{SW} changes from 600 to 800 kHz when the load changes from 200 to 900 mA. According to the measured transient waveforms, $\Delta f_{SW}/f_{SW}$ is 25% and $\Delta f_{SW}/\Delta i_{Load}$ is 285.7 kHz/A. By contrast, the optimum on-time controller ensures that the f_{SW} variation is lower than 8 kHz when f_{SW} is 2.5 MHz. Although the parasitic effects are severe, f_{SW} is nearly

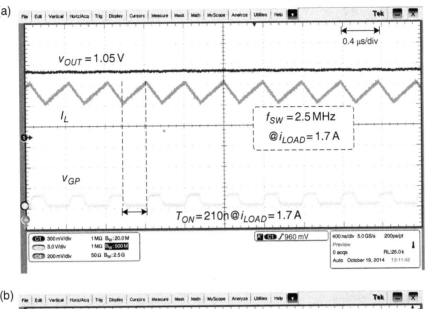

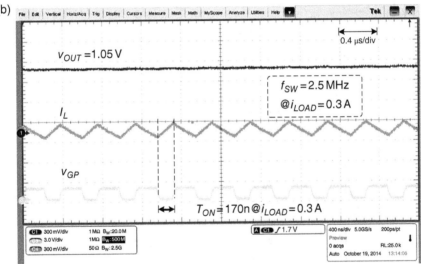

Figure 5.75 Steady state of the optimum on-time controlled buck controller: (a) $i_{Load} = 1.7$ A; (b) $i_{Load} = 0.3$ A

constant, in which $R_{on,P}$ and $R_{on,N}$ are 300 and 200 mΩ, respectively. The optimum on-time controller has good performances in $\Delta f_{SW}/f_{SW} = 0.32\%$ and $\Delta f_{SW}/\Delta i_{Load} = 5.7$ kHz/A at $f_{SW} = 2.5$ MHz. In this work, the peak efficiency is 89%, which is dominated by the on-resistance of the power stage. The values of these on-resistances are designed larger than that

Table 5.7 Comparison table of state-of-the-art COT converters

Control method	PLL-based [15]	PLL-based [12]	PLL-based [11]	DADC [13]	LCC (load sensing) [17]	SLCC [29]	Constant on-time	Optimum on-time
v_{IN}	3 V	2.7–4.5 V	2.5 V	2.7–3.3 V	3.3 V	2.7–3.6 V	3.3 V	3.3 V
v_{OUT}	1.8 V	2 V	0.7–1.8 V	0.9–2.1 V	1.2 V	1–1.2 V	1.05 V	1.05 V
L	4.7 µH	4.7 µH	1–5 µH	2.2 µH	4.7 µH	N/A	1 µH	1 µH
C_{OUT}	4.7 µF	10 µF	10 µF	4.4 µF	8.9 µF	N/A µF	4.7 µF	4.7 µF
$R_{on,P}$	N/A	N/A	N/A	N/A	N/A	N/A	300 mΩ	300 mΩ
$R_{on,N}$							200 mΩ	200 mΩ
R_{DCR}							30 mΩ	30 mΩ
f_{SW}	1 MHz	1 MHz	1 MHz	3 MHz	750 kHz	800 kHz	2.5 MHz	2.5 MHz
Δf_{SW}	15 kHz	2 kHz	5 kHz	100 kHz	84 kHz	200 kHz	900 kHz	8 kHz
$\Delta f_{SW}/f_{SW}$	1.5%	0.2%	0.5%	3.3%	11.2%	25%	36%	0.32%
Δi_{Load}	0.25 A	0.4 A	0.6 A	0.45 A	0.65 A	700 mA	1.4 A	1.4 A
$\Delta f_{SW}/\Delta i_{Load}$	60 kHz/A	5 kHz/A	8 kHz/A	222.2 kHz/A	129 kHz/A	285.7 kHz/A	642.8 kHz/A	5.7 kHz/A
Extra V_{CLK}	Required	Required	Required	Required	Not required	Not required	Not required	Not required
Maximum efficiency	95%	95.5%	93%	93%	87%	88.2%	89%	89%

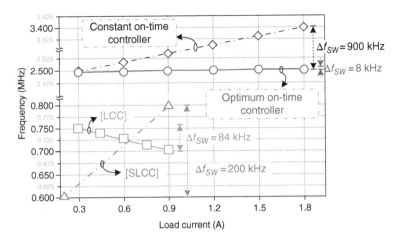

Figure 5.76 Switching frequency variation during different loading conditions

of the general design to demonstrate the performance of the optimum on-time controller. Although the power stage contains serious parasitic effects, the optimum on-time controller can achieve pseudo-constant f_{SW}, which has a competitive performance for the design applied by PLL.

References

[1] Chava, C.K. and Silva-Martinez, J. (2004) A frequency compensation scheme for LDO voltage regulators. *IEEE Transactions on Circuits and Systems I: Regular Papers*, **51**(6), 1041–1050.

[2] Sun, J. (2006) Characterization and performance comparison of ripple-based control methods for voltage regulator modules. *IEEE Transactions on Power Electronics*, **21**(2), 346–353.

[3] Redl, R. and Sun, J. (2009) Ripple-based control of switching regulators, an overview. *IEEE Transactions on Power Electronics*, **24**(12), 2669–2680.

[4] Texas Instruments (2014) 1.5- to 18-V (4.5- to 25-V bias) Input, 8-A Single Synchronous Step-Down SWIFT™ Converter, TPS53513 Datasheet, December 2014.

[5] Texas Instruments (2014) 6-A Output, D-CAP+™ Mode, Synchronous Step-Down, Integrated-FET Converter for DDR Memory Termination, TPS53317 Datasheet, January 2014.

[6] Texas Instruments (2015) TPS560200 4.5-V to 17-V Input, 500-mA Synchronous Step-Down SWIFT™ Converter with Advanced Eco-Mode™, TPS560200 Datasheet, February 2015.

[7] Chen, W.-C., Wang, C.-S., Su, Y.-P., *et al.* (2013) Reduction of equivalent series inductor effect in delay-ripple reshaped constant on-time control for buck converter with multi-layer ceramic capacitors. *IEEE Transactions on Power Electronics*, **28**(5), 2366–2376.

[8] Chen, W.-C., Huang, Y.-S., Chien, M.-W., *et al.* (2104) ±3% voltage variation and 95% efficiency 28 nm constant on-time controlled step-down switching regulator directly supplying to Wi-Fi systems. *Proceedings of the IEEE Symposium on VLSI Circuits Digest of Technical Papers*, Honolulu, HI, June 10–13, pp. 1–2.

[9] Chen, W.-C., Lin, K.-L., Chen, K.-H., *et al.* (2014) A pseudo fixed switching frequency 2 kHz/A in optimum on-time control buck converter with predicting correction technique for EMI solution. *Proceedings of the IEEE International Symposium on Circuits and Systems (ISCAS)*, Melbourne, Australia, June 1–5, pp. 946–949.

[10] Chen, H.-C., Chen, W.-C., Chou, Y.-W., *et al.* (2014) Anti-ESL/ESR variation robust constant-on-time control for DC–DC buck converter in 28 nm CMOS technology. *Proceedings of the IEEE Custom Integrated Circuits Conference (CICC)*, San Jose, CA, September 15–17, pp. 1–4.

[11] Khan, Q., Elshazly, A., Rao, S., et al. (2012) A 900 mA 93% efficient 50 μA quiescent current fixed frequency hysteretic buck converter using a highly digital hybrid voltage- and current-mode control. *Proceedings of the IEEE Symposium on VLSI Circuits Digest of Technical Papers*, Honolulu, HI, June 13–15, pp. 182–183.

[12] Lee, S.-H., Bang, J.-S., Yoon, K.-S., et al. (2015) A 0.518 mm^2 quasi-current-mode hysteretic buck DC–DC converter with 3 μs load transient response in 0.35 μm BCDMOS. *Proceedings of the IEEE International Solid-State Circuits Conference (ISSCC), Digest of Technical Papers*, San Francisco, CA, February 22–26, pp. 214–215.

[13] Su, F. and Ki, W.-H. (2009) Digitally assisted quasi-V^2 hysteretic buck converter with fixed frequency and without using large-ESR capacitor. *Proceedings of the IEEE International Solid-State Circuits Conference (ISSCC), Digest of Technical Papers*, San Francisco, CA, February 8–12, pp. 446–447.

[14] Wang, J., Xu, J., Zhou, G., and Bao, B. (2013) Pulse-train-controlled CCM buck converter with small ESR output-capacitor. *IEEE Transactions on Power Electronics*, **60**(12), 5875–5881.

[15] Huerta, S.-C., Alou, P., Oliver, J.A., et al. (2011) Nonlinear control for DC–DC converters based on hysteresis of the C_{OUT} current with a frequency loop to operate at constant frequency. *IEEE Transactions on Power Electronics*, **58**(3), 1036–1043.

[16] Cortes, J., Svikovic, V., Alou, P., et al. (2011) Accurate analysis of subharmonic oscillations of V^2 and V^2I_c controls applied to buck converter. *IEEE Transactions on Power Electronics*, **58**(3), 1036–1043.

[17] Tsai, C.-H., Lin, S.-M., and Huang, C.-S. (2103) A fast-transient quasi-V^2 switching buck regulator using AOT control with a load current correction (LCC) technique. *IEEE Transactions on Power Electronics*, **28**(8), 3949–3957.

[18] Chen, W.-C., Chen, K.-H., Wey, C.-L., et al. (2013) Dynamic bootstrap capacitance technique for high efficiency buck converter in universal serial bus (USB) power device (PD) supplying system. *Proceedings of the IEEE Asian Solid-State Circuits Conference (A-SSCC)*, Singapore, November 11–13, pp. 165–168.

[19] Su, Y.-P., Lee, Y.-H., Chen, W.-C., et al. (2013) A pseudo-noise coded constant-off-time (PNC-COT) control switching converter with maximum 16.2 dBm peak spur reduction and 92% efficiency in 40 nm CMOS. *Proceedings of the IEEE Symposium on VLSI Circuits Digest of Technical Papers*, June, pp. 170–171.

[20] Chen, W.-C., Lin, C.-C., and Chen, K.-H. (2102) Differential zero compensator in delay-ripple reshaped constant on-time control for buck converter with multi-layer ceramic capacitors. *Proceedings of the IEEE International Symposium on Circuits and Systems (ISCAS)*, Seoul, May 20–23, pp. 692–695.

[21] Chen, W.-C., Chi, K.-Y., Lin, C.-C., et al. (2012) Reduction of equivalent series inductor effect in delay-ripple reshaped constant on-time control for buck converter with multi-layer ceramic capacitors. *Proceedings of the IEEE Energy Conversion Congress and Exposition (ECCE)*, Raleigh, NC, September 15–20, pp. 755–758.

[22] Erickson, R.W. and Maksimovic, D. (2001) *Fundamentals of Power Electronic*, 2nd edn. Kluwer Academic Publishers, Norwell, MA.

[23] Li, P., Bhatia, D., Lin, X., and Bashirullah, R. (2011) A 90–240 MHz hysteretic controlled DC–DC buck converter with digital phase locked loop synchronization. *IEEE Journal of Solid-State Circuits*, **46**(9), 2108–2119.

[24] Li, P., Lin, X., Hazucha, P., et al. (2009) A delay locked loop synchronization scheme for high-frequency multiphase hysteretic DC–DC converters. *IEEE Journal of Solid-State Circuits*, **44**(11), 3131–3145.

[25] Zheng, Y., Chen, H., and Leung, K.N. (2012) A fast-response pseudo-PWM buck converter with PLL-based hysteresis control. *Transactions on Very Large Scale Integration (VLSI) System*, **20**(7), 1167–1174.

[26] Lee, K.-C., Chae, C.-S., Cho, G.-H., and Cho, G.-H. (2010) A PLL-based high stability single-inductor 6-channel output DC–DC buck converter. *Proceedings of the IEEE International Solid-State Circuits Conference (ISSCC), Digest of Technical Papers*, San Francisco, CA, February 7–11, pp. 200–201.

[27] Hsieh, C.-Y. and Chen, K.-H. (2008) Adaptive pole-zero position (APZP) technique of regulated power supply for improving SNR. *IEEE Transactions on Power Electronics*, **23**(6), 2949–2963.

[28] Shih, C.-J., Chu, K.-Y., Lee, Y.-H., et al. (2103) A power cloud system (PCS) for high efficiency and enhanced transient response in SoC. *IEEE Transactions on Power Electronics*, **28**(3), 1320–1330.

[29] Tsai, C.-H., Chen, B.-M., and Li, H.-L. (2016) Switching frequency stabilization techniques for adaptive on-time controlled buck converter with adaptive voltage positioning mechanism. *IEEE Transactions on Power Electronics*, **31**(1), 443–451.

6

Single-Inductor Multiple-Output (SIMO) Converter

6.1 Basic Topology of SIMO Converters

Conventionally, n parallel DC/DC buck converters are most commonly used to generate n supplying voltages v_{O1}–v_{On}, as illustrated in Figure 6.1. High power efficiency is achieved through distributive voltage/current levels because of the inherent characteristics of the DC/DC buck converter. However, this topology requires n inductors and n buck converter chips on the PCB, and the volume and cost of tablets and/or portable devices are largely increased. To suppress the volume of the power management unit (PMU), n parallel low dropouts (LDOs) are cascaded after one DC/DC buck converter as shown in Figure 6.2, because only one inductor, one buck converter chip, and n small LDO chips are required. Nevertheless, large dropout voltages across the LDOs occur when the corresponding output voltage is low, which deteriorates the power efficiency, and the charge in the battery runs out in a short time. The solution in Figure 6.2 is not efficient for tablets or portable devices.

Instead, a SIMO DC/DC converter is able to simultaneously generate n supply voltages, as shown in Figure 6.3, by adopting only one inductor and one SIMO converter chip [1, 2]. The advantages are that the PCB area and cost are largely suppressed. v_{O1}–v_{On} can be well regulated by properly allocating the energy stored in the inductor to each output voltage. Therefore, SIMO DC/DC converters are more attractive in applications that need multiple supply voltages.

6.1.1 Architecture

The simplified architecture of a SIMO converter, which can be divided into power stage and controller, is illustrated in Figure 6.4. Similar to DC/DC buck converters, the high-side power MOSFET M_H, low-side power MOSFET M_L, and inductor L control acquire energy from the input source v_{IN} and store the energy in the inductor. In SIMO converters, n additional switches

---

Power Management Techniques for Integrated Circuit Design, First Edition. Ke-Horng Chen.
© 2016 John Wiley & Sons Singapore Pte Ltd. Published 2016 by John Wiley & Sons Singapore Pte Ltd.

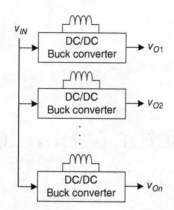

Figure 6.1 n supply voltages are generated by n parallel DC/DC buck converters

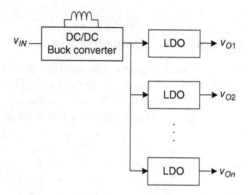

Figure 6.2 n supply voltages are generated by cascading one DC/DC buck converter and n parallel LDOs

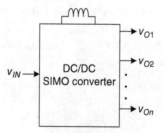

Figure 6.3 n supply voltages are generated by one SIMO converter

are adopted to appropriately allocate the stored energy to n outputs v_{O1}–v_{On}. Therefore, n different voltage levels can be generated to meet the requirements of a variety of applications.

A SIMO controller must control the turning on/off periods of all the switches. The duty cycle control signals D, $(1-D)$, and D_1–D_n must be generated to control the power switches M_H, M_L

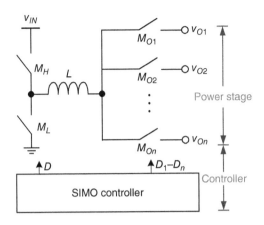

Figure 6.4 Simplified architecture of the SIMO converter.

and the output switches M_{O1}–M_{On}, respectively, for the voltage regulation of each output. The SIMO converter design is divided into two parts: the power stage and the controller design. At the power stage, $2n$ paths are formed by the combinations of M_H, M_L, and M_{O1}–M_{On}. Thus, different arrangements of the energy path sequence are realized for different design targets [3]. Moreover, additional auxiliary switches can also be included to improve the performance of SIMO converters, especially for high power-conversion efficiency. Most importantly, selecting an adequate power switch control method considering high power-conversion efficiency is crucial. As we know, trade-off designs exist among the performance of SIMO converters. The power switch control method in controller designs can determine the duty cycles of all the switches in a pre-designed energy path sequence or load-dependent energy path sequence. How to choose control methods, such as the ripple- and error-based control, is discussed below.

6.1.2 Cross Regulation

Since all outputs share only one inductor with each other in the SIMO converter, interference among them occurs only occasionally. Accumulated or insufficient energy in a single inductor causes the phenomenon of cross regulation in SIMO converters [4]. As depicted in Figure 6.5, when the load current variation occurs at v_{O2}, while i_{O1} and i_{O3}–i_{On} remain constant, v_{O2} naturally results in undershoot and overshoot at heavy and light loads, which are similar to a transient response in a single-output buck converter. However, unintended voltage variations also occur in v_{O1} and v_{O3}–v_{On}. These unintended voltage variations at victim outputs without any load variations are called "cross regulation." Cross regulation can be defined as the ratio of voltage variation at the output without any load variations to load current variation at a certain output, as expressed in Eq. (6.1):

$$\text{Cross regulation} = \frac{\Delta v_{Oj}}{\Delta i_{Ok}} (\text{mV/mA}), j \neq k \qquad (6.1)$$

An instantaneous load current variation in a certain output breaks the steady-state balanced energy delivery sequence of the SIMO converter. The stored energy in the inductor no longer

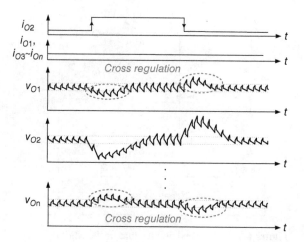

Figure 6.5 Cross-regulation phenomenon in SIMO converters

satisfies the instantaneous energy demand or release. Raising or pulling the inductor current level to its new balanced level within a short period is impossible because of limited bandwidth. Therefore, voltage variations occur at all outputs and cause cross regulation at victim outputs. Although the voltage variation can be recovered through the feedback loop of each output, the settling time becomes longer than that in the single-output buck converter because the chain reaction among all the outputs prolongs the recovery time. The design challenge in SIMO converters is to minimize cross regulation and enhance supply quality in the PMU.

6.2 Applications of SIMO Converters

SIMO converters are widely used to minimize the size of PMUs in different applications. These applications can be classified into Soc and portable electronic systems. Soc integrates the whole system in a single chip to achieve high integrations. Advanced processes are usually adopted, such as the 65 nm or 28 nm CMOS process. Portable electronic systems, for instance tablet applications, require a large driving capability and constitute the whole system in a PCB board.

6.2.1 System-on-Chip

If the sub-modules in a Soc can simply be classified into analog and digital circuits, then the simplified single-inductor dual-output (SIDO) converter can provide suitable and independent supply voltages to these sub-modules. However, if the demanded supply voltage range is wide, then the SIMO converter may be more suitable than the SIDO converter. Similarly, the reduction of one inductor is area efficient, and the PCB area is effectively reduced because a conventional PMU requires more than two off-chip inductors to generate dual switching regulator (SWR) output voltages [5–7]. Low output voltage ripple, minimized cross regulation, and high power-conversion efficiency are essential design issues for one SIDO converter. Figure 6.6 shows the SIDO step-down converter in the Soc integration. The embedded power switches

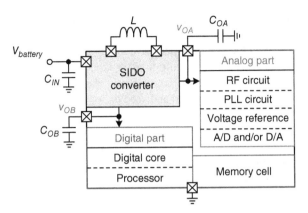

Figure 6.6 Simple illustration of the SIDO step-down converter in the SoC integration, which can provide distinct analog and digital power supplies to the analog and digital parts, respectively

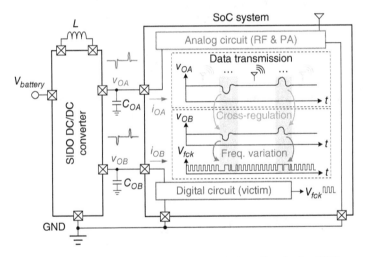

Figure 6.7 Detailed description of the transient cross-regulation effect in the SIDO converter and the performance deterioration in SoC applications

in the SIDO converter can deliver energy from $V_{battery}$ to both off-chip capacitors C_{OA} and C_{OB} through an off-chip inductor. Therefore, the SIDO converter generates two output voltages, v_{OA} and v_{OB}, to supply the analog and digital parts, respectively. In particular, the PMU implemented by the SIDO converter in the Soc system can be properly guaranteed. However, several design challenges in the SIDO or SIMO converters should be prudently considered to prevent the deterioration of the Soc performance. The well-known design challenge is the reduction of cross regulation. When any one of the outputs undergoes a sudden loading current change, cross regulation occurs at the rest of the outputs because the accumulated inductor current occurs in the single off-chip inductor.

The transient cross regulation in the SIDO converter affects the PMU and the performance of the powered circuits in Soc applications. Figure 6.7 shows the transient cross regulation effect of the step-down SIDO converter in Soc applications. The analog supply voltage source v_{OA} powers the RF and PA circuits, whereas the digital supply voltage source v_{OB} supplies the digital circuits. During the data transmission period, a large power requirement causes a sudden loading current increase at i_{OA}, thereby resulting in an expected voltage drop at v_{OA}. Simultaneously, v_{OB} is also affected by the voltage variation even under constant loading condition. The reason is that only one inductor is disturbed and used to deliver energy to the loading change at i_{OA}. The unexpected sudden voltage variation at v_{OB} degrades the performance of the digital circuits in Soc. Transient cross regulation varies the frequency of the signal V_{fck} and even induces abnormal operation in the system processor or other digital circuits. Therefore, how to minimize cross regulation in the SIDO converter becomes a necessary research topic.

6.2.2 Portable Electronics Systems

With the development of commercial portable electronics, small volume, light weight, and long usage time have become urgent requirements for consumers. The PMU has a pivotal role in portable electronics because the sub-circuits of different function blocks all require a high-quality voltage supply to ensure their performance. A low-cost PMU with a small PCB area is desirable to minimize the cost of portable electronics [34, 37].

The typical architecture of a commercial tablet is illustrated in Figure 6.8 [4, 8, 9]. Except for the LED driver, which supplies the backlights and a pair of positive/negative supply voltages for the gate driver, the panel requires several supply voltages with different levels, which are implemented by several DC/DC buck converters in commercial tablets. Here, four DC/DC buck converter chips with four inductors form the power solutions of this panel, which requires

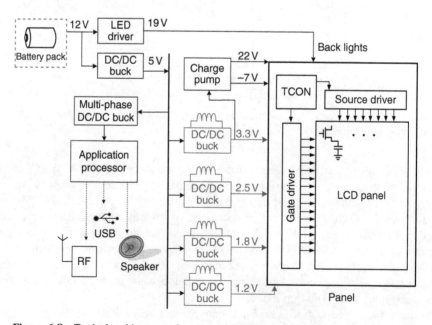

Figure 6.8 Typical architecture of commercial tablet and its power management unit

four different output voltages ranging from 1.2 to 3.3 V. The timing control unit (TCON) requires three supply voltages of 3.3, 1.8, and 1.2 V, while the gate driver and source driver demand voltages of 2.5 and 1.8 V, respectively. Using four individual buck converters, the performance requirements of each circuit block can easily be satisfied by separately designed converters.

However, the disadvantage of adopting four individual DC/DC buck converters is the large PCB area occupation. Multiple inductors dramatically increase the cost as well. The main concern here is how to design a PMU such that a small volume, light weight, and long usage time are achieved and no flicker effect occurs on the display in commercial tablets.

6.3 Design Guidelines of SIMO Converters

6.3.1 Energy Delivery Paths

Energy delivery arrangements at the power stage are highly related to the performance of SIMO converters, especially in consideration of cross regulation. Thus, energy delivery schemes need to be considered carefully in order to guarantee adequate energy supply for multiple outputs, so as to minimize cross regulation. Additionally, the DC level of the inductor current has to be sufficient to meet the total loading conditions of all outputs. Some prior representative designs for SIMO converters are discussed and compared as follows.

6.3.1.1 Constant-Charge Auto-Hopping (CCAH)

A simple method of minimizing cross regulation in SIMO converters is to separate the energy delivery of each output using zero inductor current. Therefore, the DCM control method can ensure the return of the inductor current to zero at the end of each PWM switching cycle. Cross regulation can effectively be reduced, but the other converter performances are sacrificed because of the limited DCM bandwidth.

Based on the DCM operation, CCAH aims to minimize cross regulation with unbalanced loads at $v_{O1}-v_{On}$ [10]. Figure 6.9 shows the topology and operating waveforms. Figure (b) shows that in case I the energies requested by each output, which are indicated by E_1-E_n, respectively, are close to one another. The energy stored in the inductor L is sequentially allocated to $v_{O1}-v_{On}$. At $t = 0$, which is the start of a switching cycle, energy S_1 is on and S_2-S_n are off to transfer energy to v_{O1}. Until the inductor current reaches zero, which indicates that the energy transferred to v_{O1} is sufficient, S_2 is on and S_1 and S_3-S_n are off to transfer energy to v_{O2}. When all the outputs obtain sufficient energy, all the switches S_H, S_L, and S_1-S_n are turned off with zero inductor current. The energy allocation resumes at $t = T_s$. With DCM control, the transference of energy is ordered with the switching period $T = T_s$. The zero current region, which exists inherently in DCM control, can act as a buffer region to address energy variations during the transient and minimize cross regulation.

When an unbalanced load occurs (i.e., large load current difference between outputs), CCAH adjusts the switching period to n times T_s based on load condition $T = kT_s$, where k is an integer. Using the constant charge concept, the average energy transferred to each output remains constant regardless of switching frequency. In case II, when the energy requested by v_{O1} is considerably large, the time of E_1 is significantly extended to obtain enormous energy. This process delays the energy attainment of the other outputs; thus, the energy allocation sequence cannot

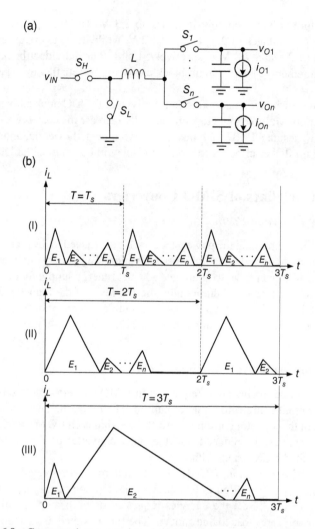

Figure 6.9 Constant-charge auto-hopping: (a) topology; (b) operating waveforms

be achieved in a switching frequency. Therefore, CCAH extends the switching frequency to $2T_s$ in case II. Case III shows that CCAH also extends the switching frequency to $3T_s$ when E_2 is considerably large.

Even CCAH can adapt to a wide range of loads and unbalanced loads, thereby delaying the energy attainment results in cross regulation. The switching frequency decreases with an increase in load current. Obvious disadvantages are large switching power loss at light loads and large ripple at heavy loads.

6.3.1.2 Pseudo-Continuous-Conduction Mode (PCCM)

The PCCM operation in [11] forces the inductor current back to the predefined DC level I_{DC} at the end of each PWM switching cycle. A non-zero inductor current can have advantages, including improved driving capability, lower output voltage ripples than that of the DCM, and higher bandwidth if compensated by Type II compensator.

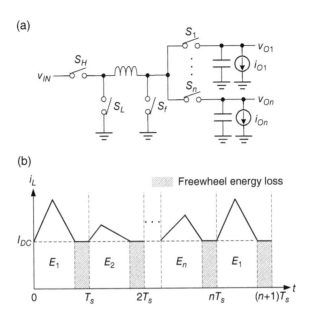

Figure 6.10 Pseudo-continuous conduction mode: (a) topology; (b) operating waveforms

Figure 6.10(a) shows a brief illustration of the classic PCCM control method in SIMO DC/DC buck converters. Except for the switches S_H, S_L, and S_1–S_n, which are required in the basic topologies of SIMO converters in Figure 6.4, a freewheel switch S_f in Figure 6.10(a) is additionally adopted to achieve PCCM operation. The operating waveforms of the inductor current are depicted in Figure 6.10(b). In the first switching cycle (from 0 to T_S), energy E_1 is allocated to v_{O1} through (S_H, S_1) and (S_L, S_1) for inductor current charging and discharging, respectively. When the inductor current returns to I_{DC}, S_L and S_F form the freewheel path until the switching cycle ends, and a constant inductor current level can be maintained by the shorted inductor because of the freewheel switch. Next, v_{O2}–v_{On} obtain energies E_2–E_n in the subsequent switching cycles in sequence. After n switching cycles, energy is allocated to v_{O1} again.

The insertion of a freewheel period to separate each output can be viewed as an energy buffer. Once the load current at one of the n outputs varies, the extension or compression of the freewheel period can achieve the voltage regulation of the load-varied output without affecting the other outputs. Therefore, cross regulation can be eliminated if a sufficient freewheel period can be maintained. In addition, the insertion of a freewheel period can simplify the compensation network. The system order is reduced from two (in voltage-mode control) to one that is similar to that of the DCM operation because the inductor current is reset to a predefined DC level in each switching period. The compensation for the PCCM operation becomes simpler. A Type II or PI compensator can be used to increase system stability.

Unfortunately, some inherent disadvantages that exist in PCCM are not suitable for battery-powered applications. The first crucial and obvious disadvantage is the high conduction power loss during the freewheel period. To ensure regulation, I_{DC} is always higher than the total loading current of all the outputs because no energy is transferred to the outputs in freewheel periods. During freewheel periods, the multiplication of the square of a high-value I_{DC} and the turn-on resistances of S_L and S_f causes large conduction loss and greatly degrades power efficiency. Although a large freewheeling power switch can alleviate conduction power loss, a large silicon area occupation drastically increases the cost. For heavy load current endurance and low on-resistance, an unlimited increase in the area of S_f is impossible. Moreover, the design of the energy buffer region becomes more complicated in the PCCM if an adaptive and adequate I_{DC} is dynamically adjusted according to load conditions. If I_{DC} is too low, a large voltage ripple and unregulated problems may occur. If I_{DC} is too high, a large conduction power loss deteriorates the power conversion efficiency. The second fatal disadvantage is the large output voltage ripple. The output voltage ripple characteristic of the PCCM is similar to that of DCM because the peak inductor current is determined by the loading current. In particular, the output voltage ripple increases drastically at heavy loads. The efficiency deterioration in the PCCM makes it unsuitable in tablets or battery-powered applications.

6.3.1.3 Adaptive Energy Recovery Control (AERC)

To simultaneously achieve low cross regulation and suppress freewheel power loss, the AERC technique is proposed in [12]. Instead of freewheel duration, an energy recovery duration is constructed in the AERC technique. The AERC creates an energy recovery duration, which is the feedback controlled by the load-dependent duty. This recovery duration is responsible for decoupling the sub-channels and behaves similarly to a buffer region in the transient condition. As illustrated in Figure 6.11(a), the switch S_{DR} is introduced to form an energy recovery path. The operating waveforms are illustrated in Figure 6.11(b). At the beginning of a switching cycle, energy is allocated to each output in sequence. After all the outputs obtain adequate energy, the energy recovery duration is activated in the remaining switching cycle. The

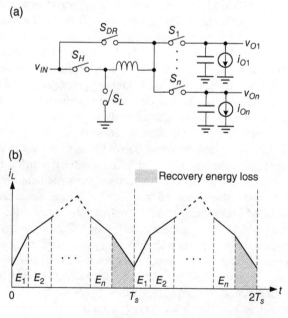

Figure 6.11 Adaptive energy recovery control: (a) topology; (b) operating waveforms

inductor current flows through S_L, L, and S_{DR} back to the input source for energy recovery. In other words, sufficient energy stored in the inductor can face any sudden load changes from all the outputs.

Both the energy recovery duration in AERC and the freewheel duration in PCCM act as energy buffer regions to minimize cross regulation. The decision of the DC inductor level in AERC is not required, in contrast to the PCCM technique. Under different loading conditions, the energy recovery duration is automatically determined once the energy allocations in all the outputs are complete. Although energy can be recovered back to the input source for energy reuse, the turn-on resistance of S_L and S_{DR} still causes large conduction power loss. Thus, the problem of deteriorated power-conversion efficiency still cannot be solved by such an active energy recovery technique. In other words, the AERC technique can ensure controller stability and reduce cross regulation at the cost of power-conversion efficiency.

6.3.1.4 Energy-Conservation Mode (ECM) Control

From the above discussions, inserting an energy buffer into the energy control sequence can minimize cross regulation at the cost of power efficiency. However, high efficiency for long usage time in portable and wearable electronics is especially important. The ECM control removes the buffer region and rearranges the energy paths according to all the output loading conditions. Thus, high efficiency and low cross regulation can be achieved concurrently because the freewheel stage is deleted [13, 14].

Figure 6.12(a) illustrates the topology of the ECM-controlled SIMO converter, which does not include extra switches in the basic SIMO converter shown in Figure 6.4. As mentioned earlier, this topology contains $2n$ energy paths, which include n inductor charging paths

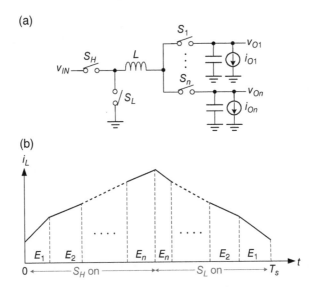

Figure 6.12 ECM control: (a) topology; (b) operating waveforms

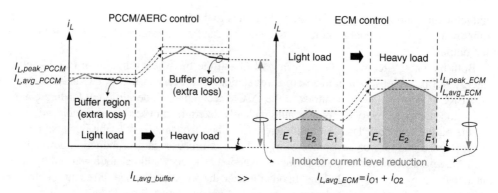

Figure 6.13 Comparison of the PCCM/AERC and ECM controls in a two-output DC/DC converter

through the high-side switch S_H, and n inductor discharging paths through the low-side switch S_L. By combining these paths with all positive and negative slopes in one switching cycle, the inductor current waveform of ECM control is created, as illustrated in Figure 6.12(b). When S_H is on and S_L is off, the inductor current is charged, and the energy is allocated from the first to the last output in sequence. When S_H is off and S_L is on, the inductor current is discharged, and the energy is allocated from the last to the first output in sequence. From another point of view, the ECM control method uses the superposition of n inductor currents in different current levels. The energy levels E_1–E_n are superposed from the bottom to the top layer to construct the inductor current waveform.

Taking a two-output DC/DC converter, that is, a SIDO converter as an example, Figure 6.13 clearly compares the energy loss in PCCM/AERC with that in ECM control. In PCCM or AERC control, the buffer region not only causes extra loss but also largely increases the inductor current level. The increase in current level results in a large conduction power loss in all the power switches of the SIMO converters and further degrades power efficiency. By contrast, ECM control simultaneously removes the buffer region and reduces the inductor current level. Obviously, the average inductor current level in PCCM/AERC, I_{L,avg_buffer}, is much higher than that in ECM control, I_{L,avg_ECM}. The advantage of ECM control is that I_{L,avg_ECM} is equal to the total load current. In other words, increasing the inductor current is not necessary, and conduction power loss can be reduced.

6.3.1.5 Buck and Boost SIMO Converter

Figure 6.14 shows the structure of the SIDO converter, which achieves dual buck and boost outputs with one single inductor utilization. In order to minimize the number of power switches to obtain silicon cost saving, the energy delivery paths for dual outputs must be arranged carefully. There are three main power switches and one freewheel switch in the power stage. Both of the output conditions can be fed back to the controller by the EAs. The full-range current-sensing circuit is used to derive the complete inductor current information to achieve duty cycle modulation with the charge-reservation methodology. The control logic can generate the control signals for the power switches. This technique implements similar current-mode control for

Single-Inductor Multiple-Output

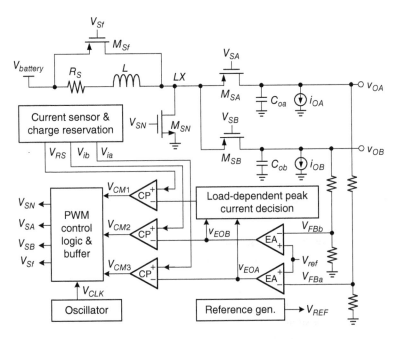

Figure 6.14 SIDO converter with arranged energy paths and charge-reservation control technique for one buck and one boost output voltage [3]

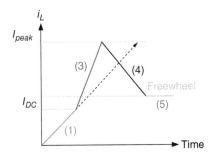

Figure 6.15 The energy delivery methodology for the SIDO converter with one buck and one boost output voltage [3]

a low output voltage ripple. Although both feedback paths need a PI compensator, a current-mode PI compensator can be used to simplify and reduce the design and the cost [33, 36].

Figure 6.15 shows the energy delivery paths of the inductor current to deliver energy from the input battery voltage source to both of the outputs. Path 1 delivers energy to the buck output and the inductor current slope is positive. After it reaches a satisfactory current level for the buck output, the energy path will switch to path 3 to charge the inductor current to the desired peak value, which is determined by both feedback paths. Thus, the inductor is always kept below the calculated peak value. After path 3, the stored energy starts to transfer to the boost output through path 2. Finally, the freewheel operation appears at the end of the switching cycle

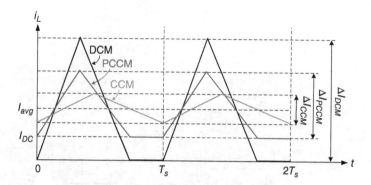

Figure 6.16 Comparison of ripple performances in the DCM, CCM, and PCCM controls

until the next PWM switching cycle starts. Therefore, the energy distribution for the dual outputs can be well arranged. In contrast, although this structure can minimize the number of power switches, at least the boost output needs to have a larger loading current compared with the buck output to obtain a negative inductor current slope. That is to say, the boost operation is used to release the inductor current when the discharging operation is activated. However, this topology is not suitable in PMU designs for Soc, because the loading current criteria cannot be met all the time. In other words, the low-side power switch cannot be removed simply for cost reduction because this will lead to a stability problem if considering all overloading current ranges.

6.3.1.6 Comparison of Ripple Performance

As introduced in the preceding sections, different energy control methods result in different ripple performances (Figure 6.16). The inductor current returns to zero at the end of each switching cycle in DCM control. A large current ripple is required to maintain the average inductor current at I_{avg}. To decrease the output ripple and achieve small cross regulation at the same time, the inductor current returns to a predefined DC value I_{DC} at the end of each switching cycle in PCCM control. However, power efficiency is sacrificed since a large conduction loss occurs during the freewheel period. To further decrease the output ripple and maintain high power efficiency, a continuous-conduction mode (CCM) can be adopted with adequate energy allocation. The relationship of output ripple between three control modes is $\Delta I_{DCM} > \Delta I_{PCCM} > \Delta I_{CCM}$.

6.3.1.7 Inductor Ringing Suppression

At light load or with DCM-controlled SIMO converters, inductor ringing suppression can help suppress the EMI problem. Figure 6.17 illustrates the block diagram of inductor ringing suppression. The switch S_{RS} and the inductor ringing suppression logic are introduced to help suppress the ringing phenomenon. Once the inductor current reaches zero, all output switches are turned off to prevent the negative inductor current. Figure 6.18(a) shows that v_X starts ringing when all the switches are off because of the residual energy in the inductor. This ringing

Single-Inductor Multiple-Output

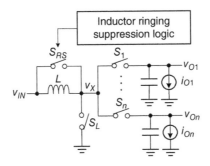

Figure 6.17 Block diagram of the inductor ringing suppression

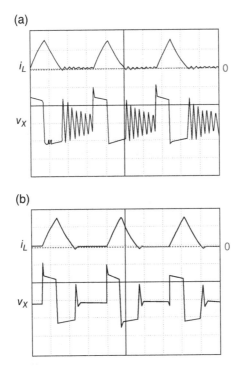

Figure 6.18 (a) Unclamped and (b) clamped waveforms of inductor current ringing [15]

degrades the EMI performance. Figure 6.18(b) shows that, with S_{RS} turned on during the zero inductor current period, the ringing effect can be clamped to suppress EMI.

6.3.2 Classifications of Control Methods

From the above discussions, the inductor current waveform and energy allocation sequence can be determined with different energy path designs. Next, how to determine the switching points from sequence to sequence is introduced. This is similar to the duty cycle generation in buck

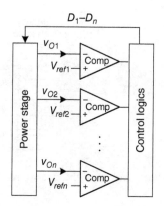

Figure 6.19 Circuit blocks in ripple-based control technique

converters. The control methods introduced in Chapter 3 can be adopted with different combinations.

6.3.2.1 Pure Ripple-Based Control

Pure ripple-based control, which directly compares the output voltage with reference voltages, is the simplest method of generating a duty cycle in buck converters. In a pure ripple-based controlled SIMO converter, as shown in Figure 6.19, multiple output voltages v_{O1}–v_{On} are compared with their individual reference voltages V_{ref1}–V_{refn}, respectively. At the beginning of a switching frequency, the energy is transferred to v_{O1}. Until v_{O1} reaches V_{ref1}, which indicates that the energy transferring to v_{O1} is sufficient, the energy starts to transfer to v_{O2} by turning off S_1 and turning on S_2. When all the outputs obtain a sufficient amount of energy, one complete switching cycle ends.

With this control method, energy is transferred to the outputs in sequence. The switching frequency varies greatly with output loading conditions. A PLL is usually introduced to achieve fixed-frequency operation [16, 17]. Given that the switching frequency is locked to a predetermined reference frequency, the output voltage of the CP in PLL is used to control the peak inductor current. If the switching frequency is too low, the SIMO converter decreases the peak value of the inductor current. By contrast, if the switching frequency is too high, the SIMO converter increases the peak value of the inductor current to achieve frequency locking. The pseudo-constant-frequency PWM control can be implemented. To obtain improved noise immunity in duty cycle determination, large ESR output capacitors are adopted. As a result, a large output ripple is a major disadvantage in ripple-based control. In addition, voltage regulation performance is not good without the help of EAs because of the DC deviation from the reference voltage.

6.3.2.2 Constant-Frequency Ripple-Based PWM Control

Although the ripple-based control technique is simple, adopting PLL greatly increases the silicon area and design complexity. The traditional PWM control with a clock signal is another available control method. The constant-frequency ripple-based PWM control is illustrated in

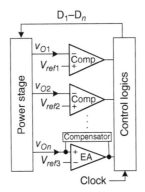

Figure 6.20 Circuit blocks of constant-frequency ripple-based PWM control

Figure 6.20 [18]. The first $n-1$ outputs adopt ripple-based control in sequence, which is the same as the operation in pure ripple-based control. Once v_{On-1} obtains sufficient energy, the remaining energy is transferred to v_{On} during the remainder of the switching frequency. However, different from v_{O1}–v_{On-1}, v_{On} is controlled by an EA. The EA amplifies the difference between v_{On} and V_{refn} to control the peak inductor current. If v_{On} is lower than V_{refn}, the EA controls and increases the peak inductor current. If v_{On} is higher than V_{refn}, the EA controls and decreases the peak inductor current for voltage regulation. With the help of the EA, the regulation performance of v_{On} is ensured. Still, v_{O1}–v_{On-1} have poor regulation performances because of the comparator-based control. In addition, adjusting the peak inductor current for different output loading conditions must be achieved by varying the output of the EA. v_{On} must be influenced once transient loading occurs at any one of the outputs. This indicates that serious cross-regulation problems occur at v_{On}. Thus, v_{On} is not suitable for supplying sensitive blocks although it is controlled by EA.

6.3.2.3 Error-Based PWM Control

For better regulation performance at all outputs, n EAs are adopted, as shown in Figure 6.21 [13, 14, 19, 20]. v_{O1}–v_{On} are fed back to n EAs to generate individual error signals. Either voltage- or current-mode control can be adopted in error-based PWM control SIMO converters. If current-mode control is used, the inductor current information i_L is required to help determine the duty cycles. Compensation networks are required in each EA. By contrast, a voltage-mode controlled SIMO converter occupies a larger PCB area, which is not suitable for small and low-cost portable and wearable electronic applications.

6.3.2.4 Freewheel Current Feedback Control

Freewheel current feedback control is required in the PCCM control. Figure 6.22 illustrates the block diagram of the freewheel current feedback control loop. Two current feedbacks containing the freewheel current i_{fw} and the inductor current i_L are required. The reference freewheel

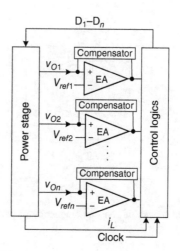

Figure 6.21 Circuit blocks of error-based and current-mode PWM control

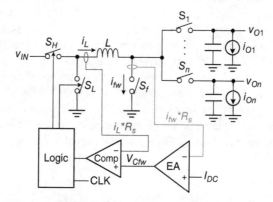

Figure 6.22 Block diagram of the freewheel current feedback control loop [21]

current I_{DC} is preset to define the DC current during the freewheel period. The average freewheel current i_{fw} is sensed ($i_{fw}*R_s$) and sent to the EA. The difference between $i_{fw}*R_s$ and I_{DC} is amplified to generate the error signal v_{cfw}. Thereafter, v_{cfw} can be viewed as the v_c peak current-mode buck converter. By sensing the inductor current and comparing v_{cfw} with i_L*R_s, the peak inductor current can be determined by controlling both S_H and S_L. At the same time, i_{fw} can be adjusted to I_{DC}.

I_{DC} can be adjusted based on different load conditions to minimize the switching loss during the freewheel duration. As the load decreases, I_{DC} can be decreased to degrade the energy stored in the inductor. At considerably light loads, when I_{DC} reaches zero, the PCCM operation is the same as DCM control. As the load increases, I_{DC} can be raised to store more energy in the inductor, thereby facilitating both the instant energy request and the transient response performance.

6.3.3 Design Goals

The major advantage of SIMO converters is their compact size in PMU design, because only one inductor and one SIMO chip are required. However, given the complexity of SIMO design, PMUs with multiple SWRs usually have better performance. Improving these characteristics in SIMO converters is especially important. For portable and wearable applications, several design goals exist, including power efficiency, load current range, output voltage ripple, cross regulation, and transient response, to meet portable and wearable requirements.

6.3.3.1 Improvement of Power-Conversion Efficiency

Two main factors that determine the power-conversion efficiency in SIMO converters are energy path design and power stage switch control. As discussed in Section 6.3.1, the insertion of energy buffer regions, such as ACRE and PCCM techniques, largely decreases power efficiency. Therefore, energy buffer regions are avoided in portable and wearable PMU designs.

A large number of power switches in a SIMO converter, which is usually implemented by P-type or N-type MOSFETs, is highly related to power efficiency. The turn-on resistance of power switches determines the conduction loss and affects the power efficiency, especially at heavy loads. In tablet applications, a relatively large supply voltage range is also required. However, a low turn-on resistance and a high power efficiency cannot be maintained when multiple voltages have wide output voltage ranges in conventional SIMO designs. One of the main design goals here is how to properly control power switches.

6.3.3.2 Reduction in the Number of Power Switches

The number of power switches should be minimized to get the cost reduction in silicon area, since a large number of power switches occupy a large silicon area. When considering the reduction in both conduction loss and switching loss at the power stage to enhance the power-conversion efficiency, the energy delivery paths at the power stage of the SIDO converter need to be well arranged. That is to say, minimizing the number of power switches simply for cost reduction may cause an increase in power loss. A trade-off may exist between the cost reduction in silicon area and the reduction in power loss. The reader should consider both of these at the same time. One freewheel power switch is sometimes used for easy stability consideration [16]. However, it causes an increase in both the cost and the power loss. In other words, the reduction in the freewheel power switch can reduce the cost and the power loss. It is meaningful to carry out the reduction in a number of power switches.

6.3.3.3 Extension of Load Current Range and Load Difference Among Different Outputs

The SIMO converter must cover all possible load current ranges to guarantee correct operating functions. In tablet applications, multiple supply voltages for T_{CON} and source drivers may operate with a large load current difference, thereby making the regulation for all outputs difficult in SIMO converters because all the outputs share one inductor at the same time.

6.3.3.4 Reduction in Output Voltage Ripple

The SIMO converter provides multiple supply voltages for sub-circuits in tablet systems. However, the quality of the output voltages needs to be guaranteed. The output voltage ripple of the PMU is a concern because of the noise-sensitive sub-circuits in tablet systems. In SIMO converters, a small output ripple with a large load current difference between multiple outputs is especially difficult to achieve. A straightforward way of suppressing the voltage ripple is through a small ESR value. By adopting an output capacitor with a small ESR value, the voltage ripple, overshoot, and undershoot are suppressed. However, noise immunity in ripple-based control is poor. The error-based PWM control with good regulation performance is a suitable choice for tablet applications.

6.3.3.5 Reduction of Cross Regulations

Cross regulation is an inherent problem in SIMO converters because energy allocations to all outputs are stored in a single inductor. The inductor current cannot be changed rapidly during transient response. One load variation in a certain output may influence the other outputs. As discussed in Section 6.3.1, several methods are adopted to minimize cross regulation at the cost of power-efficiency performance. However, this conflicts with the high-efficiency design goal in tablet applications. Therefore, the mitigation of interference between all outputs is a concern in SIMO control methodology designs.

6.3.3.6 Improvement of Transient Response

The load transient response is the basic requirement of power management design. SIMO converters are required to have the capability to recover output voltages from overshoot and undershoot within an acceptable time, because the load current of each output may suddenly change. In addition, the load transient response is used to demonstrate system stability to further guarantee stable operation. Thus, a fast load transient is also an important design goal in SIMO converters.

6.4 SIMO Converter Techniques for Soc

6.4.1 Superposition Theorem in Inductor Current Control

Figure 6.23 shows the four distinct energy delivery paths in the power stage of the SIDO converter. The dual step-down operation is derived with an input battery voltage $V_{battery}$ of 3.3 V and nominal output voltages, v_{OA} and v_{OB}, of 1.8 and 1.2 V, respectively. Four distinct energy paths can be composed to guarantee the energy transfer function and regulated output voltages. Figure 6.23(a), (c) shows that path 1 and path 3 are regarded as the inductor charging paths,

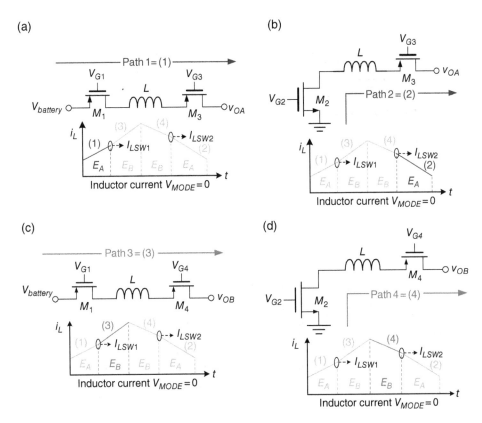

Figure 6.23 Four distinct energy delivery paths at the power stage: (a) inductor current charging path for v_{OA}; (b) inductor current discharging path for v_{OA}; (c) inductor current charging path for v_{OB}; (d) inductor current discharging path for v_{OB} [13]

with positive slopes to deliver energy to v_{OA} and v_{OB}, respectively. Figure 6.23(b), (d) shows that path 2 and path 4 are considered to be the inductor discharging paths, with negative slopes to transfer energy to v_{OA} and v_{OB}, respectively. When these four energy delivery paths are combined, the energy distribution of the dual outputs can be achieved in the CCM operation.

Based on the characteristics of the CCM operation, the final state of the inductor current level is equal to the initial state in one switching period at steady state. However, the inductor current waveform as depicted in Figure 6.23 is a specific combination of energy delivery paths in steady state, controlled by the mode decision signal V_{MODE}. V_{MODE} determines two distant methodologies to achieve the operation in the SIDO converter. As a result, to minimize both transient and steady-state cross regulations induced by the output loading current steps and the large loading current difference between the dual outputs, respectively, the dual-mode energy delivery methodology can appropriately arrange the energy delivery paths and the inductor current level. Thus, the transient cross regulation can be reduced by rapidly adjusting the inductor current when the load transient response occurs, and the steady-state cross regulation can be minimized simultaneously with the output voltage ripple in the SIDO converter.

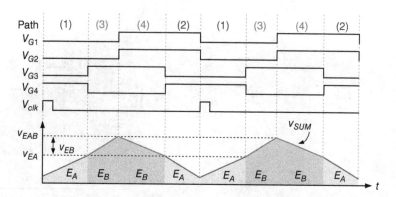

Figure 6.24 The timing diagram of the ECM control

Figure 6.24 shows the timing diagram of the ECM control with order of path 1, path 3, path 4, and path 2. This ordered sequence is circulated by the triggering of the system clock, V_{clk}. The error signal v_{EA} determines the transitions from path 1 to path 3 and from path 4 to path 2. That is, path 1 and path 2 are decided by v_{SUM} and v_{EA}, respectively, for the output v_{OA}. Similarly, path 3 and path 4 are decided by v_{SUM} and v_{EB}, respectively, which is the difference between v_{EAB} and v_{EA}. The ECM control achieves separated dual step-down operations by the inductor current superposition scheme. Both of the error signals, as well as the energy demands of the dual output, can be modulated within one PWM switching cycle. Thus, each output receives the power form input battery in each switching period. Besides, each power switch, M_1 to M_4, would switch twice in one switching cycle time with the ordered energy delivery paths.

6.4.2 Dual-Mode Energy Delivery Methodology

Figure 6.25 illustrates the dual-mode energy delivery methodology controlled by the mode signal V_{MODE}. If the energy delivered to the output of v_{OA} is smaller than that of v_{OB}, then V_{MODE} is set to low. The energy delivery paths follow the order of path 1, path 3, path 4, and path 2. Therefore, energy is delivered to v_{OA} at the beginning of every PWM switching cycle that is triggered by the system clock V_{clk}. By contrast, V_{MODE} is set to high when the output v_{OA} has a larger energy request than v_{OB}. Therefore, the energy delivery paths follow the order of path 3, path 1, path 2, path 4, and commence when the energy delivered to v_{OB} is synchronous with V_{clk}. Switches among the four energy delivery paths are also determined by the intersections of the summing signal v_{SUM} with v_{EAB}, v_{EA}, and v_{EB}. In this case, the intersection of v_{EAB} and v_{SUM} determines the peak inductor current to optimize the stored energy in the inductor compared with prior studies [11], because v_{EAB} is the summation of v_{EA} and v_{EB} to represent the exact energy requirement of the dual outputs. Utilizing the superposition theorem in the SIDO converter with the current-programmed control is similar to the combination of the two separate inductor current levels in two single-output buck converters. Additionally, v_{EA} determines the transition points from path 1 to path 3 and from path 4 to path 2 when V_{MODE} = 0. E_A indicates the provision of the energy for v_{OA}, and is delivered at the period with lower inductor current value. E_B indicates the energy provision for v_{OB}, and is located at the period

Single-Inductor Multiple-Output

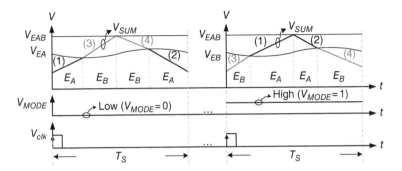

Figure 6.25 Illustration of the dual-mode energy delivery methodology

that contains the peak inductor current. Moreover, the energy transition points from path 3 to path 1 and from path 2 to path 4 are decided by the error signal v_{EB} when $V_{MODE} = 1$. E_B is derived at the period with lower inductor current value, whereas E_A is obtained near the peak inductor current. Either of the energy operation modes achieves the energy delivery scheme with the current-programmed control through the determination of V_{MODE}.

Figure 6.26 shows the detailed LV-EDC control scheme. In the ECM control methodology, each compensation enhancement multi-stage amplifier (CEMA) in the energy distribution regulation amplifier (EDRA) circuit acts as EA to feed back each output loading condition and to generate the error signal v_{EA} or v_{EB}. The CEMA circuit is designed to use the low-voltage multi-stage structure presented in [6] to guarantee high system loop gain for good output voltage regulation. The energy modulation circuit is composed of three individual comparators and the energy path logic, which can carry out the control duty cycles for each distinct output voltage.

Figure 6.26(a) shows the precise control scheme when the signal V_{MODE} is set to low. Figure 6.25 shows that with the selection of the ordered energy delivery path 1-3-4-2, v_{EAB} and v_{EA} are compared with v_{SUM} through the comparators $comp_1$ and $comp_2$, respectively. Subsequently, the comparison output signals v_{CAB} and v_{CA} are sent to the energy path logic to determine the control signals for turning the power switches on or off. Figure 6.26(b) shows that v_{EAB} and v_{EB} are compared with v_{SUM} through the comparators $comp_1$ and $comp_3$, respectively, to generate the ordered energy delivery path 3-1-2-4 when V_{MODE} is set to high. The energy path logic can determine the 4-bit control signal, V_{CTL}, with the synchronization clock signal V_{clk} to realize the energy delivery paths. The dual-mode energy delivery methodology leads to the distinct energy delivery scheme at the power stage corresponding to different output loading current conditions. To further enhance both transient and steady-state cross regulations, the energy delivery paths must be optimized in accordance with the relationship of the two outputs. The energy operation mode and the energy delivery paths must be adjusted accordingly to obtain a suitable energy delivery scheme for the dual outputs.

6.4.3 Energy-Mode Transition

The output voltage ripple is constantly limited to one certain value (e.g., 5% of the rated value) to avoid influencing the performance of a few noise-sensitive analog sub-circuits in the Soc. Equations (6.2) and (6.3) illustrate the respective output voltage ripples Δv_{OA} and Δv_{OB} of the

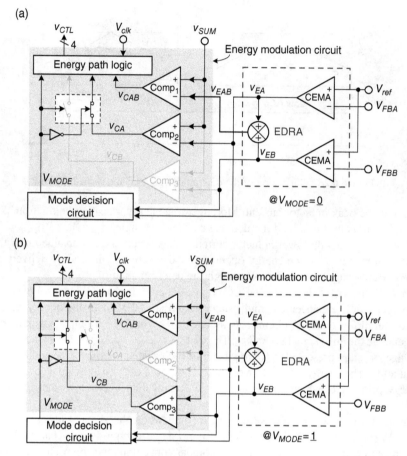

Figure 6.26 Control scheme of the LV-EDC circuit: (a) operation of $V_{MODE} = 0$; (b) operation of $V_{MODE} = 1$

two outputs v_{OA} and v_{OB}, respectively, in the SIDO converter, where R_{LA} and R_{LB} are the equivalent load resistances at v_{OA} and v_{OB}, respectively; T_{on_M3} and T_{on_M4} indicate the turned-on periods of the power switches, M_3 and M_4, respectively; R_{ERA} and R_{ERB} are the ESR on the output capacitors C_{OA} and C_{OB}, respectively.

The first term appearing in both Eqs. (6.2) and (6.3) contains information on the output voltages, loading current conditions, and output capacitors. Design specifications define these values. Therefore, the inductor current I_{LSW} in Eq. (6.4) flowing through each switch, which is the maximum value of the currents I_{LSW1} and I_{LSW2}, as shown in Figure 6.23, determines the output voltage ripples. I_{LSW1} and I_{LSW2} represent the exact inductor current values at the energy path transition point, where v_{SUM} and v_{EA} or v_{SUM} and v_{EB} intersect in different energy operating modes.

$$\Delta v_{OA} \cong \frac{v_{OA}}{C_{OA}R_{LA}} T_{on\_M4} + R_{ERA} I_{LSW} \qquad (6.2)$$

$$\Delta v_{OB} \cong \frac{v_{OB}}{C_{OB}R_{LB}} T_{on\_M3} + R_{ERB}I_{LSW} \qquad (6.3)$$

where

$$I_{LSW} = \max\{I_{LSW1}, I_{LSW2}\} \qquad (6.4)$$

The power switches M_3 and M_4 are used to allocate energy to the dual outputs; however, both switches cannot simultaneously receive energy because of the usage of the one off-chip inductor. The characteristic of the discontinuous inductor current is certainly derived from both outputs; thus, the output voltage ripple is seriously affected by its ESR on the output capacitor and I_{LSW}. The solution to reducing the output voltage ripple is to decrease the value of I_{LSW}. Therefore, for a given material-dependent ESR value, minimizing the current I_{LSW} can facilitate the reduction of the output voltage ripples and steady-state cross regulation. Although the first term in both Eqs. (6.2) and (6.3) cannot be alleviated because of the design specifications, the energy-mode transition operation can help obtain a low I_{LSW} to reduce the output voltage ripple. The superposition theorem can achieve the SIDO operation without the need for the freewheel stage, thereby reducing the precise inductor current level to obtain low I_{LSW} and output voltage ripples.

However, the output voltage ripple may be increased by an increase in I_{LSW} when a large loading current difference occurs between the dual outputs. A high output voltage ripple is not acceptable, and this is the design goal of the wide loading current range for all multiple outputs. For example, if i_{OA} is considerably larger than i_{OB} and V_{MODE} is set to low, then the area E_A, as shown in Figure 6.25, occupies most of the energy transition period in one switching cycle. Therefore, both I_{LSW} and the output voltage ripple increase, which is treated as the phenomenon of steady-state cross regulation. Steady-state cross regulation can be decreased by adjusting the combination and duration of the energy delivery paths based on the relationship among multiple output loading current conditions.

For a given low V_{MODE}, the increase in i_{OB} leads to an increase in the inductor current with an extension of the energy delivery period for v_{OB} to acquire more power (Figure 6.27). The energy delivery mode must be maintained because of the remained loading current at v_{OA}. Thus, the energy path transition point is set to the lower side of the inductor current to prevent v_{OA} from being overcharged, so as to minimize the steady-state cross regulation. On the contrary, once i_{OA} increases considerably, the energy operating mode changes from $V_{MODE} = 0$ to $V_{MODE} = 1$ to achieve a better energy delivery scheme. That is, the inductor delivers energy to the light-load output during the period of a lower inductor current, but transfers energy to the heavy-load output within the duration of the peak inductor current. A similar operation is achieved when V_{MODE} is set to high at the beginning of the load transient response. As a result, the energy-mode transition operation ensures the receipt of energy within the periods of higher and lower inductor current levels for heavy-load and light-load outputs, respectively. The operation can avoid overcharging during the load transient period. The voltage regulations of the dual outputs can be assured with the reduction of output voltage ripples, as well as transient and steady-state cross regulations.

Figure 6.28 shows the steady-state cross regulation with the corresponding output voltage ripple. Figure 6.27, the energy path transition points of the two outputs, which are determined by v_{EA} when $V_{MODE} = 0$ or v_{EB} when $V_{MODE} = 1$, are set to the lower inductor current level.

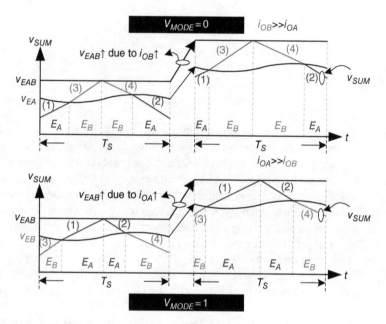

Figure 6.27 Energy-mode transition operation during the load transient response

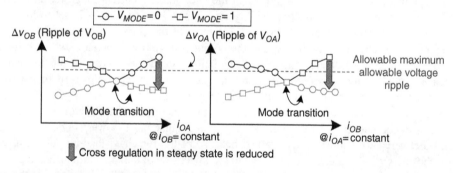

Figure 6.28 Illustrations of steady-state cross regulation with the corresponding output voltage ripple

If i_{OA} increases and i_{OB} keeps constant, V_{MODE} would be set high and the output voltage ripple is properly suppressed by the energy-mode transient operation. Thus, the output voltage ripples are limited within an allowable ripple range. This result indicates that the required energy for v_{OA} is delivered from the period with a low inductor current level to the period with a high inductor current level as i_{OA} increases. Similarly, the V_{MODE} signal would be set from high to low when i_{OB} increases continuously, but i_{OA} is maintained. Hence, the output ripple of v_{OA} can also be generated within an allowable ripple range in the steady-state operation.

The transient cross regulation is also enhanced by both the current-programmed energy delivery methodology and the energy-mode transition operation. Output loading variations can respond rapidly to their error signals to modulate the inductor current because of the current-programmed scheme. The energy-mode transition operation is also activated when one of the outputs derives a loading current step.

If the loading current in one output changes from light to heavy, then energy is provided to the relatively light load output at the beginning of each switching cycle to minimize transient cross regulation before delivering energy to the heavy load output. Therefore, the relatively light load output caused by the no-load variation does not have a large voltage drop during the load transient period because it derives a constant energy supply during the load transient period. Accordingly, energy in the single inductor is properly distributed during the load transient period to minimize transient cross regulation. In conclusion, the energy control sequence is one of the control methods that seriously affects cross regulation. This factor should consider the relative energy request from multiple outputs. Thus, the energy control sequence becomes more complex if the number of outputs increases considerably. This case is the reason that the number of outputs in the SIMO converter is always limited to four at most. Four different output sources with the help of multiple LDO regulators can implement a high-performance PMU in Soc applications.

6.4.4 Automatic Energy Bypass

In the SIDO/SIMO converter designs, changing the control method from PWM to PFM control is possible if one or more outputs have relatively light loading current condition. The PFM control method has the advantage of high efficiency at light loads. To extend the PFM concept in the SIDO/SIMO converter, the automatic energy bypass (AEB) mechanism is used to further enhance the power-conversion efficiency and ensure voltage regulation in steady state. The AEB mechanism reduces the number of energy delivery paths to lower both switching and conduction power losses at the power stage without sacrificing output voltage regulation. Therefore, when any one of the multiple outputs has a relatively decreasing loading current condition, the existing energy delivery paths need to be bypassed if the load decreases continuously.

For a given $V_{MODE} = 0$, Figure 6.29 shows that the energy delivery paths 1-3-4-2 are automatically bypassed, although the PWM operation with a constant switching frequency is still applied to the rest of the outputs. For example, when i_{OA} decreases but i_{OB} remains constant, the decreasing value at i_{OA} reduces the loading current-dependent error signal v_{EA} caused by the characteristics of the current-programmed control. Consequently, path 1 is bypassed first to reduce the duration of energy delivery to v_{OA}. The remaining three energy delivery paths can ensure the voltage regulation of all outputs. The light-load output is not overcharged in

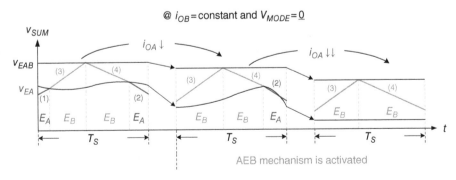

Figure 6.29 Operation of the AEB mechanism when i_{OB} is constant and V_{MODE} is low

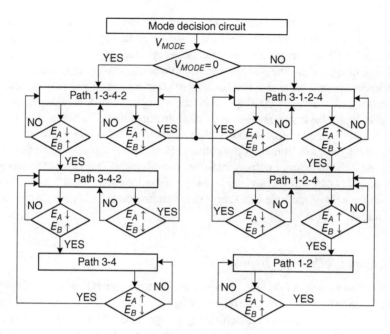

Figure 6.30 Flow chart of both energy-mode transition operation with the implemented AEB mechanism

steady state, and the output voltage ripples of all the outputs are still maintained within an allowable range. When i_{OA} decreases continuously, that is, to the ultra-light-load or no-load conditions, the energy delivery paths for v_{OA} are excluded. The operation at i_{OA} changes from PWM to PFM control. Meanwhile, the output voltage of the two outputs can be maintained even when a large loading current difference occurs among the two outputs. Consequently, the high inductor current level that results from the heavy-load output does not cause overcharge at the ultra-light-load output because of the AEB mechanism. For a given $V_{MODE} = 1$, a similar operation is also activated corresponding to the continuous decrease at i_{OB}.

The AEB mechanism is achieved particularly with implementation of the energy-mode transition operation in the SIDO converter. The reason is that the energy-mode transition operation can set the light-load output to obtain the energy during the period of the small inductor current level. Figure 6.30 shows the flowchart for both energy-mode transition operation and the AEB mechanism. The PMU in Soc applications becomes more flexible because no specific loading current restriction exists among all the supplying outputs. Moreover, the steady-state cross regulations are alleviated because the problem of overcharging is effectively eliminated by the PFM-like AEB control mechanism.

6.4.5 Elimination of Transient Cross Regulation

Although the AEB mechanism can reduce cross regulation, transient cross regulation still exists because the energy imbalance cannot be solved immediately. A possible solution is to extend the bandwidth; however, this strategy costs power and decreases the phase margin. Therefore,

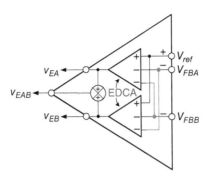

Figure 6.31 Modified EDRA with EDCA to eliminate transient cross regulation in the SIDO DC/DC converter

another control path is established locally to eliminate the disturbance from any of the outputs. For example, to properly eliminate the transient cross regulation in the SIDO DC/DC converter, the energy prediction function needs to be utilized by the modified EDRA as the EA (as shown in Figure 6.25), which is derived from CEMA. The multi-stage structure implemented by a cascading function can allow the circuit to operate under low supply voltage in advanced nanometer technology, such as 65 nm CMOS technology. However, to alleviate transient cross regulation, the prediction function must be adopted with the energy distribution correlation amplifier (EDCA), which is implemented in the modified EDRA, as shown in Figure 6.31, and behaves as the EA to monitor the output voltage condition. The EDCA can correlate both outputs to eliminate transient cross regulation in case of any sudden loading current change occurring in either of the outputs.

Figure 6.32 illustrates the energy delivery scheme by the energy mode of path 1-3-4-2 during the load transient response. Parts (a) and (b) show the respective excess energy at v_{OB} and the insufficient energy at v_{OA} when the step-up loading current change occurs at v_{OA} and v_{OB}, respectively. The excess energy at v_{OB}, which is shown as the shadowed area, results in voltage overshoot at v_{OB} even when v_{OB} has a constant loading current condition. The cross regulation causes the overshoot effect at v_{OB}. By contrast, the insufficient energy at v_{OA}, which is shown as the shadowed area, results in a voltage drop at v_{OA} even when v_{OA} has a constant loading current condition. The cross regulation causes the undershoot effect at v_{OA}. Any loading current variation that occurs at any one of the multiple outputs disturbs all the outputs because it disturbs the inductor current. In the chain reaction, the energy delivery sequence passes the disturbance to the other outputs, thereby necessitating the cross regulation to be alleviated in order to receive high-quality multiple outputs.

Similarly, Figure 6.33 depicts an illustration of the transient cross regulations in the case of the energy path 3-1-2-4. When the load current i_{OB} increases, the undesirable voltage variation occurs at v_{OA} because of the changed energy distribution scheme. The insufficient energy at v_{OB} also results in an undesirable drop when the load current i_{OA} increases. Excess or insufficient energy degrades the supply quality, which indicates that the transient cross regulation needs to be properly eliminated. Table 6.1 summarizes the transient cross regulations.

To effectively eliminate the transient cross regulation, energy prediction should be employed to prevent disturbing the other outputs. Therefore, if the inductor current can be locally

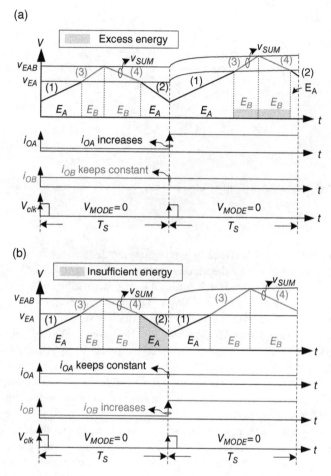

Figure 6.32 Illustration of transient cross regulation in the case of the energy path 1-3-4-2 and $V_{MODE} = 0$: load step at (a) v_{OA} and (b) v_{OB}

adjusted, then energy can be properly allocated to the other outputs that remain at constant loading current condition.

Figure 6.34(a) shows the energy prediction methodology in the case of the energy path 1-3-4-2. The error signal v_{EB} can be modulated even if the output v_{OB} remains at constant loading condition. v_{EAB} does not become considerably high to cause the overshoot effect because of the adjustment from v_{EB}. Similarly, Figure 6.34(b) shows that the energy prediction function is also activated when the load current i_{OB} increases. The error signal v_{EA} can be varied accordingly so that the energy distribution for v_{OA} can be kept constant without being affected by the disturbance from v_{OB}. Therefore, the undershoot effect does not occur at v_{OA} compared with the condition shown in Figure 6.32(b). Consequently, the energy prediction function can help prevent the output from being affected by the disturbance from the output that has sudden loading current variation. Hence, transient cross regulation is minimized locally by the energy prediction function because the inductor current is locally adjusted to its adequate value.

Single-Inductor Multiple-Output

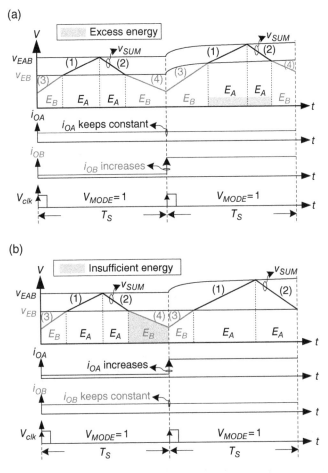

Figure 6.33 Illustration of the transient cross regulation in the case of the energy path 3-1-2-4 and $V_{MODE} = 1$: load step at (a) v_{OB} and (b) v_{OA}

Table 6.1 Summary of transient cross regulation

Operating mode V_{MODE}	Output loads		Error signal			Energy for output	
	i_{OA}	i_{OB}	v_{EA}	v_{EB}	v_{EAB}	E_A	E_B
$V_{MODE} = 0$ (path 1-3-4-2)	I	C	I	C	I	Insufficient	Excess (CR)
	D	C	D	C	D	Excess	Insufficient (CR)
	C	I	C	I	I	Insufficient (CR)	Insufficient
	C	D	C	D	D	Excess (CR)	Excess
$V_{MODE} = 1$ (path 3-1-2-4)	I	C	I	C	I	Insufficient	Insufficient (CR)
	D	C	D	C	D	Excess	Excess (CR)
	C	I	C	I	I	Excess (CR)	Insufficient
	C	D	C	D	D	Insufficient (CR)	Excess

I: increase, D: decrease, C: constant, CR: transient cross regulation.

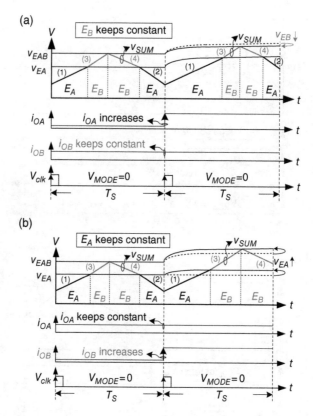

Figure 6.34 Methodology for eliminating transient cross regulation in the case of the energy path 1-3-4-2 and $V_{MODE} = 0$: load step at (a) v_{OA} and (b) v_{OB}

Figure 6.35 shows the energy prediction methodology in the case of the energy path 3-1-2-4. Figure 6.35(a) shows that, when a load transient response occurs at v_{OB}, the transient cross regulation effect can effectively be suppressed because v_{OA} can be supplied by constant energy during the load transient period of v_{OB}. Figure 6.35(b) illustrates that a similar operation is also activated to eliminate the transient cross regulation effect when a load transient occurs at v_{OA}. Accordingly, both the error signals, v_{EA} and v_{EB}, can be locally adjusted based on the voltage variation of each output. This implementation guarantees constant energy distribution for the output with unchanged loading current condition. The undesired voltage drop caused by the transient cross regulation is successfully eliminated.

6.4.6 Circuit Implementations

6.4.6.1 Pre-regulator

In advanced nanometer CMOS technology, the high-input battery voltage $V_{battery}$ cannot connect directly to low-voltage core devices because of insufficient voltage stress. The simple but inefficient design aims to use a high-voltage input/output (I/O) device to implement the whole

Single-Inductor Multiple-Output

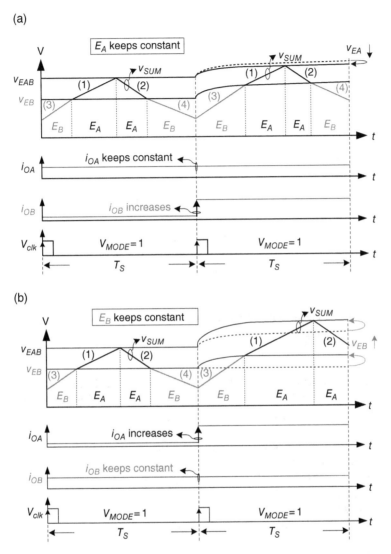

Figure 6.35 Methodology for eliminating transient cross regulation in the case of the energy path 3-1-2-4 and $V_{MODE} = 1$: load step at (a) v_{OB} and (b) v_{OA}

controller. However, the disadvantage of using this device includes dealing with a large silicon area and increasing cost. Nevertheless, if using the core devices to implement the controller is necessary, then a pre-regulator is needed to convert the high-input voltage to a low-supply voltage to prevent damaging the core devices. Figure 6.36(a) shows that the overall efficiency depends on the efficiency η_{SIMO} from the battery voltage being converted to the multiple outputs and on the efficiency $\eta_{pre-regulator}$ from the battery being converted to the pre-regulator output. To supply a regulated and noiseless power to the SIDO/SIMO controller, a high-efficiency pre-regulator is used as the design object. In commercial products, the pre-regulator

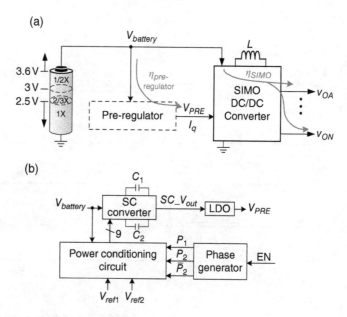

Figure 6.36 (a) Pre-regulator in the SIDO/SIMO design and (b) pre-regulator designed by combining the SC circuit and LDO regulator

is served by an LDO circuit. The advantages of using an LDO circuit include its simple structure and small silicon area, but a serious disadvantage is its poor efficiency if the pre-regulator output voltage V_{PRE} is low [22–24].

Battery voltage gradually deteriorates with usage time. Therefore, an SC converter can be used to provide large and various step-down conversion ratios for high conversion efficiency. The advantage of using an SC converter includes its simple structure, but the disadvantage is the output switching voltage ripple. Thus, the cascaded LDO circuit is selected to suppress the noise generated from the SC converter and ensure a stable and regulated supply voltage to drive the SIDO/SIMO controller. Figure 6.36(b) shows the SC pre-regulator design, which is controlled by the power conditioning circuit and the phase generator to get a low output voltage V_{PRE}.

Figure 6.37 illustrates the detailed configuration of the SC converter with cascaded LDO circuit. The SC converter is controlled by the power conditioning circuit and the phase generator. Battery voltage continues to decrease with an increase in usage time. The power conditioning circuit can determine the adaptive conversion ratio of the SC converter based on the decreasing trend of the high-input battery voltage $V_{battery}$. R_1 and R_2 are 400 and 100 kΩ, respectively. The reference signals of V_{ref1} and V_{ref2} generated from the bandgap reference circuit are 0.5 and 0.6 V, respectively. The adaptive conversion ratio aims for high power-conversion efficiency even if the battery voltage continues to decrease. The decoder can generate the gate control signals S_1–S_9 for the SC converter through the factor control signals V_{T1} and V_{T0}, and the phase clock signals P_1, P_2, and $\bar{P}_2$ from the phase generator. The power-conditioning circuit requires that the high input battery voltage $V_{battery}$ be automatically scaled down to a low voltage SC_V_{out} based on the predetermined factors of 1/2 or 2/3.

Single-Inductor Multiple-Output

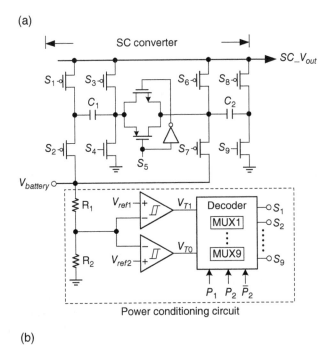

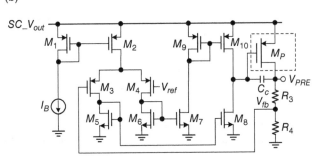

(c)

		S_1	S_2	S_3	S_4	S_5	S_6	S_7	S_8	S_9
Common phase		1	0	0	0	1	1	0	0	0
Gain phase	1/2	0	1	1	1	1	0	1	1	1
	2/3	0	1	1	0	0	1	1	1	1
Auto-bypass function		0	0	1	0	1	0	0	1	0

Figure 6.37 (a) Power conditioning circuit and SC converter with cascaded LDO circuit in (b); (c) truth table for the control signals

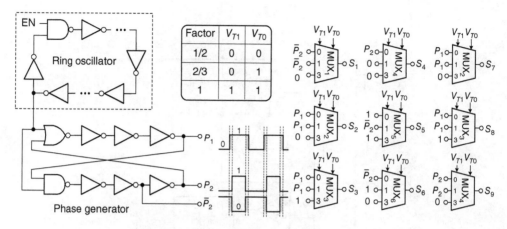

Figure 6.38 Phase generator in the pre-regulator

The auto-bypass function also disables the SC converter and directly connects the battery voltage $V_{battery}$ to the SC_V_{out} for a high pre-regulator efficiency when $V_{battery}$ is lower than 2.5 V. However, the efficiency is high without any conversion power loss when directly connected to $V_{battery}$. Figure 6.37 lists the operation of the SC converter for the gain and common phases under different adaptive conversion ratios. This mechanism allows the pre-regulator to enhance the conversion efficiency over a wide range of input battery voltage. In Eq. (6.5), M represents the adaptive conversion ratio of the SC converter:

$$\eta_{PRE} = \eta_{SC} \times \eta_{LDO} \approx \frac{SC\_V_{out}}{M \cdot V_{IN}} \times \frac{V_{PRE}}{SC\_V_{out}} = \frac{V_{PRE}}{MV_{IN}} \qquad (6.5)$$

The LDO circuit in the pre-regulator is compensated with a small on-chip output capacitor of 0.1 pF. This would also increase the power supply rejection (PSR) from the high input battery voltage for the LV-EDC controller.

Figure 6.38 depicts the phase generator. The phase clock generated by the ring oscillator is designed with a dead-time mechanism produced by a simple logic scheme to prevent leakage in the SC converter. The multiplexer determines the gate control signals for the switches in the SC converter by the factor control signals, V_{T1} and V_{T0}. Thus, all the switches in the SC converter are maintained at the off state to eliminate leakages of charge sharing during the phase-exchange period. Consequently, the conversion efficiency of the pre-regulator can be further enhanced.

Figure 6.39 shows the experimental steady-state waveforms of the pre-regulator. The output voltage SC_V_{out} of the SC converter in the pre-regulator is derived with a ripple of approximately 30 mV. The output voltage ripple at the output of the pre-regulator V_{PRE} can be suppressed to 10 mV because of the cascaded LDO regulator used as the SC converter post-regulator. The constant supply voltage for the LV-EDC can guarantee the converter performance, and the pre-regulator efficiency is constantly kept above 50% because of the power conditioning circuit.

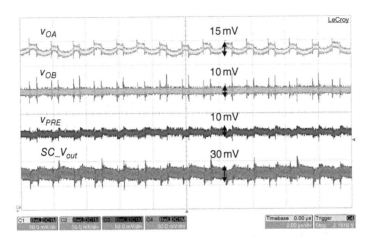

Figure 6.39 Measured results of the pre-regulator

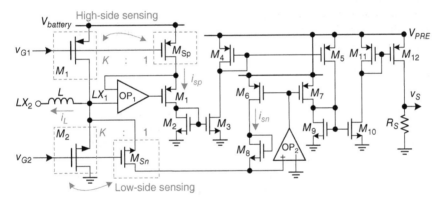

Figure 6.40 Full-range inductor current-sensing circuit

6.4.6.2 Full-Range Inductor Current Sensor

Figure 6.40 shows the full-range current-sensing circuit that implements the current-programmed control. In the SIDO converter, the current-sensing circuit is utilized to generate the full-range current-sensing signal v_s and to generate the control duties for the dual outputs, v_{OA} and v_{OB}. i_L is the inductor current, and K is the current-sensing ratio of both M_1 to M_{Sp} and M_2 to M_{Sn}. The four transistors are all implemented by the I/O devices to tolerate high voltage stress. The transistor M_{Sp} produces the sensing current during the turn-on period of the high-side power switch M_1. The source-to-drain voltages of M_1 and M_{Sp} are approximately equal because of the closed-loop control by the operational amplifier OP_1. Accordingly, the current i_{sp} flowing through M_{Sp} becomes proportional to the current flowing through M_1, thereby achieving high-side inductor current sensing. Similarly, M_{Sn} produces the inductor current information during the low-side switch turn-on period. The closed-loop control by the operational amplifier OP$_2$ conducts a sensing current i_{sn} flowing through the transistor M_6 and the sensing transistor M_{Sn}. Therefore, the full-range inductor current-sensing signal v_s in Eq. (6.6) across the sensing resistor R_s is generated by the summation of the two sensing current signals, i_{sp} and i_{sn}:

$$v_s = (i_{sp} + i_{sn})R_S = \frac{i_L}{K}R_s \tag{6.6}$$

Therefore, the full-range inductor current-sensing signal can be obtained as v_s to monitor the instantaneous inductor current information and to achieve the current-programmed operation in the SIDO DC/DC converter.

6.4.6.3 Voltage Summation Circuit

To achieve the ECM control methodology with the current-programmed control in the SIDO DC/DC converter, the voltage summation circuit shown in Figure 6.41 can facilitate realizing the voltage signal summation function. The supply voltage of the voltage summation circuit is supplied by the voltage V_{PRE} from the on-chip pre-regulator because of low-voltage operation. Noiseless and stable driving quality can guarantee the proper functions of the analog sub-circuits. The dual voltage-to-current conversion structures generate the currents i_{S1} and i_{S2} using the operational amplifiers OP$_1$ and OP$_2$, to enable the current conversion function. Two distinct input voltages, v_{IN1} and v_{IN2}, conduct two different summation currents i_{S1} and i_{S2}, respectively, through the resistors R_1 and R_2, respectively. If the resistors R_1, R_2, and R_3 are designed with the same value, then the direct summation function of the input voltages, v_{IN1} and v_{IN2}, can be realized. This scenario can ensure the superposition operation for the ECM control and the current-programmed operation. Sub-harmonic compensation still needs to be utilized in the current-programmed control methodology for stable operation. The summing function of the voltage summation circuit is expressed as:

$$v_{OUT} = (i_{S1} + i_{S2}) \cdot R_3 = \left(\frac{v_{IN1}}{R_1} + \frac{v_{IN2}}{R_2}\right) \cdot R_3 \tag{6.7}$$

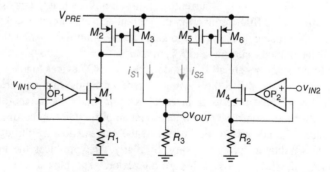

Figure 6.41 Schematic diagram of the voltage summation circuit

6.4.6.4 Mode Decision Circuit

The EA output signals v_{EA} and v_{EB} contain the output loading information because of the current-programmed control. Figure 6.42(a) shows that the mode-detection circuit decides the energy operating mode appropriately for the dual-mode energy delivery methodology. The mode-detection circuit comprises the level-shift structures, a common-gate amplifier, and two debounce cells. The transistors M_1 to M_4 are used in the shift-up structures, whereas M_5 to M_8 act as the shift-down operations. The common-gate amplifier also responds faster than the conventional operation amplifier without consuming enormous amounts of power. The hysteresis buffer and the debounce cells Db_1 and Db_2 (that delay the mode decision signal) are derived from the amplifier output to guarantee a smooth and stable energy mode transition

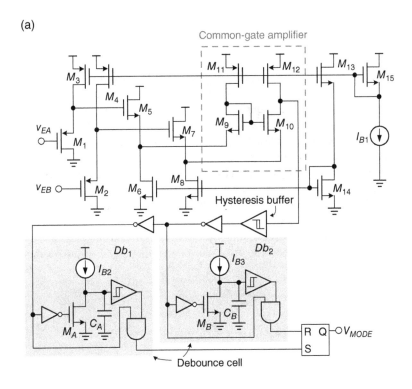

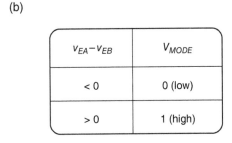

Figure 6.42 (a) Schematic diagram of the mode detection circuit. (b) Table of V_{MODE}

operation. See the attached table in Figure 6.42(b). The mode-decision circuit determines the energy operating mode according to the output loading current conditions.

In addition, the cascade level-shift structure is implemented to cope with the wide-range error signals to achieve correct operation. Because of the low-voltage operation in the LV-EDC circuit, the error signal is produced with a low voltage value, especially at light-load conditions, owing to the current-programmed scheme. The implementation of the conventional comparator structure would suffer from the large signal delay, with the compared results for the mode decision and the energy-mode transition operation being deteriorated. Moreover, the input of the common-gate amplifier cannot connect directly to both v_{EA} and v_{EB} because the additional bias current can affect the voltage level of the error signals, which might cause abnormal operation.

6.4.6.5 Energy Path Logic

Figure 6.43 shows the energy path logic in the energy modulation circuit. To achieve the dual-mode energy delivery methodology, a 4-bit signal V_{CTL} is produced as the control signal for power switches. v_{CA}, v_{CB}, and v_{CAB} derived from the outputs of the comparators in the energy modulation circuit (as shown in Figure 6.26) can be used to determine the energy path transition point. Decoder$_0$ and Decoder$_1$ are used to generate the corresponding energy delivery paths of $V_{MODE} = 0$ and $V_{MODE} = 1$, respectively. Finally, the output signal of the multiplexer V_{CTL} is selected through the V_{MODE} signal to realize the dual-mode energy delivery methodology. Figure 6.44 shows the timing diagram of the energy path logic.

Therefore, all the energy delivery paths at the power stage can be indicated through the energy path logic in LV-EDC. The ECM control methodology ensures that each output voltage can receive energy in every PWM switching cycle based on their loading current conditions. This scenario also assists in ensuring that the energy-mode transition operation and the

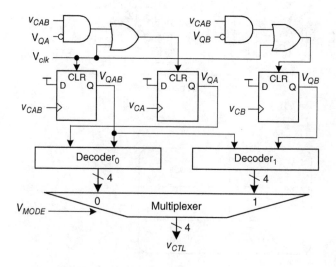

Figure 6.43 Schematic diagram of the energy path logic

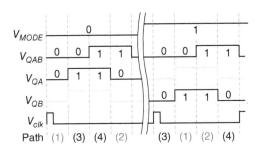

Figure 6.44 Timing diagram of the energy path logic

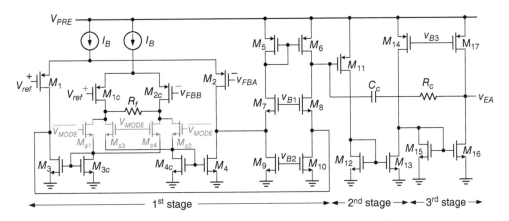

Figure 6.45 Schematic diagram of EDCA

automatic energy-bypass mechanism will enhance the driving performance to simultaneously achieve transient and steady-state cross regulation minimization.

6.4.6.6 Energy Distribution Correlation Amplifier

To achieve the energy prediction function to eliminate transient cross regulation in the SIDO DC/DC converter, the EDCA is shown in Figure 6.45 to achieve the EDRA (as shown in Figure 6.26). The EDRA functions as an EA to modulate both output voltages to realize the energy delivery function. The EDCA is also implemented with the multi-stage structure for the cascade function, because of the low-voltage operation in the controller. The EDCA solves the problem of insufficient voltage headroom and provides satisfactory voltage gain to meet the specifications of output voltage regulation. The on-chip capacitor and resistor, C_c and R_c, respectively, generate the compensation pole and compensation zero to ensure system stability.

With the three-stage structure of the EDCA, the energy prediction function is realized at the first stage. Hence, both the feedback signals, v_{fbA} and v_{fbB}, can affect the dual error signals, v_{EA} and v_{EB}. Figures 6.32 and 6.33 show that once either of the outputs has a loading current variation, its voltage variation can simultaneously modulate both error signals to reduce transient

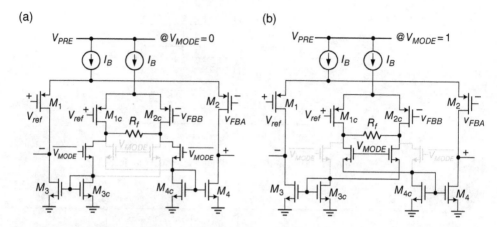

Figure 6.46 Operation of EDCA with the distinct energy modes for transient cross regulation elimination: (a) $V_{MODE} = 0$ and (b) $V_{MODE} = 1$.

cross regulation. The auxiliary switches, M_{s1} to M_{s4}, and the resistor, R_f, can control the energy prediction function. Switches M_{s1} to M_{s4} are controlled by the signal V_{MODE} to realize proper energy distribution with distinct energy modes. The resistor R_f controls the factors of the energy prediction function. Different output conditions, such as input voltages, output voltages, and loading current variations, may request the distinct value of R_f to achieve a perfect transient cross regulation elimination.

Figure 6.46 depicts the detailed operation of the energy prediction function in the EDCA. With the energy-mode indication signal V_{MODE}, the switches can be turned on or off to change the energy correlation at the distinct energy modes. Thus, the error signal, which is derived from the constant loading output, is also varied to ensure the receipt of unchanged energy in one PWM switching cycle. Figure 6.46(a), (b) shows the operation that reveals the methodology for properly modulating the error signals in case of any sudden loading current change at each output. Therefore, the energy delivery schemes shown in Figures 6.34 and 6.35 can be properly realized to eliminate the transient cross regulation.

The load transient response is the basic operation in the power management module. With sudden output load variations, the output voltage is derived as the undershoot or overshoot voltage when the output load current activates a step-up or step-down response, respectively. Therefore, power management needs to enlarge the delivery power to compensate for the voltage variation and lead to the voltage recovery operation during the load transient response.

Figure 6.47 shows the measured load transient response with an input voltage of 3.3 V, and the dual output voltages of 1.8 and 1.2 V. To eliminate transient cross regulation, the energy prediction function can help ensure the adequate delivery of energy to the output without negatively affecting the load variation. Figure 6.47(a) shows that when i_{OA} has a 240 mA load step but i_{OB} is unchanged, a 50 mV transient cross regulation is yielded at v_{OB} without using the energy prediction function. The energy prediction function can help ensure the unchanged energy distribution for the output that remains at constant load condition. Figure 6.47(b) shows that the undesirable voltage variation with the constant load output can effectively be eliminated. Similarly, Figure 6.48 shows that the transient cross regulation effect can also effectively

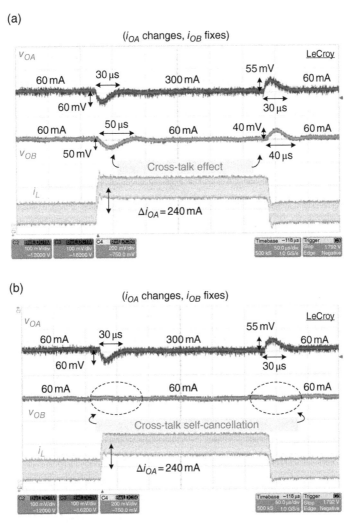

Figure 6.47 Measured load transient response when the load current variation occurs at i_{OA}: (a) without and (b) with the energy prediction function

be eliminated when i_{OB} has an instant change of 240 mA but i_{OA} retains the same load. The energy prediction function can also help eliminate the transient cross regulation in the SIDO DC/DC converter.

6.4.7 Experimental Results

6.4.7.1 Die Photo

Figure 6.49 shows that two SIDO DC/DC converters with the ECM control methodology fabricated in a 65 nm CMOS process demonstrate a few critical specifications. The off-chip

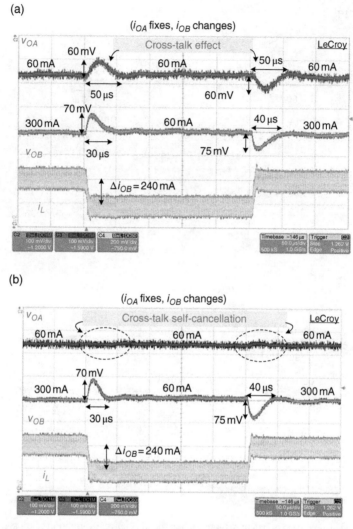

Figure 6.48 Measured load transient response when the load current variation occurs at i_{OB}: (a) without and (b) with the energy prediction function

inductor and off-chip capacitors for the dual output are 4.7 and 4.7 µF, respectively. Both chips have four embedded power switches. The pre-regulator and the LV-EDC circuit help obtain the correct control functions with utilization of the core devices. The low-voltage supply operation can reduce the power consumption, but may encounter a few design difficulties that increase the design complexity. Nevertheless, the 65 nm implementation presents one proper integration function in the Soc applications.

Figure 6.49(a) shows the fabricated chip that features the ECM control methodology for delivering energy to both outputs within one PWM switching cycle [13, 14]. The chip micrograph occupies an active core area of 1.44 mm². The advantage of the ECM control is its

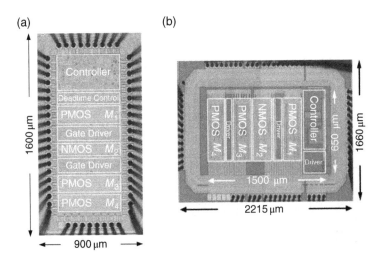

Figure 6.49 Chip micrographs of two SIDO converters fabricated using 65 nm CMOS process: (a) features the ECM control methodology; (b) features the transient and steady-state cross-regulation minimization

capacity to enhance power-conversion efficiency. Figure 6.49(b) shows the fabricated chip that features the reduction in transient and steady-state cross regulations [25]. The chip micrograph occupies an active core area of 1.28 mm². The mode transition operation and the energy automatic bypass mechanism help guarantee voltage regulation at both output nodes without any loading current restriction.

6.4.7.2 Measured SIDO Steady-State Operation

The SIDO converter with LV-EDC has two outputs, v_{OA} and v_{OB}, with values of 1.8 and 1.2 V, respectively. Figure 6.50 shows the measured steady-state operation with a loading current of 100 mA at i_{OA} and i_{OB}. Figure 6.50(a) demonstrates the ordered energy delivery paths 1-3-4-2 in sequence when V_{MODE} is set to low. The average inductor current is 200 mA, which is equal to the summation of the two output loads. The output voltage ripples are derived with 20 and 22 mV at v_{OA} and v_{OB}, respectively. Figure 6.50(b) shows the energy delivery scheme with paths 3-1-2-4 when V_{MODE} is set to high. The output voltage ripples are kept within 24 and 20 mV. Thus, the energy delivery scheme can be achieved with either of the two energy operation modes when no large load difference exists between the dual outputs.

Figure 6.51 shows the measured waveforms with i_{OA} and i_{OB} of 120 and 60 mA, respectively, with distinct operating modes. Figure 6.51(a) shows the measured results by forcing V_{MODE} to be low. The light-load output v_{OB} is the obtained energy within the period containing the peak inductor current. Therefore, the voltage ripples of v_{OA} and v_{OB} are 25 and 40 mV, respectively, indicating that the light-load output suffers from steady-state cross regulation because of incorrect utilization of the energy operating mode. Figure 6.51(b) shows that, with the mode decision circuit operation, the voltage ripple of v_{OB} is reduced to 20 mV, whereas the voltage ripple of v_{OA} is retained at 25 mV when V_{MODE} is changed to high. This result demonstrates the

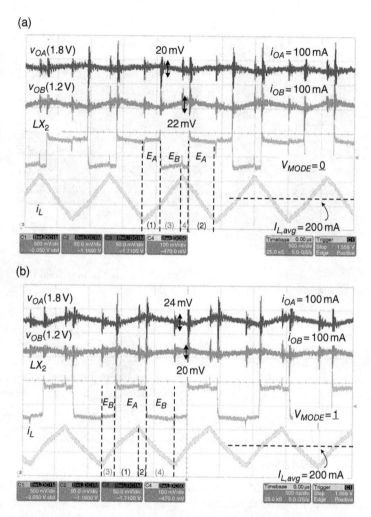

Figure 6.50 Measured steady-state operation with both i_{OA} = 100 mA and i_{OB} = 100 mA when v_{IN} is 3.3 V: (a) V_{MODE} = 0 with path 1-3-4-2 and (b) V_{MODE} = 1 with path 3-1-2-4

suppression of the steady-state cross regulation if the energy operating mode is applied to the SIDO converter.

Figure 6.52 shows the measured AEB mechanism. When V_{MODE} is set to low, a decrease in i_{OA} activates the AEB mechanism to bypass the energy delivery path, enhancing efficiency and ensuring voltage regulation. Figure 6.52(a) shows that the energy path is reduced to path 3-4-2 in one PWM switching cycle when i_{OA} and i_{OB} are 50 and 100 mA, respectively. Therefore, the switching power loss is reduced, whereas voltage regulation is still guaranteed. Figure 6.52(b) also shows the ultra-light-load condition of v_{OB}. Through the AEB mechanism, the SIDO converter can operate single step-down operation for v_{OA} unless v_{OB} requires energy replenishment. Consequently, no minimum loading restriction occurs at any output in the SIDO converter, which can definitely realize dual independent supply power in Soc applications.

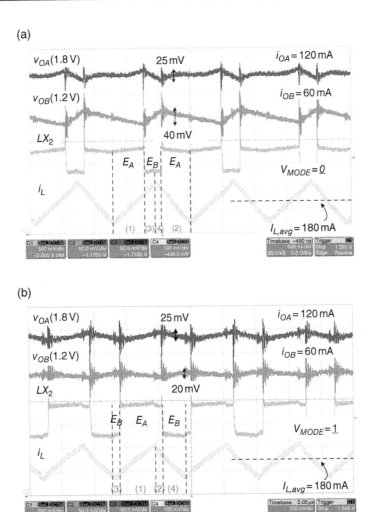

Figure 6.51 Measured steady-state operation with both $i_{OA} = 120$ mA and $i_{OB} = 60$ mA when v_{IN} is 3.3 V: (a) $V_{MODE} = 0$ with path 1-3-4-2 and (b) $V_{MODE} = 1$ with path 3-1-2-4

6.4.7.3 Energy-Mode Transition

Figure 6.53 shows the measured load transient response. The load currents are initially set to 60 mA for each output at $V_{MODE} = 0$. When i_{OA} changes abruptly from 60 to 240 mA, the energy operating mode is changed to obtain a better energy delivery scheme and realize the energy delivery period of v_{OA}, including the peak inductor current. Subsequently, when i_{OB} increases from 60 to 240 mA and i_{OA} is maintained at 240 mA, the energy operating mode is again changed by the energy-mode transient operation. Therefore, both transient and steady-state cross regulations are minimized. The largest voltage variation is approximately 100 mV and the voltage recovery time is shorter than 30 μs. The final state of the measured load transient response is particularly identical to that in the initial state. Thus, the final state derives the

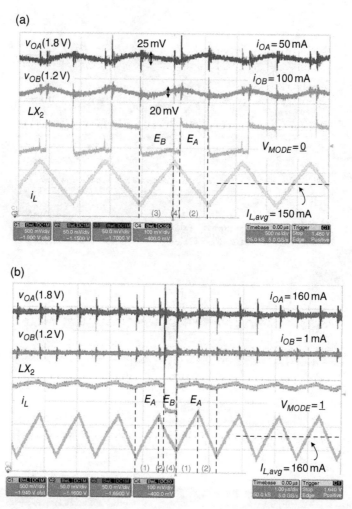

Figure 6.52 Measured AEB mechanism under different load conditions when v_{IN} is 3.3 V: (a) i_{OA} = 50 mA and i_{OB} = 100 mA with V_{MODE} = 0 and (b) i_{OA} = 160 mA and i_{OB} = 1 mA with V_{MODE} = 1

60 mA loading current at each output but operates with the distinct energy operating mode compared with the initial state. This result is caused by the hysteresis buffer in the mode decision circuit. Nevertheless, either of the energy operating modes can be utilized when the two output loads are close to the measured results in Figure 6.50.

6.4.7.4 Measured Line Transient Response and DVS Operation

Figure 6.54 shows the measured line transient response of the SIDO DC/DC converter. The LV-EDC controller is supplied by the pre-regulator, which provides a constant low supply voltage from the input battery supply voltage. Thus, a sudden variation in the battery voltage does

Single-Inductor Multiple-Output 393

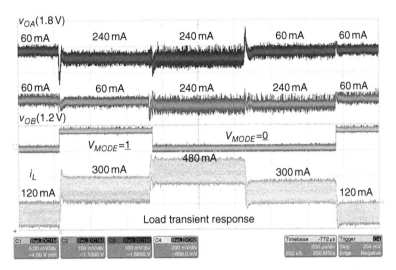

Figure 6.53 Measured load transient response if the energy-mode transition operation is applied to the SIDO converter

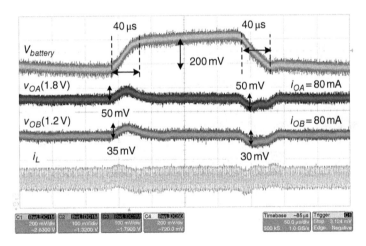

Figure 6.54 Measured line transient response of a 0.2 V voltage step at $V_{battery}$ with i_{OA} = 80 mA and i_{OB} = 80 mA

not result in incorrect functioning of the LV-EDC. The analog circuit can retain its proper operation without leading to the problem of insufficient voltage headroom. Therefore, when $V_{battery}$ has a voltage step of 200 mV within 40 μs, the dual output voltages derive the induced voltage variations, which are less than 50 mV. LV-EDC has the capacity to ensure closed-loop operation to guarantee the regulated output voltages.

In Soc applications, Figure 6.55 shows the measured DVS operation for high-efficiency power management. The voltage divider of the output v_{OB} can be adjusted in order to obtain

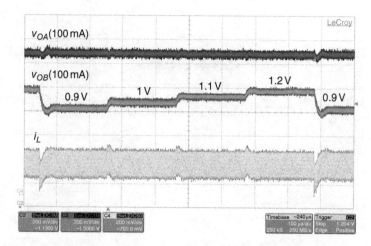

Figure 6.55 Measured DVS operation of v_{OB} in the SIDO DC/DC converter

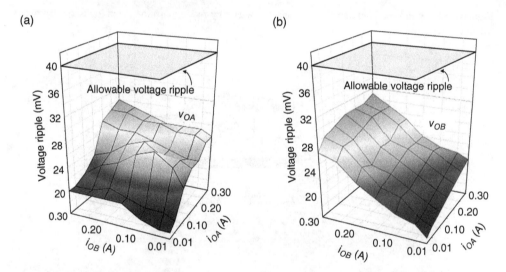

Figure 6.56 Measured output voltage ripple: output ripple of (a) v_{OA}; (b) v_{OB}

the proper supply voltage level for the Soc applications. That is to say, when the Soc system operates at distinct operating modes, the supply voltage can be well modulated in order to minimize the power consumption in standby mode while enhancing the performance (such as the data transmission operation). The SIDO converter can activate the DVS operation to adjust the supply voltage level of v_{OB} in the range of 0.9–1.2 V. Besides, the constant supply voltage v_{OA} achieves the proper driving operation for analog circuits in Soc applications.

6.4.7.5 Measured SIDO Output Voltage Ripple and Power Conversion Efficiency

Figure 6.56 shows the measured output voltage ripple. The steady-state output voltage ripple can be suppressed below the allowable value through the energy-mode transition operation and the AEB mechanism. This process demonstrates the steady-state cross-regulation

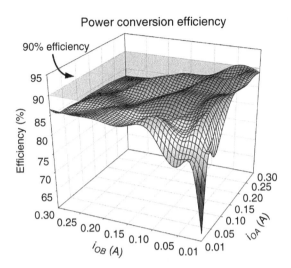

Figure 6.57 Measured power-conversion efficiency with $v_{IN} = 3.3$ V, $v_{OA} = 1.8$ V, and $v_{OB} = 1.2$ V

minimization. Figure 6.57 shows the power-conversion efficiency. The LV-EDC circuit with a 1 V operation consumes 60 μA to achieve correct operation. Most of the power consumption is derived from the conduction power loss and switching power loss in the four main power switches. The efficiency is above 80% at medium and heavy loads, and reaches a peak value of 91% because of the AEB mechanism. Consequently, the highly efficient and wide-load-range SIDO converter can be a good candidate for PMU design in Soc applications.

6.4.7.6 Design Summary and Comparisons

Table 6.2 lists detailed design specifications. The nominal output voltages are 1.8 and 1.2 V. The steady-state cross regulation is successfully suppressed, and the transient cross regulation eliminated to an unnoticeable result. The peak power conversion efficiency is 91%, which can be kept above 80% of the output loading current condition.

Table 6.3 shows a comparison of other SIDO (SIMO) methodologies. The SIDO DC/DC converter achieves energy delivery function with ECM control, which does not need the freewheel stage to stabilize the system. With the unnoticeable transient cross regulation, the quality of the supply function can be guaranteed for Soc applications.

Both SIDO and SIMO converters are used in the PMU design of Soc applications. This process is based on the previously discussed design goals, including the reduction in number of power switches, reduction in both conduction and switching power losses, extension of output loading current range, improvement of load transient response, reduction of output voltage ripple, reduction of both transient and steady-state cross regulations, and improvement of power-conversion efficiency. These design goals cannot be achieved simultaneously; thus, a multidimensional trade-off should be implemented based on the specifications. Most importantly, readers need to realize the advantages and disadvantages of each control technique

Table 6.2 Design specifications

Technology	65 nm CMOS process
Input voltage	2.7–3.6 V
Inductor/DCR	4.7 µH/200 mΩ
Switching frequency	1 MHz
Chip size	Figure 6.49 (a) 1.44 mm^2; (b) 1.28 mm^2
Outputs (nominal) (V)	$v_{OA} = 1.8$ $v_{OB} = 1.2$
Output capacitor/ESR	4.7 µF/30 mΩ 4.7 µF/30 mΩ
Transient cross regulation (load transient response)	60 mA → 240 mA Unnoticeable
	240 mA → 60 mA
	Unnoticeable 60 mA → 240 mA
	240 mA → 60 mA
Steady-state cross regulation (steady-state ripple)	With energy mode transition operation
	25 mV/120 mA 20 mV/60 mA
	Without energy mode transition operation
	25 mV/120 mA 40 mV/60 mA
Efficiency	Maximum 90% ($i_{OA} = 300$ mA, $i_{OB} = 50$ mA)

Table 6.3 Comparisons of other SIDO (SIMO) methodologies

	This work	JSSC 2007 [18]	JSSC 2009 [3]	JSSC 2011 [10]	PE 2010 [26]
Technology	65 nm CMOS	0.5 µm BiCMOS	0.25 µm CMOS	0.35 µm CMOS	0.25 µm CMOS
Supply voltage (V)	2.7–3.6	2.5–4.5	1.8–2.2	1.8–2.4	2.7–5
Switching frequency	1 MHz	700 kHz	660 kHz	1.25 MHz	1.3 MHz
Types	SIDO	SIMO	SIMO	SIDO	SIDO
Outputs	1.8 V/1.2 V	5–12 V/−9.5 V	1.25–2.25 V	3–3.6 V	1.2 V/1.8 V
Inductor (µH)	4.7	10	10	1	4.7
Output capacitor (µF)	4.7	4.7	33	4.7	47
Total load current (mA)	600	110	N/A	600	600
Transient voltage drop/settling time	100 mV/30 µs ($\Delta i_{load} = 0.18$ A)	100 mV/N/A ($\Delta i_{load} = 0.03$ A)	N/A	500 mV/100 µs ($\Delta i_{load} = 0.2$ A)	50 mV/10 µs ($\Delta i_{load} = 0.27$ A)
Transient cross regulation ($\Delta V/V$) (%)	Unnoticeable	<1	<0.35	<5	<2.7
Steady-state cross regulation (ripple) (mV)	<30	<160	<22	<160	<40
Peak efficiency (%)	90	81	93	87.8	87
Active area (mm^2)	0.975	8.7	3.78	2.21	5.29

6.5 SIMO Converter Techniques for Tablets

6.5.1 Output Independent Gate Drive Control in SIMO Converter

6.5.1.1 SIMO on Tablets

Figure 6.58 compares a conventional power solution and a SIMO converter in a PMU. In commercial tablets, four voltage levels are generated by four individual buck converters. These four buck converters require four 2 mm × 1.6 mm inductors that occupy 12.8 mm^2 of PCB area. By contrast, SIMO DC/DC buck converters possess the advantage of compact size. As shown in Figure 6.58, only one 2.5 mm × 3.2 mm inductor is required in the SIMO solution, thereby greatly decreasing the inductor area by 37.5%. In addition, the chip counts and PCB routing area further minimize the area and cost of the power unit in commercial tablets. These advantages make SIMO converters popular in PMU design.

6.5.1.2 Efficiency and Switch Control Methods

Figure 6.59 shows the simplified architecture of a SIMO converter, which supplies four different output voltages to the tablet. In contrast to a conventional buck converter, four switches

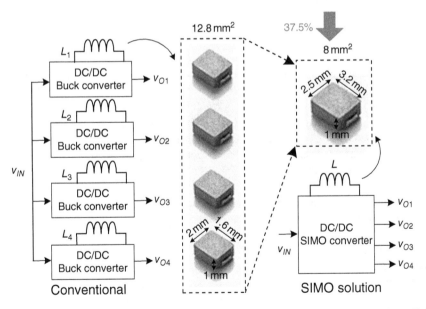

Figure 6.58 PCB inductor areas can be decreased by 37.5% by replacing conventional parallel buck converters with a SIMO converter

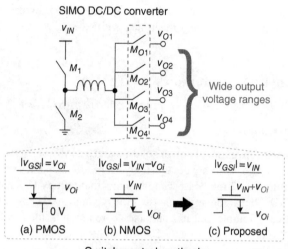

Figure 6.59 Comparison of the switch control methods in a SIMO converter

Table 6.4 Comparison of switch control methods

Switch	(a) PMOS	(b) NMOS	(c) Method in this book
Silicon area	Large	Small	Small
On-resistance	Large	Medium	Small
Gate driving voltage ($|v_{GSi}|$ where $i = 1$–4)	v_{Oi}	$v_{IN}-v_{Oi}$	v_{IN}
Efficiency v_{Oi} independent	NO	NO	YES

M_{O1}–M_{O4} are inserted additionally to allocate energy to four outputs with only one inductor. The selection of a switch control method is particularly crucial to achieve higher efficiency in SIMO converters.

Conventional switch control methods can briefly be divided into two categories: the use of (a) a P-MOSFET [12, 27] and (b) an N-MOSFET [17]. Using the P-MOSFET, the gate is connected to the ground when the switch is turned on. Thus, the gate driving voltage, which is the gate-to-source voltage v_{GSi}, is equal to the corresponding output voltage v_{Oi}. If the output voltage is low, then the low gate driving voltage results in high on-resistance and degrades the efficiency. By contrast, the gate is connected to v_{IN} to turn on the N-MOSFET switch. Given that v_{IN} is the highest voltage in the SIMO converter, connecting the gate to v_{IN} results in lower on-resistance. According to different output voltage levels, the gate driving voltage is equal to $v_{IN}-v_{Oi}$. A large on-resistance and a low power efficiency occur if the output voltage is high in the N-MOSFET, in contrast to the P-MOSFET.

A comparison of these switch control methods is summarized in Table 6.4. Gate driving voltages, on-resistances, and efficiencies depend on output voltages in both (a) and (b). Thus, high efficiency during a large output voltage range for tablet applications cannot be achieved with conventional switch control methods. Furthermore, the N-MOSFET, which inherently possesses a lower on-resistance and a smaller silicon area than the P-MOSFET, is suitable to use as a power switch. Therefore, a switch control method for N-MOSFETs with a constant gate driving voltage during a large output voltage range is highly desirable.

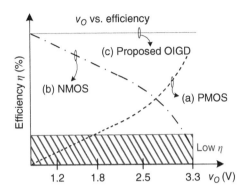

Figure 6.60 Efficiency comparisons of different switch control methods during a large output

Efficiency comparisons of different switch control methods are illustrated in Figure 6.60. In conventional control methods (a) and (b), the overall power-efficiency performance cannot satisfy the tablet requirements when multiple outputs are requested in a wide range, such as 1.2–3.3 V. For example, the use of P-MOSFET and N-MOSFET switches results in a low efficiency of 64% and 70% at outputs of 1.2 and 3.3 V, respectively, because of the low gate driving voltage and large on-resistance. This implies that the conventional usage of P-MOSFET and N-MOSFET switches is inappropriate in SIMO converters if the output range is wide [12, 17, 27].

A useful switch control method is output independent gate drive (OIGD) control for all N-MOSFET switches. The gate driving voltage can be kept constant regardless of the output voltage level. In addition, the internal body control (IBC) prevents the formation of body diodes and blocks the leakage current between the drain and the source of the power MOSFET. High power efficiency can be achieved during wide output voltage ranges to prolong the usage time of commercial tablets.

6.5.1.3 Output Independent Gate Drive Control

Figure 6.61 depicts the architecture of the SIMO DC/DC converter with the OIGD technique. Similar to conventional SIMO converters, M_1 and M_2 control the inductor charging and discharging phases, respectively, while M_{O1}–M_{O4} alternately deliver energy to v_{O1}–v_{O4} within one switching cycle. For the voltage regulations, v_{O1}–v_{O4} are fed back to the SIMO controller to generate duty control signals D, $1 - D$, and D_1–D_4 for M_1, M_2, and M_{O1}–M_{O4}, respectively. In achieving low on-resistance and high efficiency, which are independent of the output voltage, OIGD techniques are implemented to control each N-type power MOSFET M_{O1}–M_{O4}. Each OIGD control includes two major blocks: a boosted gate (BG) circuit and an IBC circuit.

Boosted Gate Circuit
In generating a constant gate driving voltage independent of v_{O1}–v_{O4} in OIGD control, a BG circuit is designed to drive the gates of M_{O1}–M_{O4}. In obtaining the lowest on-resistance power switches, the highest voltage in the SIMO buck converter v_{IN} is chosen to drive the gates. In other words, each gate-to-source voltage v_{GSi} is equal to the input voltage v_{IN} when the power

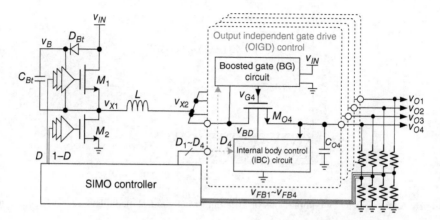

Figure 6.61 Architecture of the SIMO with the OIGD technique

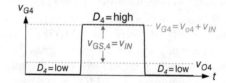

Figure 6.62 The concept of OIGD control

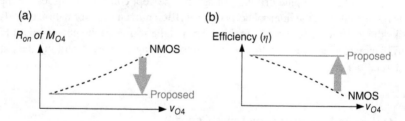

Figure 6.63 Improved performance of the SIMO converter with OIGD in terms of (a) on-resistance and (b) efficiency

switches are turned on. To achieve this scenario, each gate voltage is boosted to the value of $v_{Oi} + v_{IN}$, where $i = 1-4$.

As illustrated in Figure 6.62, we take the fourth output as example. v_{G4} is connected to the ground to fully turn off M_{O4} when the duty cycle D_4 of v_{O4} is logic 0 (low). When D_4 is logic 1 (high), the OIGD control boosts v_{G4} to $v_{O4} + v_{IN}$ to fully turn on M_{O4}. At this time, v_{G4} is equal to v_{IN}, which is always constant and independent of v_{O4}. The performance improvements of the SIMO converter with OIGD are shown in Figure 6.63. In a conventional switch control method with N-MOSFET, whose gate driving voltage is equal to $v_{IN} - v_{Oi}$, the on-resistance increases as v_{O4} increases. Thus, the power efficiency deteriorates dramatically according to v_{O4}. The OIGD facilitates the output voltage-dependency problem. Briefly, the OIGD control makes the gate drive voltage independent of all multiple output voltages, and efficiency degradation never occurs at any value of the output voltage.

Internal Body Control

The gate voltages of M_{O1}–M_{O4} are boosted to $v_{Oi} + v_{IN}$ to achieve the output-independent characteristic. Over-voltage occurs between the gate and the body terminals at a low-voltage 5 V N-MOSFET. Thus, the OIGD control adopts the IBC circuit to facilitate this problem.

As shown in Figure 6.61, the IBC circuit controls the body voltage according to each output voltage. When M_{Oi} turns on with its gate boosted to $v_{Oi} + v_{IN}$, the body is biased by v_{Oi}. Therefore, the gate-to-body voltage is equal to v_{IN}, which is limited in the normal operating voltage range of low-voltage 5 V devices. The gate-to-body overstress problem does not occur. Moreover, the body effect is eliminated by biasing the body to v_{Oi} because the source and the body of M_{Oi} are at the same voltage level. Without the body effect, the on-resistance is further minimized to improve power efficiency.

6.5.1.4 Deadtime Overstress Recycling

In the absence of deadtime, simultaneously turning on any two switches in M_{O1}–M_{O4} results in the leakage of energy from one output to the other. This deteriorates the regulation performance and efficiency. Thus, the deadtime between any two switches in M_{O1}–M_{O4} is necessary to avoid redundant energy transfer loss.

During the deadtime of M_{O1}–M_{O4}, these switches are turned off. Nevertheless, the continuity characteristic of the inductor current i_L charges up the parasitic capacitor of v_{X2}, C_p, as depicted in Figure 6.64, because all the leakage current paths and body diode are blocked. Overstress voltage inevitably occurs at v_{X2}, thereby causing damage to M_{O1}–M_{O4} and other control circuits. Excessive charge accumulation results in high overstress voltage, especially at heavy loads, indicating high inductor current levels. The low-power deadtime overstress recycling (DOR) technique not only limits the maximum voltage of v_{X2} and prevents M_{O1}–M_{O4} from being overstressed, but also recycles the extra charge stored at v_{X2} back to the input source

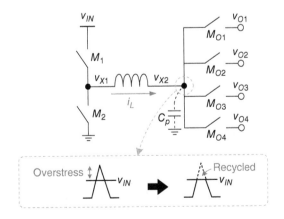

Figure 6.64 Deadtime overstress problem

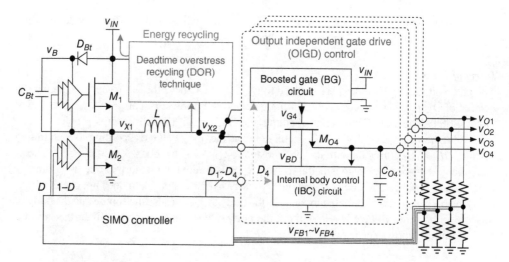

Figure 6.65 Overall architecture of the SIMO with OIGD and DOR techniques

during deadtime periods. High recycling efficiency improves the SIMO efficiency and battery longevity of tablets.

Figure 6.65 illustrates the overall architecture of the SIMO with OIGD and DOR techniques. The DOR technique constructs an energy-recycling path to retrieve the overstress energy from v_{X2} to v_{IN}. An accurate detection circuit is designed with the DOR technique to turn on the recycling path once v_{X2} exceeds v_{IN}, thereby preventing a leakage current when overstress does not occur. In addition, a low-power design is required in DOR to increase the energy-recycling efficiency.

High efficiency during wide output voltage ranges and prevention of overstress problems by energy recycling are achieved simultaneously with OIGD and DOR techniques. The OIGD and DOR techniques are in the power stage without affecting the energy path sequence. These two techniques can be applied to any conventional SIMO converter with all the N-MOSFET switches. Designing an appropriate control method for tablet applications is desirable.

6.5.1.5 Universal Controller Applications

The OIGD and DOR techniques resolve output-dependent efficiency and deadtime overstress problems at the power stage. The performances of the SIMO converter in terms of output voltage accuracy, cross regulation, and output voltage ripple are highly related to the controller design. The OIGD and DOR techniques can fortunately combine universal control methods in prior arts according to different applications.

The rippled-based PWM control with ordered power distribution is a good choice to simplify the controller design and cost [18]. As illustrated in Figure 6.66, three feedback voltages v_{FB1}–v_{FB3} are directly compared with the reference voltage V_{ref} by three individual comparators. As shown in Figure 6.67, the duty cycles D_1–D_3 can be determined simply when v_{Oi} reaches V_{ref}. At the beginning of each switching cycle, M_{O1} is turned on to deliver energy to v_{O1}. Once v_{O1} reaches V_{ref}, M_{O2}, M_{O3}, and M_{O4} turn on in an ordered sequence, which

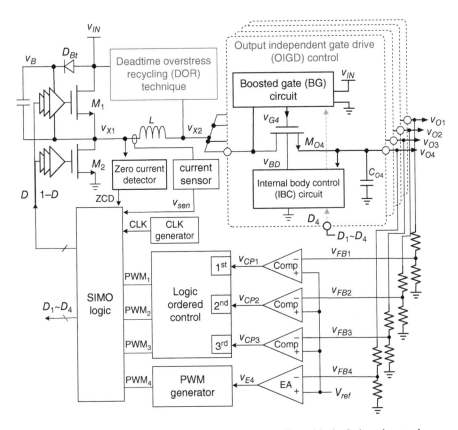

Figure 6.66 Implementation of the SIMO controller with ripple-based control

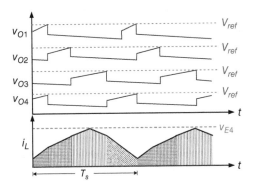

Figure 6.67 Timing diagram of the SIMO converter

simplifies the logic circuit designs. In controlling the energy in the inductor, an EA, instead of a comparator, is connected to v_{FB4} to amplify the difference between v_{FB4} and V_{ref}. That is, the error signal v_{E4} is used to determine the peak inductor current. However, when the load changes, the errors between v_{FBi} and V_{ref} of all the outputs accumulate to the EA and adjust

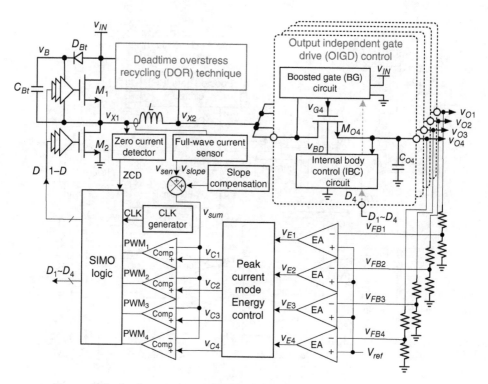

Figure 6.68 Implementation of the SIMO controller with the EA-based control

the peak inductor current. Unfortunately, cross regulation in v_{O4} is especially serious. Without the help of the EA at v_{O1}–v_{O3}, the accuracy of the output voltage cannot be ensured when a large load current difference occurs among the four outputs. Furthermore, a large ESR capacitor is required to improve the noise immunity when comparing v_{FBi} and V_{ref}. Large voltage ripples and cross regulations cause a flicker effect on the panel and degrade the display performance.

The current-mode control, which achieves a small output voltage ripple and a high output voltage accuracy, is appropriate for tablet applications [13, 14, 19]. Figure 6.68 shows the architecture of a SIMO converter with current-mode control. Four output voltages v_{O1}–v_{O4} are fed back (v_{FB1}–v_{FB4}) to four EAs for good voltage regulation performance. Four error signals v_{E1}–v_{E4} are generated by amplifying the difference between v_{FB1}–v_{FB4} and the voltage reference V_{ref}. Through the peak current-mode energy control block, v_{E1}–v_{E4} are summed to generate the energy control signals v_{C1}–v_{C4}. By comparing v_{C1}–v_{C4} with v_{SUM}, which is the summing signal of the current sense v_{sen} and the slope compensation v_{slope}, the duty cycles D, $1 - D$, and D_1–D_4 for M_1, M_2, and M_{O1}–M_{O4}, respectively, are generated with the help of the SIMO logic and CLK generator.

The timing diagram of the SIMO converter is depicted in Figure 6.69. v_{C4}, which determines the peak current value, generates D to control the inductor current charging and discharging time. v_{C1}–v_{C3} adequately distribute the inductor energy through D_1–D_4 to each output. Since output voltage ripples are not used to determine duty cycles, a large voltage ripple for better noise immunity is not required. A small output voltage ripple can be achieved in the peak current-mode control with a small ESR-value output capacitor. Moreover, high voltage

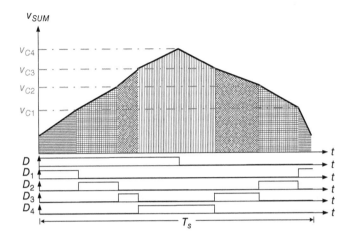

Figure 6.69 Timing diagram of the SIMO converter

Table 6.5 Performance comparison of the SIMO controllers

Control method	Ripple based	Current mode	Ripple based + OIGD&DOR	Current mode + OIGD&DOR
Voltage ripple	Large	Small	Large	Small
Cross regulation	v_{O4} is large	Small	v_{O4} is large	Small
Voltage accuracy	Low	High	Low	High
Efficiency	Low	Low	High	High

accuracy can be achieved using EAs in each output (Table 6.5). Adopting a conventional ripple-based control, large voltage ripple, low voltage accuracy, and series cross regulation at v_{O4} degrade the performance of the display panel. A ripple-based control SIMO converter is not suitable for supplying tables. By contrast, small voltage ripple, low cross regulation, and high voltage accuracy can be achieved by an adequately designed current-mode control. Nevertheless, low efficiency shortens the usage time of tablets if power switches are not controlled appropriately. The power consumption at the power stage is significantly decreased with the OIGD and DOR techniques. Consequently, the embodied OIGD and DOR techniques in the current-mode control allow commercial tablet applications to achieve high performance and high efficiency under all load conditions.

6.5.2 CCM/GM Relative Skip Energy Control in SIMO Converter

Commercial tablets require multiple supplies with a relatively high load current difference, which causes a large ripple and serious cross regulation in the SIMO DC/DC converter. Maximizing the load current difference among all the outputs, while simultaneously reducing the

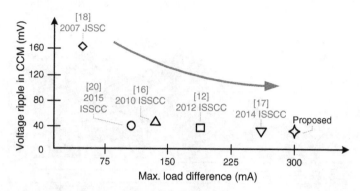

Figure 6.70 Comparison of CCM voltage ripple and maximum load difference performances in prior arts and the proposed work

voltage ripple of each output, is urgently required by tablets. SIMO converters need to provide multiple outputs for different functional blocks that need versatile load current conditions. Thus, designing the SIMO control method to achieve these goals is crucial.

Figure 6.70 gathers the performance of SIMO converters in prior arts. Although the ordered power-distributive control (OPDC) [18] provides a simple control method for multiple outputs, serious cross regulation and a large output ripple of 160 mV occur, thereby degrading power performance. Meanwhile, the PLL-based SIMO converter [16] can simplify frequency compensation networks and ease the EMI filter design with a fixed frequency control. However, its driving capability is not sufficient for tablet applications. To further minimize cross regulation and voltage ripple, an AERC [12] introduces an energy buffer region. The energy buffer region unfortunately causes low efficiency because of the large conduction loss, which dissipates battery energy in a short time. Energy must be properly allocated to all outputs in one switching cycle in the absence of any energy buffer duration. The ripple-based adaptive off-time control [28] and error-based control [17] still cannot simultaneously achieve high driving capability and small output ripple for tablets.

Therefore, a control method is presented here to achieve small output ripple and high power efficiency with high driving capability. The SIMO DC/DC converter with continuous conduction mode/green mode relative skip energy control (CCM/GM RSEC) can minimize the variation in output voltage with a large current difference between the multiple outputs, either in steady state or transient time. Using the current-mode control [13, 14, 25], high output voltage accuracy and small output ripple can be achieved simultaneously.

6.5.2.1 Absolute Skip Energy Control

Except for power-efficiency improvement techniques in the power stage as discussed in Section 6.5.1, control methods are also related to power efficiency. To further prolong the usage time of tablets, an absolute skip methodology, which simply depends on load conditions, is commonly utilized in single-output converters to decrease the switching loss at light loads [29, 30]. In other words, the energy transfer to the output is skipped in certain switching cycles when the load current becomes light. The number of skipped switching cycles increases as the load current decreases.

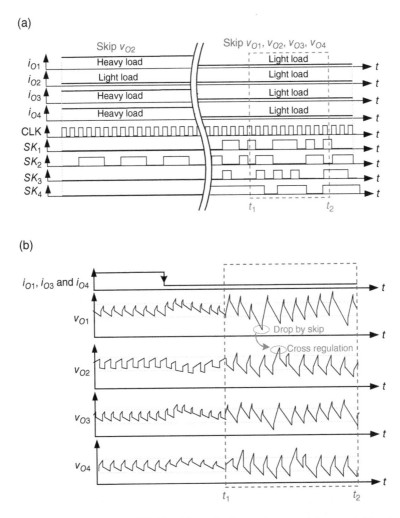

Figure 6.71 (a) Timing diagram of SIMO with an absolute skip methodology and (b) output-voltage waveforms with the absolute skip function during t_1–t_2

By directly determining an absolute skip methodology for each output in SIMO converters, the skip timing diagram is established, as shown in Figure 6.71(a), where i_{O1}–i_{O4} are the load currents of v_{O1}–v_{O4}, respectively, and SK_1–SK_4 indicate the corresponding skip statuses, respectively. Based on the absolute skip method, when v_{O1}, v_{O3}, and v_{O4} are at heavy loads and v_{O2} is at light loads, v_{O2} is skipped periodically. When all the outputs are at light loads, v_{O1}–v_{O4} are skipped alternately and irregularly, and no noise is injected, which significantly increases the voltage ripples. This occurs because all the outputs share the same switching cycle in SIMO converters. Thus, skip activation in a certain output necessarily influences the energy distribution of the other outputs. As shown in Figure 6.71(b) during t_1–t_2, this phenomenon may cause a large voltage ripple and noise, which affects the operating function of tablets. Furthermore, efficiency is not evidently improved under all conditions at the cost of ripple

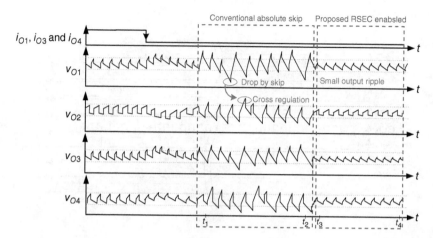

Figure 6.72 Comparison of the output-voltage waveforms of the conventional absolute skip function during t_1–t_2 and that of the RSEC technique during t_3–t_4

performance because the efficiency of the SIMO converter includes load conditions of all output voltages, and skipping one or two outputs may not significantly increase efficiency. These observations indicate that the absolute skipping method is not suitable for SIMO converters.

6.5.2.2 Relative Skip Energy Control (RSEC)

Multiple outputs share one inductor in the SIMO converter. Thus, determining the skip status based on the absolute energy of each output is unsuitable. However, the relative energy between multiple outputs can account for the total energy of all the outputs. Hence, determining the skip status based on relative energy results in better supply qualities. Figure 6.72 shows output voltage waveforms with conventional absolute skip energy control during t_1–t_2 and that of the RSEC technique during t_3–t_4. Both skip methods can achieve good voltage regulation. However, a large voltage ripple occurs in the absolute skip method. The skip-induced drop and cross regulation can be eliminated by employing the RSEC.

A CCM/GM RSEC in the SIMO DC/DC converter is shown here. The RSEC identifies the trade-off between efficiency and voltage ripple in both CCM and GM at high accuracy in the high-performance mode and energy saving in the sleeping mode of tablets, respectively. Unnecessary skip activation is avoided to substantially suppress voltage ripple and cross regulation. Adequate energy skipping is determined based on the relative energy between each output. Thus, energy is properly distributed in a relatively wide voltage and load current range. Good load regulation, high power capability, low noise, and ripple supply are achieved.

6.5.2.3 Architecture of SIMO Converter with CCM/GM RSEC

Based on the summary in Table 6.5, using current-mode control results in small voltage ripple, low cross regulation, and high voltage accuracy; it is therefore suitable for tablet applications. Current-mode control is adopted with its timing diagram, as depicted in Figure 6.69. Figure 6.73 shows the architecture of the SIMO DC/DC converter with CCM/GM RSEC. To simplify the illustration, the circuit blocks of the OIGD and DOR techniques are not shown. Energy is

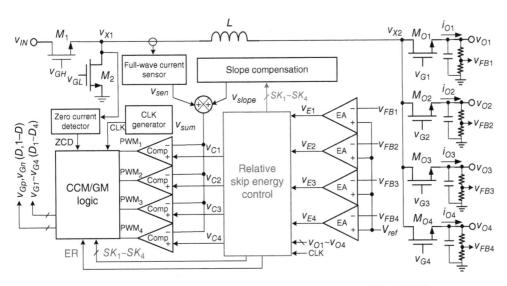

Figure 6.73 Architecture of the SIMO converter with CCM/GM RSEC

appropriated to four outputs v_{O1}–v_{O4}, through the power stage, which contains the high- and low-side power MOSFETs M_1 and M_2, one single inductor L, and four time-multiplexing power switches M_{O1}–M_{O4}. Four output voltages are fed back to their corresponding EAs to improve the voltage regulation performance. The error signals v_{E1}–v_{E4} are generated by amplifying the difference between the feedback voltages v_{FB1}–v_{FB4}, respectively, and the reference voltage V_{ref}. The RSEC circuit manipulates the error signals v_{E1}–v_{E4} to generate the energy control signals v_{C1}–v_{C4}.

A full-wave current sensor is required to develop the current-mode-controlled SIMO DC/DC converter. Slope compensation is required to prevent sub-harmonic oscillation. The total signal v_{SUM} is generated by adding the current sense signal V_{sen} and the compensation slope v_{slope}. By comparing v_{C1}–v_{C4} with v_{SUM}, the duty cycle control signals for all power MOSFETs are generated to achieve Figure 6.69. Here, the duty cycles M_1 and M_2 are defined as D and $1 - D$, respectively, similar to those in conventional buck converters. The duty cycles of M_{O1}–M_{O4}, which indicate the on-time periods of M_{O1}–M_{O4} in a switching cycle, are defined as D_1–D_4, respectively.

The SIMO converter enters into its green mode (GM) at very light loads. Once the zero current detector (ZCD) identifies the zero inductor current, all the power MOSFETs are turned off to avoid the negative inductor current. The energy request signal ER, which indicates that energy is insufficient in one of the outputs, activates the switching activities of all the power MOSFETs again for voltage regulation.

6.5.2.4 Relative Skip Energy Control

To address the disadvantages of the absolute skip method, the concept of CCM/GM RSEC is illustrated in Figure 6.74 [19]. The four cases indicated by the horizontal axis – (a) heavy, (b) medium, (c) light, and (d) very light loads in the total output power of the SIMO converter – are shown in the histogram. E_1–E_4, which represent the output energies of v_{O1}–v_{O4}, respectively,

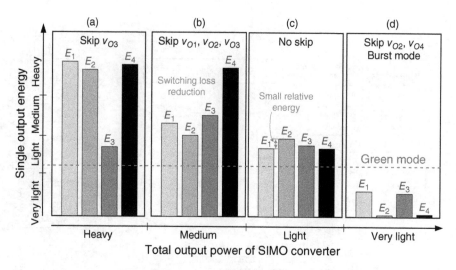

Figure 6.74 Concept of the CCM/GM RSEC

correspond to the vertical axis. In case (a), although E_3 is at light loads, the total output power of the SIMO converter is at heavy loads. This indicates that the total output power of the SIMO converter is not directly related to the individual energy of each output. Thus, skip activation cannot simply be determined based on the absolute energy of the single output. Instead, the relative energy between v_{O1}–v_{O4} is used to identify the skip status.

The RSEC skips the output because its energy is relatively smaller than that of the others. In case (b), as shown in Figure 6.74, v_{O1}–v_{O3} are at medium loads, and v_{O4} is at heavy loads. Although v_{O1}–v_{O3} are not at light loads, E_1–E_3 are skipped because E_4 is substantially higher. Thus, the switching loss can be significantly decreased. In case (c), the values of E_1–E_4 are close to each other at light loads. However, no skipping function occurs in the RSEC technique. Compared with the conventional absolute skip method, where the skip function happens in all the outputs, a large skip-induced voltage ripple occurs in the proposed method. The skipped energy is imposed on the other outputs in a switching cycle. Hence, the excessive energy also increases the skipping frequency and induces larger voltage ripples. However, RSEC can address this problem. When the output power is further decreased in case (d), the RSEC enters the GM to decrease switching loss and improve efficiency. However, RSEC skips the relatively low energy outputs v_{O2} and v_{O4}.

The energy allocation and operation of the SIMO converter with the RSEC technique are illustrated in Figure 6.75. The total inductor current is indicated by v_{SUM}. To develop a current-mode-controlled SIMO converter, the output currents of v_{O1}–v_{O4} are stacked from the bottom to the top layer in a switching cycle. In the bottom layer, which is located between v_{Et} and v_{C1}, the energy is transferred to v_{O1}. By disregarding the other layers, the duty cycle determination, which compares v_{SUM} with v_{C1}, can be considered a peak current-mode buck converter. Similarly, the second, third, and top layers, which are located between v_{C1} and v_{C2}, v_{C2} and v_{C3}, and v_{C3} and v_{C4}, respectively, transfer energy to v_{O2}, v_{O3}, and v_{O4}, respectively. Accordingly, the energy control signals v_{C1}–v_{C4} are generated by superposing v_{E1}–v_{E4}. The duty cycles of M_{O1}–M_{O4} in Figure 6.73, which indicate the energy allocation, are determined by comparing v_{C1}–v_{C4} with v_{SUM}.

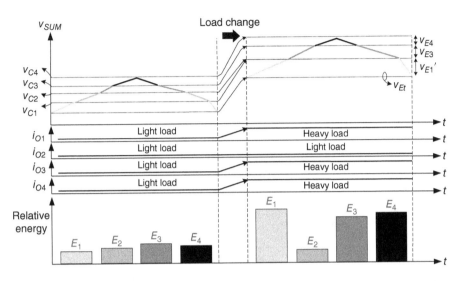

Figure 6.75 Energy allocation and operation of the SIMO converter with the RSEC technique

When the relative energies among the four outputs are low, the inductor current is delivered to v_{O1}–v_{O4} in sequence without any skip activations, as shown in the left-hand side of Figure 6.75. If one of the outputs requires a relatively low energy, its energy layer is extracted from the energy stack. For instance, if v_{O2} has a relatively low energy request, the second energy layer is not shown. No energy is transferred to v_{O2} in this switching cycle. By comparing v_{C1}–v_{C4} with v_{SUM}, D_2 is always zero, and M_{O2} is not automatically turned on in this switching cycle.

In addition, the RSEC technique is essential. Without any skip function, regulation cannot be guaranteed in case of the occurrence of any large load current difference among the four outputs. As the load current decreases in one output, its duty cycle must ideally be reduced to obtain less energy for voltage regulation. However, as shown in Figure 6.76, the minimum on-time is limited by process parameters, such as the speed of logics and comparators and the propagation delay of drivers. The minimum on-time limits the smallest load current difference and degrades voltage regulation. The output voltage with a relatively low energy increases and becomes unregulated. The RSEC technique overcomes the limitation of the load current difference by properly skipping the energy-excessive output in the energy delivery sequence. The equivalent minimum on-time is decreased to zero, and the output is well regulated, as illustrated in Figure 6.76. Hence, without any significant improvement in efficiency, E_3 is skipped to ensure regulation in case (a), as shown in Figure 6.74.

Figure 6.77 illustrates the circuit block diagram of the RSEC. To develop the current-mode-controlled SIMO converter, RSEC achieves superposition by generating the energy control signals v_{C1}–v_{C4}. The duty cycles of M_{O1}–M_{O4} indicated in Figure 6.73 are determined accordingly.

When the skip function is not activated, the values of v_{E1}–v_{E4} are added to determine the energy layer shown in Figure 6.75 and generate v_{C1}–v_{C4}. For example, v_{C2} is the total of v_{E1} and v_{E2}. The energy layer of v_{O2} is located between v_{C1} and v_{C2} with a height equal to v_{E2}.

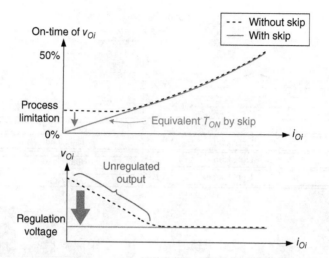

Figure 6.76 Unregulated problem occurs without any skip mechanism

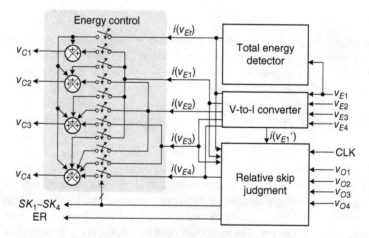

Figure 6.77 Circuit block diagram of the RSEC

In summary, v_{C1}–v_{C4} can be expressed in Eq. (6.8) and the difference in energy control signals, which indicates the energy layer of each output, is identified, as shown in Eq. (6.9):

$$v_{Cj} = \sum_{i=1}^{j} v_{Ei}, \ j = 1, \ldots, 4 \tag{6.8}$$

$$v_{C(j+1)} - v_{C(j)} = v_{E(j+1)}, \ j = 1, \ldots, 3 \tag{6.9}$$

Except for v_{O1}, each energy layer of v_{O2}–v_{O4} is represented by Eq. (6.9) when $j = 1$–3, respectively. To judge the skip status, the energy layer of v_{O1} is defined as:

$$v'_{E1} = v_{E1} - v_{Et} \tag{6.10}$$

To achieve Eqs. (6.8) and (6.9), v_{E1}–v_{E4} are converted to the current signals $i(v_{E1})$–$i(v_{E4})$ using the voltage-to-current (V-to-I) converter and are injected into the energy control block for summing. In the RSEC technique, the relative judgment circuit generates SK_1–SK_4 to determine the skipping status of each output according to the error signals v_{O1}–v_{O4} and the total energy v_{Et}. The total energy detector calculates the valley value of v_{SUM} to present v_{Et} in each switching cycle. The energy request signal ER indicates the status of the deficient energy in the GM operation. The RSEC technique can operate for either the CCM under heavy/medium loading condition or for GM under very light loading condition [35, 38].

Continuous-Conduction Mode
The energy control signal paths, energy stack, and relative skip judgment are shown in Figure 6.78. As shown in Figure 6.78(a), the relative energy of each output $v_{E1}'(= v_{E1} - v_{Et})$, v_{E2}, v_{E3}, and v_{E4} is higher than the skip threshold v_{SK}. Thus, no skipping occurs. The energies of

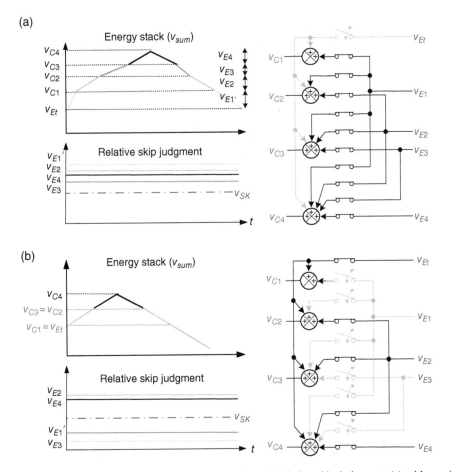

Figure 6.78 Energy control signal paths, energy stack, and relative skip judgment: (a) with no skip; (b) with v_{O1} and v_{O3} skipped

v_{O1}–v_{O4} are stacked from the bottom to the top layer. v_{C1}–v_{C4} are composed of the sums of v_{E1}–v_{E4}. When both E_1 and E_3 are relatively lower than the others, E_1 and E_3 are removed from the energy stack, as indicated in Figure 6.78(b). Hence, v_{E1} and v_{E3} are not included in the energy control signals. The energy layers of E_1 and E_3 are eliminated by setting $v_{C1} = v_{Et}$ and $v_{C3} = v_{C2}$, respectively. v_{E1} and v_{E3} are not involved in v_{C1}–v_{C4}. Considering different load conditions, the formula of v_{C1}–v_{C4} can be modified from Eqs. (6.8)–(6.11) according to SK_1–SK_4. Similarly, the difference in energy control signals is modified from Eqs. (6.9)–(6.12). To achieve this, the switches in the energy control are controlled by the skip status.

$$v_{Cj} = v_{Et} \cdot SK_1 + \sum_{i=1}^{j} v_{Ei} \cdot \overline{SK_i}, \; j = 1, \ldots, 4 \qquad (6.11)$$

$$v_{C(j+1)} - v_{C(j)} = v_{E(j+1)} \cdot \overline{SK_{(j+1)}}, \; j = 1, \ldots, 3 \qquad (6.12)$$

Green Mode

The SIMO operating in GM for power saving is shown in Figure 6.79 when the sum of all the output powers is at very light loads. When the inductor current reaches zero, all the power MOSFETs M_1, M_2, and M_{O1}–M_{O4} are turned off during t_0–t_1 and t_4–t_5. Thus, no switching operation occurs, and no switching power loss is consumed. No energy is transferred to the outputs, so that the output voltages decrease continuously, thereby increasing v_{E1}–v_{E4} and v_{C1}–v_{C4}. Until one of the four output voltages becomes too low, the energy request signal ER remains high. This step activates the switching operation to allocate energy to the energy-deficient output. For example, during t_2–t_3, which contains two switching cycles, v_{O3} requires lower energy. Thus, SK_3 indicates a skip activation in the second switching cycle. With the obtained energy, v_{C1}–v_{C4} decrease to stop the switching operation again, when the

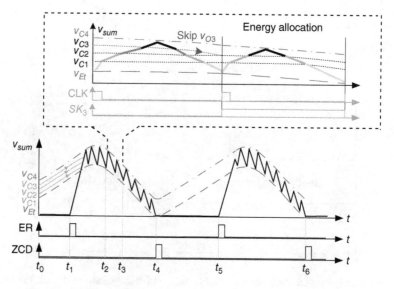

Figure 6.79 Operating waveforms of the RSEC technique in the GM

zero inductor current is detected. Hence, the ZCD signal is high. Moreover, once the tablet shifts from the sleeping mode, the lack of energy significantly increases v_{C1}–v_{C4}. Therefore, the SIMO converter automatically and smoothly enters the CCM.

6.5.3 Bidirectional Dynamic Slope Compensation in SIMO Converter

Slope compensation is necessary in the current-mode control to avoid sub-harmonic oscillation and ensure stable operation [31]. In current-mode buck converters, the optimal compensation slope is determined by the output voltage [32, 39]. Similarly, the optimal compensation slope of a SIMO converter is determined based on the four output voltages. However, when skipping occurs, the predetermined compensation slope may instantly become too large or too small according to the skipped output voltages. An insufficient compensation slope causes sub-harmonic oscillation, as shown in Figure 6.80(a), whereas overcompensation slows down the transient response, as indicated in Figure 6.80(b).

The compensation slope is generally discussed in relation to the rising edge of the slope signal. Thus, the compensation slope is crucial in the inductor charging phase, which is the duration of time when the inductor current has a positive slope. However, when the compensation slope is in the inductor discharging phase, which is the duration of time when the inductor current has a negative slope, the performance of current-mode buck converters is not affected. In SIMO converters, one switching cycle is divided into several slots to provide energy to each output. In summary, the compensation slope throughout the whole switching period must be considered in SIMO converter designs.

6.5.3.1 Conventional Slope Compensation in SIMO Converters

Slope compensation is crucial for the stable operation of current-mode controlled DC/DC converters. The compensation slope with a slope of m_a is required during the inductor charging phase (D). To prevent sub-harmonic oscillation, m_a must follow Eq. (6.13) [32]:

$$m_a \geq \frac{1}{2}m_2 \quad (6.13)$$

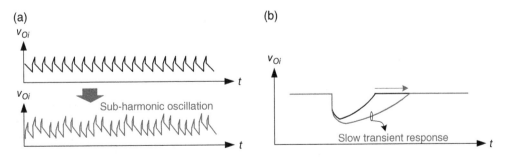

Figure 6.80 Fixed compensation slope in SIMO converters may cause (a) sub-harmonic oscillation when a skip function occurs or (b) slow transient response with an overcompensation slope

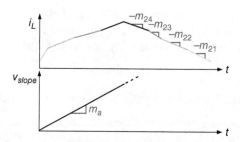

Figure 6.81 Realizing the slope compensation in the SIMO converter is directly similar to that in a single-output buck converter

In other words, the minimum value of the compensation slope must be larger than half of the inductor current discharging slope m_2, which is expressed as:

$$m_2 = \frac{v_O}{L} \tag{6.14}$$

In the current-mode SIMO converter, the discharging inductor current is divided into four segments to allocate energy to four outputs. Four discharging slopes m_{21}–m_{24} are shown:

$$m_{2i} = \frac{v_{Oi}}{L}, \text{ where } i = 1-4 \tag{6.15}$$

By directly conceptualizing the buck converter slope compensation in SIMO converters, the compensation slope must conform to all the outputs with different values. To prevent sub-harmonic events under all conditions, the maximum output voltage $v_{O,max}$ among four outputs is chosen in the worst case. The waveform of the compensation slope with a slope of m_a, which follows Eq. (6.16), is shown in Figure 6.81:

$$m_a = \frac{v_{O,max}}{L} \tag{6.16}$$

However, as mentioned in Figure 6.80, sub-harmonic oscillation or slow transient response may occur with different skip statuses under all load conditions.

6.5.3.2 Bidirectional Dynamic Slope Compensation

Instead of conventional slope compensation, a new bidirectional dynamic slope compensation is presented here. Bidirectional dynamic slope compensation contains multiple-segment slope compensation and falling slope calibration during the inductor charging and discharging phases, respectively. Sub-harmonic oscillation during all the skipping periods SK_1–SK_4 and an invalid pulse for energy delivery are eliminated in current-mode SIMO DC/DC converters.

From Sections 6.5.1–6.5.3, the SIMO converter designs both the power stage and the controller. The power stage includes the OIGD and DOR techniques, and the controller consists of

Single-Inductor Multiple-Output

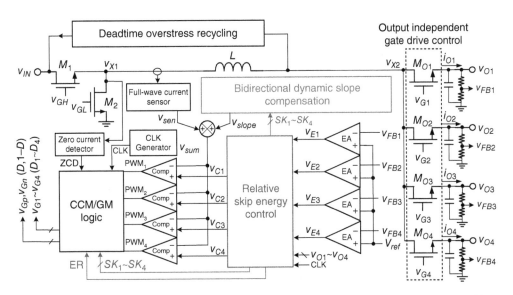

Figure 6.82 Complete SIMO converter and proposed techniques in the power stage and controller

the complete RSEC and bidirectional dynamic slope compensation. Figure 6.82 illustrates the complete architecture of the SIMO converter.

Multiple-Segment Slope Compensation

$v_{O,max}$ is adopted for m_a calculation to ensure stable operation in all cases. However, overcompensation occurs in other output voltages. In particular, when $v_{O,max}$ is skipped, the compensation slope is unnecessarily large and degrades the transient response. Thus, multiple-segment slope compensation can achieve adequate slope compensation for any combination of output voltages and loading current conditions. As depicted in Figure 6.83(a), the compensation slope during the inductor charging phase is divided into multiple segments to address each output voltage. For example, when energy is transferred to v_{O1}, the compensation slope is determined by m_{21}, which is related only to v_{O1}. The rule of avoiding sub-harmonic oscillations is consequently modified as Eq. (6.17) from Eq. (6.13):

$$m_{ai} \geq \frac{1}{2} m_{2i}, \text{ where } i = 1-4 \qquad (6.17)$$

When the skip function is activated, the corresponding compensation segment is skipped to avoid overcompensation. The optimal compensation slope is achieved by multiple-segment slope compensation. As illustrated in Figure 6.83(b), if v_{O2} is skipped in a certain switching cycle, then the corresponding compensation slope m_{a2} disappears in the multiple-segment compensation sequence, thereby instantly achieving adequate compensation.

Falling Slope Calibration

Except for the above-mentioned compensation slope required during the inductor charging phase, the slope signal v_{slope} during the inductor discharging phase is highly related to energy

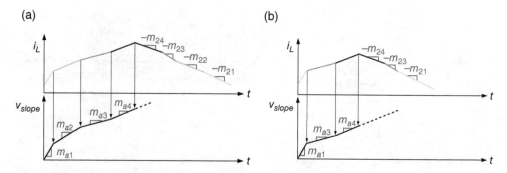

Figure 6.83 Multiple-segment slope compensation during the inductor charging phase (a) without any skip activation and (b) with the skip activation of v_{O2}

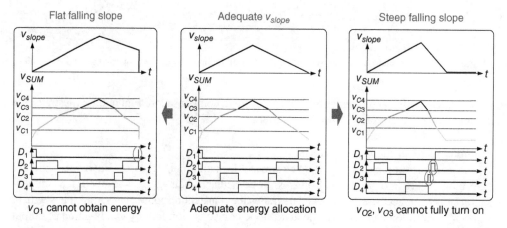

Figure 6.84 Slope compensation signal v_{slope} during the inductor discharging phase with different falling slopes

delivery and allocation in SIMO converters. In buck converters, v_{slope} during the inductor discharging phase is not involved in the duty cycle determinations. Hence, the simplest way is to reset v_{slope} to zero at any time in the inductor discharging phase before the next switching cycle starts. However, the duty cycle determinations (D_1–D_4) for energy allocation continue during both the inductor charging and discharging phases in SIMO converters. Hence, v_{slope} must be well designed in both directions.

In Figure 6.84, if v_{slope} is reset to zero by an arbitrary steep falling slope, then the waveform of the total signal v_{SUM} is distorted and deviates from the inductor current waveform. By comparing v_{C1}–v_{C4} with v_{SUM} to generate D_1–D_4, respectively, short pulses may occur, thereby degrading energy allocations. As depicted on the right-hand side of Figure 6.84, a short pulse occurs in D_2 and D_3 during the inductor discharging phase. With a short pulse, the time for the power MOSFETs M_{P2} and M_{P3} to be fully turned on before the end of the short pulse is not enough. Thus, energy cannot be transferred to v_{O2} and v_{O3}, but switching loss increases significantly. The steep falling slope for v_{slope} is evidently inadequate. However, if v_{slope} is reset

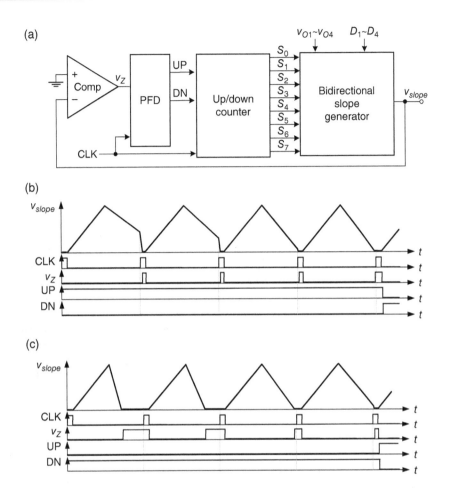

Figure 6.85 Bidirectional slope compensation: (a) block diagram; (b) calibration processes with a flat falling slope; (c) steep falling slope

to zero by a flat falling slope, then D_1 may disappear during the inductor discharging phase, as illustrated on the left-hand side of Figure 6.84. Thus, v_{O1} cannot obtain adequate energy to regulate the output voltage.

For adequate energy allocations, the slope signal v_{slope} must be adjusted according to its peak value and the inductor discharging time. Setting v_{slope} to zero at the end of every switching cycle is important. Thus, a falling slope calibration is proposed to track v_{slope} to zero by modulating the falling slope.

The block diagram of the bidirectional slope compensation that achieves multiple-segment slope compensation and falling slope calibration is shown in Figure 6.85(a). The slope signal v_{slope} is compared with the ground to generate v_Z and indicate when v_{slope} reaches zero. The PFD generates the signals UP and DN when v_Z lags and leads the clock signal CLK, respectively. Through the up/down counter, the slope generator can dynamically adjust the falling slope of v_{slope} by eight levels, S_0–S_7. Calibration processes with flat falling slope and steep

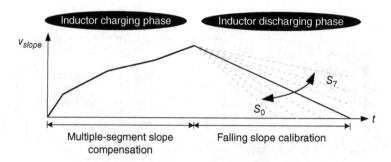

Figure 6.86 Operating waveforms of the bidirectional slope compensation

slope are illustrated in Figure 6.85(b) and (c), respectively. CLK resets v_{slope} at the end of each switching cycle to ensure normal operation. In Figure 6.85(b), v_Z lags CLK, and the UP signal is activated to continuously increase the falling slope when it is flat. Thus, v_{slope} cannot fall to zero at the end of the switching cycle. After several switching cycles, an optimal falling slope can be calibrated. However, if the falling slope is too steep, as shown in Figure 6.85(c), v_Z leads CLK, and the DN signal is activated to continuously decrease the falling slope.

The overall waveforms of the bidirectional slope compensation are depicted in Figure 6.86. During the inductor charging phase, v_{slope} increases with multiple segments according to the values of v_{O1}–v_{O4} and the skipping status. During the inductor discharging phase, a three-bit adjustable falling slope calibrates the value of v_{slope} to zero at the end of a switching cycle to prevent an invalid short pulse in any energy delivery procedure.

6.5.4 Circuit Implementations

6.5.4.1 OIGD and DOR Techniques

Boosted Gate Circuit and Internal Body Control Circuit
The circuit implementation of the OIGD control, which includes an N-type power MOSFET M_{Oi} for the ith output (i = 1–4) and the corresponding control circuits, is illustrated in Figure 6.87. The OIGD control contains BG and IBC circuits. The BG circuit generates a constant gate driving voltage to achieve output voltage-independent power efficiency, as depicted in Figures 6.62 and 6.63. However, the IBC circuit prevents the gate-to-body overstress problem and the body effect by controlling the body voltage of M_{Oi}.

Figure 6.87(a) shows the OIGD when each power N-MOSFET is in the off state ($D_i = 0$). To fully turn off M_{Oi}, its gate v_{Gi} is connected to the ground through M_7. The voltage across the boosted capacitor C_{Bi} is simultaneously charged to v_{IN} through M_1 and M_2. By contrast, Figure 6.87(b) shows the OIGD when each power N-MOSFET is in the on state ($D_i = 1$). In generating a constant gate-to-source voltage, M_4–M_6 are turned on to construct a path to control the gate voltage v_{Gi}. At this time, the bottom plate of C_{Bi} is connected to v_{X2}, which approaches v_{Oi} when M_{Oi} is turned on. Therefore, v_{Gi} is boosted to $v_{Oi} + v_{IN}$ to decrease the on-resistance of M_{Oi}. v_{GSi} is equal to v_{IN} regardless of output voltage levels. In other words, the OIGD control forces the gate driving voltage to be constantly equal to v_{IN}. Thus, the on-resistance is kept low for high efficiency regardless of different output voltage levels.

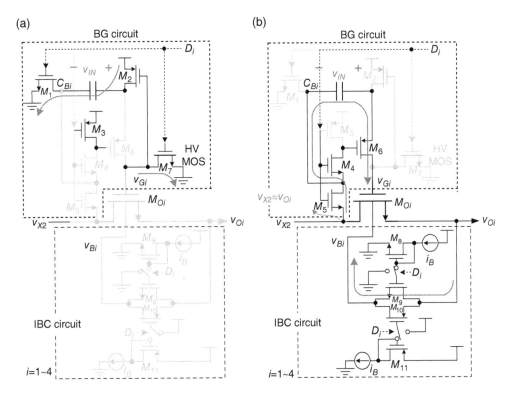

Figure 6.87 Circuit implementation of the OIGD control, including the BG and IBC circuits

v_{Gi} is zero when each power N-MOSFET is off ($D_i = 0$), which does not cause overstress between the gate and the body of M_{Oi}. Body control is not required, and IBC is turned off. With the boosted voltage at v_{Gi} during the on state, the IBC circuit biases v_{Bi} to v_{Oi} to avoid overstress and body effects. However, directly connecting v_{Bi} to v_{Oi} once M_{Oi} is turned on causes the occurrence of an in-rush current. Therefore, IBC charges v_{Bi} to v_{Oi} by a constant current. Considering a wide output voltage range from v_{O1}–v_{O4}, (M_8, M_9) and (M_{10}, M_{11}) form N-MOSFET and P-MOSFET current mirror pairs, which are controlled by the bias current I_B to bias v_{Bi} at low and high v_{Oi}, respectively.

M_7 is the only high-voltage device in OIGD control. To sustain the BG voltage between the drain to the source of M_7, which is higher than the maximum endurable voltage of a low-voltage device, a lightly doped drain MOSFET is required. However, the gate voltage controlled by the duty cycle D_i is still a low-voltage signal.

Another issue is that the body of M_{Oi} floats when M_{Oi} and the IBC circuit are off, as shown in Figure 6.87(a). The body is generally required to connect to the ground when M_{Oi} is off to prevent current leakage caused when the body diode is turned on. However, the body charge is discharged to the ground and charged again when M_{Oi} is turned on, which consumes extra charge from v_{Oi} in each switching activity. Instead of floating the body of M_{Oi}, the body charge current can automatically be restored to the other outputs. However, the leakage current between the drain and the source still does not occur. For instance, if v_{O3} has the

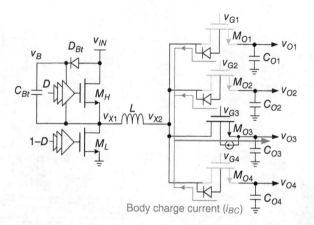

Figure 6.88 Body charge current recovery (where v_{O3} has the lowest value among the four outputs)

lowest value among the four outputs, as illustrated in Figure 6.88, then the bodies of M_{O1}, M_{O2}, and M_{O4} float during the deadtime period. Until M_{O3} is turned on, v_{X2} descends to a lower level, which is approximately equal to v_{O3}. The body-to-drain parasitic diodes turn on to discharge the bodies of M_{O1}, M_{O2}, and M_{O4}. The discharging phenomenon forms the body charge current I_{BC}, which retrieves v_{O3} for output loading. Compared with the condition when the body is connected to ground, a floating body can decrease energy dissipation and recover such dissipation to the outputs.

Deadtime Overstress Recycling

During the deadtime period, the OIGD control turns off four power MOSFETs M_{O1}–M_{O4}. The continuity of the inductor current accumulates charges at v_{X2} and causes overstress voltage because no path exists to release the surplus inductor current. v_{X2} increases to a value higher than v_{IN}, particularly at heavy loads. The DOR technique recycles energy at v_{X2} by returning M_{DR} back to v_{IN}, and to prevent the overstress problem.

Figure 6.89(a) illustrates the low-power circuit implementation of the DOR technique, where M_{DR} establishes an energy recycling path between v_{X2} and v_{IN}. The resistors R and R_1, capacitor C, switch S_1, transistors M_5–M_6, and FSM form the control of the energy recycling path. M_1–M_4 form the body control of M_{DR}. R and C can rapidly reflect the variation at v_{X2} to v_{Tri}. M_5–M_6 act as an inverter that controls the gate of M_{DR} and v_{GR}, according to the value of v_{Tri}. R_1 in the series of M_5 is introduced to prevent malfunction caused by noise. S_1 turns off during non-deadtime periods, so that the energy recycling path is not required. If any kind of noise is injected into v_{Tri}, then R_1 can prevent v_{GR} from being dragged to zero through M_5, even when the recycling control circuit is triggered abnormally. During deadtime periods, once v_{Tri} triggers the inverters M_5–M_6 to turn on M_{DR}, the FSM immediately turns on S_1 to connect v_{GR} to zero without flowing through R_1. Thus, the recycling path is rapidly constructed to prevent overstress and at the same time achieve energy recycling. Moreover, M_1–M_4 can properly bias the body voltage v_{BR} of M_{DR} to avoid leakage current during deadtime periods and periods when one of M_{O1}–M_{O4} turns on. During a non-deadtime period, v_{IN} is higher than v_{X2}, and M_1 and M_2 are turned off and on to bias v_{BR} at v_{IN}. During a deadtime period, once v_{X2} is higher

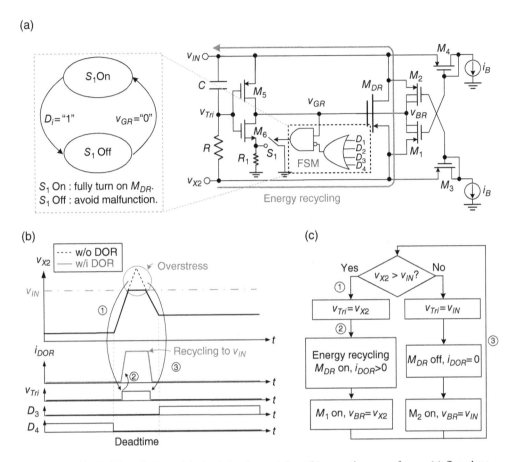

Figure 6.89 DOR technique: (a) circuit implementation; (b) operating waveforms; (c) flowchart

than v_{IN}, M_1 and M_2 turn on and off to bias v_{BR} at v_{X2}. In other words, v_{BR} is always biased at the highest voltage determined by the highest voltage of the drain and source terminals to avoid turning on the body diode and leakage current.

The operating waveforms are illustrated in Figure 6.89(b). During a deadtime period, once v_{X2} exceeds v_{IN}, v_{Tri} rapidly increases to establish the energy recycling path. Thus, v_{X2} is clipped to v_{IN} because the overstress energy depicted by the black dashed line is converted to the recycling current i_{DOR} and flows back to v_{IN}. The reliability of the power switches M_{O1}–M_{O4} and the BG circuit increases because the maximum value of v_{X2} is limited to v_{IN}. M_{DR} turns off once v_{X2} becomes smaller than v_{IN}. The flowchart of the DOR technique is illustrated in Figure 6.89(c). First, the relationship of v_{X2} and v_{IN} is used to assess the operating state of the DOR technique. If v_{X2} is smaller than v_{IN}, v_{Tri} is equal to v_{IN}, indicating that the recycling path is not required and $i_{DOR} = 0$. v_{BR} is biased at v_{IN} to avoid the body diode being turned on. If v_{X2} is larger than v_{IN}, v_{Tri} is equal to v_{X2}, confirming that the recycling path is constructed with i_{DOR}, which is larger than zero. Energy is retrieved from v_{X2} to v_{IN} to prevent overvoltage. v_{BR} is biased at v_{X2} to avoid turning on the body diode.

6.5.4.2 Relative Skip Energy Control Technique

The circuit implementations of the RSEC, as illustrated in Figure 6.77, are described in this section. The total energy detector and V-to-I converter, which help determine the superposition in the current-mode SIMO converter, are illustrated in Figure 6.90. To generate the skip status under all load conditions, circuit relative skip judgment is implemented, as depicted in Figure 6.91. The energy control circuit that achieves superposition and generates the energy control signals v_{C1}–v_{C4} is illustrated in Figure 6.92. Equations (6.11) and (6.12) are realized through this energy control circuit.

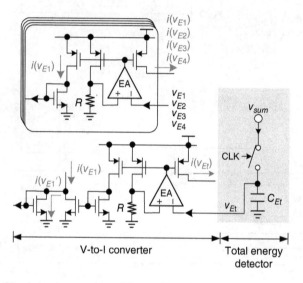

Figure 6.90 Circuit implementation of the total energy detector and four V-to-I converters

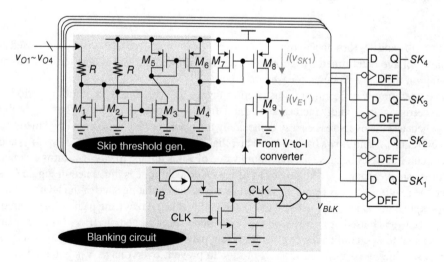

Figure 6.91 Circuit implementation of the relative skip judgment circuit

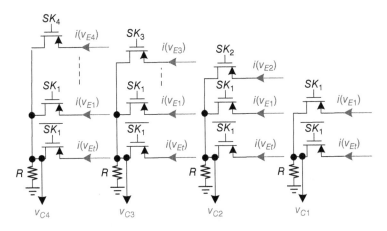

Figure 6.92 Circuit implementation of the energy control circuit

Total Energy Detector and V-to-I Converter

To easily achieve superposition in energy control, the error signals v_{E1}–v_{E4} are converted to the current signals $i(v_{E1})$–$i(v_{E4})$ using four V-to-I converters in Figure 6.90. According to different skip statuses, the combination of v_{E1}–v_{E4} can be achieved by adding the corresponding currents. To determine Eq. (6.10) to judge the skip status of v_{O1}, the total energy v_{Et} is required to help generate $v_{E1}'(= v_{E1}-v_{Et})$. v_{Et} is equal to the valley of v_{SUM}. Hence, the total energy detector can identify v_{Et} at the end of each switching cycle.

Five V-to-I converters are contained in circuit implementations, which generate $i(v_{E1})$–$i(v_{E4})$ and $i(v_{Et})$, respectively. A V-to-I converter includes an EA, resistor R, and current mirrors. Using the EA, the five voltages are distributed across the resistors and generate the corresponding currents by a factor of $1/R$. Through the current mirrors, the current signals are sent to the energy control circuit to achieve superposition. The total energy detector is formed using a sample-and-hold circuit, which contains a switch and a sampling capacitor C_{Et}. At the end of each switching cycle, the CLK signal turns on the sampling switch and samples the valley of v_{SUM} to obtain v_{Et}. Similarly, v_{Et} is converted to the current signal $i(v_{Et})$. A relative skip judgment $i(v_{E1}')$, which represents the energy layer of v_{O1}, is derived by subtracting $i(v_{Et})$ from $i(v_{E1})$.

Relative Skip Judgment

The relative skip judgment circuit shown in Figure 6.91 is developed to generate the skip status SK_1–SK_4 for each output under all load conditions. The first skip threshold voltages v_{SK1}–v_{SK4} of the four outputs are generated. By comparing v_{SK1}, v_{SK2}, v_{SK3}, and v_{SK4} with v_{E1}', v_{E2}, v_{E3}, and v_{E4}, respectively, SK_1, SK_2, SK_3, and SK_4 can be obtained. To speed up the comparison process and prevent the comparators occupying a large area, current comparators that simply require current mirrors are adopted. Thus, the skip threshold voltages v_{SK1}–v_{SK4} are also converted to currents $I(v_{SK1})$–$I(v_{SK4})$.

In generating duty cycles, which compare v_{SUM} and v_{C1}–v_{C4}, the comparison delays differ because the slope of v_{SUM} varies with v_{IN} and v_{OUT} at different outputs. To compensate for the comparison delay of each output, v_{IN} and v_{O1}–v_{O4} are involved to help determine the skip thresholds $I(v_{SK1})$–$I(v_{SK4})$. By comparing the skip thresholds from $I(v_{SK1})$ to $I(v_{SK4})$ with each

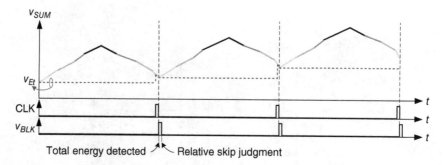

Figure 6.93 Operating waveforms of the RSEC technique

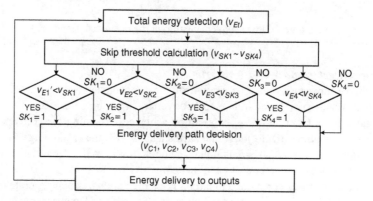

Figure 6.94 Flowchart of the RSEC technique

energy layer from $i(v_{E1}')$ to $i(v_{E4})$, respectively, the skip status from SK_1 to SK_4 is determined after a blanking time of the CLK signal.

Energy Control
As illustrated in Figure 6.92, the energy control circuit is realized by collecting the current signals from the five V-to-I converters according to the different skip statuses generated by the relative skip judgment circuit. To generate the energy control signals v_{C1}–v_{C4}, four identical resistors R, which are the same as those in the V-to-I converters, are adopted for the current collection. According to the switches controlled by the skip status, the voltage signals of v_{C1}–v_{C4} are reconstructed from the current signals. Finally, Eqs. (6.11) and (6.12) can be realized.

The operating waveforms and flowchart of the RSEC technique are shown in Figures 6.93 and 6.94, respectively. When CLK is logic high, v_{Et} tracks v_{SUM} and holds its valley value once the next cycle starts. The skip threshold voltages are determined using the skip threshold generator. After a blanking time, SK_1–SK_4 are determined according to v_{SK1}–v_{SK4} and v_{E1}'–v_{E4} and are stored in the DFFs at the negative edge of the signal V_{BLK} in Figure 6.91. The energy delivery path is determined by v_{C1}–v_{C4}, which are generated by SK_1–SK_4. Finally, energy can be properly transferred to four outputs using the relative skip mechanism.

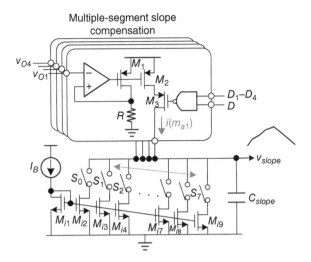

Figure 6.95 Circuit implementation of the bidirectional slope generator

6.5.4.3 Bidirectional Dynamic Slope Compensation

Bidirectional Slope Generator

The block diagram of the bidirectional slope compensation is illustrated in Figure 6.85(a), which includes a comparator, PFD, up/down counter, and bidirectional slope generator. The comparator, PFD, and up/down counter are realized by the conventional basic architectures to decrease the design complexity and the occupied area.

Figure 6.95 shows the circuit implementation of the bidirectional slope generator, which simultaneously performs multiple-segment slope compensation and falling slope calibration. During the inductor charging time, multiple-segment slope compensation generates the highest four compensation segments according to the corresponding output voltages, D_1–D_4, and SK_1–SK_4. Thus, Eq. (6.17) is used to ensure stable operation and prevent sub-harmonic oscillation. In the inductor current discharging phase, the falling slope calibration activates and determines the optimal falling slope. The falling slope is divided into three bits, which are implemented by eight bias currents, M_{i2}–M_{i9}. If the falling slope is too flat, the calibration loop in Figure 6.95 turns on more bias currents from S_0 to S_7 in sequence during the calibration process. However, if the falling slope is too steep, the calibration loop turns off the bias currents from S_7 to S_0 in sequence during the calibration process. Finally, the optimal slope can be derived when v_{slope} reaches zero at the end of each switching cycle.

6.5.5 Experimental Results

6.5.5.1 Chip Micrograph

In this chapter, two chips are implemented to verify the OIGD and RSEC techniques and bidirectional slope compensation. Figure 6.96 shows the first realization of the SIMO converter with the CCM/GM RSEC technique and bidirectional slope compensation. This chip was fabricated in a 0.18 μm CMOS process with an active area of 2.24 mm². For simplicity, M_{O1}–M_{O4} are

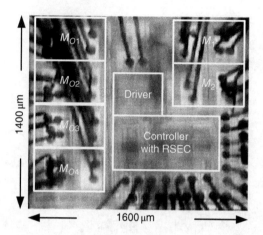

Figure 6.96 Chip micrograph with the RSEC technique and bidirectional slope compensation

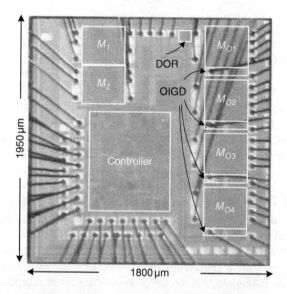

Figure 6.97 Chip micrograph with the OIGD and RSEC techniques and bidirectional slope compensation

implemented by the PMOS conventional switch control as illustrated in Figure 6.59(a). To further verify the OIGD and DOR techniques, the complete SIMO converter was fabricated in a 0.25 μm CMOS process with an active area of 3.51 mm^2, as shown in Figure 6.97.

In both chips, the current-mode control with a 1 MHz switching frequency is used to minimize cross-regulation and provide good line/load regulation. The output filter inductor and capacitors are 4.7 μH and 4.7 μF, respectively. The input voltage can range from 3.6 to 5 V, and the four output voltages are 1.2, 1.8, 2.5, and 3.3 V.

6.5.5.2 Measured Results of Techniques in Power Stage Design

The measured results of the OIGD technique are shown in Figure 8.3. To clearly demonstrate the OIGD technique, the top and bottom plates of C_{Bi} are illustrated at the same time. At v_{O1} = 3.3 V, the measured waveforms are as shown in Figure 6.98(a). When M_{O1} is off, the bottom plate of C_{Bi} is connected to the ground, and the voltage across C_{Bi} is charged to v_{IN}. Here, v_{IN} is equal to 4.2 V. When M_{O1} is on, the bottom plate of C_{Bi} is connected to v_{X2}, which approximates to v_{O1}. Thus, the top plate of C_{Bi} is boosted to $v_{O1} + v_{IN}$ ($\approx$7.5 V), and the gate of M_{O1} is controlled. Similarly, at v_{O4} = 1.2 V, the measured waveforms are as shown in Figure 6.98(b). The top plate of C_{Bi} is again boosted to $v_{O4} + v_{IN}$ ($\approx$5.4 V), and the gate of M_{O4} is controlled. Regardless of the multiple output voltages, the gate drive voltages are constant and equal to v_{IN} (= 4.2 V) to ensure high efficiency over a wide range of output voltages.

Figure 6.99 describes the DOR technique. During deadtime, v_{X2} is charged up because of the continuity of the inductor current and the lack of a current release path. Once v_{X2} is increased

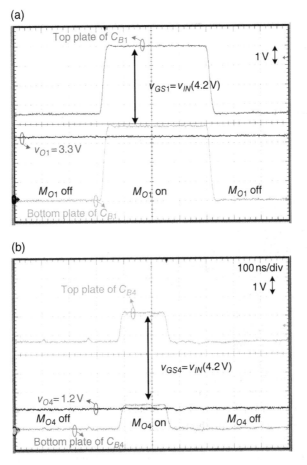

Figure 6.98 Measured results of the boosted gate in the OIGD technique at (a) v_{O1} = 3.3 V; (b) v_{O4} = 1.2 V

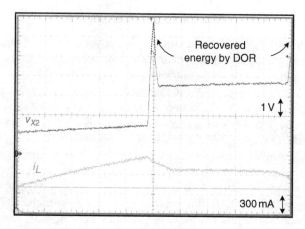

Figure 6.99 Measured results of the DOR technique in the worst case when $v_{IN} = 5$ V

to v_{IN}, the recycling path in DOR is activated to limit v_{X2} near v_{IN} by retrieving the overstress energy to v_N. Figure 6.99 shows the worst case of overvoltage when $v_{IN} = 5$ V. In addition, the inductor current is twisted during deadtime because of the increase in v_{X2}.

6.5.5.3 Measured Results of Techniques in Controller Design

Figure 6.100 indicates the measurement results when i_{O1}, i_{O2}, i_{O3}, and i_{O4} are 50, 50, 300, and 50 mA, respectively. Under this load condition, v_{O4} requires a relatively low energy and a small duty cycle. Without a skip function, v_{O4} is not regulated because the minimum duty cycle of D_4 is limited by process limitations, which include comparator and logic delay, as illustrated in Figure 6.76. Only one inductor is adopted in the SIMO converter; hence, the adequate energy allocation in v_{O4} also affects the regulation of the other outputs. As observed in Figure 6.100(a), oscillation occurs in v_{O2} as well.

When the skip function is activated, all the outputs are regulated, as shown in Figure (b). The energy transfer to v_{O4} is skipped occasionally to achieve adequate energy. The skipped energy is inevitably reallocated to v_{O1}–v_{O3}, thereby slightly increasing their voltage ripples. However, with the RSEC technique, the maximum output ripple is kept below 17 mV.

A comparison of the conventional absolute skip method and the RSEC technique is presented in Figure 6.101. With a large load current difference ($i_{O1} = i_{O2} = i_{O4} = 50$ mA and $i_{O3} = 300$ mA), the conventional absolute skip method frequently skips v_{O4} and induces a large output voltage ripple of 36.2 mV. SK_4 shows that the absolute skip method allocates a skipping time of v_{O4} in an extremely chaotic manner. By contrast, the RSEC technique intermittently skips v_{O4}; as shown in Figure 6.101(b), SK_4 occasionally skips v_{O4} ($SK_4 = 1$) and transfers slight skipping energy to the other outputs. When skipping occurs, the inductor current decreases slightly to obtain less energy. With the RSEC technique, the voltage ripple can be suppressed from 36.2 to 12.6 mV.

Figure 6.102 shows a comparison of the energy allocations with different slope compensation methods. In Figure 6.102(a), a fixed falling slope results in an invalid pulse, higher switching loss, and deterioration of power distribution. Thus, the duty cycles of the four outputs are

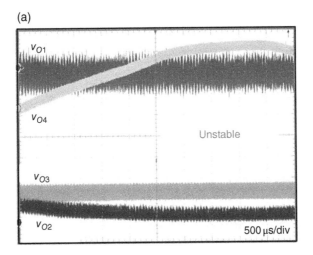

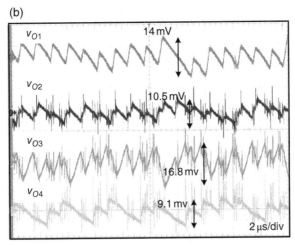

Figure 6.100 $i_{O1} = i_{O2} = i_{O4} = 50$ mA and $i_{O3} = 300$ mA (a) with and (b) without the skip function

slightly different in the switching cycles even if no skipping occurs. This results in inductor current variation without skip activations. However, the short invalid pulse for energy allocation is avoided with the bidirectional slope compensation in Figure 6.102(b). The eight segments effectively transfer energy to the four outputs with stable operations.

6.5.5.4 Measured SIMO under Different Load Conditions

The measured transient response in the first implementation chip is shown in Figure 6.103. When i_{O3} changes from 50 to 300 mA, the voltage ripples of all the outputs are slightly increased. This occurs because the low relative energy in v_{O4} induces skip activities in v_{O4},

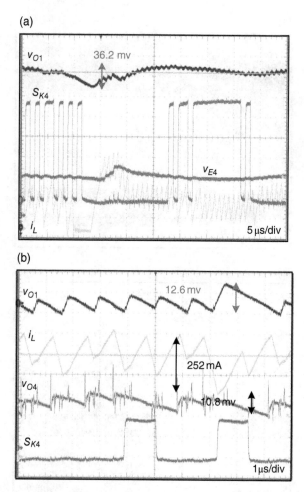

Figure 6.101 Comparison of (a) the conventional skip and (b) RSEC with $i_{O1} = i_{O2} = i_{O4} = 50$ mA and $i_{O3} = 300$ mA

which is shown in SK_4. With the RSEC technique, the overshoot/undershoot voltage and cross regulation are kept at 27 and 10.8 mV, respectively, because the energy is properly controlled.

The measured transient response in the complete implementation chip is shown in Figure 6.104. When the OIGD and DOR techniques are introduced, the transient performances do not vary significantly with the same control method, which includes RSEC and bidirectional slope compensation. When $i_{O1} = i_{O2} = i_{O4} = 50$ mA and i_{O3} changes from 300 to 50 mA and vice versa, the overshoot/undershoot, cross regulation, and ripple in steady state are kept at 16 and 10 mV, respectively.

At very light loads, the RSEC SIMO converter enters GM to reduce the switching loss. The measured waveform is shown in Figure 6.105. Once the inductor current reaches zero, the switching activities in all the switches are paused. The switching loss is decreased to ensure light load efficiency because no switching activities occur in some switching cycles. Without

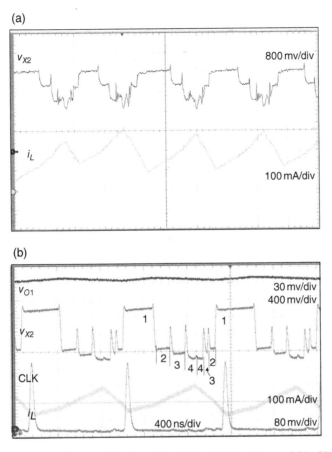

Figure 6.102 Comparison of energy allocations (a) with fixed falling slope and (b) with the bidirectional slope compensation

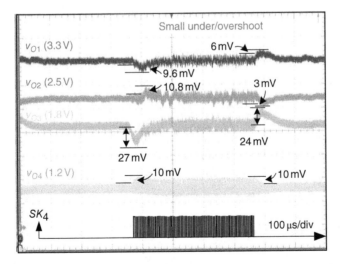

Figure 6.103 Load transient of the RSEC technique in chip 1 when $i_{O1} = i_{O2} = i_{O4} = 50$ mA and $i_{O3} = 50$–300 mA

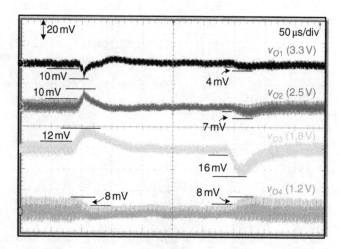

Figure 6.104 Measured transient response in chip 2 when $i_{O1} = i_{O2} = i_{O4} = 50$ mA and i_{O3} changes from 300 to 50 mA and vice versa

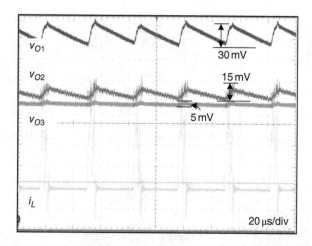

Figure 6.105 RSEC in GM when $i_{O1} = 8.3$ mA, $i_{O2} = 6$ mA, and $i_{O3} = i_{O4} = 1$ mA

acquiring energy, all outputs drop because the energy stored in the output capacitors are consumed by the loads until one of the outputs requires energy. The switching activity starts again to maintain voltage regulations. However, v_{O3}, which is an energy-excessive output, is skipped occasionally. With GM, the efficiency at light load can be further enhanced.

6.5.5.5 Statistics of Performances

Figure 6.106 shows the selection of the skip threshold voltage V_{SK}. With a higher V_{SK}, all outputs are inclined to skip frequently. This induces a lower switching loss and a higher power

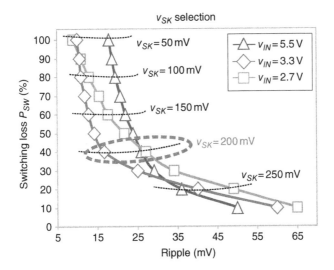

Figure 6.106 Optimization between switching loss and voltage ripple

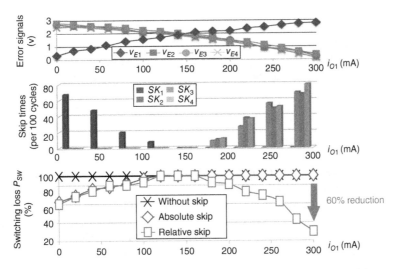

Figure 6.107 Statistics of error signals, skip number distributions, and switching power loss suppression

efficiency but causes a large voltage ripple. By contrast, a lower V_{SK} keeps the voltage ripple smaller at the cost of power efficiency. The optimal skip threshold voltages of each output are determined by the trade-off between large switching loss and small voltage ripple.

The statistical data of error signals, skip number distributions, and switching power loss suppression with v_{O2}–v_{O4} at medium loads is shown in Figure 6.107. When v_{O1} is at medium loads, the relative energies of all the outputs are close. Therefore, the error signals v_{E1}–v_{E4} have similar levels in equally allocating energy to the four outputs with nearly the same duty

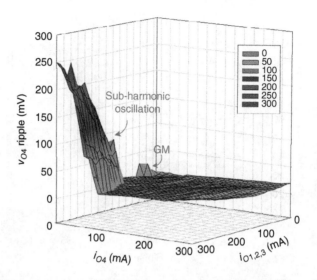

Figure 6.108 Ripple performance of fixed slope compensation under versatile loading current conditions when $v_{O1} = 3.3$ V, $v_{O2} = 2.5$ V, $v_{O3} = 1.8$ V, and $v_{O4} = 1.2$ V

cycles D_1–D_4. Under this condition, no skipping occurs because of the small relative energies between the four outputs. When v_{O1} is at light loads, the conventional absolute skip method starts to skip v_{O1} because the absolute energy of V_{O1} is low. Similarly, the RSEC skips v_{O1} because of its relatively low energy. The skipping times of v_{O1} increase, and the switching loss decreases as the load current of v_{O1} decreases. When v_{O1} is at heavy loads, v_{O2}–v_{O4} are skipped alternately according to the RSEC technique because their energies are relatively low. This skip activation reduces the switching loss by a maximum of 60% compared with the conventional absolute skip method.

Figure 6.108 shows the ripple performance of fixed slope compensation under versatile loading current conditions when $v_{O1} = 3.3$ V, $v_{O2} = 2.5$ V, $v_{O3} = 1.8$ V, and $v_{O4} = 1.2$ V. When i_{O1}–i_{O4} are at light loads, the SIMO converter enters the GM. The ripple increases slightly because the switching operation is paused in some switching cycles. As the total load currents increase in CCM, the larger inductor current flowing through the ESR of the output capacitors results in the maximum voltage ripple of 40 mV at maximum loading condition. With the decrease in i_{O4}, the skipping time of v_{O4} increases. With fixed slope compensation, subharmonic oscillation happens, thus deteriorating the voltage ripple performance. The multiple-segment slope compensation facilitates this problem and keeps the voltage ripple within 50 mV under all load conditions.

Figure 6.109 shows a comparison of the efficiencies of the conventional N-type power MOSFET and the OIGD and DOR techniques over a wide range of output voltages. The gate driving voltage decreases with a higher output voltage level. Thus, the turn-on resistance increases with the N-MOSFET. The power efficiency is reduced to 70%. Proper switch control can achieve low turn-on resistance under all output voltage conditions with the OIGD and DOR techniques. The efficiency is always kept near 90% at a total output current of 0.7 A. With the SIMO converter, the efficiency is improved by 19.9% at an output voltage of 3.3 V.

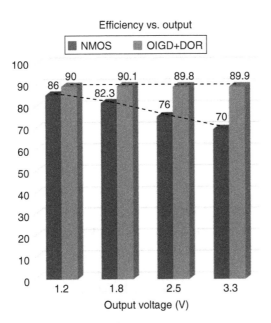

Figure 6.109 Efficiency comparison of the conventional N-type power MOSFET and the OIGD and DOR techniques over a wide range of output voltages

Figure 6.110 shows a comparison of the conventional P-type, N-type power MOSFET and the OIGD and DOR techniques. For tablet applications, v_{O1}, v_{O2}, v_{O3}, and v_{O4} are 3.3, 2.5, 1.8, and 1.2 V, respectively. The output voltages have a wide range; thus, the conventional switch control with P-MOSFET and N-MOSFET exhibits low on-resistance. The P-MOSFET, which has a higher on-resistance, results in the lowest efficiency in full load ranges. The OIGD and DOR techniques with N-MOSFET, which inherently has lower on-resistance, can greatly improve the power efficiency. The efficiency comparison shows that the peak efficiency of the P-MOSFET, N-MOSFET, and N-MOSFET with OIGD and DOR techniques are 85, 87, and 90%, respectively.

Figure 6.111 shows the recycling performance during deadtime. To retrieve more energy, low-power DOR circuits are designed, as illustrated in Figure 6.89(a). Low-power v_{X2} detection and body control are designed to prevent body diode leakage. Recycling efficiency is defined as the recycled energy/total energy loss to quantize the retrieved energy. As shown in Figure 6.111, 95% of the recycling efficiency is achieved by recycling 20 μW and consuming 1 μW in the DOR circuit.

6.5.5.6 Experimental Setup of Tablet System

Figure 6.112(a) demonstrates that a 12-inch tablet is well powered by the SIMO converter, which provides four voltages of 1.2, 1.8, 2.5, and 3.3 V. The small output ripple and low cross regulation can remove any flicker effect. The whole tablet system, which includes touch, display panel, CPU, and speaker, is constructed as illustrated in Figure 6.112(b). The panel, which

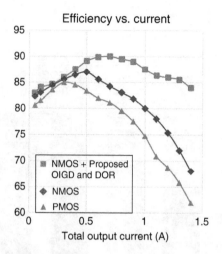

Figure 6.110 Efficiency comparisons of the conventional P-type, N-type power MOSFET and OIGD and DOR techniques for tablet applications

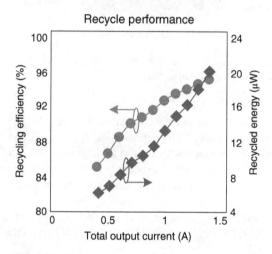

Figure 6.111 Recycling performance during deadtime

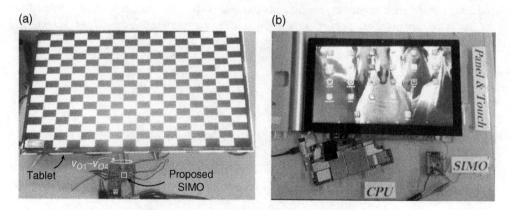

Figure 6.112 (a) 12-Inch tablet powered by the SIMO converter and (b) the whole tablet system, including the SIMO converter

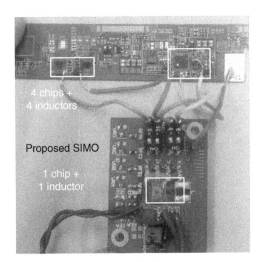

Figure 6.113 Comparison of the original power solution and the SIMO converter

is stacked and aligned to touch, is powered by the SIMO converter. Different apps can be executed on the tablet with CPU control.

Figure 6.113 shows the zoomed-in original power solution of a commercial tablet and provides a comparison with that of the SIMO converter. On the one hand, in the original power solution, four DC/DC buck converter chips and four inductors occupy a large PCB area. On the other hand, a tablet with the SIMO converter uses only one SIMO converter chip and one inductor. The PCB area and cost are significantly minimized for lighter and more compact tablet designs.

6.5.5.7 Design Summary and Comparisons

The SIMO converter is compared with previous converters; the comparisons can be divided into control methodology design and power stage design.

A comparison of the control methodology design and performance with that in prior converters for the first implementation chip is shown in Table 6.6. Instead of employing the conventional absolute skip method, the RSEC technique can properly judge the skip status under all load conditions for adequate energy allocations. Thus, the smallest voltage ripple and good regulation performance are achieved in prior converters. Moreover, the SIMO converter can provide the largest total load current and load current difference because of the current-mode control with skip function. Cross regulation is minimized to 0.0432 mV/mA.

A comparison of the power stage design and performance of the converter with that of previous converters for complete implementation of the chip is shown in Table 6.7. Compared with the conventional use of P-type and N-type MOSFETs, the OIGD technique can maintain low on-resistance over a wide range of output voltages. Therefore, the efficiency performance is independent of the output voltage level. The highest peak efficiency of the SIMO converter can reach 90%. With energy recycling, the AERC [12] mitigates the efficiency degradation caused by an intentional insertion of the buffer region. However, only 80% of the energy

Table 6.6 Comparison of the control methodology design and performance with that for the first implementation chip

	Method in this book	[16]	[12]	[18]
Control methodology	RSEC technique	PLL based + ripple based	Current mode + charge control	Current mode + ripple based
Skip method	Relative	Absolute	Absolute	Absolute
Input voltage (V)	2.7–5.5	5	3.4–4.3	2.5–4.5
Output voltage (V)	0.6–5	1–3	1.2–2.8	7.5–10.2, −9.5
Maximum CCM voltage ripple (mV)	<40	<50[a]	<40	<160
Maximum total load current (A)	1.2	0.78[a]	1[a]	0.11
Maximum load current difference (mA)	300	130[a]	200[a]	30
Cross regulation (mV/mA)	0.0432	0.93	0.067	1.5
Load regulation (mV/mA)	0.02	0.93	N/A	1.5
Switching frequency	1.1 MHz	2 MHz	1.2 MHz	700 kHz
Maximum efficiency (%)	85.2	N/A	83.1	80.8
Power density @I_{max} (W/mm^2)	1.34	0.17[a]	1.2	N/A

[a] Estimated by measurement waveforms.

Table 6.7 Comparison of the power stage design and performance of the proposed converter with that in prior arts.

	Method in this book	[12]	[16]	[17]
Switch	NMOS with OIGD control	PMOS	NMOS	NMOS
Energy recycling method	DOR	AERC[a]	X	X
Recycling efficiency (%)	95	<80[b]	X	X
Input voltage (V)	3.6–5	3.4–4.3	5	2.7–5
Output voltage (V)	1.2–3.3	1.2–2.8	1–3	0.6–1.8
Maximum efficiency (%)	90	83.1	N/A	87
Max. CCM ripple (mV)	<40	<40	<25	<30
Cross regulation (mV/mA)	0.04	0.067	N/A	0.04
Load regulation (mA/mA)	0.02	N/A	N/A	N/A
Switching frequency (MHz)	1	1.2	2	1
Power density @I_{max} (W/mm^2)	1.13	1.2	N/A	0.4

[a] Adapted energy recovery control.
[b] Estimated by conduction loss.

recycling efficiency is achieved because the recovery current flows through two power MOSFETs with large conduction loss. By contrast, the DOR technique retrieves a maximum of 95% of the inherent energy loss at v_{X2} because of the low-power design of the recycling detection circuit and recycling path.

References

[1] Fan, S., Xue, Z., Lu, H., et al. (2011) Area-efficient on-chip DC–DC converter with multiple-output for biomedical applications. *IEEE Transactions on Circuits and Systems I*, **61**(11), 1671–1680.

[2] Kim, J., Kim, D.S., and Kim, C. (2013) A single-inductor eight-channel output DC–DC converter with time-limited power distribution control and single shared hysteresis comparator. *IEEE Transactions on Circuits and Systems I*, **60**(12), 3354–3367.

[3] Huang, M.-H. and Chen, K.-H. (2009) Single-inductor multi-output (SIMO) DC–DC converters with high light-load efficiency and minimized cross-regulation for portable devices. *IEEE Journal of Solid-State Circuits*, **44**(4), 1099–1111.

[4] Lee, Y.-H., Fan, M.-Y., Chen, W.-C., et al. (2011) A near-zero cross-regulation single-inductor bipolar-output (SIBO) converter with an active-energy-correlation control for driving cholesteric-LCD. *Proceedings of the IEEE Custom Integrated Circuits Conference (CICC)*, San Jose, CA, September 19–21, pp. 1–4.

[5] Qiu, Y., Chen, X., and Liu, H. (2010) Digital average current-mode control using current estimation and capacitor charge balance principle for DC–DC converters operating in DCM. *IEEE Transactions on Power Electronics*, **25**(6), 1537–1545.

[6] Lee, Y.-H., Wang, S.-J., Yang, Y.-Y., et al. (2009) A DVS embedded power management for high efficiency integrated SoC in UWB system. *Proceedings of the IEEE Asian Solid-State Circuits Conference*, Taipei, November 16–18, pp. 321–324.

[7] Cannizzaro, S.O., Grasso, A.D., Mita, R., et al. (2007) Design procedures for three-stage CMOS OTAs with nested-Miller compensation. *IEEE Transactions on Circuits and Systems I: Regular Papers*, **54**(5), 933–940.

[8] Texas Instruments (2013) TI Tablet Solutions, http://www.ti.com/lit/sl/slyy028d/slyy028d.pdf (accessed November 15, 2015).

[9] Samsung Electronics (2013) 2-Chip Display Driver Architecture for Tablet Display, https://www.google.com.tw/url?sa=t&rct=j&q=&esrc=s&source=web&cd=1&ved=0CBwQFjAA&url=http%3A%2F%2Fwww.edn.com%2FPdf%2FViewPdf%3FcontentItemId%3D4425798&ei=-1VdVdvFEYOU8QXR8IGgAw&usg=AFQjCNFRdXsioiH31xKr7UjFbN0mkHNRKg&sig2=OgaHn_OoGISZNBXQlCE95A&bvm=bv.93756505,d.dGc&cad=rja (accessed November 15, 2015).

[10] Jing, X., Mok, P.K.T., and Lee, M.C. (2011), A wide-load-range constant-charge-auto-hopping control single-inductor dual-output boost regulator with minimized cross-regulation. *IEEE Journal of Solid-State Circuits*, **46**(10), 2350–2362.

[11] Ma, D., Ki, W.-H., and Tsui, C.-Y. (2003) A pseudo-CCM/DCM SIMO switching converter with freewheel switching. *IEEE Journal of Solid-State Circuits*, **38**(6), 1007–1014.

[12] Kuan, C.-W. and Lin, H.-C. (2012) Near-independently regulated 5-output single-inductor DC–DC buck converter delivering 1.2 W/mm^2 in 65 nm CMOS. *Proceedings of the IEEE International Solid-State Circuits Conference (ISSCC), Digest of Technical Papers*, San Francisco, CA, February 19–23, pp. 274–276.

[13] Lee, Y.-H., Chen, K.-H., Lin, Y.-H., et al. (2010) An interleaving energy-conservation mode (IECM) control in single-inductor dual-output (SIDO) step-down converters with 91% peak efficiency. *Proceedings of the IEEE Symposium on VLSI Circuits (VLSIC), Digest of Technical Papers*, Honolulu, HI, June 16–18, pp. 57–58.

[14] Lee, Y.-H., Yang, Y.-Y., Wang, S.-J., et al. (2011) Interleaving energy-conservation mode (IECM) control in single-inductor dual-output (SIDO) step-down converters with 91% peak efficiency. *IEEE Journal of Solid-State Circuits*, **46**(4), 904–915.

[15] Jung, S.-H., Jung, N.-S., Hwang, J.-T., et al. (1999) An integrated CMOS DC–DC converter for battery-operated systems. *Proceedings of the Power Electronics Specialists Conference (PESC)*, Charleston, SC, August, pp. 43–47.

[16] Lee, K.-C., Chae, C.-S., Cho, G.-H., and Cho, G.-H. (2010) A PLL-based high-stability single-inductor 6-channel output DC–DC buck converter. *Proceedings of the IEEE International Solid-State Circuits Conference (ISSCC), Digest of Technical Papers*, San Francisco, CA, February 7–11, pp. 200–201.

[17] Lu, D., Qian, Y., and Hong, Z. (2014) An 87%-peak-efficiency DVS-capable single-inductor 4-output DC–DC buck converter with ripple-based adaptive off-time control. *Proceedings of the IEEE International Solid-State Circuits Conference (ISSCC), Digest of Technical Papers*, San Francisco, CA, February 9–13, pp. 82–83.

[18] Le, H.-P., Chae, C.-S., Lee, K.-C., et al. (2007) A single-inductor switching DC–DC converter with five outputs and ordered power-distributive control. *IEEE Journal of Solid-State Circuits*, **42**(12), 2706–2714.

[19] Su, Y.-P., Lin, C.-H., Yang, T.-F., et al. (2014) CCM/GM relative skip energy control in single-inductor multiple-output DC–DC converter for wearable device power solution. *Proceedings of the IEEE Asian Solid-State Circuits Conference (A-SSCC)*, KaoHsiung, November 10–12, pp. 65–68.

[20] Jung, M.-Y., Park, S.-H., Bang, J.-S., et al. (2015) An error-based controlled single-inductor 10-output DC–DC buck converter with high efficiency at light load using adaptive pulse modulation. *Proceedings of the IEEE International Solid-State Circuits Conference (ISSCC), Digest of Technical Papers*, San Francisco, CA, February 22–26, pp. 222–223.

[21] Woo, Y.-J., Le, H.-P., Cho, G.-H., et al. (2008) Load-independent control of switching DC–DC converters with freewheeling current feedback. *Proceedings of the IEEE International Solid-State Circuits Conference (ISSCC), Digest of Technical Papers*, San Francisco, CA, February 3–7, pp. 446–447.

[22] Hazucha, P., Moon, S.T., Schrom, G., et al. (2007) High voltage tolerant linear regulator with digital control for biasing of integrated DC–DC converters. *IEEE Journal of Solid-State Circuits*, **42**(1), 66–73.

[23] Lin, Y.-H., Zheng, K.-L., and Chen, K.-H. (2008) Power MOSFET array for smooth pole tracking in LDO regulator compensation. *IEEE Transactions on Power Electronics*, **23**(5), 2421–2427.

[24] Milliken, R.J., Silva-Martínez, J., and Sanchez-Sinencio, E. (2007) Full on-chip CMOS low-dropout voltage regulator. *IEEE Transactions on Circuits and Systems I: Regular Papers*, **54**(9), 1879–1890.

[25] Lee, Y.-H., Huang, T.-C., Yang, Y.-Y., et al. (2011) Minimized transient and steady-state cross regulation in 55-nm CMOS single-inductor dual-output (SIDO) step-down DC–DC converter. *IEEE Journal of Solid-State Circuits*, **46**(11), 2488–2499.

[26] Xu, W., Li, Y., Gong, X., et al. (2010) A dual-mode single-inductor dual-output switching converter with small ripple. *IEEE Transactions on Power Electronics*, **25**(3), 614–623.

[27] Wang, S.-W., Cho, G.-H., and Cho, G.-H. (2012) A high-stability emulated absolute current hysteretic control single-inductor 5-output switching DC–DC converter with energy sharing and balancing. *Proceedings of the IEEE International Solid-State Circuits Conference (ISSCC), Digest of Technical Papers*, San Francisco, CA, February 19–23, pp. 276–277.

[28] Huang, H.-W., Chen, K.-H., and Kuo, S.-Y. (2007) Dithering skip modulation, width and dead time controllers in highly efficient DC–DC converters for system-on-chip applications. *IEEE Journal of Solid-State Circuits*, **42**(11), 2451–2465.

[29] Tsai, J.-C., Huang, T.-Y., Lai, W.-W., and Chen, K.-H. (2011) Dual modulation technique for high efficiency in high-switching buck converters over a wide load range. *IEEE Transactions on Power Electronics*, **58**(1), 1671–1680.

[30] Kwan, H.-K., Ng, D.C.W., and So, V.W.K. (2013) Design and analysis of dual-mode digital-control step-up switched-capacitor power converter with pulse-skipping and numerically controlled oscillator-based frequency modulation. *IEEE Transactions on Very Large Scale Integration System*, **21**(11), 2132–2140.

[31] Yan, Y., Lee, F.C., and Mattavelli, P. (2013) Comparison of small signal characteristics in current mode control schemes for point-of-load buck converter applications. *IEEE Transactions on Power Electronics*, **28**(7), 3405–3414.

[32] Erickson, R.W. and Maksimovic, D. (2001) *Fundamentals of Power Electronics*, 2nd edn. Kluwer Academic Publishers, Norwell, MA.

[33] Chen, K.-H., Chang, C.-J., and Liu, T.-H. (2008) Bidirectional current-mode capacitor multipliers for on-chip compensation. *IEEE Transactions on Power Electronics*, **23**(1), 180–188.

[34] Patounakis, G., Li, Y.W., and Shepard, K.L. (2004) A fully integrated on-chip DC–DC conversion and power management system. *IEEE Journal of Solid-State Circuits*, **39**(3), 443–451.

[35] Alimadadi, M., Sheikhaei, S., Lemieux, G., et al. (2007) A 3 GHz switching DC–DC converter using clock-tree charge-recycling in 90 nm CMOS with integrated output filter. *Proceedings of the IEEE International Solid-State Circuits Conference (ISSCC), Digest of Technical Papers*, San Francisco, CA, February 11–15, pp. 532–533.

[36] Hsieh, C.-Y. and Chen, K.-H. (2008) Adaptive pole-zero position (APZP) technique of regulated power supply for improving SNR. *IEEE Transactions on Power Electronics*, **23**(6), 2949–2963.

[37] Ma, F.-F., Chen, W.-Z., and Wu, J.-C. (2007) A monolithic current-mode buck converter with advanced control and protection circuit. *IEEE Transactions on Power Electronics*, **22**(5), 1836–1846.

[38] Mulligan, M.D., Broach, B., and Lee, T.H. (2007) A 3 MHz low-voltage buck converter with improved light load efficiency. *Proceedings of the IEEE International Solid-State Circuits Conference (ISSCC), Digest of Technical Papers*, San Francisco, CA, February 11–15, pp. 528–529.

[39] Ridley, R.B. (1991) A new, continuous-time model for current-mode control. *IEEE Transactions on Power Electronics*, **6**(2), 271–280.

7

Switching-Based Battery Charger

7.1 Introduction

Portable devices have become very popular and ubiquitous electrical commercial products in the market. Laptops, smartphones, tablets, and recently wearable devices present a common feature: they all require a battery. Batteries allow products to become portable. Without a battery, these products completely lose their convenience. Thus, battery chargers have become an important and well-developed technology in analog circuit designs. In this chapter, the basic concepts of switching-based battery chargers, from their fundamental control to some advanced applications, are introduced.

Lithium-ion (Li-ion) batteries are the mainstream battery choice for laptops and smartphones because this type of battery presents the highest energy density (capacity per unit volume) among the available commercial battery technologies. This feature is of great importance for portable devices requiring small sizes and long standby times. High cell voltages allow battery packs to include only one cell, thereby simplifying battery design. Such battery packs are the most-often used for current smartphone applications. Li-ion batteries also provide low maintenance, which is an advantage that most chemical batteries cannot achieve, no memory effects, and a scheduled charging cycle that can affect battery life.

Despite their many obvious benefits, Li-ion batteries also present a number of drawbacks, the most obvious of which is the need for a protection circuit to maintain safe operation. Thus, a well-designed charging method for Li-ion batteries is essential.

The battery-charging technique has been well developed and is fairly mature. The standard charging topology for Li-ion batteries is well known to involve constant current/constant voltage (CC/CV) mode control. Under this control, the CC mode charges the battery with a constant current when the voltage is low. This constant current is ideally set to 1 C, which is a charging current rate equal to the ampere-hour rating of the cell. However, in practical designs, researchers have suggested that this be set to below 0.8 C to protect and prolong the battery

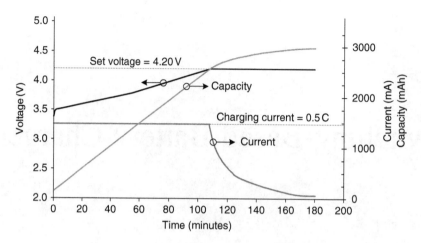

Figure 7.1 Charging characteristic of a Li-ion battery

life. While charging the battery in CC mode, the battery voltage increases to reach the set voltage. For a one-cell Li-ion battery pack, this set voltage is within 4.1–4.3 V based on the capacity and manufacturing specifications. The typical set voltage is 4.2 V ± 50 mV. The accuracy of the set voltage is relatively important. The battery should not be charged above this set voltage or the battery may be damaged and the battery life will be shortened. Furthermore, the battery cannot be fully charged if the set voltage is too low.

Besides CC/CV control, another control mode is available to protect Li-ion batteries. Overcharging and deep discharge problems are two main issues related to Li-ion batteries. If the battery voltage is at low levels (nominally lower than 2.5 V), the battery is in deep-discharge condition. This problem does not often occur because every smartphone has built-in battery protection, wherein the phone shuts down if the state of charge (SOC) of the battery is lower than 2–3%. This phenomenon often occurs in batteries that have not been used for a long period of time because the battery has a self-leakage current that leads to complete discharge. Under this condition, the charging current begins from a low level to prevent damage because of rapid increases in charging current. This pre-charge stage is usually referred to as trickle-current control.

Figure 7.1 shows a typical CC/CV charging procedure under 0.5 C charging current rate and 4.2 V set voltage. The charge starts from the CC mode with a constant charging current. Considering that the battery voltage is below the set voltage, the charging current charges the battery and increases the voltage to reach the set voltage. The SOC of the battery reaches 65–70% after the CC mode is complete, which means that the battery voltage has reached the set voltage. Given that the battery voltage reaches the upper limit, the battery charger enters the CV mode of control. To keep a constant voltage, the charging current begins to decrease to prevent damage caused by overcharged voltages. Full charge occurs when the battery voltage reaches the set voltage and the charging current drops to the minimum value, which means that the charging current cannot decrease any further. Increasing the charging current does not significantly reduce the full charging time but lets the battery voltage reach the set voltage faster. However, the saturation time, which is defined as the time required to decrease the charging current in the CV mode, also increases. The saturation time can be explained by the internal ESR of the

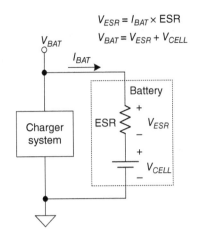

Figure 7.2 ESR effect of a Li-ion battery

battery, the effects of which are illustrated in Figure 7.2. The measured battery voltage includes the voltage drop in the ESR, which is affected by the charging current and the real voltage of the battery cell. In the CC mode, the battery reaches the set voltage at 4.2 V with only around 65% SOC. The 35% uncharged capacity is attributed to the voltage drop in the ESR. The charging current must decrease to avoid overstressing the battery voltage, thereby lowering the voltage drop in the ESR and maintaining the total battery voltage at the set voltage. The battery cell is actually fully charged until the voltage drop in the ESR becomes negligible. The increased charging current quickly charges the battery to around 65% SOC. However, fully charging the Li-ion battery is not necessary based on its characteristics because of the increase in voltage stress on the battery. Providing a lower set voltage or decreasing the saturation time prolongs the battery life but shortens the standby time. Major commercial products promote the maximum standby time by setting the maximum affordable set voltage.

In recent applications, battery-powered portable devices require multiple DC/DC converters to supply many different chip sets and functionalities. Fast- and high-efficiency battery charging is essential to meet complex power supply requirements. Therefore, embedded power management systems (PMSs) are often demanded by portable devices to fulfill different power requirements, as shown in Figure 7.3(a). An energy delivery controller in the PMS is used to arrange power paths to optimally distribute energy to each block. Afterward, multiple DC/DC converters convert different supplies for various function blocks, while a charger circuit manages the battery-charging procedure. Thus, stable and continuous energy from the power supply and battery for multiple DC/DC converters can provide adequate and distributive voltage and current to different functional blocks in portable devices. LDO regulator-based or switching regulator-based structures are commonly used because of the design of the charger systems, as illustrated in Figure 7.3(b). LDO-based chargers present the advantages of freedom from ripples, compactness, and high accuracy, but suffer from low efficiency in the case of large dropout voltages [1]. By contrast, a switching-based charger can guarantee high efficiency over wide input and output voltage ranges [2].

A simplified architecture of a switching-based charger is shown in Figure 7.4. The battery charger system operates under different power delivery conditions according to different

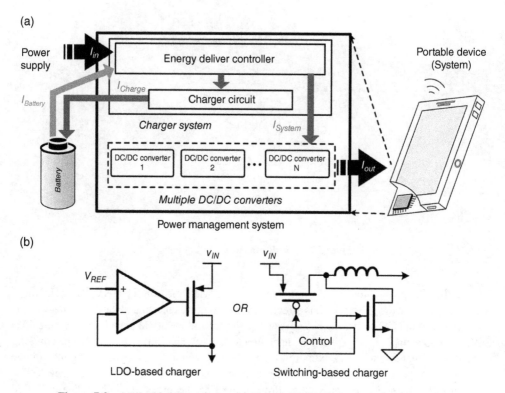

Figure 7.3 (a) Architecture of an embedded PMS. (b) Charger structures in PMS

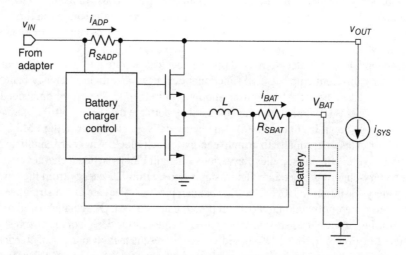

Figure 7.4 Simplified architecture of a switching-based battery charger

connections between the system, input power source, and battery. The management of a switching-based battery charger is called automatic energy delivery control (AEDC) and is presented in [2]. The portable system can be supplied by batteries or external power sources (e.g., adaptor, universal serial bus (USB), etc.). According to different power conditions, the AEDC

Switching-Based Battery Charger

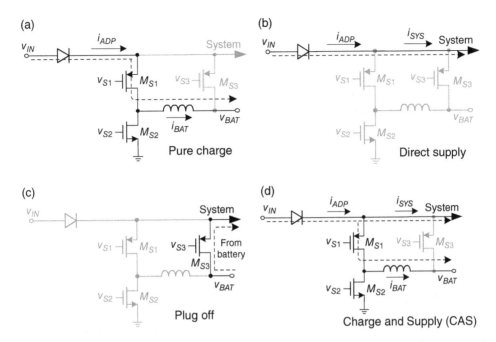

Figure 7.5 Power delivery paths in the proposed charger system: (a) pure charge state; (b) direct supply state; (c) plug off state; (d) CAS state

must select a suitable power delivery path through the power stage to balance the input energy and system loading. If the external power sources have enough energy to drive the loading system, battery charging can then proceed with rated charging. However, if the energy source is insufficient to satisfy both charging and loading requirements, then the charging process varies with the system loading.

Four operational states are present to satisfy all power conditions: pure charge state, direct supply state, plug off state, and charge and supply (CAS) state. Figure 7.5 shows the power path in each state, which corresponds to different input and output relationships. The input power source generally has a maximum power limitation. Thus, the first priority of the charging system is to keep the subsequent system working normally. According to the constraints of the input power and system loading information, the energy delivery control circuit in the AEDC generates gate control signals to induce the power stage and charger system to execute a suitable power state. The charger circuit then collaborates with the AEDC circuit and simultaneously adjusts the charging current for battery charging. A detailed description of each operational state is provided below.

7.1.1 Pure Charge State

When the loading system is shut down or disconnected from the charger system, the only load is the battery. The energy from the input power source is completely delivered to the battery and constitutes a pure charging state. Meanwhile, the charging current is limited by the rated charging current of the battery.

7.1.2 Direct Supply State

The loading system can operate normally even if the battery is removed when the input power source is connected to the charger system. That is, the input power source can provide energy directly to the system. This phenomenon may also occur when the battery is fully charged. The charger system keeps monitoring the battery status. The charging state quickly switches to the CAS state when the battery is connected to the charger system once more, or requires energy.

7.1.3 Plug Off State

The portable system uses the battery as the power supply once the input power source is disconnected. Thus, the charger system must provide a power delivery path from the battery to the loading system. Moreover, the battery should be disconnected from the loading system once the battery voltage reaches its lower limit to protect the battery from being over-discharged.

7.1.4 CAS State

As the development of CPUs has moved toward multi-core applications, the required power level has also increased because of the requirements for improved dynamic performance for fast processing of multiple complicated tasks. This increase in power may enable the power from the input power source to exceed the maximum power limit. Therefore, the CAS state is presented to prevent this phenomenon. While the CPU is performing multiple tasks, the required power also increases. Once the required power exceeds the limit that the input power source can provide, the controller decreases the charging current to the battery, thereby reducing the total power from the input power source.

To illustrate the operation of the CAS state, the maximum power provided by the input power source is defined as $P_{IN,max}$, the power consumption of the loading system is P_{system}, and the battery charging power is P_{charge}. This state can be divided into two cases according to the relationships among $P_{IN,max}$, P_{system}, and P_{charge}. First, if $P_{IN,max} \geq P_{system} + P_{charge}$, the input power source can afford the total power that is consumed by the loading system and the battery charger. By contrast, if the total power consumption exceeds the maximum power, that is, $P_{IN,max} < P_{system} + P_{charge}$, the power consumption of the system is more prioritized than its charging function. The proposed CAS state reduces the charging current to satisfy the power demand for maintaining the loading system operation. Thus, the input power source can avoid overloading by adjusting the battery charging power from P_{charge} to P'_{charge}, that is, $P_{IN,max} = P_{system} + P'_{charge}$. In other words, the charging current becomes I'_{BAT} as determined by the CAS state expressed in Eq. (7.1):

$$I'_{BAT} = \frac{V_{IN}}{V_{BAT}} \times (I_{ADP,max} - I_{SYS}) \qquad (7.1)$$

In Figure 7.6, the CAS state shows that the charging current is correspondingly limited to the I_{SYS}. Therefore, the charger system is unable to charge the battery with the rated current, and the charging current is a distributed result. The AEDC must carefully control the charging current

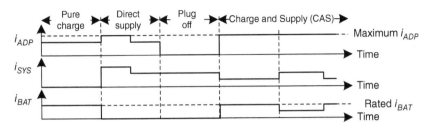

Figure 7.6 Current delivery to the input power source, the battery, or the loading system is decided by the AEDC

to extract the remaining power from the $P_{IN,max}$. The AEDC keeps I_{ADP} below its maximum value without affecting the requirements of the loading system.

The trend in the latest commercial products is geared toward prolonging the standby time with high computation performance. Switching-based battery chargers are a more fascinating topology for the current trend because these devices feature advantages such as high efficiency and flexibility of power delivery. The following sections introduce a general design procedure for analyzing a typical battery charger that includes small signal analysis, closed-loop modeling, and demonstration setup through a simple SPICE tool. Two advanced applications, namely the turbo-boost technique and the continuous built-in resistance detection (CBIRD) technique, are also illustrated.

7.2 Small Signal Analysis of Switching-Based Battery Charger

Small signal models are essential in analyzing the behavior of switching-based battery chargers. The equivalent model of a typical switching-based battery charger in [3] is shown in Figure 7.7. To derive the equivalent small signal model, some basic concepts from [4], which introduces a typical buck small signal model, are helpful. Symbols with capital letters in this section, such as V_g in Figure 7.8, are used for DC biasing values or large signal modeling. By contrast, symbols with small letters, such as $\hat{d}$ in Figure 7.8, represent the small signal perturbation or source. Considering the well-modeled small signal model of a buck in [4], as shown in Figure 7.8, the equivalent small signal model of a switching-based battery charger can be substituted directly into this model. Given that the switching-based battery charger can be treated as a buck convertor but with a different output filter stage and loading, small signal modeling of the switching-based battery charger can be achieved using the model in Figure 7.9.

As illustrated in Section 7.1, a typical switching-based battery charger is controlled by the CC/CV mode, which depends on the SOC of the battery. Thus, the first step in analyzing the small signal of a switching-based battery charger is to construct the transfer function of the CC and CV modes. The small signal perturbations $\hat{v}_g$ and $\hat{d}$ denote the small signal perturbation of the line (power source) and control signal (duty), respectively, as shown in Figure 7.9. Provided that the final target of small signal modeling is derived from a closed-loop transfer function, the modeling focuses on the perturbation of control signal $\hat{d}$, which directly affects the closed-loop response. As a result, the perturbation of line $\hat{v}_g$ is neglected to simplify the modeling, as shown in Figure 7.10. According to the small signal model in Figure 7.10, the transfer function from

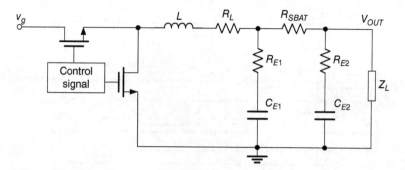

Figure 7.7 The equivalent model of a typical battery charger

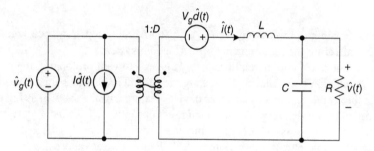

Figure 7.8 Equivalent small signal mode of a typical buck converter

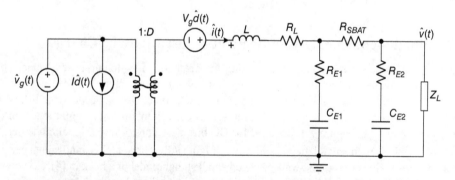

Figure 7.9 Equivalent small signal mode of a typical battery charger

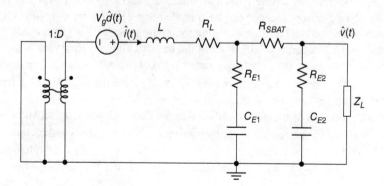

Figure 7.10 Equivalent small signal mode of the CC and CV modes

control signal $\hat{d}$ to output voltage $\hat{v}$, which represents the transfer function of the CV mode, can be derived as in Eqs. (7.2) and (7.3):

$$G_{vd} = \frac{\hat{v}_o}{\hat{d}} = V_g \cdot Z_L \cdot \frac{(\alpha_1 s + 1)(\alpha_2 s + 1)}{\beta_1 s^3 + \beta_2 s^2 + \beta_3 s + \beta_4} \quad (7.2)$$

$\alpha_1 = R_{E1} \times C_{E1}$

$\alpha_2 = R_{E2} \times C_{E2}$

$\beta_1 = L \times C_{E1} \times C_{E2} \times (R_{E1} \times R_{E2} + R_{E1} \times Z_L + R_{E2} \times Z_L + R_{SBAT} \times C_{E2} + R_{SBAT} \times Z_L)$

$\beta_2 = R_L \times C_{E1} \times C_{E2} \times (R_{E1} \times R_{E2} + R_{E1} \times Z_L + R_{E2} \times Z_L + R_{SBAT} \times C_{E2} + R_{SBAT} \times Z_L)$

$\quad + L \times (R_{E1} \times C_{E1} + R_{E2} \times C_{E2} + Z_L \times C_{E1} + Z_L \times C_{E2} + R_{SBAT} \times C_{E1})$

$\quad + R_{E1} \times R_{E2} \times Z_L \times C_{E1} \times C_{E2} + R_{E1} \times R_{E2} \times R_{SBAT} \times C_{E1} \times C_{E2} \quad (7.3)$

$\quad + R_{E1} \times R_{SBAT} \times Z_L \times C_{E1} \times C_{E2}$

$\beta_3 = L + R_L \times (R_{E1} \times C_{E1} + R_{E2} \times C_{E2} + Z_L \times C_{E1} + Z_L \times C_{E2} + R_{SBAT} \times C_{E1})$

$\quad + R_{E1} \times C_{E1} \times Z_L + R_{E1} \times R_{SBAT} \times C_{E1} + R_{E2} \times Z_L \times C_{E2} + R_{E2} \times R_{SBAT} \times C_{E2}$

$\quad + R_{SBAT} \times Z_L \times C_{E2}$

$\beta_4 = R_L + R_{SBAT} + Z_L$

The Bode plot of Eq. (4.2) is shown in Figure 7.11. Here, the Bode plot is generated through MATLAB with the coefficients in [3], as also shown in Table 7.1. The output loading Z_L

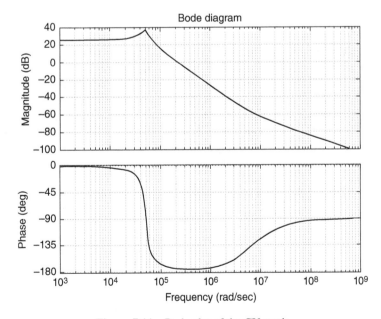

Figure 7.11 Bode plot of the CV mode

Table 7.1 Parameters of the Bode plot

V_g	19 V	R_{E1}	10 mΩ
L	10 μH	R_{E2}	10 mΩ
C_{E1}	20 μF	R_L	20 mΩ
C_{E2}	20 μF	Z_L	2.25 Ω
R_{SBAT}	10 mΩ	I	4 A

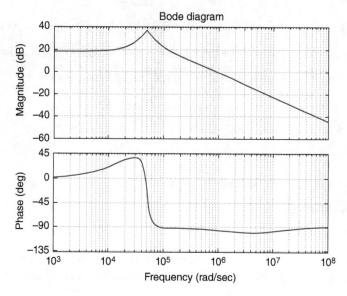

Figure 7.12 Bode plot of the CC mode

illustrated in Table 7.1 is a resistive load, but the target is to develop a battery charger model. A resistive load is used to simplify the modeling. Development of the closed-loop small signal model is demonstrated in the next section, and the Li-ion battery equivalent model is introduced and used in the analysis as a complete model. While some differences are noted in the Li-ion battery equivalent model, in practical design, the stability is not affected by this difference.

Following the same modeling as in Figure 7.10, the transfer function of the CC mode can be derived as the transfer function from control signal $\hat{d}$ to the charging current $\hat{i}$ sensed by R_{SBAT}, which is derived in Eq. (7.4). The coefficients β_1–β_4 and α_1 are similar to those shown in Eq. (7.3). The Bode plot of Eq. (7.4) is shown in Figure 7.12, and the relevant coefficients are listed in Table 7.1.

$$G_{id} = \frac{\hat{i}}{\hat{d}} = V_g \cdot \frac{(\alpha_1 s + 1)(\alpha_3 s + 1)}{\beta_1 s^3 + \beta_2 s^2 + \beta_3 s + \beta_4} \quad (7.4)$$

$$\alpha_3 = (R_{E2} + Z_L) \times C_{E2} \quad (7.5)$$

Besides the CC/CV mode, the CAS state must also be analyzed. The control loop of the CAS state is different from the CC/CV mode; hence, discussion of the small signal model of the CAS state is necessary. The small signal model in Figure 7.10 must be adjusted to model the

Switching-Based Battery Charger

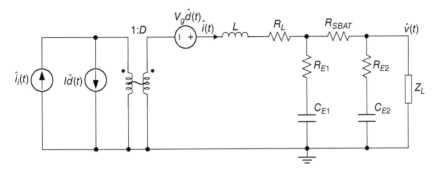

Figure 7.13 Equivalent small signal mode of the CAS state

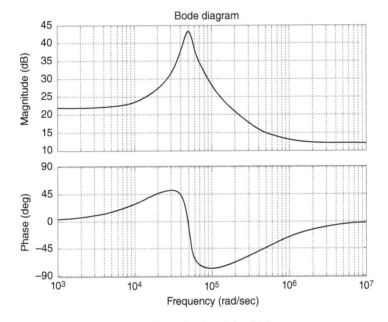

Figure 7.14 Bode plot of the CAS state

CAS state, which monitors the total power from the power source. Thus, the transfer function of the CAS state is observed from the control signal $\hat{d}$ to the input current $\hat{i}_i$. The perturbation from the line, which demonstrates the power source, is removed to simplify the modeling in Figure 7.10. To calculate the input current, an extra small signal source must be added, as shown in Figure 7.13. The control-to-input-current transfer function in the CAS state can be modeled as in Eq. (7.6) through this adjustment. Figure 7.13 demonstrates two sources with control signal $\hat{d}$, which are $I\hat{d}$ and $V_g\hat{d}$. Providing a superposition theory leading to a transfer function with two zeros and three poles is the general solution to this situation. Equation (7.6) is a more direct approach to calculate these two sources using Kirchhoff's current law at the input node. However, Eq. (7.6) still represents a property of superposition by the plus symbol. If we simplify these two terms in Eq. (7.6), it will have the same result as providing the superposition theory. The coefficients β_1–β_4 are identical to those shown in Eq. (7.3). The Bode plot of Eq. (7.6) is shown in Figure 7.14, with the coefficients in Table 7.1.

$$G_{iid} = \frac{\hat{i}_i}{\hat{d}} = I + V_g \cdot \frac{\gamma_1 s^2 + \gamma_2 s + 1}{\beta_1 s^3 + \beta_2 s^2 + \beta_3 s + \beta_4} \qquad (7.6)$$

where $\gamma_1 = C_{E1} \times C_{E2} \times (R_{E1} \times R_{E2} + R_{E1} \times Z_L + R_{E2} \times Z_L + R_{SBAT} \times C_{E2} + R_{SBAT} \times Z_L)$

and $\gamma_2 = C_{E1} \times (R_{E1} + R_{SBAT} + Z_L) + C_{E2} \times (R_{E2} + Z_L)$
$\qquad (7.7)$

7.3 Closed-Loop Equivalent Model

The closed-loop model is based on the transfer functions of the CC, CV, and CAS loops derived in Section 7.2. Figure 7.15 shows the closed-loop controller circuits with feedback circuits of CV, CC, and CAS loops, error amplifier with compensation circuits, and duty control block. These three loops do not dominate the loop at the same time, as only one loop dominates the system according to the status of the battery and system.

The simplified controller block diagram is shown in Figure 7.16 to analyze the closed-loop model. The transfer functions of the power stages G_{vd}, G_{id}, and G_{iid} have been derived in Section 7.2. $H(s)$, $G_C(s)$, and FM, respectively, represent the feedback sensor gain, the transfer function of the error amplifier with a compensator, and the transfer function from the control voltage to the duty cycle.

The closed-loop model can be separated into three parts: the CV, CC, and CAS loops. Analysis begins from the CV loop. As the transfer function G_{vd} has been demonstrated in Eq. (7.2), the current focus is the feedback sensor gain and compensation. The CV loop can be treated as a voltage-mode control buck converter. Thus, the transfer function of the feedback sensor gain for the CV loop is composed of the feedback voltage divider, which is normally achieved by

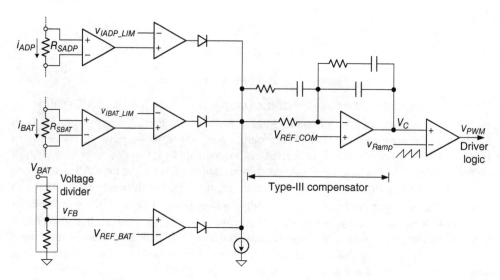

Figure 7.15 Simplified controller blocks of the battery charger

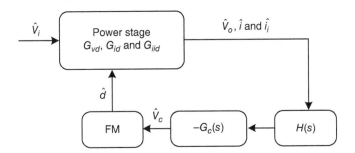

Figure 7.16 Simplified closed-loop block diagram

resisters and a voltage amplifier, which amplifies the sensed feedback voltage. The following transfer functions illustrated in this section are derived using the parameters in the demonstrated simulation. Readers should design the actual value by considering their own design issues. The parameters in this section may not fit all conditions.

The transfer function of the feedback gain $H(s)$ for the CV loop is shown in Eq. (7.8), and g_{rd} and g_{va} represent the resistor divider and voltage amplifier gains, respectively.

$$H(s) = g_{rd} \times g_{va} \qquad (7.8)$$

The PWM control duty is generated by comparing the control voltage $\hat{v}_c$ with the ramp signal. As the ramp peak value V_{peak} is decided, the FM can be derived in Eq. (7.9) as a transfer function from the control voltage to the duty signal:

$$FM = \frac{1}{V_{peak}} \qquad (7.9)$$

The Bode plot in Figure 7.11 illustrates the transfer function of the CV mode, with a double pole peaking at around 40 kHz. This peak is also discovered at the power stage of the typical voltage-mode control buck converter, because of the LC double pole of the output filter. According to this relation, the compensation of the CV loop can be designed directly by following the compensation method of a typical voltage-mode control buck converter, which is known as Type-III compensation. In fact, the CV loop is not the only loop to use Type-III compensation – the CC and CAS loops do so as well. Based on the results of their frequency responses, the Bode plots of these three loops show the same phenomenon: a double pole peaking around 40 kHz. Thus, a Type-III compensator is a wise choice for compensation.

Figure 7.17 shows a typical analog Type-III compensator that requires six passive components. The transfer function of the Type-III compensator is shown in Eq. (7.10). As shown in the derived transfer function, the Type-III compensator has three poles and two zeros. One pole is located at the origin to let the compensator act as an integrator at low frequency and provides high gain for the whole loop. The two zeros must be placed below the cutoff frequency to compensate the integrator and the double pole, which leads to excessive phase lag. The rest of the two poles are adopted to eliminate high-frequency noise. These two poles ensure that the magnitude decreases continuously after the crossover of 0 dB.

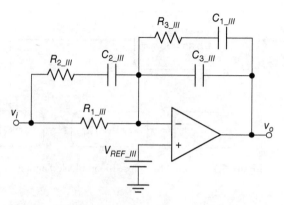

Figure 7.17 Type-III compensator

Table 7.2 Parameters of the demonstrated Type-III compensator

R1_III	200 kΩ	C1_III	2000 pF
R2_III	7.5 kΩ	C2_III	130 pF
R3_III	20 kΩ	C3_III	51 pF

$$G_C(s) = \frac{1}{R_{1\_III} \times (C_{1\_III} + C_{3\_III})} \times \frac{[s(R_{3\_III} \times C_{1\_III}) + 1][s((R_{1\_III} + R_{2\_III}) \times C_{2\_III}) + 1]}{s[s(R_{2\_III} \times C_{2\_III}) + 1]\left[s\left(R_{3\_III} \times \frac{C_{1\_III} \times C_{3\_III}}{C_{1\_III} + C_{3\_III}}\right) + 1\right]}$$

(7.10)

$$\omega_{z1} = \frac{1}{R_{3\_III} \times C_{1\_III}}$$
$$\omega_{z2} = \frac{1}{(R_{1\_III} + R_{2\_III}) \times C_{2\_III}}$$
$$\omega_{p1} = \frac{1}{R_{2\_III} \times C_{2\_III}}$$
$$\omega_{p2} = \frac{1}{R_{3\_III} \times \frac{C_{1\_III} \times C_{3\_III}}{C_{1\_III} + C_{3\_III}}}$$

(7.11)

In practical designs, the cutoff frequency of a switching-type converter is usually set to at least 10 times less than the switching frequency. The switching frequency in the demonstrated simulation is 300 kHz. The cutoff frequency is thus suggested to fall within 10–30 kHz. The power stage of the CV loop also contributes to the two zeros, as derived in Eq. (7.2). These two zeros are higher than half of the switching frequency. Thus, the two poles adopted by the Type-III compensator must be placed below half of the switching frequency to eliminate the high-frequency noise. According to the above considerations and assuming $R_{1\_III} = 200$ kΩ, the parameters of the passive components can be calculated as shown in Table 7.2. The Bode plot of the designed Type-III compensator is illustrated in Figure 7.18. Using the above analysis, the closed-loop transfer function of the CV loop can be illustrated by Eqs. (7.2), (7.8)–(7.10). The Bode plot of the entire closed-loop gain is shown in Figure 7.19. As all Bode plots in this section are illustrated by HSPICE, example code is also provided for the reader.

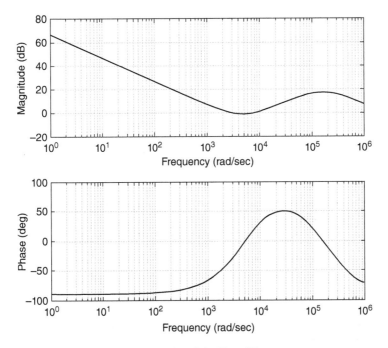

Figure 7.18 Bode plot of the Type-III compensator

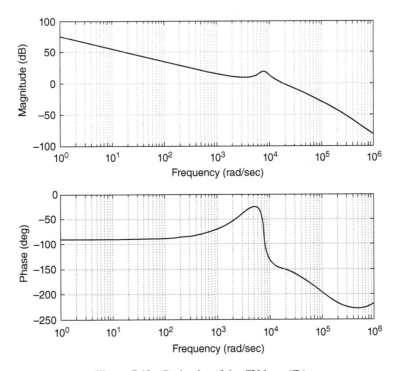

Figure 7.19 Bode plot of the CV loop (Z_L)

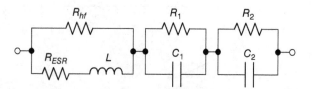

Figure 7.20 Equivalent model of a one-cell Li-ion battery model

Table 7.3 Parameters of the equivalent battery model

$R1$	13.77 mΩ	$C1$	0.337672
$R2$	47.15 mΩ	$C2$	1.79935
R_{hf}	5.8 Ω	L	0.637 μH
R_{ESR}	65.18 mΩ		

In Section 7.2, the loading of the equivalent battery charger Z_L is modeled as a resistive load to simplify the calculation. Although the modeling in Section 7.2 properly describes the behavior of the three loops, the actual small signal of the battery is still necessary in the last step of modeling. Figure 7.20 shows a typical one-cell Li-ion battery small signal model [3], with parameters provided in Table 7.3. This battery small signal model is modeled for SOC from 100 to 20%. Parameters may vary according to the different manufacturers. If readers are focused on a particular battery package, the characteristics of the battery model must be determined. In the model used in the present work, the impedance of the battery rapidly increases if the SOC is lower than 20%. This phenomenon can approximately be corrected by tripling the resistances of R_1 and R_2. By adopting this battery model, the closed-loop transfer function of the CV loop can be re-plotted as in Figure 7.21.

The other two loops, that is, the CC and CAS loops, use the same compensator and the same PWM control scheme. Thus, the only differences between the loops are the power stage, which has been derived, and the feedback sensor gain. These two loops sense the battery charging current and the adaptor (input) current by the sensing resistors R_{SADP} and R_{SBAT}, respectively. The transfer function of the feedback sensor gain can be derived as in Eq. (7.12), where R_{sense} and g_{ca} represent the sensing resistor and the gain of the current amplifier, respectively. $T_{filter}(s)$ is adopted to filter high-frequency harmonics that occur when the current-sensing topology is occupied. The simulation values are shown in Eq. (7.13). The CC and CAS loops share the same value because their sensing procedures are highly similar. The two poles of the filter are set to 60 and 150 kHz.

$$H(s) = R_{sense} \times g_{ca} \times T_{filter}(s)$$

$$R_{sense} = R_{SADP} = R_{SBAT}$$

$$T_{filter}(s) = \frac{1}{\left(\dfrac{s}{\omega_{f1}}+1\right)\left(\dfrac{s}{\omega_{f2}}+1\right)} \tag{7.12}$$

$$H(s) = 10\,\text{m} \times 40 \times T_{filter}(s) \tag{7.13}$$

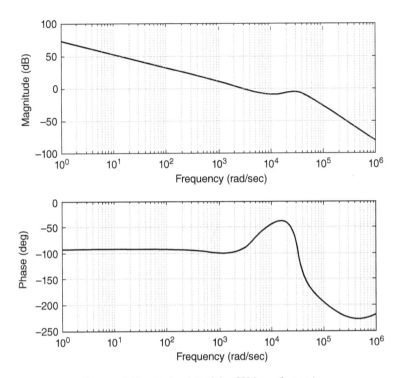

Figure 7.21 Bode plot of the CV loop (battery)

Using the above results, the closed-loop transfer function of CC and CAS loops can be illustrated by Eqs. (7.4), (7.9), (7.10), and (7.12) and by Eqs. (7.6), (7.9), (7.10), and (7.12), respectively. The Bode plots of the entire closed-loop gain are shown in Figures 7.22 and 7.23. As mentioned previously, the CV, CC, and CAS loops share the same Type-III compensator. The parameters of the compensation are designed based on the behavior of the CV loop, which differs from the CC and CAS loops. Fortunately, the position of the double poles remains basically at the same frequency. Based on the results of all three loops, the Bode plots show that the cutoff frequency of the CV loop moves to 2 kHz after adopting the battery model, whereas the cutoff frequency of the CC and CAS loops is within 10–30 kHz, which is the ideal interval for the demonstrated design. If the compensator is changed to let the cutoff frequency of the CV loop fit the ideal interval, this change may increase the cutoff frequency of the CC and CAS loops to over 30 kHz, which is an unwanted value as it is too close to the switching frequency. To share the same compensator and lower costs by reducing 18 components to 6, some compromise between the actual and ideal values must be made. In this case, although the bandwidth of the CV loop is sacrificed, the lowering of cost achieved by sharing the compensator is beneficial to the present system. However, the transfer function of the CAS loop shown in Figure 7.23 does not perfectly fit the actual response. The difference obtained is attributed to the superposition in Eq. (7.6) that cannot be emulated using HSPICE, hence there is a missing zero at high frequency. The lack of this zero does not significantly affect the analysis. Without this zero, the frequency response is poorer than the original value, which means that if the results shown in Figure 7.23 are suitable, then the actual results must be suitable as well.

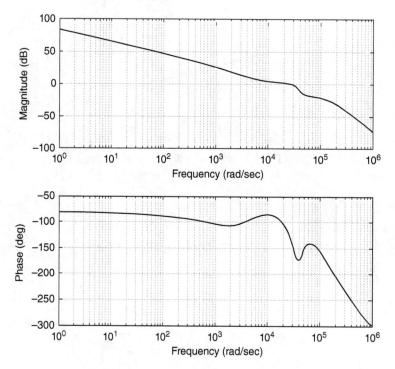

Figure 7.22 Bode plot of the CC loop (battery)

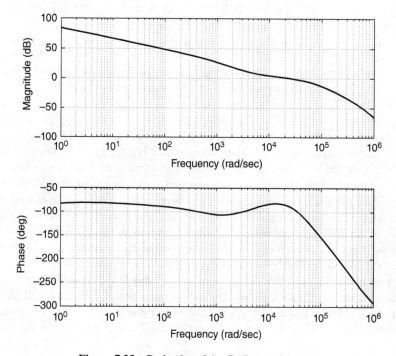

Figure 7.23 Bode plot of the CAS state (battery)

7.4 Simulation with PSIM

As the small signal of the power stage and the transfer function of the closed loop have been analyzed in Sections 7.2 and 7.3, respectively, the complete battery charger system can be set up through these analyses. In an integrated circuit, the most reliable simulating tool is HSPICE. However, constructing a fully functioning battery charger takes time and a significant amount of effort. Another issue is the simulation time: at least 12 h is necessary to simulate a well-designed battery charger that only includes the CV and CC modes through HSPICE. A faster way to understand the transient behavior of a battery charger is through a simple SPICE tool. In this section, the demonstrated simulation is simulated by PSIM.

The basic charger system shown in Figure 7.24 includes CC, CV, and CAS loops. The modeling of the battery model is noticeably replaced by a capacitor and a resister representing the

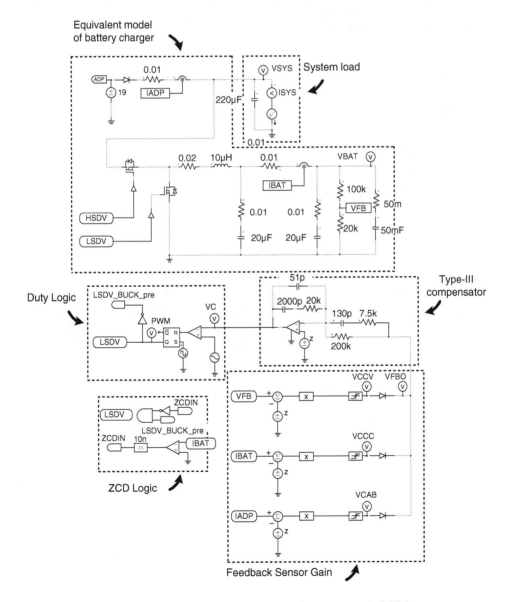

Figure 7.24 Schematic of the battery charger system in PSIM

inner impedance of the battery. The battery model change introduced in Figure 7.20 is attributed to efforts to improve the simulation time. Although the simulating tool has been replaced by a simple SPICE tool, the simulation time required to charge a real battery model is relatively long because of the large capacitance of the model. Thus, the original battery model is replaced by a smaller capacitor and an equivalent inner resistance. This change means that the simulation results do not completely fit the real battery charging behavior. However, the phenomena demonstrated in the present simulation provide readers with a preliminary concept of a battery charger with short simulation times.

The equivalent model of a switching-based charger introduced in Section 7.2 is presented in Figure 7.24, and all the parameters have been shown in Table 7.1, except for the replaced load and inner impedance, which are 50 μF and 55 mΩ, respectively. The initial voltage of the battery is set to 11.2 V and varies according to different battery packs. The system is emulated with current loading and an equivalent capacitance of 220 μF. The parameters in this chapter are only for demonstration and are not suitable for all possible cases.

The controller shown in Figure 7.24 is based on the results in Section 7.3; details are shown in Figure 7.25. The control loop of the CV loop is based on Eq. (7.8). The battery voltage V_{BAT} provides feedback in terms of a ratio of the voltage divider. This feedback voltage (V_{FB}, as shown in Figure 7.25) is subtracted by a summer, which acts like an error amplifier

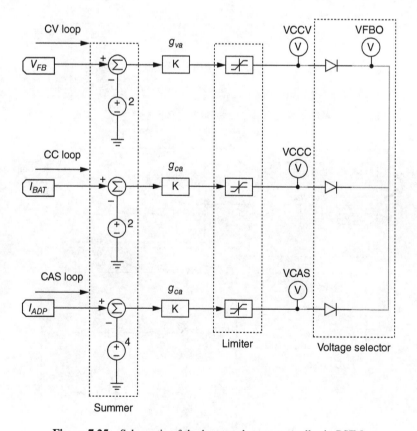

Figure 7.25 Schematic of the battery charger controller in PSIM

with unit gain. This voltage error is amplified by the designed voltage amplifier gain g_{va} and clamped by a limiter to emulate the limited voltage supply. The CC loop is based on the same procedure. The inductor current, which also represents the charging current, is sensed as I_{BAT}. The current is subtracted by a summer with the desired charging current, which is set to 2 A in the demonstration. This error will be amplified by the current amplifier gain g_{ca} and is also clamped by a limiter. The g_{va} and g_{ca} in Eqs. (7.8) and (7.12) are set to 6 and 40, respectively, and the corresponding values set in the demonstration are 600 and 40. The difference may be explained by the fact that the feedback sensor gain in Eq. (7.12) includes the sensing resistance, which often appears in a CMOS-based current-sensing circuit. The current sensor in PSIM is the actual charging current. Therefore, the sensing resistance is not included in the simulation, and the gain of the CC loop is much larger than that of the CV loop. The g_{va} increases to compensate for the difference in this sensing resistance. Another difference between Eq. (7.12) and our demonstration is the high-frequency filter. Since the current sensing in the simulation is ideal and does not lead to high-frequency harmonics, a filter need not be adopted. The feedback sensor gain of the CAS loop is identical to that of the CC loop in Eq. (7.12). The CAS loop differs from the CC loop in that the CAS loop senses the input current I_{ADP} and the reference current at the summer is the limit of the adapter.

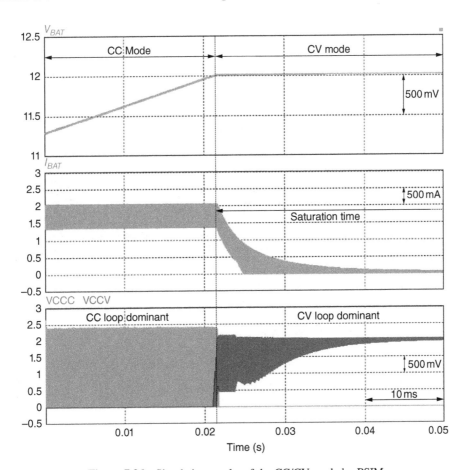

Figure 7.26 Simulation results of the CC/CV mode by PSIM

Using the three loops described above, the feedback signals of V_{BAT}, the charging current I_{BAT}, and the input current I_{ADP} are all well sensed. The three diodes subsequently act as voltage selectors that select the largest feedback signals in different states of the battery charger system. The selected feedback signal will be compensated by the Type-III compensator and create a control voltage. The control voltage will compare with the ramp signal by the comparator and create the PWM duty to the driver logic.

Figure 7.26 shows the simulation results obtained from the CC mode to the CV mode. The battery voltage is initially lower than the target set voltage, which is 12 V, in the simulation. Thus, V_{FB} remains low at the beginning of the simulation and the voltage selector, which is formed by diodes, selects the CC loop as the dominant loop (see the label in Figure 7.25, VFBO = VCCC). While the CC loop begins to charge the battery (I_{BAT} = 2 A), V_{BAT} also begins to increase until it reaches the set voltage (V_{BAT} = 12 V). V_{FB} also increases and dominates the voltage selector after reaching the target voltage at 2 V. Then, the CV loop begins to dominate the system (see the label in Figure 7.25, VFBO = VCCV), and the charging current begins to decrease. The charging current will continue to decrease, preventing the overvoltage of V_{BAT}, which is also called the saturation time in Section 7.1.

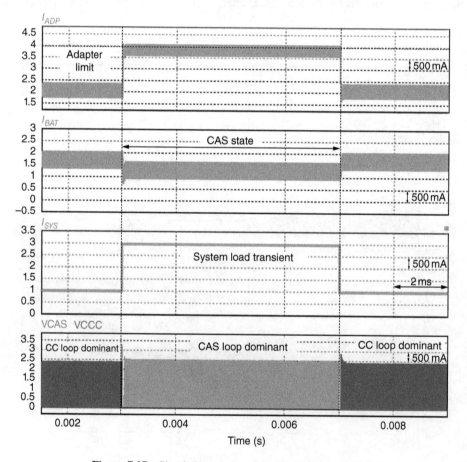

Figure 7.27 Simulation results of the CAS state by PSIM

Figure 7.27 shows the transient results of the CAS state. The system load I_{SYS} is set to feature an increased loading period. Once I_{SYS} is raised, the adapter current I_{ADP} is also raised until it reaches the pre-set adapter current limit (I_{ADP} = 4 A). Then, the CAS loop begins to dominate the system (see the label in Figure 7.25, VFBO = VCAS). The total current must be provided by the adapter below the current limit. Thus, the charging current must decrease to protect the adapter. After the system load completes the transient period, the charging current rises back to the original charging current (I_{BAT} = 2 A).

The above simulations used a simple SPICE tool to provide a brief understanding of the switched battery-charger topology. The useful parameters for the compensation and stability of the whole system are presented in a more detailed simulation using HSPICE in Sections 7.2 and 7.3. Moreover, simulation using PSIM provides good confirmation of the HSPICE simulation results. To validate the functionality of a circuit, results can be tested through PSIM.

7.5 Turbo-boost Charger

For a typical switching-based battery charger, four types of power delivery paths, namely the pure charge state, the direct supply state, the plug off state, and the CAS state, are available; these paths are shown in Figure 7.5. As stated in Section 7.1, the CAS state is adopted to satisfy the extra power consumption of a CPU with multi-core applications. However, CPU technology is advancing at a very fast pace. Increasing the CPU frequency may increase the processing speed for performing multiple tasks but exceed the thermal capability of the device for a short period of time. Moreover, increasing the CPU frequency also increases the power required by the CPU and the total power delivered from the input source. If the total input power exceeds the limitation of the input source, such as an adapter, then the adapter may crash or shut down by its own overloading protection. By adopting the CAS state, the battery charger can reduce the charging current to release the loading stress on the adapter. However, for more advanced CPU technologies, lowering the charging current may not satisfy the required power of the system. One way to solve this problem is to increase the power limitation of the adapter; this approach, however, will also increase the cost of the hardware.

A solution that provides enough power to the system and does not increase the cost of the hardware is required. The turbo-boost charger [5], which was developed by Texas Instruments, lets the input source and battery provide power to the system at the same time. In this way, the total power delivered to the system is increased without any increase in hardware cost. Essentially, the turbo-boost charger is related to a CPU technology developed by Intel called turbo-boost technology. This turbo-boost technology was developed to increase dynamic performance during multiple processing tasks with multi-core CPU applications and graphics processing units (GPUs). This technology allows the processor power to exceed the thermal design power (TDP) for a short time period and enhance the performance of the processor. For a regular adapter design, the power limitation is designed to fit the TDP level. The charger reduces the charging current while the adapter exceeds the limitation specified through the CAS state. However, once the charging current is lowered to zero, the processor must reduce its frequency to compromise the input power. Thus, a turbo-boost charger is developed to increase the total power provided to the system without compromising its performance.

The power delivery path of the CAS state and the turbo-boost charger are illustrated in Figure 7.28. The battery charger is operated as a buck converter during the CAS state since the CAS

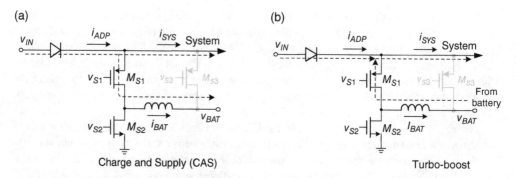

Figure 7.28 Power delivery paths of (a) the CAS state and (b) the turbo-boost

control reduces the charging current while the power required by the system is increased and the adapter reaches the power limitation. The turbo-boost charger is active while the system load is increased by a large amount, which means that the processor is operated under turbo-boost technology. The system loading is so high that even lowering the charging current to zero cannot provide enough power to the system. A straightforward way to solve this problem is to add another power path. In the present system, only one power source can provide extra power: the battery. Thus, the battery charger is operated as a boost converter, which converts the power from the battery to the system. The operation of the CAS state and the turbo-boost charger is illustrated in Figure 7.29. Since the turbo-boost function of CPUs can start at any time, the battery charger must automatically detect the load to start the turbo-boost charger.

After determining the fundamental concept of the turbo-boost charger technique, the next step is to achieve the turbo-boost charger. Texas Instruments recently released several patents describing the control methods of turbo-boost chargers [6, 7]; the details of these patents are not included here. The following analysis will focus on the conceptual design issue.

Both the CAS state and the turbo-boost charger monitor the input adapter current; however, the turbo-boost charger is designed to afford a much larger system load than the CAS state. Thus, the sensing path of the turbo-boost charger is the same as that of the CAS loop illustrated in Section 7.3. Once I_{ADP} exceeds the pre-set current limit, the battery charger will shut down the charging current and remain on standby. After some delay time, if I_{ADP} is still higher than the limit, then the charger system will confirm the start of the turbo-boost technology and enable the turbo-boost charger. To end a turbo-boost charger operation, the turbo-boost topology of the CPU is disabled, which leads to a decrease in system load and input current. Once I_{ADP} is decreased to the pre-set lower bound, the turbo-boost charger is disabled and the battery charger reverts back to a buck-type charger. However, some conditions must be considered to protect and ensure the stability of the battery charger system. For example, if the SOC of the battery is quite low, no extra power can be provided by the battery; therefore, the turbo-boost charger will be disabled. Temperature conditions and excess currents/voltages also influence the enable/disable mechanisms of the turbo-boost charger. The FSM of the turbo-boost charger is shown in Figure 7.30.

As illustrated in Figure 7.28(b), the turbo-boost charger delivers power from the battery to the system. Thus, how to control the power delivery during the turbo-boost period must be determined. The battery charger needs to control the charger system under a boost topology

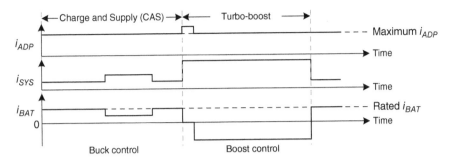

Figure 7.29 Timing diagram of the CAS state and the turbo-boost

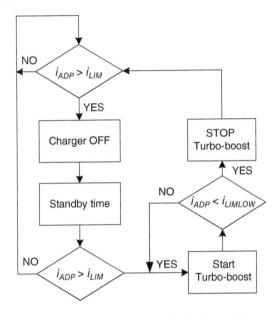

Figure 7.30 Simplified FSM of turbo-boost charger

instead of a buck topology. Essentially, both the CAS state and the turbo-boost charger control the charging current by sensing I_{BAT}. Switching the battery charger between buck and boost can be done by controlling the logic. Thus, the critical design issue is the method to use in selecting the amount of charging current during turbo-boosting. Under the CAS state, the charging current is automatically regulated by the closed loop of the CC and CAS loops. By contrast, the control method for the turbo-boost charger developed by Texas Instruments is a fairly smart and cost-efficient solution. After the charger confirms that the turbo-boost charger is enabled, the sensed charger current is locked to a desired value and the desired current is delivered to the system. The desired value is set by the user with an internal reference voltage, or the host carrying the digital message from the CPU and transferring it to a reference voltage. This control method can be classified as an open-loop control. Given that the charging current is not

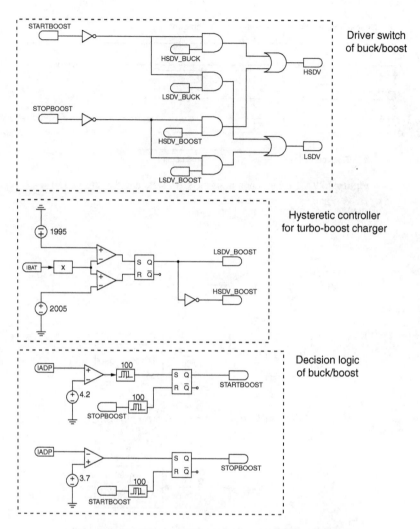

Figure 7.31 Schematic of a turbo-boost charger by PSIM

regulated by any feedback signal, the charging current will always remain at the desired value, unlike the CAS loop that the charging current will vary with the system load. Through this control method, the circuit complexity can be reduced efficiently and the sensing circuit can share with the CAS loop. Because of the open-loop control, considering the compensation, which reduces the need for extra components significantly, is no longer necessary.

The simulation of the turbo-boost charger is also established by PSIM. The basic charger is the same as those shown in Figure 7.24. The additional controllers of the turbo-boost charger are shown in Figure 7.31. The system load is set to feature a heavy-load transient period, which can emulate the behavior of the turbo-boost technology of the CPU. The simulation result is shown in Figure 7.32. While the system load increases, the input current I_{ADP} also increases. Since the increased loading is very large, I_{ADP} reaches the pre-set input limit (I_{LIM}, shown in

Switching-Based Battery Charger

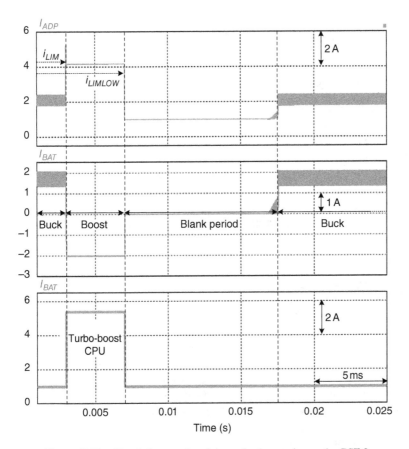

Figure 7.32 Simulation results of the turbo-boost charger by PSIM

Figure 7.30) of 4.2 A and enters the turbo-boost charger mode. The charging current I_{BAT} becomes negative, which means that the charge is delivered from the battery to the system through the boost path, as shown in Figure 7.28. As mentioned previously, the control topology during the turbo-boost period is an open-loop control. The approach here is to control I_{BAT} through a hysteretic control with a value of 2 A, as shown in Figure 7.31. Once the charger system enters into the turbo-boost charger mode, the logic will turn the driver signal from the regular charger to the hysteretic controller. After the load transient, I_{ADP} is lower than the pre-set lower limit of 3.7 A (as I_{LIMLOW}, shown in Figure 7.30), the turbo-boost charger is switched off, and the logic restores the controller to the regular charger loop. After the turbo-boost charger is switched off, a blank period ensues. This blank period is due to the CAS loop sensing I_{ADP} while the turbo-boost charger is switched on. Thus, the sensed feedback signal allows the VFBO to remain at a large value because of the capacitors in the Type-III compensator. After the logic turns the controller to the regular charger loop, the stored value in the CAS loop needs to decrease back to the regular value, thereby causing the blank period, and the charger can begin a regular function. The blank is about 1 ms in the simulation, which is a negligible time relative to the whole charging period. The simulation shown in Figure 7.31

focuses on the basic operations of the turbo-boost charger, which are simplified by neglecting the standby check and the protecting circuits.

7.6 Influence of Built-In Resistance in the Charger System

In Section 7.1, the influence of the ESR in the battery pack was shown in Figure 7.2. The ESR induces the error voltage V_{ESR}, which affects the sensed V_{BAT}. To be more accurate, the inherent resistance of the Li-ion battery charger includes several parasitic resistances (collectively called the built-in resistance (BIR), R_{BIR}), including contact resistance ($R_{CONTACT}$), fuse resistance (R_{FUSE}), PCB wire trace resistance (R_{PCB}), and ESR, as depicted in Figure 7.33. In a Li-ion battery charger system, R_{BIR} may vary from 100 to 500 mΩ according to the different types of Li-ion battery employed and the PCB layout. The BIR exerts a serious influence on the charging process. The error voltage drop V_{ESR} is enlarged to V_{BIR}, as shown in Eq. (7.14). The enlarged voltage drop may lead to a slow full charging time because the saturation time is prolonged.

$$V_{BAT} = V_{BIR} + V_{CELL}$$
$$V_{BIR} = I_{charge} \times R_{BIR} = I_{charge} \times (R_{CONTACT} + R_{FUSE} + R_{PCB} + ESR) \quad (7.14)$$

Figure 7.34 depicts the charging profile, including the typical CC and CV modes, under the BIR effect. During the charging procedure, V_{BAT} is the sum of the voltages across the BIR, V_{BIR} and V_{CELL}, which accurately represent the energy stored in the battery. However, the battery voltage can only be sensed at the point V_{BAT}. Owing to the BIR effect, early transition times occur from the CC to the CV mode. A too-early transition will cause a shortened period in the CC mode and a prolonged charging time since the charging current in the CV mode is much smaller than that in the CC mode. In other words, fast charging can be achieved if the CC-mode duration is extended.

To determine a proper transition point, named $V_{FULL,COMP}$, for fast charging, many previous works proposed cancelling the BIR effect through extending the CC charging time [1, 8]. If the BIR value can be estimated from the charging current, then the voltage drop across the BIR can be determined. Thus, an adequate transition point from the CC mode to the CV mode can be

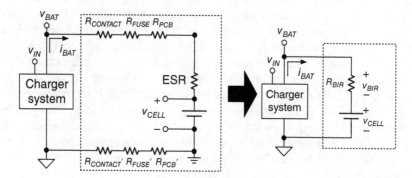

Figure 7.33 Equivalent parasitic resistances in the charger system

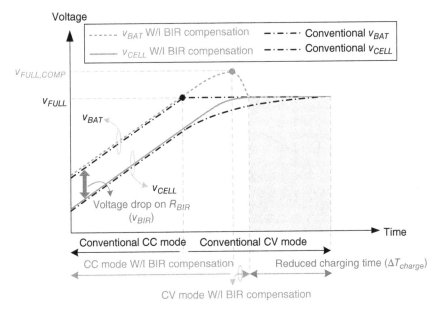

Figure 7.34 Charging profile with the BIR effect

derived precisely to extend the period of the CC mode and obtain a large energy and fast charging. By generating a compensated voltage $V_{FULL,COMP}$, instead of the conventional set voltage V_{FULL}, the CC-mode period is extended, as shown in Figure 7.34.

Prior techniques for alleviating the BIR effect were done through one-time BIR detection in the charging procedure. Unfortunately, the BIR effect is temperature-, voltage-, and environment-dependent. As shown in Figure 7.35(a), R_{BIR} in a Li-ion battery varies at different environmental conditions and especially so at different temperatures. During the charging process, large currents flow into the battery and generate a large amount of heat on the R_{BIR}. The R_{BIR} is temperature-dependent, thus, one-time detection may result in an incorrectly compensated value, and alleviation of the BIR effect is not obvious. More seriously, improperly compensated BIR effects may result in overcharging, even permanent damage, or explosion of the battery. In other words, one-time detection cannot handle uncertain environmental variations during the charging process. That is, adaptive compensation for R_{BIR} tracks the variation of temperature and achieves precise compensation, as depicted in Figure 7.35(b). Therefore, real-time detection of the BIR is needed to accurately alleviate the BIR effect [9].

As shown in Figure 7.6, I_{BAT} in the CAS state is unpredictable because I_{SYS} varies with the loading system. This variation may cause compensation of the BIR effect to become more complicated during the charging process. To compensate the BIR effect, the charger system must also take the charging current information into consideration. Conventional one-time detection methods, as shown in Figure 7.36(a), lead to incorrect predictions when the charger operates in the CAS state [1, 8], which may result in overcharging and even permanent damage or explosion of the battery. The charging current variation in the CAS mode and the BIR variation under different environmental conditions highlight the need to adjust the compensation voltage $V_{FULL,COMP}$ adaptively. This real-time detection gathers information on the recent charging current and continuously monitors the BIR value to generate an accurate and adaptively

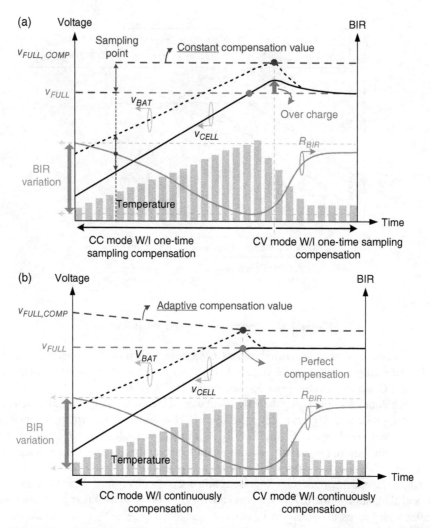

Figure 7.35 BIR effect due to BIR variation: (a) one-time BIR detection if the BIR changes; (b) real-time BIR detection if the BIR changes

compensated $V_{FULL,COMP}$. Figure 7.36(b) shows that $V_{FULL,COMP}$ follows the variation in charging current. An optimal value can be achieved to accurately compensate the BIR effect and shorten the charging time of a Li-ion battery.

7.7 Design Example: Continuous Built-In Resistance Detection

As illustrated in Section 7.6, the BIR affects the charging time. Even with the technique of compensating the BIR, one-time sampling still presents the risk of overcharging because of variations in temperature and charging current. Thus, continuous BIR detection (CBIRD) is

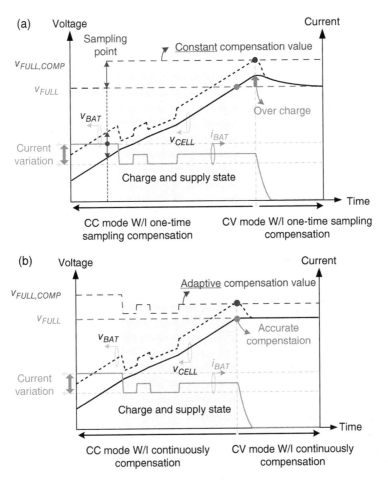

Figure 7.36 BIR effect due to charging with current variations: (a) one-time BIR detection versus charging current variation; (b) continuous BIR detection versus charging current variation

presented in [2] to eliminate the effect of the BIR and speed up the charging procedure. The proposed charging system of [2] is shown in Figure 7.37.

7.7.1 CBIRD Operation

The output voltage V_{BAT} is composed of two parts, namely the DC voltage and the AC ripple. Through a mathematical derivation, V_{BAT} can be expressed as:

$$V_{BAT} = V_{CELL} + I_{BAT} R_{BIR} \qquad (7.15)$$

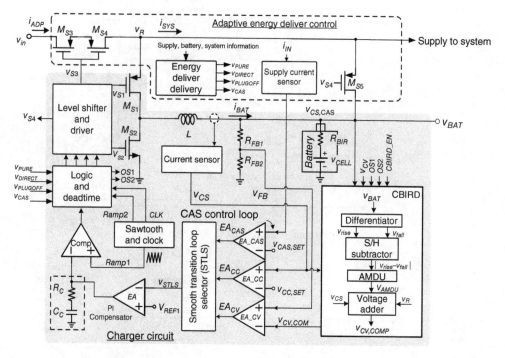

Figure 7.37 Proposed switching-based charger system of [2]

The ripple is generated by the charging current I_{BAT} flowing through R_{BIR}. The output voltage ripple on the battery can be seen approximately as ΔV_{CV} in Eq. (7.16), which is determined by the BIR under the inductor current I_{BAT}:

$$\Delta V_{CV} = I_{BAT} R_{BIR} \tag{7.16}$$

To compensate the BIR effect, ΔV_{CV} should be derived first. Then, the compensated $V_{CV,COMP}$, which is shown in Figure 7.37, can be generated continuously. The proposed detection and compensation method is not interrupted during the charging process compared with previous techniques [1]. The proposed CBIRD method derives ΔV_{CV} from the V_{BAT}'s AC ripple, which will not affect the charging process. Continuous detection can achieve adaptive compensation.

The switching-based charger operates similarly to a CCM buck converter [8]. Two inductor current slopes in the charging and discharging phases constitute the AC voltage ripple at V_{BAT}, which can be used to disclose information on the BIR. Through differentiation, the DC part of V_{BAT} is eliminated, and the slope of the voltage ripple across BIR could be derived as in Eq. (7.17), where V_R indicates the voltage value after the input voltage V_{IN} across two cascading power switches M_{S3} and M_{S4}, as shown in Figure 7.37.

$$V_{rise} = K_1 \frac{dV_{BAT}}{dt} = K_1 R_{BIR} \frac{V_R - V_{BAT}}{L}$$

$$V_{fall} = K_1 \frac{dV_{BAT}}{dt} = -K_1 R_{BIR} \frac{V_{BAT}}{L} \qquad (7.17)$$

V_{rise} and V_{fall} are the inductor current slopes in the charging and discharging phases, respectively. K_1 is the coefficient induced by the differentiator circuit. The difference between V_{rise} and V_{fall} includes information on R_{BIR}:

$$|V_{rise} - V_{fall}| = K_1 R_{BIR} \frac{V_R}{L} \qquad (7.18)$$

Moreover, the charging current I_{BAT} can be identified by the current-sensing circuit in Figure 7.37 to derive ΔV_{CV}. The sensing ratio is designed as K_2 in this work; that is, $V_{CS} = K_2{}^*I_{BAT}$. Thus, the function of ΔV_{CV} can be derived by Eq. (7.19) through the product of the current-sensing signal V_{CS} and $|V_{rise} - V_{fall}|$. Owing to the constant value of K, V_R, and L, the value of ΔV_{CV} can be calculated precisely through the estimation of V_{CS}, V_{rise}, and V_{fall}. Adding ΔV_{CV} with a scaling factor $R_{FB2}/(R_{FB1} + R_{FB2})$ to the original V_{CV}, the precise transition point $V_{CV,COMP}$ is derived as in Eq. (7.20):

$$V_{CS} \times |V_{rise} - V_{fall}| = K_2 \cdot I_{BAT} \times K_1 \cdot R_{BIR} \cdot \frac{V_R}{L} = \Delta V_{CV} \times K \frac{V_R}{L} \qquad (7.19)$$

where $\Delta V_{CV} = I_{BAT} \cdot R_{BIR}$ and $K = K_1 \cdot K_2$

$$V_{CV,COMP} = \frac{R_{FB2}}{R_{FB1} + R_{FB2}} \cdot V_{FULL,COMP} = \frac{R_{FB2}}{R_{FB1} + R_{FB2}} \cdot (V_{FULL} + \Delta V_{CV})$$
$$= V_{CV} + \frac{R_{FB2}}{R_{FB1} + R_{FB2}} \times I_{Charge} \cdot R_{BIR} \qquad (7.20)$$

The proposed CBIRD circuit is designed to implement Eqs. (7.15)–(7.20). Figure 7.38 depicts the circuit flow to realize the CBIRD function. It includes a differentiator, a sample-and-hold (S/H) subtractor, an analog multiplication/division unit (AMDU), and a voltage adder. First, the differentiator is used to obtain the AC ripple of V_{BAT} and V_{rise} and V_{fall} in Eq. (7.17) after filtering out the DC voltage. The S/H subtractor samples V_{rise} and V_{fall}, and holds their difference value in Eq. (7.18). Then, through the AMDU circuit, the multiplication function multiplies V_{CS} by $|V_{rise} - V_{fall}|$ and the division function derives ΔV_{CV} in Eq. (7.19). Through the voltage adder, ΔV_{CV} can be scaled down by the voltage divider ratio and then

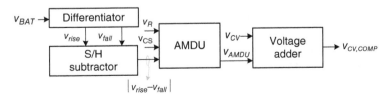

Figure 7.38 Implementation of the CBIRD circuit

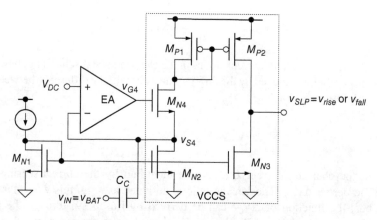

Figure 7.39 Proposed differentiator circuit

added to V_{CV} to accomplish $V_{CV,COMP}$ in Eq. (7.20). Finally, $V_{CV,COMP}$ is used as the reference voltage to dynamically adjust the mode transition point for compensating the BIR effect.

7.7.2 CBIRD Circuit Implementation

7.7.2.1 Differentiator

The differentiator in Figure 7.39 uses the coupling capacitor C_C to block the DC components and get the AC information from V_{BAT}. The error amplifier and the transistor M_{N4} maintain a DC level at the input node of V_{S4}. Therefore, only the AC voltage information is converted to a current signal through the VCCS circuit. The rising and falling voltage ripple information, V_{rise} and V_{fall}, can be obtained at the node of V_{SLP}. The differentiator outputs V_{rise} and V_{fall} correspond to the charging and discharging phases, respectively. V_{rise} and V_{fall} follow Eq. (7.17). Given that the differentiator needs a settling time to ensure that V_{SLP} changes its value between V_{rise} and V_{fall}, the sampling time of the S/H subtractor should therefore match the operation of the differentiator to obtain correct and precise V_{rise} and V_{fall}. The common mode level of V_{SLP} may vary because of the process, voltage, and temperature (PVT) variations. However, the common level will not affect the differentiation results since the only difference between V_{rise} and V_{fall} will be sampled and calculated as BIR information.

7.7.2.2 Sample-and-Hold Subtractor

In Figure 7.40, the S/H subtractor circuit samples the differentiator's outputs in charging and discharging phases. By selecting the sampling path during charging and discharging phases, the subtraction function in Eq. (7.18) can simultaneously be achieved. As shown in Figure 7.41, the sampling clocks, $OS1$ and $OS2$, control the MOSFETs M_{N1} to M_{N4} in sampling V_{SLP}. In the charging phase, M_{N1} and M_{N2} turn on, and V_{rise} is stored in C_{FLY}. In the discharging phase, M_{N3} and M_{N4} turn on, and C_{OUT} stores the value $V_{fall} - V_{rise}$, which is equal to $|V_{rise} - V_{fall}|$ according to the Kirchhoff voltage law.

Switching-Based Battery Charger

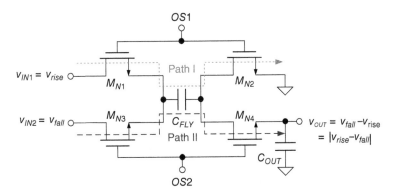

Figure 7.40 Proposed S/H subtractor circuit

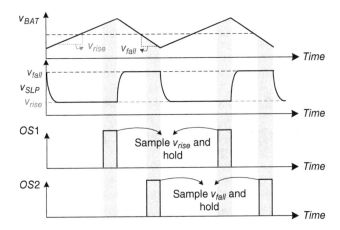

Figure 7.41 Timing diagram for the S/H subtractor circuit

7.7.2.3 Analog Multiplication/Division Unit

Figure 7.42(a) shows the basic structure and concept of the proposed AMDU circuit, which is composed of two V-to-I converters (V-to-I_1 and V-to-I_2), one comparator, and one switch S_{SW}. V-to-I_1 translates V_{IN1} to K_1I_1 to charge C_1 with the charging slope m_1 in Eq. (7.21). Similarly, V-to-I_2 translates V_{IN2} to K_2I_2 to charge C_2 with the charging slope m_2 in Eq. (7.21):

$$m_1 = \frac{I_1}{C_1} = \frac{K_1 V_{IN1}}{C_1} \text{ and } m_2 = \frac{I_2}{C_2} = \frac{K_2 V_{IN2}}{C_2} \quad (7.21)$$

K_1 and K_2 are the coefficients for V-to-I_1 and V-to-I_2, respectively. Once the voltage stored in C_1 approaches the value of V_{IN3}, the output of the comparator becomes low to turn off the switch S_{SW}. Hence, a charging time Δt related to V_{IN1} and V_{IN3} is obtained to realize the division function described in Eq. (7.22). Consequently, the charging time of C_2 can be determined by

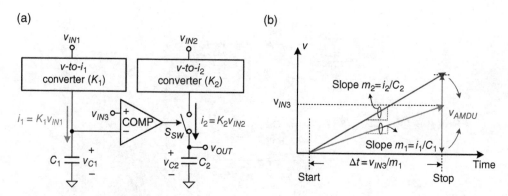

Figure 7.42 (a) Concept of the proposed AMDU circuit. (b) Timing diagram of the AMDU circuit

Δt. Finally, V_{OUT} of the AMDU can be calculated by Eq. (7.23) to complete the function. The timing diagram of the AMDU is depicted in Figure 7.42(b).

$$\Delta t = \frac{V_{IN3}}{m_1} = \frac{C_1 V_{IN3}}{K_1 V_{IN1}} \tag{7.22}$$

$$V_{OUT} = m_2 \times \Delta t = \left(\frac{K_2 V_{IN2}}{C_2}\right) \times \left(\frac{C_1 V_{IN3}}{K_1 V_{IN1}}\right) = \frac{V_{IN2} \times V_{IN3}}{V_{IN1}} \tag{7.23}$$

The AMDU circuit is illustrated in Figure 7.43(a). The cascode current mirror is used to avoid channel length modulation and ensure accuracy. S_1 and S_2 are switches used to reset the voltage on the capacitors after each calculation. By setting the same value and careful layout matching for the two V-to-I converters and capacitors, $C_1 = C_2$ and $K_1 = K_2$ can be guaranteed. The AMDU circuit output voltage is simply composed of V_{IN1}, V_{IN2}, and V_{IN3}, as shown in Eq. (7.23). Consequently, V_{AMDU} in Eq. (7.24) can be derived by substituting the parameters V_{CS}, $|V_{rise} - V_{fall}|$, and V_R for V_{IN1}, V_{IN2}, and V_{IN3} in Eq. (7.23). To get ΔV_{CV}, V_{AMDU} should be modified by the voltage divide ratio through the voltage adder to correctly reflect the BIR effect:

$$V_{AMDU} = V_{CS} \times \frac{\left(|V_{rise} - V_{fall}|\right)}{V_R} = \Delta V_{CV} \times \frac{K}{L} \tag{7.24}$$

7.7.2.4 Voltage Adder

The voltage adder is depicted in Figure 7.43(b), and includes a unit-gain buffer, a V-to-I circuit, and a cascode current mirror. The unit-gain buffer EA2 is used to keep V_{CV}. The V-to-I circuit transfers V_{AMDU} to a current signal for easy signal processing. Afterward, the current signal is

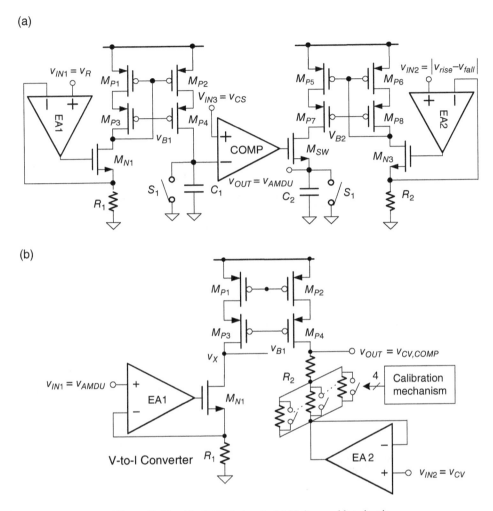

Figure 7.43 (a) AMDU circuit. (b) Voltage adder circuit

converted back to a voltage signal across the resistor R_2 through the current mirror. R_1 and R_2 are used to provide a scaling ratio β:

$$V_{CV,COMP} = V_{CV} + \beta V_{AMDU} = V_{CV} + \Delta V_{CV} \text{ where } \beta = \frac{R_2}{R_1} = \frac{R_{FB2}}{R_{FB1} + R_{FB2}} \cdot \frac{L}{K} \qquad (7.25)$$

The relationship between the voltage divider, the inductor value, and the constant K in Eq. (7.19) can be established by setting the value of β, which can be adjusted by tuning the value of R_2 through 4-bit calibration digital codes to get a precise result. To solve the unavoidable mismatch in circuits, a trimming process is added to our design to improve performance. At the beginning of system operation, a known ESR value is used to adjust the value of the V-to-I resistor in the AMDU output. Thus, non-ideal effects can effectively be eliminated.

Besides, because of the large difference between V_X and $V_{CV,COMP}$, the cascode current mirror, M_{P1}–M_{P4}, is used to avoid channel length modulation and ensure high accuracy in this operation. Finally, the voltage adder implements Eq. (7.25) and generates the compensated value $V_{CV,COMP}$. This will be used as a new reference voltage for the charger transiting from CC to CV mode.

The BIR compensation shortens the charging time, but may induce a stability issue on the transition between CC and CV modes. Most previous work does not discuss this issue [1], which is related to the release of ΔV_{CV}. When the charging mode enters the CV mode, ΔV_{CV} should be removed from $V_{CV,COMP}$. Sudden changes in the transition voltage result in rapid transitions or even oscillations among different charging modes. Under the proposed CBIRD, ΔV_{CV} releases smoothly and appropriately. According to Eq. (7.20), ΔV_{CV} includes information on the charging current. Once the charger system enters the CV mode, the charging current I_{BAT} is gradually reduced by the CV loop regulation. Therefore, ΔV_{CV} decreases smoothly with decreasing inductor current and $V_{CV,COMP}$. Hence, a smooth transition from the CC mode to the CV mode can be achieved.

7.7.3 Experimental Results

To validate the CBIRD circuit, a test signal emulating the voltage ripple is used as the input voltage V_{BAT}, as shown in Figure 7.44. The voltage ripple is about 20 mV when the rated charging current is 2 A and the BIR is equal to 100 mΩ. Likewise, the voltage ripple is 60 mV when the BIR is 300 mΩ. According to Eq. (7.25), ΔV_{CV} should be 100 and 300 mV, respectively, because the feedback resistor ratio m is 0.5. In the measurement results, the values of ΔV_{CV} are

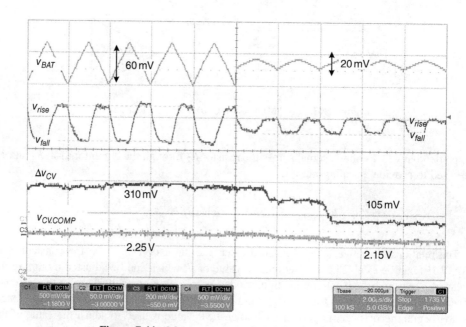

Figure 7.44 Measurement results of the CBIRD circuit

Switching-Based Battery Charger

105 and 310 mV, respectively. Thus, the actual values for $V_{CV,COMP}$ when V_{CV} is 2.1 V are 2.25 and 2.15 V, respectively. The error percentage of $V_{CV,COMP}$ is less than 0.5% in the proposed technique.

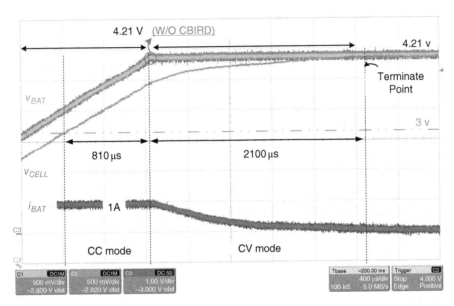

Figure 7.45 Charging process during the CC and CV modes without the CBIRD technique

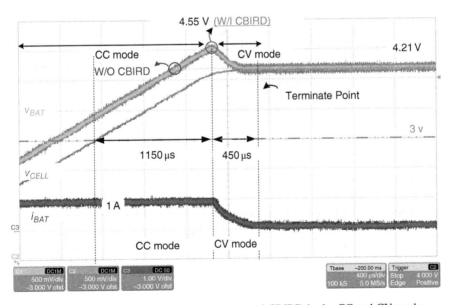

Figure 7.46 Charging process with the proposed CBIRD in the CC and CV modes

Figure 7.45 illustrates the conventional charging process without the proposed CBIRD technique. The charging current in the CC mode is set to 1 A, and the full voltage is 4.2 V. The STLS circuit ensures loop stability between the transitions of the two modes. However, transitioning from the CC mode to the CV mode under the control of the charger is too early because of the BIR effect. Consequently, the charging time is prolonged because of the shortened period of the CC mode. As shown in Figure 7.45, the charger transits from the CC mode to the CV mode when V_{BAT} reaches its rated voltage of 4.2 V. The real voltage of the battery, V_{BATO}, is not fully charged, thus the charger continuously delivers energy to the battery with a decreasing charging current in the CV mode. The charger requires 2100 μs in the CV mode until the charging process terminates.

By contrast, Figure 7.46 shows the charging process with the CBIRD function. The measurement conditions are identical to those presented in Figure 7.45. Because of the CBIRD technique, the transition point of the CV mode is modified from 4.21 to 4.55 V. Therefore, the operating period in the CC mode is extended by nearly 350 μs. In other words, the BIR effect is compensated by the CBIRD technique. Compared with the conventional charger in Figure 7.45, the proposed charger could reduce the charging time of 1300 μs. In the measurement results, the emulated capacitor charges from 0 V. However, in practical applications, the lower limit of the Li-ion battery voltage is often set to 3 V to protect the battery from being overdischarged. The charging time in the CC mode from 3 to 4.2 V is approximately 800 and 1150 μs without and with the proposed CBIRD technique, respectively. That is to say, the proposed charger reduces the total charging time by 44.8% from 2900 to 1600 μs.

References

[1] Lin, C.-H., Hsieh, C.-Y., and Chen, K.-H. (2010) A Li-ion battery charger with smooth control circuit and built-in resistance compensator for achieving stable and fast charging. *IEEE Transactions on Circuits and Systems I (TCAS-I)*, **57**(2), 506–517.

[2] Huang, T.-C., Peng, R.-H., Tsai, T.-W., *et al.* (2014) Fast charging and high efficiency switching-based charger with continuous built-in resistance detection and automatic energy deliver control for portable electronics. *IEEE Journal of Solid-State Circuits*, **49**(7), 1580–1593.

[3] Texas Instruments (2006) Closed-Loop Compensation Design of a Synchronous Switching Charger Using bq2472x/3x, SLUA371, Application Report.

[4] Erickson, R.W. and Maksimovic, D. (2001) *Fundamentals of Power Electronics*, 2nd edn. Kluwer Academic Publishers, Norwell, MA.

[5] Texas Instruments (2012) Turbo-Boost Charger Supports CPU Turbo Mode, SLYT448, Analog Applications Journal.

[6] Ye, M., Stair, R., Chen, S., *et al.* (2012) Control method of hybrid power battery charger. US Patent No. 20120139345 A1, filed May 12, 2011 and issued June 7, 2012.

[7] Ye, M., Qian, J., Chen, S., and Stair, R. (2012) Method for limiting battery discharging current in battery charger and discharger circuit. US Patent No. 20120139500 A1, issued June 7, 2012.

[8] Saint-Pierre, R. (2000) A dynamic voltage-compensation technique for reducing charge time in Li-ion batteries. *Proceedings of the 15th Annual Battery Conference on Applications and Advances*, Long Beach, CA, January 11–14, pp. 179–184.

[9] Peng, R.-H., Tsai, T.-W., Chen, K.-H., *et al.* (2013) Switching-based charger with continuously built-in resistor detector (CBIRD) and analog multiplication-division unit (AMDU) for fast charging in Li-ion battery. *Proceedings of the 39th European Solid-State Circuits Conference (ESSCIRC)*, Bucharest, September 16–20, pp. 157–160.

8

Energy-Harvesting Systems

8.1 Introduction to Energy-Harvesting Systems

Research on energy-harvesting applications has gained importance in the last decade [1–10]. The most commonly discussed applications include wireless sensor nodes for healthcare, embedded or implanted sensor nodes for medical applications, tire pressure-monitoring systems for automobiles, battery-charging devices for long-sustainability systems, security or guard systems for homes, and environmental condition-monitoring systems. Figure 8.1 [11] shows a wireless healthcare bio-wireless sensor network (WSN) in a hospital used to monitor the vitals of patients [12]. Each patient wears several sensor nodes in several small monitoring systems, such as an electrocardiogram (ECG). Each system is responsible for specific physiological signals. The central health server can collect the sensed data through wireless communication.

The popularity and widespread use of WSNs may be ascribed to several factors. The first factor is the progress of silicon technology in micrometer and even down to nanometer scales. According to Moore's law [13], the size of transistors on integrated circuits follows a decreasing trend of 0.7× every one and a half years. The gate count on a chip, as well as the operating speed and processing performance, increase significantly. The supply voltage on a silicon chip is also greatly reduced because of thinner oxides in advanced nanometer technologies. Smaller transistors or passive components result in lower parasitic effects and, thus, further reduce power loss. For a scale reduction by factor α ($\alpha > 1$), the power consumption of a shrunk circuit that performs the same task can be reduced by $(1/\alpha)^3$ [1]. The second factor is the progress of RF technology. RF technologies, especially low-power transmission networks such as Zigbee [14] and Bluetooth low energy (BLE) [15], facilitate data transmission and WSN construction. The third factor is the system integration of SoC and a heterogeneous system. A highly integrated SoC greatly decreases the requirements of external components, thereby reducing the volume and weight of the entire system. Heterogeneous integration technologies integrate

---

Power Management Techniques for Integrated Circuit Design, First Edition. Ke-Horng Chen.
© 2016 John Wiley & Sons Singapore Pte Ltd. Published 2016 by John Wiley & Sons Singapore Pte Ltd.

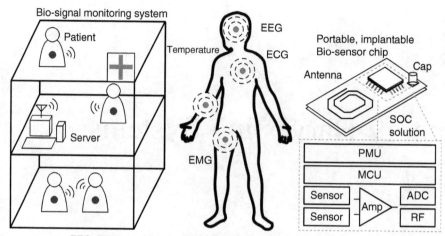

Figure 8.1 Wireless healthcare bio-WSN system

EEG: Electroencephalogram; ECG: Electrocardiogram; EMG: Electromyogram; PMU: Power management unit; MCU: Micro control unit.

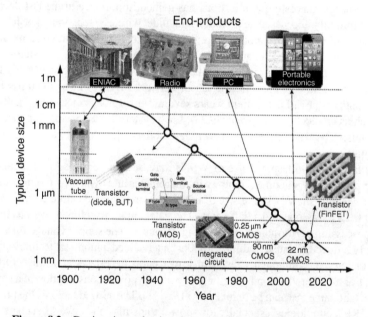

Figure 8.2 Device size reduction through silicon integration technology

micro-electro-mechanical systems (MEMSs), bio-medical or chemical sensors, display devices, and micro-electronic circuits through such technologies as through-silicon via (TSV) and three-dimensional (3D) IC packaging [16]. Consequently, highly integrated systems greatly reduce sensor node size. Figure 8.2 shows the chronological development of size and power reduction because of advanced silicon and system integration technologies.

Energy-Harvesting Systems

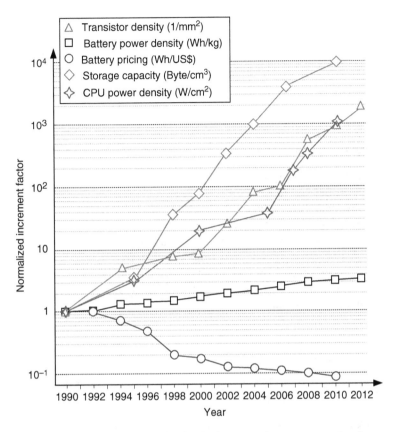

Figure 8.3 Improvement of technologies versus battery energy density

In contrast to rapidly growing silicon and SoC technologies, energy storage technologies do not develop at the same rate, as shown in Figure 8.3 [3]. Developments in battery technologies primarily focus on mass production and packaging, but the materials used in batteries improve slowly. Thus, the prices of batteries drop significantly, but energy density at the same volume increases slowly. Such limitation somehow limits the lifetimes of battery-based portable electronics or sensors.

Technological developments allow low energy consumption, compact size, high performance, and wireless communication in sensor networks or portable electronics. However, operating lifetime remains a challenge. If the quantity and energy density of energy storage devices cannot be improved significantly, the system lifetime can also be prolonged by recharging energy storage devices to allow the system to self-power during normal or standby operation [6–10]. Harvesting energy from the environment is the most suitable choice for the aforementioned demands.

Energy harvesting involves energy source analysis, storage devices, energy conversion circuit design, and material physics [1–9]. Recent developments include components and devices at micro- and macro-scales, encompassing materials, electronics, and integration. The popularity of energy-harvesting technologies has benefited from low-power electronic technology.

Owing to the low-power requirements, low harvested energy in microwatts can be used as energy. Low-power requirements also reduce the size of harvesting source that is capable of generating sufficient energy. An energy-harvesting system has three important components: energy converter (energy source/generator), harvesting circuit, and energy storage device. Energy sources and generators are introduced in Section 8.2. Some specific circuits for different applications are classified and discussed in Section 8.3.

8.2 Energy-Harvesting Sources

As shown in Figure 8.4, the purpose of energy harvesting is to reuse or recycle energy from the environment and artificial facilities or creatures. Energy sources are divided into several types: kinetic energy, thermal energy, and electromagnetic radiation [1, 5].

Kinetic energy is one of the most readily available energy sources from living creatures and the environment. This section provides a brief explanation and the operating principles of conversion of the electrical energy converted from kinetic energy by different transducers. The main principle of kinetic energy harvesting is the displacement of a moving part or the mechanical deformation of a structure. Displacement or deformation can be converted to electrical energy through different methods. Commonly used kinetic energy transducers, namely

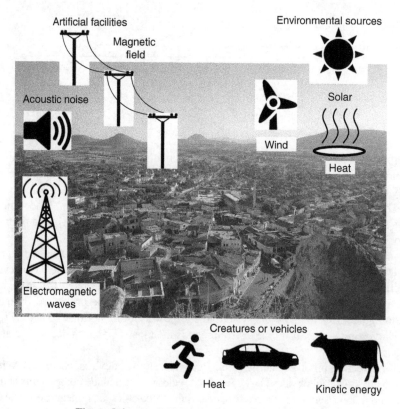

Figure 8.4 Available energy from the environment

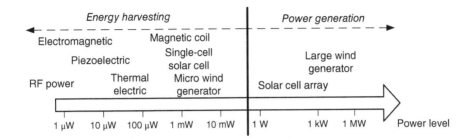

Figure 8.5 Energy scales of various sources

magnetic induction, piezoelectric, and electrostatic transducers, are explained in the succeeding Sections 8.2.1–8.2.3, respectively [1, 5, 17–20]. Wind is also a kinetic energy-induced source obtained from air movement. Wind power transducers, which convert wind power into electrical energy, are introduced in Section 8.2.4 [1, 21–24].

Thermal energy is a widely studied and commercialized energy for low-power electronic devices; thermoelectric systems can generate energy at a scale smaller than 100 μW [1, 5, 25–29]. Thermal energy-harvesting devices can derive thermal energy from different sources, such as humans, animals, machines, and other natural sources. A thermoelectric generator (TEG) basically consists of a thermocouple, which comprises p-type and n-type semiconductors and produces an electrical current proportional to the temperature difference between hot and cold junctions. TEGs are discussed in detail in Section 8.2.5.

Electromagnetic radiation, either in the form of light, magnetic field, or RF microwave, may come from natural or artificial sources. Energy transmission in these sources is wireless and does not necessitate machinery. Solar cell [1,5,30–33], magnetic field [1,5,34], and RF power [1,5,35–38] are discussed in Sections 8.2.6–8.2.8, respectively.

The corresponding energy scales of different energy sources are shown in Figure 8.5. The following subsections introduce eight energy-harvesting transducers of the most commonly harnessed energy sources and describe their respective behaviors. Table 8.1 shows a comparison and the characteristics of different energy sources.

8.2.1 Vibration Electromagnetic Transducers

Vibration occurs in almost all dynamic systems. Electromagnetic transducers are used for harvesting kinetic (vibration) energy. A general model of kinetic energy conversion to electrical power in a vibrating mass is provided in Figure 8.6. This simple equivalent model based on linear system theory was proposed by Williams and Yates [39]; the mathematical description of the model is as follows:

$$M\Delta Z'' + (b_E + b_M)\Delta Z' + K\Delta Z = -M\Delta Y'' \tag{8.1}$$

In Eq. (8.1), M is the weight of the mass, ΔZ is the movement of the mass, ΔY is the input displacement, K is the elasticity coefficient of the spring, b_E is the electrical damping coefficient, b_M is the mechanical damping coefficient, $\Delta Z'$ and $\Delta Z''$ are the first-order and

Table 8.1 Power scales of different sources under given space constraints

	Estimated power (cm^3 or cm^2)	Output type	Output range
Solar cell	**10 μW to 15 mW** (Outdoors: 0.15–15 mW) (Indoors: <10 μW)	DC	1 V (single cell)
Kinetic energy	**<1–200 μW** (Piezoelectric: ~200 μW) (Electrostatic: 50–100 μW) (Electromagnetic: <1 μW)	AC	Tens of volts (peak voltage)
Thermal	**15 μW** (10°C gradient)	DC	10–100 mV
RF power	**<1–300 μW** (Source dependent)	AC	Hundreds of microvolts
Wind power generator	**<10–1000 μW**	AC	Generator dependent several volts, that is, approximately tens of volts
Magnetic coil	N/A	AC	Coil dependent several volts, that is, approximately tens of volts

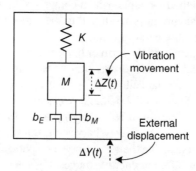

Figure 8.6 Schematic of a generic vibration converter

second-order differential results of ΔZ, respectively. $\Delta Z'$ and $\Delta Z''$ represent the velocity and the acceleration of the mass, respectively. $\Delta Y''$ represents the acceleration of the input displacement. The model is based on force balance. The energy transferred from the oscillating mass and converted to electricity behaves like a linear damper to the mass spring system. The kinetic power converted to electricity in the system is equal to the decreased power from mechanical vibration. The electrically induced force is $b_E \Delta Z$. Power can be defined as the product of force and velocity. For example, with the integration of force $b_E \Delta Z$ and velocity $\Delta Z'$ by Laplace transform and mathematical derivation, the magnitude of the output power can be defined as in Eq. (8.2). In this equation, ζ_T is the combined damping ratio of ζ_E and ζ_M (where ζ_E is the equivalent electrical damping ratio and ζ_M is the equivalent mechanical damping ratio); A is the acceleration magnitude of input vibrations, and ω is the frequency of the driving vibration [17,18].

Energy-Harvesting Systems

$$|P| = \frac{M\zeta_E A^2}{4\omega\zeta_T^2}, \text{ where } \zeta_T = \zeta_E + \zeta_M \qquad (8.2)$$

Equation (8.2) shows the relationship of mass and the acceleration magnitude and frequency of vibration. As shown in Eq. (8.2), a higher mass and a higher acceleration have more output power. Owing to the nature of vibration, a higher vibration frequency results in a decreased acceleration if the spring of a damping system similar to the system illustrated in Figure 8.6 has constant elasticity coefficient K. This premise suggests that the transducer should be designed to resonate at the lowest fundamental frequency in the input spectrum rather than at higher harmonics. The loss in the mechanical structure is denoted by ζ_M and should be as low as possible. Finally, the amount of power is linearly proportional to the mass. Thus, the converter design should allow for the largest possible mass within the space constraints.

By using input vibrations and applying Faraday's law, electromagnetic power conversion is obtained from the relative motion of an electrical conductor in a magnetic field. According to Faraday's law, the variation in a magnetic flux through an electrical circuit creates an electric field. This flux variation can be realized with a moving magnet, wherein the flux is linked with a fixed coil or with a fixed magnet with a flux linked to a moving coil. The first configuration is preferred to the second one because the electrical wires are fixed. A simple example of a linear electromagnetic generator is shown in Figure 8.7 [8,40,41].

The voltage on the coil is determined by Faraday's law, as expressed in Eq. (8.3), where ε is the induced electromagnetic field and Φ_B is the magnetic flux:

$$\varepsilon = -\frac{d\Phi_B}{dt} \qquad (8.3)$$

The open-circuit voltage V_{Out}, which is generated by the linear electromagnetic generator, is defined by Eq. (8.4), where N is the number of turns in the coil, B is the magnetic strength of the magnet, A_{Coil} is the cross-sectional area of the coil, and v is the velocity of the magnet as it travels through the coil. If the generator is loaded with resistance, power is extracted from the generator and a current will flow in the coil. This current creates its own magnetic field, which opposes the field inducing it. The interaction between the field caused by the induced current and the field from the magnets produces a force that opposes the motion.

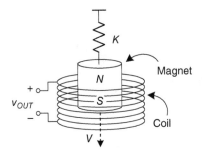

Figure 8.7 Electromagnetic generator

$$V_{Out} = NBA_{Coil}v \tag{8.4}$$

The output voltages of electromagnetic transducers are low, in the range of 100 mV. The main drawback of electromagnetic transducers is their difficulty to be integrated into electronic and micro systems. Low energy density also limits the application of electromagnetic transducers.

8.2.2 Piezoelectric Generator

In 1880, the Curie brothers discovered the piezoelectric effect in quartz crystals. Generally, the piezoelectric effect can be defined as the conversion of mechanical energy into electrical energy (direct effect) or the conversion of electrical energy into mechanical energy (inverse effect). Certain materials suffer from electrical polarization proportional to the applied strain when subjected to mechanical strain. Figure 8.8 shows a schematic of a piezoelectric cantilever [1,3,5].

The constitutive equations for a piezoelectric material are given in Eqs. (8.5) and (8.6), where δ is the mechanical strain, σ is the mechanical stress, Y is the modulus of elasticity [3], d is the piezoelectric strain coefficient, D is the electrical displacement (charge density), E represents the electric field, and ε_P is the dielectric constant of the piezoelectric material:

$$\delta = \frac{\sigma}{Y} + dE \tag{8.5}$$

$$D = \varepsilon_P E + d\sigma \tag{8.6}$$

Equation (8.5) is a combination of Hooke's law, represented by the term dE, and the piezoelectric coupling, denoted by the term σ/Y. Similarly, Eq. (8.6) is a combination of Gauss's law for electricity, represented by the term $\varepsilon_P E$, and the piezoelectric coupling, denoted by $d\sigma$. Piezoelectric coupling provides a two-way conversion between mechanical energy and electrical energy. The electric field across the material affects its mechanics, whereas stress in the material affects its dielectric properties.

Piezoelectric devices are commonly assumed to provide high voltages and low currents. However, voltage and current levels depend on the physical implementation and the particular

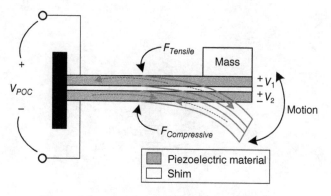

Figure 8.8 Schematic of a piezoelectric cantilever

electrical load circuit used. In reality, designing a system that produces voltages and currents in a useful range is fairly easy. Experimental results show that a conversion voltage in the range of several volts and a conversion current on the order of tens to hundreds of microamperes are achievable. The advantage of piezoelectric conversion is the direct generation of appropriate voltages. Owing to the high energy density, piezoelectric generators are considered to be a relatively good option for vibration energy harvesting compared with electromagnetic and electrostatic energy generators.

8.2.3 *Electrostatic Energy Generator*

The working principle of electrostatic generators indicates that the moving part of the transducer moves against an electrical field, thus generating energy [1,3,5]. A rectangular parallel plate capacitor is used to illustrate the principle of electrostatic energy conversion. The voltage V across the capacitor is expressed by Eq. (8.7), where Q is the charge on the capacitor, d is the gap or distance between plates, A is the area of the plate, and ε_0 is the dielectric constant of the free space. Figure 8.9 shows a schematic of an electrostatic generator.

$$V = \frac{Qd}{\varepsilon_0 A} \tag{8.7}$$

The capacitance is defined by:

$$\frac{Q}{V} = C = \frac{\varepsilon_0 A}{d} \tag{8.8}$$

If the charge on the plates is held constant, the voltage can be increased by reducing the capacitance. A voltage increase by capacitance reduction can be achieved either by increasing the distance d between plates or by reducing A. The energy stored in the capacitor is expressed by Eq. (8.9). If the voltage is held constant, the charge can be increased either by reducing d or increasing A. Increasing the voltage or charge increases the energy stored in the capacitor. Thus, energy transfer schemes can be classified as either charge- or voltage-constrained conversion; both energy transfer schemes are adopted in various works [1].

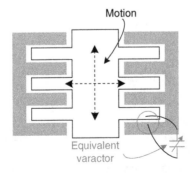

Figure 8.9 Schematic of electrostatic generator

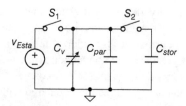

Figure 8.10 Electrostatic conversion circuit model

$$E_{stat} = \frac{1}{2}QV = \frac{1}{2}CV^2 = \frac{Q^2}{2C} \quad (8.9)$$

The main disadvantage of electrostatic generators is the voltage source requirement to initiate the conversion process because the capacitor must be charged to an initial voltage. Figure 8.10 shows an example of an electrostatic conversion circuit [42]. A V_{Esta} voltage source is used as the supply source to charge the equivalent varactor C_V. The maximum voltage $V_{CV,max}$ on the varactor according to the capacitance difference in each vibration is shown in Eq. (8.10), where C_{max} and C_{min} are the maximum and minimum capacitances of the varactor C_V, respectively, and C_{par} is the parasitic capacitance of C_V:

$$V_{CV,max} = V_{Esta} \cdot \frac{(C_{max} + C_{par})}{(C_{min} - C_{par})} \quad (8.10)$$

In each vibration cycle, the energy generated by the electrostatic generator E_{stat} is shown as:

$$E_{stat} = \frac{1}{2}V_{Esta}^2 \cdot (C_{max} - C_{min}) \cdot \left(\frac{C_{max} + C_{par}}{C_{min} + C_{par}}\right) = \frac{1}{2}V_{CV,max} \cdot V_{Esta} \cdot (C_{max} - C_{min}) \quad (8.11)$$

C_{max} and C_{min} can vary from hundreds to several pico-farads. The considerable difference between C_{max} and C_{min} results in a large variation in $V_{CV,max}$. Such a variation could reach hundreds of volts at a 1 V supply of V_{Esta}. The high output voltage places a constraint on device selection. An additional disadvantage of electrostatic generators is the possibility of the capacitor plates coming into contact with each other suddenly, thereby resulting in a short circuit. Thus, mechanical stops must be included in the design. Despite its limitations, electrostatic generators can still be considered a good option for SoC applications because of their potential to be integrated into silicon processes through MEMs technology.

8.2.4 Wind-Powered Energy Generator

For decades, wind generators (WGs) have been a well-known renewable energy source utilized in large-scale electric power generation in grid-connected applications. In recent years, micro-WGs, as shown in Figure 8.11, have been used as energy-harvesting sources for indoor or autonomous monitoring systems as these power sources become more popular [1,21–24]. Despite the demand for growth in harvesting, not all locations have strong winds. Thus, the challenge in designing a micro-WG is to make it work even at low wind speeds.

Energy-Harvesting Systems

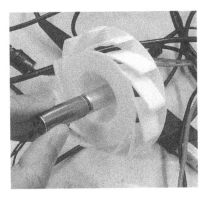

Figure 8.11 Micro-wind generator

The harvested power P_{WT} captured by a wind turbine is related to blade shape, pitch angle, and radius speed. Equation (8.12) defines P_{WT} and shows the related parameters. In Eq. (8.12), ρ is air density (typically 1.25 kg/m³); $T_p(\lambda, \beta)$ is the wind-turbine power transfer function, which varies with the design; β is the pitch angle (degrees); λ is the tip-speed ratio; R is the blade radius (m); and V is the wind speed (m/s).

$$P_{WT} = \frac{1}{2}\pi\rho T_p(\lambda,\beta)R^2V^3 \qquad (8.12)$$

In Eq. (8.12), the coefficient of T_p is limited to 0.59 according to the Betz limit [23]. The tip-speed ratio λ in Eq. (8.13) is the ratio between the rotational speed of the tip of a blade and the actual velocity of the wind, V, where Ω is the WG rotor speed of rotation (rad/s).

$$\lambda = \frac{\Omega R}{V} \qquad (8.13)$$

Wind turbines convert the kinetic energy of the generated wind because the principle of wind power is the same as that of normal electricity generation. By multiplexing the efficiency of the generator, η_G, the total power, P_{WG}, produced by the WG is given as:

$$P_{WG} = \eta_G P_{WT} \qquad (8.14)$$

The WG power coefficient is not constant but maximized for a tip-speed ratio $\lambda_{optimal}$. A specific point exists at which the WG output power is maximum at a certain rotation speed. The WG power curves at various wind speeds are shown in Figure 8.12 [21]. Figure 8.12 shows that the value of the optimal tip-speed ratio is constant for all maximum power points (MPPs). As shown in Eq. (8.15), the speed of WG rotation is related to wind speed, where Ω_n is the optimal rotation speed at a certain wind velocity V_n:

$$\Omega_n = \lambda_{optimal}\frac{V_n}{R} \qquad (8.15)$$

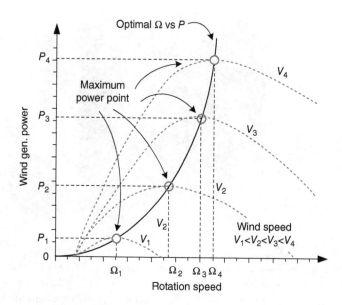

Figure 8.12 Rotation speed vs. generated power

Different rotation speeds at the same wind speed result in different generated powers. The loading condition influences the operating speed of a WG because the current generated by the generator produces a torque that is the inverse of the torque of the wind turbine. The balance between the input torque and the inverse torque can be achieved through load control to extract maximum power. Wind power depends on environmental conditions. Micro-WGs can generate several milliwatts of power, which is relatively large in the scope of energy harvesting. Through continuous harvesting, the power generated by a micro-WG is capable of fully meeting the power requirement of a monitor sensor. However, the drawback of wind power is the large size of WGs. Reducing the size of WGs is difficult because of the wind power density and wind turbine mechanism. Wind power is considered a competitive option for monitoring environmental factors.

8.2.5 Thermoelectric Generator

The thermoelectric effect is the direct conversion of temperature differences to electric voltage and vice versa. A thermoelectric device produces voltage when the temperature on one side is different from that on the other side. Conversely, when a voltage is applied to a device, a temperature difference is generated [1,5].

Thomas Johann Seebeck discovered that a compass needle was deflected by a closed loop formed by two metals in which a temperature difference existed in the junctions. This effect is attributed to the metals responding differently to the temperature difference, creating a current loop and a magnetic field. Seebeck did not realize that the magnetic field was induced by an electric current. This phenomenon was called the thermomagnetic effect until Danish physicist Hans Christian Ørsted corrected the mistake and renamed the phenomenon "thermoelectricity."

Energy-Harvesting Systems

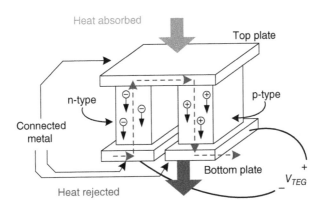

Figure 8.13 Schematic of a thermoelectric generator

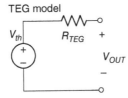

Figure 8.14 Model of a thermoelectric generator

Following developments in the study of thermoelectricity, current thermoelectric generators (hereafter TEGs) are currently solid-state devices. Figure 8.13 shows a schematic of a TEG. Charge carriers in metals and semiconductors are free to move, and resemble gas molecules that carry charge and heat. When a temperature gradient exists in a material, the mobile charge carriers at the hot end preferentially diffuse to the cold end. The collection of charge carriers results in a net charge and an electrostatic potential buildup. This mechanism is the basis of thermoelectric power generation.

Figure 8.13 shows a semiconductor thermoelectric couple, or a thermocouple consisting of n-type (containing free electrons) and p-type (containing free holes) thermoelectric elements [25–29]. The electron and hole carriers flow in opposite directions and therefore constitute a net current. The best thermoelectric materials are heavily doped semiconductors. The output voltage of TEGs is in the range of several millivolts, and to achieve reasonable output voltages a large number of thermocouples are required. These thermocouples are placed electrically in series and thermally in parallel. Figure 8.14 shows an equivalent model of a TEG, wherein V_{th} represents the voltage generated by the TEG and R_{TEG} is the inner resistance of the TEG.

The temperature difference generates a voltage V_{TEG} as defined in Eq. (8.16), where α is the Seebeck coefficient and ΔT is the temperature difference from the top plate to the bottom plate. V_{TEG} is the open-circuit voltage generated in the TEG. Heat flow drives the electrical current, which also determines the power output:

$$V_{TEG} = \alpha \Delta T \tag{8.16}$$

According to Carnot's theorem [1], the energy transfer between heat reservoirs has a limited efficiency, which is defined by Eq. (8.17). This limiting value, η_{th}, is called the Carnot cycle efficiency because it is the ideal efficiency. T_h and T_c are the temperatures on the top plate and the bottom plate, respectively:

$$\eta_{th} = 1 - \frac{T_c}{T_h} \tag{8.17}$$

TEG converts heat (Q) into electrical power (P) at efficiency η:

$$P = \eta Q \tag{8.18}$$

Efficiency η is not a constant value because it varies with the temperature difference $T_h - T_c$ between thermoelectric plates. The efficiency of a TEG increases nearly linearly with temperature difference. This effect is due to the characteristic of TEGs and all heat engines, that is, efficiency is limited by the Carnot cycle. Equation (8.19) briefly describes the efficiency of TEGs, where η_r is the reduced efficiency, the efficiency relative to the Carnot efficiency:

$$\eta = \Delta T \frac{\eta_r}{T_h} \tag{8.19}$$

The energy transmission efficiencies of TEGs are often lower than 10%. The rejected heat must be removed through a heat sink, which requires an additional heat-dissipation mechanism and a larger area. TEGs are silent, reliable, and scalable. They are extremely suitable for small distributed power generation in energy-harvesting applications, such as bio-WSNs.

8.2.6 Solar Cells

Solar energy generation or photovoltaics (PVs) are a method of generating electrical power by converting solar radiation directly into electricity using semiconductors that exhibit the PV effect [1,5]. Solar energy generators are a mature technology for large-scale energy generation. PV systems generate energy in milliwatts to megawatts. The generated electricity can be used for numerous applications, such as water heaters, signal lamps, and grid-connected PV systems, as shown in Figure 8.15 [43]. The application of PVs in portable products, such as calculators, watches, or monitoring sensor nodes, is a valid option under appropriate conditions.

In an outdoor environment on a sunny midday, the power density of the solar radiation on the earth's surface is about 100 mW/cm^3, which is approximately three times as high as that of other harvesting sources. However, solar energy is not particularly attractive for indoor environments because the power density drops to as low as 10–20 µW/cm^3. Silicon solar cells are a mature technology that can be divided into two major types: single-crystal silicon and thin-film polycrystalline. The efficiency of single-crystal silicon cells ranges from 12 to 25%. Thin-film polycrystalline and amorphous silicon solar cells are also commercially available and have lower cost and efficiency than single-crystal silicon cells. Table 8.2 shows the output power under various conditions of a single-crystal silicon solar cell with 15% efficiency. The power density of solar radiation decreases approximately as $1/d^2$, as expected, where d is the distance from the light source.

Energy-Harvesting Systems

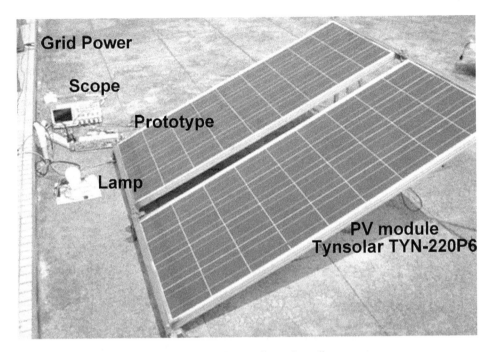

Figure 8.15 Outdoor solar cell

Table 8.2 Solar cell power under various conditions

Conditions	Outdoor, midday	Outdoor, overcast	10 cm from 60 W bulb	38 cm from 60 W bulb	Indoor lighting
Power ($\mu W/cm^3$)	15 000	750	5000	550	6.5

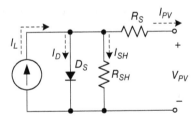

Figure 8.16 Modeling of a solar cell

A model of a solar cell is provided in Figure 8.16 [30–33]. Two parasitic resistances, that is, a parallel shunt resistance R_{SH} and a series resistance R_S, and a parallel diode are included in this model. The output current of the PV cell, I_{PV}, is expressed as in Eq. (8.20), where I_L represents the current generated by the solar cell, I_D represents the voltage-dependent diode current, and I_{SH} represents the current lost due to shunt resistances. The diode current I_D is modeled using the Shockley equation for an ideal diode, where n is the diode ideality factor (normally between 1 and 2 for a single-junction cell), I_0 is the saturation current, V_T is the thermal voltage, k is Boltzmann's constant (1.381×10^{-23} J/K), and q is the elementary charge. For an ideal cell, R_{SH} is infinite and provides no current leakage, whereas R_S is zero. No voltage drop was observed in V_{PV}:

$$I_{PV} = I_L - I_D - I_{SH} = I_L - I_0 \left(e^{\frac{q(V_{PV} + I_{PV} R_S)}{nkT}} - 1 \right) - \frac{V_{PV} + I_{PV} R_S}{R_{SH}} \quad (8.20)$$

The current generated by solar cells during operation is shared by the power losses in internal resistances and diodes. The characteristic curves of solar cell panels, which include the different irradiation levels and different ambient temperatures, are shown in Figure 8.17. In Eq. (8.21), the open-circuit voltage, V_{OC}, and the short-circuit current, I_{SC}, multiplied by the fill factor, FF_N, represent the efficiency of cell η. The power is zero at I_{SC} and V_{OC}. The voltage and current at the MPP are denoted by V_{MP} and I_{MP}, respectively. P_{in} represents the input irradiation power. As shown in Eq. (8.22), FF_N is defined as the ratio of the realistic MPP $P_{max,n}$ and the theoretical power $P_{T,n}$, which is the product of I_{SC} and V_{OC}:

$$\eta = \frac{I_{SCN} \cdot V_{OCN} \cdot FF_N}{P_{in}} \quad (8.21)$$

$$FF_N = \frac{P_{max,n}}{P_{T,n}} = \frac{I_{MPN} \cdot V_{MPN}}{I_{SCN} \cdot V_{OCN}} \quad (8.22)$$

The maximum available power provided by the solar cell, P_{max}, is related to the ambient temperature and irradiation level, as shown in Figures 8.17 and 8.18 [43]. Given their high energy density, PV cells are a good choice of energy source in places with sufficient light. However, the output power of PV panels varies significantly with the output voltage and current. Thus, commercial PV inverter products usually use a maximum power point tracking (MPPT) technique to achieve superior energy extraction from PV panels.

8.2.7 Magnetic Coil

Electromagnetic transducers use kinetic energy to induce magnetic-field-flux variation on a fixed coil to generate energy. Aside from the moving magnet, the alternating current on a power

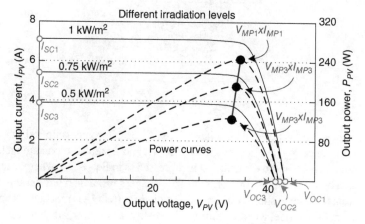

Figure 8.17 Characteristic curves of a solar panel at different irradiation levels

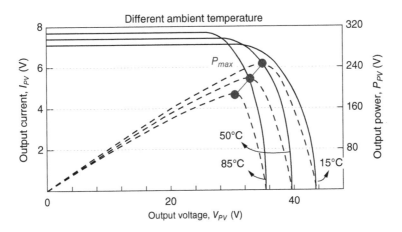

Figure 8.18 Characteristic curves of a solar panel at different temperatures

line also generates a time-variant magnetic field. The magnitude of the magnetic field variation is proportional to the current on the power wire; a large current with a large magnetic flux can generate considerable energy [1,5].

Current sensors, such as a Rogowski coil and current transformers (CTs), which are based on Faraday's law of induction, are conventionally used to provide inherent electrical isolation between the output signal and the current measured [34]. Figure 8.19 shows a schematic of a Rogowski coil, where r is the radius of the coil and V_{Rcoil} is the open-circuit voltage. The path integral of magnetic flux density B inside the coil is expressed as in Eq. (8.23), where μ_0 is the permeability of the free space and I_{AC} is the current in the power wire. Current I_{AC} flows through an enclosed area surrounded by a curve denoted by C:

$$\oint_C \vec{B} \cdot d\vec{l} = \mu_0 I_{AC} \tag{8.23}$$

If the cross-sectional diameter of the Rogowski coil is smaller than its radius r, then the magnetic flux density B can be simplified as:

$$B = \frac{\mu_0 I_{AC}}{2\pi r} \tag{8.24}$$

By applying Faraday's law of induction, the output voltage of the Rogowski coil undergoing a change in current I_{AC} can be obtained using Eq. (8.25). A is the cross-sectional area of the coil body, which is formed by the windings, k is the integration constant, and N is the number of turns. Voltage V_{Rcoil} is proportional to the derivative of I_{AC} plus the initial voltage $V_{Rcoil}(0)$:

$$V_{Rcoil} = -N\frac{d\phi}{dt} = \frac{\mu_0 NA}{2\pi r} \cdot k \int_t \frac{dI_{AC}}{dt} \cdot dt + V_{Rcoil}(0) = \frac{-\mu_0 NA}{2\pi r} \cdot k \cdot I_{AC} + V_{Rcoil}(0) \tag{8.25}$$

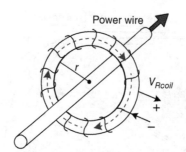

Figure 8.19 Schematic of a Rogowski coil

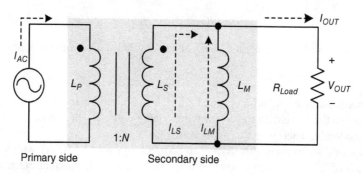

Figure 8.20 Model of a CT

CTs, which also utilize Faraday's law of induction to measure currents, are similar to a Rogowski coil. The construction of a CT is basically the same as that of a Rogowski coil, but a core material with high relative permeability is inserted into the former. The output of a Rogowski coil is a voltage that is proportional to the derivative of the primary current. The output of a CT is a sensing current on secondary winding, I_{Out}. Thus, the output of a CT is loaded with a sense resistor, R_{Load}. The current through R_{Load} generates a magnetic flux that counters the flux generated by the primary current. The CT model with sense resistor R_{Load} depicted in Figure 8.20 [34] neglects stray inductances, core losses, and winding resistances. However, such a CT can still provide sufficient insight into the CT operation principle. Figure 8.20 shows that I_{AC} represents the sensing current from the appliance. L_P is the primary-side inductance, L_S is the inductance induced by the primary side, and L_M is the magnetizing inductance. Thus, I_{LM} is influenced by the voltage across L_M and I_{LS} is induced by I_{AC}; I_{Out} can be expressed as the summation of I_{LM} and I_{LS}:

$$I_{Out} = I_{LS} + I_{LM} = \frac{I_{AC}}{N} - \frac{1}{L_M} \int_t V_{Out} \cdot dt \tag{8.26}$$

In current sensing, R_{Load} should be very low to reduce the current I_{LM} and enhance the sensing accuracy. By contrast, the most important issue in energy-harvesting applications is the transfer of maximum energy from a CT. The loaded resistor should match the equivalent resistor of a CT to extract the maximum power. Monitor sensors placed on the power lines of an electrical appliance can use a CT as power source.

8.2.8 RF/Wireless

The concept of wireless energy transmission is not new. Approximately 100 years ago, Nikola Tesla attempted to transmit low-frequency energy over long distances. In the 1950s, the rectification of microwave signals was proposed and researched in the context of high-power beaming. Figure 8.21 shows an experiment on wireless power transmission [44]. Near-field wireless energy transmission is omnipresent. All passive radio frequency identification (RFID) tags function on the same principle.

Wireless power transmissions can be classified into two categories: near field and far field [1,5,35–38]. These are the regions of the electromagnetic field around a radiation-emitting object, such as a transmitting antenna. Near field is normally within a distance of a few wavelengths. The distance beyond the near field is called the far field. The near field produces electromagnetic induction and electric charge effects on the electromagnetic field. A near-field transmission involves energy transfer effects that couple directly to the receivers near the antenna. Thus, a near-field transmission functions like a transformer, which draws more power at the primary circuit if power is drawn from the secondary circuit. By contrast, a far-field transmission constantly draws the same energy from the transmitter regardless of the receiving condition at the receiving part.

However, energy-harvesting applications collect ambient energy from the environment. The principle of near-field transmission, which consumes more power for transmitters, is related more to power transmitting than power harvesting. Thus, RF-harvesting applications are mainly focused on far-field transmission. Numerous potential RF sources, such as broadcast radio, mobile telephony, and wireless networks, exist in populated areas.

Equation (8.27) shows the difference between the received power versus the transmitted power. In Eq. (8.27), P_{rad} is the power incident on the node, P_S is the radiative power, λ is the wavelength, and R is the distance between the reader and the node:

$$P_{rad} = \frac{P_S \lambda^2}{4\pi R^2} \qquad (8.27)$$

The available power decreases rapidly with an increase in propagation distance. In reality, the transmitted power drops quickly at a rate faster than $1/R^2$ in an indoor environment. A more

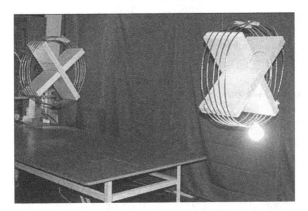

Figure 8.21 RF power transmission: MIT demonstration first showed how electricity can be wirelessly transferred to a device

likely figure is $1/R^4$ [35–38]. The target of RF energy harvesting is to collect as much RF as possible from disparate sources and convert the collected RF into useful energy. Energy collection is based on the antenna. However, a few challenges to far-field transmission exist. First, the received energy is very low, only on the nano-watt scale. Both high radiation frequency and the passive circuit structure of most RF-harvesting circuits are detrimental to efficiency. A high input frequency hinders utilization of the active circuit to enhance efficiency, and the power consumption of a high-speed control circuit is correspondingly high. Second, energy collection is achieved by the antenna. Each antenna has its own characteristic impedance and matching frequency. Thus, harvesting multiple frequencies from different sources with the use of a single antenna is difficult. The energy levels presented by RF energy sources are too low to fully provide all requirements of present electronic devices.

8.3 Energy-Harvesting Circuits

8.3.1 Basic Concept of Energy-Harvesting Circuits

Energy-harvesting transducers transfer energy from the sources described in Section 8.2. These sources are environmentally dependent, and the intensities of light, heat, magnetic field, and vibration are neither constant nor predictable. To extend the sustainability of a system, one should observe the level of the available energy and balance it with the power consumption of the system. Environmental characteristics define the possible energy-harvesting source. After the selection of the energy-harvesting source through the appropriate selection of a power converter topology, the harvested energy can be used efficiently [1–9,25,30].

In conventional portable electronics as well as in sensor nodes, batteries are the most commonly used power source as they are the most cost-effective choice. Aside from cost, batteries have the advantages of widespread availability, high reliability, mature mass-produced technology, minimal to zero environmental calibration requirement, ease of use (no need for thermal, vibrational, or photonic exposure beyond the sensor goals), and less energy-conversion overhead (the battery is the voltage source). However, battery replacement is a major issue for some sensing applications. Sometimes, battery volume may induce a problem when the sensor nodes have volume constraints.

Despite the power dissipation of circuits, a battery self-discharges by 0.1–5% per month because of its chemistry. The self-discharge rate depends on the material of the battery. High-temperature exposure tends to increase the self-discharge rate of batteries. Regardless of the cost, the suitability of an application for environmental energy harvesting is determined based on two factors: lifetime and environment. The lifetime of a system with a battery supply (under certain size limitations) can be estimated. System lifetime can be considered a benchmark for the lifetime of a harvesting supply. Environmental conditions define the available power and decide what kind of energy source should be harnessed. Considering the use of an energy-harvesting system, the overall performance needs to outperform a battery solution in terms of energy density, power density, and/or cost. Typically, the niche for energy harvesting is in long-lived applications in which the energy density is critical; the location of the sensor nodes may not be reachable and the replacement operations may be too numerous to perform [1].

In recent years, the application of energy-harvesting systems has focused mainly on WSNs. To define a power supply structure, the power requirement of a sensor node in a WSN, as shown in Figure 8.22 [3,4], should be analyzed. The power demands of the sensor nodes may therefore be raised to five or six orders, from 100 nW (e.g., 1 V and 100 nA of load power) during the sleeping mode to over 100 mW during the RF-transmitting mode. Under the variable conditions listed in Table 8.3, the limited and unstable environmental energy can hardly offset

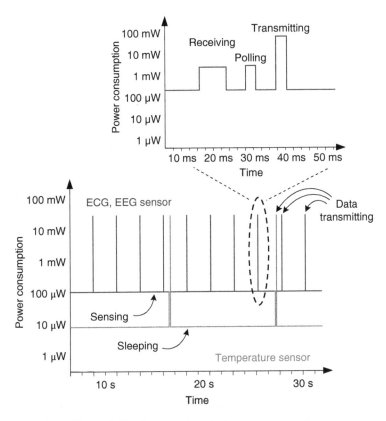

Figure 8.22 Power status of wireless sensor nodes

Table 8.3 Power status and operating mode of wireless sensor nodes

	Power consumption	Period (s)	Demand
Sleeping	10^{-1}–10^2 µW	10^{-1}–10^4	The minimum power to enable the circuit to wake up when events occur
Sensing	10 µW–10 mW	10^{-4}–10^4	Readout, conversion, and storage signals from the sensor
Receiving	10^{-1}–10 mW	10^{-3}–1	Listening to data packets or commands from the server
Polling	10 µW–10^2 mW	10^{-5}–10	Signal processing, which may be combined with sensing circuit operation
Transmitting	1–10^2 mW	10^{-6}–1	Sending of status, data, or command to the server

the power consumption of the sensor nodes. In fact, the output power of state-of-the-art harvesting technologies is not yet commensurate with high-power-consuming loads, such as in the case of power amplifiers and their associated antennae. The propagated energy of the RF communication signal decreases by the square of the distance traveled. The power requirement of PAs unavoidably increases as the telecommunication distance increases.

From the information in Table 8.3 and the characteristics of WSNs, if the sensor nodes rely only on the harvested energy, then two problems may be encountered. First, the harvested energy may not be sufficient to supply the dynamic system loading. The average power derived from the harvesting source may be higher than the average power requirement of the sensor node, but the large transient power requirement is not affordable. Second, even if the harvested energy is sufficient for the dynamic system loading, a storage device is still needed for the high efficiency demand. Energy efficiency, which is related to the MPPT, will be described in Section 8.4.

A hybrid-mode power supply system is recommended to overcome the disadvantages of an energy-harvested and battery-supplied system. The use of a rechargeable battery is a preferred method for the harvested energy. The battery supplies high power (up to several milli-watts) during a short period of time (during the receiving, transmitting, and polling modes). For the rest of the time, the energy harvester charges the battery with a trickle current. Figure 8.23 shows a system diagram of a conventional serial harvesting system. The AC/DC to DC/DC converter stage deals with the input source and generates a charging current for the storage device. The DC/DC converter stage performs a regulation function. This structure conforms to the ideal scenario and is commonly used in many applications. As the input energy has to be converted twice, the drawback of the two-stage approach is its low efficiency. To further optimize the conversion efficiency and reduce energy loss, the additional conversion loss induced by the two-stage structure can be reduced if the energy is transferred directly to the output load.

A system diagram of a parallel harvesting system is shown in Figure 8.24. The parallel structure aims to improve the conversion efficiency and retain the charging function. The parallel structure takes a primary path to convert energy directly to the load with the use of one-stage conversion to avoid additional conversion losses. The secondary path transfers the redundant energy to the storage device and performs the charging function when the energy source generates more energy than the load required. A DC/DC converter is placed after the battery to supply the primary path when the energy source is not fully adequate for the load.

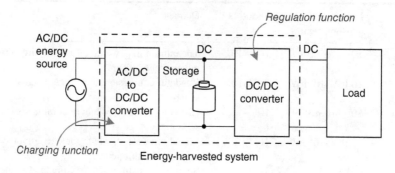

Figure 8.23 System diagram of a conventional serial harvesting system

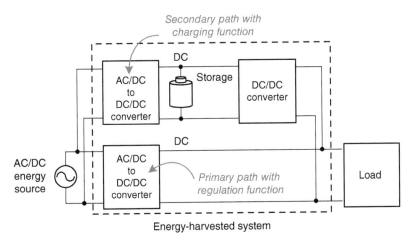

Figure 8.24 System diagram of a parallel harvesting system

However, the parallel structure introduces more design challenges than the conventional serial structure. First, the primary path AC/DC to DC/DC converter must have the ability to regulate the output voltage. Second, an energy distribution scheme is required to allocate the input energy to two paths accurately because the primary and secondary paths share the same energy source. Third, the extra circuits increase the overhead costs and the area.

Many harvesting circuits are similar to conventional power converters. Some of the main differences between an energy-harvesting circuit design and a conventional DC/DC or AC/DC converter design are summarized as follows:

1. The amount of energy from a source is limited. Thus, the loading condition of a system has to meet the characteristics of the energy-harvesting source.
2. An energy source is not a fixed voltage or a current source; it is a load or an environmental variant.
3. The input voltage may be very low or may vary across a wide range.
4. The operating voltage and the current have to be defined at the MPP.

In the following subsection the AC/DC to DC/DC converter, which can be used in Figures 8.23 and 8.24 with different functions and characteristics, will be introduced.

8.3.2 AC Source Energy-Harvesting Circuits

The design of an energy-harvesting circuit with an AC source has several objectives. The AC input must be rectified according to the DC value; otherwise, the energy harvested can hardly be used or stored. The magnitude of the AC source in a harvesting system varies correspondingly with environmental conditions. Thus, to protect the circuit from over-voltage damage, voltage limitations for some AC applications are required. The two-stage approach to AC source harvesting, which is depicted in Figure 8.23, is a mainstream approach and has been presented in previous publications. Depending on the input voltage and the current range,

AC/DC conversion needs to convert the input voltage to a proper voltage level for back-end usage. In the following subsections, the circuit description proceeds from the passive rectifier to the active rectifier. Afterward, some structures, which have combined the AC/DC to DC/DC converter and the DC/DC voltage regulator in recent years, are described. A presentation of the design concerns, with pros and cons, as regards each circuit follows shortly.

8.3.2.1 Full/Half Bridge Rectifier

The most straightforward and robust AC/DC converter is the conventional full bridge rectifier, which is shown in Figure 8.25. A passive full bridge rectifier needs no control and is very useful in harvesting applications. The harvesting system may run out of power and may have to be fully turned off to save energy. Passive components have no static power consumption, except when leakage occurs. Sometimes the input voltage under low-power conditions can hardly reach the turn-on voltage of the full bridge rectifier. Even if the input voltage is sufficient, the dropout voltages on the diodes seriously influence power efficiency when the input voltage is low. Schottky diodes with low forward voltage can improve the efficiency but suffer from reverse current leakage and reverse recovery leakage.

Charge Pump Rectifier
A voltage doubler charge pump, as shown in Figure 8.26, is similar to a full bridge rectifier [45,46]. As high-frequency-switching operation is power consuming and difficult to control

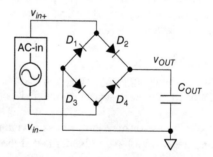

Figure 8.25 Full bridge rectifier

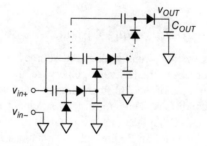

Figure 8.26 Voltage doubler charge pump

accurately, a fully passive structure has been used in high-frequency applications, such as RFID powering. The input voltage is automatically pumped to a higher level through cascading. Cascading is very useful for sources with relatively low output voltages. However, diode loss remains a serious issue for this circuit. If the diode forward voltage is V_D, then the output voltage is similar to that described in Eq. (8.28) for an N-stage charge pump:

$$V_{OUT} = N(V_{in} - 2V_D) + V_D \qquad (8.28)$$

Each cascading stage suffers from dropout voltages. The higher the cascading stages, the larger the power losses that occur in the diodes. A considerable number of components results in a decrease in cost or area efficiency.

Active Rectifier
Active diode rectifiers control transistors as switches. The large diode forward voltage is replaced by the voltage drop of the transistor. Figure 8.27 shows a fully synchronous active rectifier. Two comparators are used to compare the voltage between the terminals, that is, V_{in+} and V_{in-}, and the output voltage V_{OUT}. When the terminal voltage is higher than V_{OUT}, the corresponding path is turned on to perform rectification. The minimum input voltage requirement constructed by the diode forward voltage can be released, and the diodes suffer from energy loss as well. However, the trade-off for the aforementioned advantages is the control circuit requirement. The control circuits must be powered first to turn on the active rectifier. Thus, the circuit cannot be fully turned off and cannot start up when the stored energy is exhausted. These trade-offs are not beneficial to high system sustainability.

MOSFET Rectifier with an Active Diode
Figure 8.28 shows a combination of passive and active rectifiers [47]. MOSFETs M_1 to M_4 act as a rectifier to select the proper turn-on path and translate AC input into DC output. The active diode is placed between the MOSFET rectifier and the output storage device to prevent a reverse current when V_{OUT} is higher than the terminal voltage of the AC input. This cascade structure has a forward voltage of one transistor threshold voltage (V_{th}). A standard diode

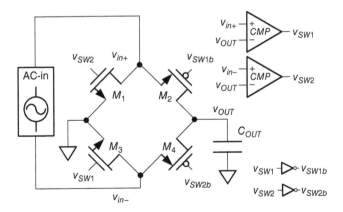

Figure 8.27 Fully synchronous active rectifier

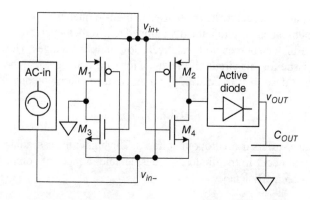

Figure 8.28 Passive MOS rectifier with active diode

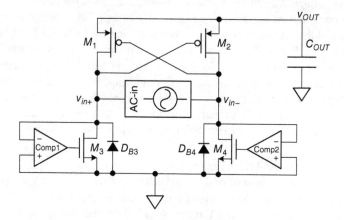

Figure 8.29 Diode-connected P-MOSFET with active N-MOSFET as rectifier

forward voltage is about the same as V_{th}. Compared with a conventional full bridge rectifier, a MOSFET rectifier with an active diode allows for one V_{th} of the forward voltage to be saved. Furthermore, the body diode of the MOSFET switch in the active diode can be used as a forward diode during start-up status. Thus, the combined structure can passively and automatically start up without any need for a power-on scheme even under zero energy condition.

Cross-Coupled P-MOSFET with Active N-MOSFET as Rectifier

Figure 8.29 shows another structure which uses diode-connected P-MOSFET and comparator-based N-MOSFET switches as the rectifier [48–50]. This structure has the same functions as those of the passive MOSFET rectifier with an active diode, but experiences less conduction loss for two reasons. First, the power path passes through two transistors, which is one less than the structure illustrated in Figure 8.28. Second, the N-MOSFET switch in Figure 8.29 has less average turn-on resistance than the N-MOSFET in the rectifier in Figure 8.28, because the gate of the N-MOSFET is controlled by the active circuit.

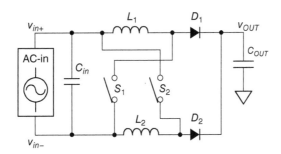

Figure 8.30 Dual-boost converter for AC/DC conversion

Dual-Boost AC/DC Converter

The aforementioned rectifier circuits only perform AC/DC conversion without any regulation capability. The output voltage level of the rectifier is determined by the input AC voltage. Thus, a rectifier that can simultaneously rectify and regulate the output voltage through the same circuit is a good option. Figure 8.30 shows the dual-boost converter for AC/DC conversion [51]. Two sets of boost converters operate alternately when the polarity of the AC input changes. The boost converter can be turned off when V_{OUT} exceeds the rated value. Thus, if the input energy is high enough to fully support the output loading, then the dual-boost structure has the ability to regulate V_{OUT} and can supply direct to the back-end system.

However, a trade-off occurs between zero energy start-up and regulation. The regulation function relies on the fully turned-off system to limit the energy being transferred. If the system is not powered first, the energy is prohibited from the harvesting source. The aforementioned passive rectifier as voltage doubler in Figure 8.26 could serve as a complement to help address this problem.

In terms of efficiency, the voltage doubler charge pump is not a good choice. However, given its fully passive and voltage-boost characteristics, the voltage doubler charge pump is a good option for a start-up circuit. If the harvesting system starts from zero energy, the voltage doubler charge pump can provide a voltage that will drive the control circuit in the start-up state. Afterward, a more efficient converter can be used instead of the charge pump. Many harvesting systems need this kind of "handover" scheme to start up the system from zero energy and improve the efficiency when normal operation is available. The handover structure can extend the sustainability of the system to a theoretically unlimited range. The system can operate again as long as adequate energy comes from the harvester.

Single-Inductor Dual-Boost Converter

Figure 8.31 is a modified type of the dual-boost structure in Figure 8.30 [52]. By arranging the switches, that is, S_1 and S_2, and the diodes, both boost converters share an inductor. This approach reduces bulky external components while maintaining the same functionality of the dual-boost structure.

Single-Inductor Buck/Boost Converter

Different energy sources have different output voltages. For general-purpose power conversion, Figure 8.32 shows a single-inductor buck/boost converter [53]. The buck/boost structure

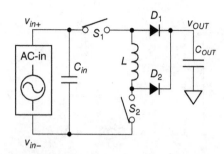

Figure 8.31 Single-inductor dual-boost converter

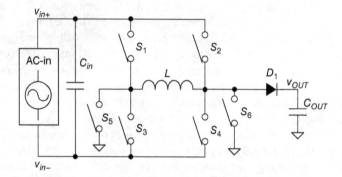

Figure 8.32 Single-inductor buck/boost converter

has the ability to up-convert or down-convert the input voltage, which provides more flexibility to the selection of energy-harvesting sources. The input voltage level is not limited by the converter structure. The diodes in Figures 8.30–8.32 can be replaced by active diodes to further enhance conversion efficiency.

Energy sources may not have the capability to store electric charge. In switch-based harvesting circuits, sudden current extraction from the inductor causes a considerable voltage drop in the energy-harvesting source. Sometimes the voltage drop is too large, so that it could cause reverse leakage or influence the normal operation of the converters. Thus, an input capacitor C_{In} is necessary for inductor-based converters. Energy from the energy source can be stored in C_{In} as a buffer to maintain a relatively stable voltage.

Summary of AC-Source Energy-Harvesting Circuits

Figure 8.33 shows a diagram summarizing and comparing the characteristics and functions of the structures discussed in this subsection. The rectifiers and converters are listed in the order of high-power loss to low-power loss and divided by different function requirements. Figure 8.33 provides simple instructions to help filter the possible structures and proceed with the harvesting circuit design.

Energy-Harvesting Systems

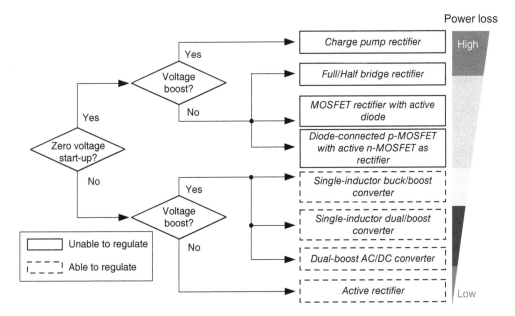

Figure 8.33 Function selection and characteristics of different AC/DC structures

8.3.3 DC-Source Energy-Harvesting Circuits

In energy harvesting, the number of AC sources is larger than that of DC sources. The most widely discussed DC harvesting sources include TEGs and solar cells. The goal of a DC harvesting circuit is to convert the input voltage to a proper and regulated voltage level for battery charging or system supply. DC-source harvesting circuits are simpler than AC-source harvesting circuits. The former are similar to conventional DC/DC converters. DC/DC harvesting circuits can also be used as DC/DC converters that perform voltage regulation, as in Figures 8.23 and 8.24. Some concerns about converters in energy-harvesting applications will be discussed in Section 8.4.

8.3.3.1 Low-Dropout Regulator

The most attractive features of an LDO regulator are low noise, compact size, and fast transient response. Figure 8.34 shows the circuit of an LDO regulator. From Figure 8.34, an LDO regulator is clearly quite simple if no special specification is required. With fewer external components, the LDO regulator is efficient in terms of area and cost. Without switching in the power transistors and the high loop gain, LDO demonstrates great noise immunity, which is especially important for implanted biomedical signal-sensing applications. For example, sensing systems using small physiological signals, such as EEG, are noise sensitive and volume limited. For such applications, LDO is a good option as a power management unit.

LDO has the following drawbacks: (i) low efficiency and (ii) performance limited to stepdown conversion. The low supply voltage generated by some harvesting sources also limits the use of LDO, especially when a battery-charging function is required. The efficiency of LDO

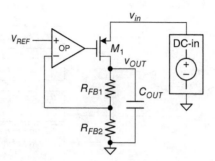

Figure 8.34 LDO regulator

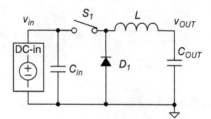

Figure 8.35 Buck converter

depends on the difference in input and output voltages. When the input voltage is high and the output is low, most of the energy is wasted on the power transistor because of the large dropout voltage. However, the low quiescent power of DC/DC voltage regulators makes linear regulators a good choice for low-power systems.

Buck Converter
Similar to LDOs, buck converters can provide step-down conversion. Figure 8.35 shows the power stage of a standard buck converter [54]. The characteristics of a switching converter allow for high efficiency but result in switching noise.

Boost Converter
Energy-harvesting sources, such as TEGs, have a very low DC voltage. A single solar cell provides a higher voltage of up to 1 V (open circuit). However, such a voltage is still not sufficient to charge batteries. Boost converters have been a straightforward solution, commonly used in many previous works [54]. Figure 8.36 shows a schematic of a boost converter. Boost converters have the advantage of high efficiency and output voltage regulation capability. Boost converters can function as a battery charger or even directly supply the load system.

Buck/Boost Converter
As shown in Figure 8.32, a buck/boost converter provides more flexibility to both input and output voltage ranges [54]. Figure 8.37 shows a schematic of a non-inverting buck/boost

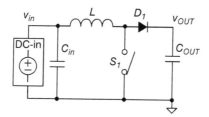

Figure 8.36 Boost converter

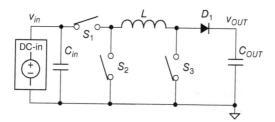

Figure 8.37 Buck/boost converter

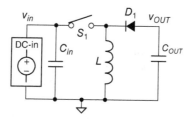

Figure 8.38 Inverting buck/boost converter

converter. The drawbacks of this converter are efficiency degradation caused by the extra pass transistors and the driving losses induced by extra switches.

Inverting Buck/Boost Converter

Figure 8.38 shows a schematic of an inverting buck/boost converter [54]. Inverting buck/boost converters have fewer switches, which results in higher efficiency. As the output voltage is inverted, the controller and the gate driving circuit for switches have to be designed especially. Such specification results in an increase in production costs.

Charge Pump

Aside from inductor-based converters, capacitor-based converters, such as charge pumps, are also widely used in energy-harvesting circuits. Figure 8.39 shows a 2× boost charge pump circuit. If the circuit is cascaded, a high-voltage boost ratio can be achieved [55,56]. Charge pumps can also perform voltage regulation through a closed-loop control.

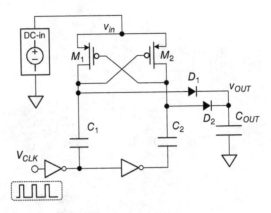

Figure 8.39 Charge pump step-up converter

Summary of DC-Source Energy-Harvesting Circuits
DC/DC harvesting circuits have the same design goals as those of conventional DC/DC converters. However, quiescent power is a more pressing issue for harvesting circuits than for conventional DC/DC converters. DC/DC converters may need sub-circuits to provide a reference voltage or a bias current to provide regular support to the DC/DC converter. These sub-circuits, such as bandgap voltage reference, comparators, or amplifiers, consume considerable power. The power loss may reach up to hundreds of microwatts. For DC/DC converters, which convert a large output power or require a fast transient response, the large power consumption caused by the controller is necessary. By contrast, the power consumption of low-input power applications, such as harvesting functions, might not be affordable or may even exceed the possible derived input power. Thus, one of the challenges in designing DC/DC harvesting circuits is how to design an ultra-low-power control circuit and operation scheme.

8.4 Maximum Power Point Tracking

8.4.1 *Basic Concept of Maximum Power Point Tracking*

Harvesting sources, such as PV cells, TEGs, vibration, or magnetic coils, have their own electrical and mechanical properties. The output voltage, current, and power of each source are influenced by the loading effect. Under a steady environmental condition, each source has a corresponding output voltage and current value to generate maximum output power. The point at which the maximum output power is generated is selected as the MPP, and the equivalent load at that point should be defined and tracked through MPPT control [1, 23, 30].

Using the equivalent model of solar cells in Figure 8.16 as reference, the current I_L generated by the solar cells is assumed constant in steady irradiation conditions. If the load on the output voltage V_{PV} is light, V_{PV} rises to a higher voltage, which induces current leakage on the parasitic diode D_S and the resistor R_{SH}. Conversely, if the load is heavy, V_{PV} drops to a low level, and most of the power is wasted on the series resistor R_S. Thus, identifying the MPP is important to extract the maximum possible power under the same environmental conditions. Many harvesting systems utilize MPPT circuits to enhance efficiency and derive maximum output power

[1,21–23,25,35]. However, the following should be noted regarding a harvesting system with the MPPT function:

1. The characteristic of the energy sources must be known before applying the MPPT method.
2. The power consumption of the MPPT circuit is critical for numerous low-power applications.
3. MPPT control fixes the input source at its maximum output energy.
4. A storage device is necessary for harvesting systems that adopt the MPPT function.
5. Continuous tracking is necessary to overcome environmental condition variations. MPP varies under different environmental conditions. The harvesting circuits should control the power-delivery condition to ensure that the energy-harvesting system operates at its MPP.

The MPPT scheme is a method of tracking input power. If a harvesting system supplies voltage directly to the load system, the MPPT function hinders the harvesting system from regulating the output voltage at the same time unless the input power is always the same as the load requirement, which is unlikely. Thus, some general ideas and methods for implementing the MPPT control are presented in the following subsections.

8.4.2 Impedance Matching

Many harvesting sources have complex behaviors or internal equivalent circuit models. Prior studies, for example [23,25,52,57], have presented numerous optimized designs for different sources and characteristics; these circuits can only be used on the specific sources they are designed for. The impedance-matching method is the most popular and important approach for a general-purpose MPPT. Figure 8.40 shows the Thevenin equivalent circuit of an energy-harvesting source modeled as an ideal voltage source, V_{EQ}, with series resistance R_S. Generally, all sources can be modeled as in Figure 8.40, with different Thevenin equivalent impedance even if the impedance is not pure resistance and includes inductance and capacitance. However, the idea of matching the load impedance with the inner impedance is still valid and provides tracking instructions when MPPT is desired.

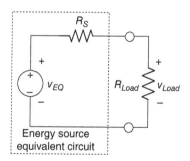

Figure 8.40 Thevenin equivalent circuit of energy source

The power P_{Load} on the output loading R_{Load} is shown as:

$$P_{Load} = V_{Load}^2 R_{Load} = \left(\frac{V_{EQ}}{R_S + R_{Load}}\right)^2 R_{Load} \qquad (8.29)$$

Impedance-matching theory implies that the system achieves maximum power transfer when the loading impedance is equivalent to the inner impedance. The maximum output power transferred to the load, $P_{Load,max}$, is expressed as in Eq. (8.30). Matching efficiency is defined as the ratio of the power in the load to the maximum output power, as shown in Eq. (8.31):

$$P_{Load,max} = \frac{V_{EQ}^2}{4R_{Load}} \qquad (8.30)$$

$$\frac{P_{Load}}{P_{Load,max}} = \frac{4}{2 + \dfrac{R_{Load}}{R_S} + \dfrac{R_S}{R_{Load}}} \qquad (8.31)$$

Figure 8.41 shows matching efficiencies at different impedance errors. An impedance error represents the mismatch percentage compared with the inner impedance of the harvesting source. When the load impedance is perfectly matched with the inner impedance, the error percentage is zero and the matching efficiency is unity. If 90% matching efficiency is desired, the endurable load impedance error ranges from −48 to +93%. This large endurable range indicates that even if a large error percentage occurs, the output power remains very close to the MPP.

8.4.3 Resistor Emulation

An equivalent resistance of the converter can be obtained if the ratio of the input voltage to the average input current is considered, regardless of whether switching or linear power converters are adopted. This scheme is called resistance emulation [23,25,30,58]. The equivalent resistance can be derived by calculating the average input current during the operation of a converter.

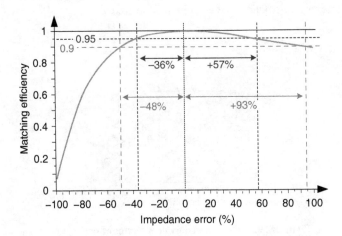

Figure 8.41 Matching efficiency versus impedance error

Energy-Harvesting Systems

For example, if the buck/boost converter in Figure 8.37 with constant switching frequency PWM control is operated in DCM, the charging and discharging phases of the buck/boost converter are as shown in Figure 8.42. The behavior of the inductor current when the input voltage changes is shown in Figure 8.43.

The buck/boost converter only connects to the input voltage source in the charging phase, where I_{Peak} is the peak current in the charging phase. The average input current of the converter, $I_{in,avg}$, is defined by Eq. (8.32), where D is the duty cycle, T is the time of a switching period, and L is the inductor used in the converter:

$$I_{in,avg} = \frac{I_{Peak} \cdot D}{2T} = \frac{V_{in} T \cdot D^2}{2L} \qquad (8.32)$$

The input voltage V_{in} is divided by $I_{in,avg}$ to derive the equivalent resistance R_{eq} as shown in Eq. (8.33), which is related to D and T:

$$R_{eq} = \frac{V_{in}}{I_{in,avg}} = \frac{2L}{T \cdot D^2} \qquad (8.33)$$

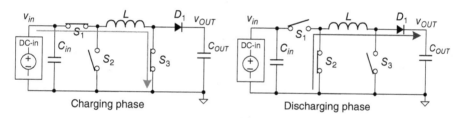

Figure 8.42 The buck/boost converter operation

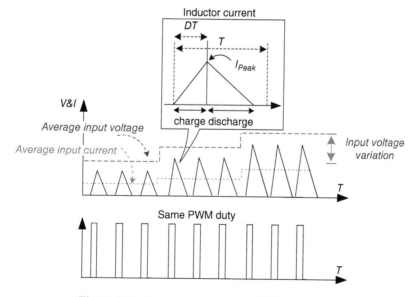

Figure 8.43 Inductor current under DCM operation

The duty or switching frequency can be used to adjust R_{eq} through PWM or PFM control. Thus, the converters can be considered a tunable resistance to match the inner impedance and obtain maximum output power.

Several considerations should be noted regarding the resistance emulation method. Converter parameters should be selected appropriately. R_{eq} should fit the inner impedance of the harvesting source. The coverage range of R_{eq} should be designed according to the target source. If a switching converter is used as the emulated resistor, a low switching frequency will induce a large switching current ripple. If the driving capability of the energy source is insufficient to sink the current, then the terminal of the source will drop significantly. The large voltage variation in some energy sources, such as solar cells, which have complex equivalent internal models, influences the output power condition. Even if the equivalent resistance is the ratio of the average input voltage to the current, the large voltage variation may result in additional power loss or power conditions deviating from the MPP.

8.4.4 MPPT Method

Different harvesting sources require different tracking methods to achieve the MPP. The resistor emulation introduced in Section 8.4.3 enables MPPT controllers to control the energy sources to achieve energy at their respective MPPs. In resistor emulation, the MPPT controller adjusts the resistor emulation converter control factors, such as the duty or the frequency. The adjustment changes the emulated loading to match the characteristics of the harvesting source. The MPPT method is the procedure used to perform adjustment.

8.4.4.1 Mountain-Climbing (Perturb and Observe) Method

The commonly used MPPT method for harvesting sources with complex models and behaviors is the mountain-climbing method, which is also called the perturb and observe (P&O) method [23,30]. This method is applied to solar cells because of the complex relationship between their operating environment and the maximum power. The MPPT operation on a solar cell is illustrated in Figure 8.44. The output power is measured after each adjustment. If the power

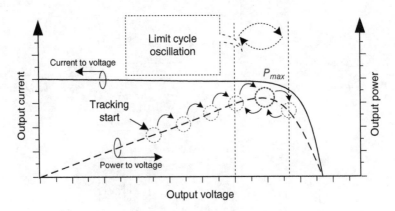

Figure 8.44 MPPT of solar cells

increases, then further adjustment in the same direction is performed until the power no longer increases. This method is so named because it is similar to mountain climbing and it depends on the rise of the power versus voltage curve below the MPP. Once the adjustment exceeds the MPP, the output power will fall below the MPP. The adjustment turns to the opposite direction and traces back to the MPP. The tracking operation goes back and forth around the MPP. To track the output power continuously, adjustment should be performed regularly to deal with environmental variations. Thus, if the environmental condition is stable, the adjustment is in a limiting cycle oscillation.

Figure 8.45 shows a flowchart of MPPT. In some applications, if the MPP curve has more than one peak value, the mountain-climbing method may suffer from a local optimization problem. This indicates that the tracked MPP is not the global optimization in the MPP curve but the local optimization. The tracking flow cannot detect whether it is located at the global optimization or at the local optimization. P&O is the most commonly used MPPT method because of its ease of implementation and flexibility with different applications.

To ensure the appropriateness of adjustments, the mountain-climbing method should constantly sample and monitor the output power condition. Sampling and monitoring are the most challenging parts of this method. A set of data converters is conventionally used to sample the output current and voltage in an electricity-generating solar cell array. Such data conversion and calculation is also applied to solar energy-harvesting applications, as shown in Figure 8.46. Information on the sampled current and power is calculated by the processor and then stored in the memory to control the tracking strategy. This method is very straightforward and has the advantage of high accuracy. The same method can also be applied to the output part. In monitoring the output current and voltage, the output power also represents the status of the input source, as shown in Figure 8.47 [43].

However, continuous data conversion and digital signal processing with storage are power consuming. For low-power energy-harvesting applications, the considerable power consumed by the tracking method significantly influences the derived output power. Some low-input power applications use the characteristics of the switching power converter to monitor the power status without complex data conversion and calculation. Figure 8.48 shows the peak

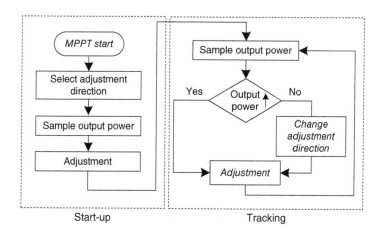

Figure 8.45 MPPT flowchart

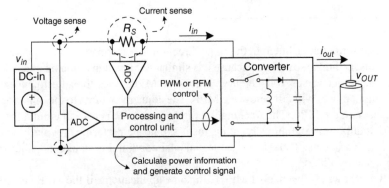

Figure 8.46 Input power-monitoring scheme

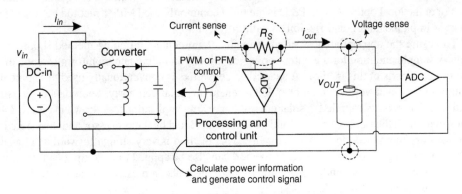

Figure 8.47 Output power-monitoring scheme

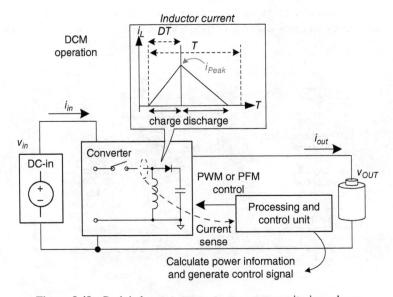

Figure 8.48 Peak inductor current output power-monitoring scheme

inductor current-monitoring method [53,59]. If the power converter operates in the DCM, the energy stored in the inductor of each cycle will be delivered completely to the output, which means that the output power is proportional to the peak current. By using a current-sensing circuit in the power converter, the peak current, I_{Peak}, can be used as an index to verify the power condition. The aforementioned power-sensing methods can only be used under certain converter structures and operating modes.

8.4.4.2 Open-Circuit Test Method

For energy sources with fixed built-in resistances, the MPP is proportional to the open-circuit voltage. The most commonly discussed energy source that applies the open-circuit test method is a TEG [25,27]. Figure 8.49 shows the current I_{TEG} and the power P_{TEG} over the voltage of the TEG. The same idea of impedance matching is applied in this method, that is, maximum power exists when the inner resistance is the same as the load resistance. According to voltage divider theory, the MPPT voltage is approximately half the open-circuit voltage.

Figure 8.50 shows a block diagram of a harvesting system that applies the open-circuit test method. As the name of the method implies, power monitoring and sampling are achieved by opening the circuit from the harvesting source during the sampling period. A comparison of the sampled voltage with the closed-circuit voltage is achieved under operation.

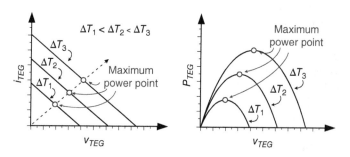

Figure 8.49 The current I_{TEG} and the power P_{TEG} over the voltage of the TEG

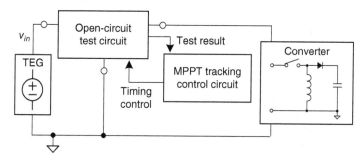

Figure 8.50 Open-circuit test method and system diagram

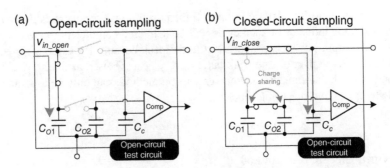

Figure 8.51 Circuit implementation of (a) open-circuit sampling and (b) closed-circuit sampling

Owing to the limited driving capability of the energy source and the operating speed of the converter in both open- and closed-circuit sampling, the output voltage of the energy source requires transient time to achieve the steady-state voltage. The recovery time is source dependent and should be considered in the sampling timing control. The implementation of the open-circuit test method is shown in Figure 8.51. The circuit operations during open-circuit sampling and closed-circuit sampling are shown in Figure 8.51(a) and (b), respectively. C_{O1} and C_{O2} have the same capacitance values and perform charge sharing to derive half of the open-circuit voltage, V_{in_open}. The closed-circuit voltage, V_{in_close}, is compared with the divided V_{in_open} to define the power condition. The converter adjusts the emulated resistance according to the comparison result. If the divided V_{in_open} is higher than V_{in_close}, then the emulated resistance is extremely high. By contrast, if V_{in_open} is lower than V_{in_close}, then the emulated resistance is extremely low. The open-circuit test method is also used in solar cell MPPT, but it adopts a different divider ratio of 0.7. The MPP of a solar cell is often near 0.7 of its open-circuit voltage. Thus, compared with the slow mountain-climbing procedure, the open-circuit test can approach the MPP relatively quickly.

8.4.4.3 Iterative-Based MPPT

The mountain-climbing method saves and compares the power status after each sampling and adjustment. The iterative-based MPPT method shifts the comparison from the previous sampled power status to a dynamic target power through the following procedure. First, the target power traces the present sampled power status. The target power level is then promoted and the converted power is adjusted to approach the target. The procedure is repeated, and the power target and the sampled power status iteratively track each other to reach the MPP [53].

In this method, the DCM in Figure 8.48 is taken as a converter example and the PWM duty is utilized as the control factor. Figure 8.52 shows the tracking operation. A target power represented by a voltage signal, V_{Target}, is compared with the peak of the current sense signal V_{CS}. At the beginning, V_{Target} is set higher than V_{CS}. The time-out period, T_O, is designed as the settling time for the converter and the energy-harvesting source. If V_{Target} and V_{CS} encounter each other within the time-out period, V_{Target} will be set to a lower level. Through several comparisons, V_{Target} will approach V_{CS}. Meanwhile, V_{Target} represents the recent power status. Afterward, V_{Target} is set to a slightly higher level as a new target for V_{CS}. The MPPT controller adjusts

Energy-Harvesting Systems

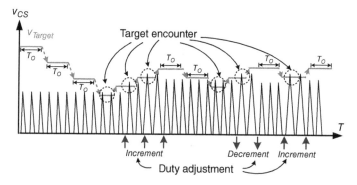

Figure 8.52 Target power and sampled power status iteratively track each other

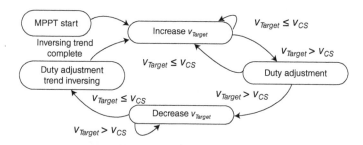

Figure 8.53 FSM of iterative-based tracking

the PWM duty and checks the V_{CS} variation. The adjustments may cause the output power to either rise or fall. V_{CS} reaching V_{Target} before the time-out period indicates that the output power is rising and the duty-adjusting trend is correct. V_{Target} shifts to a higher level and is set to a new target. The adjusting trend continues in the same direction. However, if V_{CS} does not reach V_{Target}, an incorrect adjusting trend occurs. After the time-out period, the MPPT controller lowers the V_{Target} to find the V_{CS} and to change the adjusting trend. In conclusion, V_{Target} is always set to a slightly higher target compared with the recent input power condition. Figure 8.53 shows the FSM of the iterative-based MPPT control.

References

[1] Priya, S. and Inman, D.J. (2009) *Energy Harvesting Technologies*. Springer-Verlag, New York.
[2] Yang, Y., Lambert, F., and Divan, D. (2007) A survey on technologies for implementing sensor networks for power delivery systems. *Proceedings of the IEEE Power Engineering Society General Meeting*, June, pp. 1–8.
[3] Roundy, S.J. (2003) Energy scavenging for wireless sensor nodes with a focus on vibration to electricity conversion. PhD thesis, University of California, Berkeley, CA.
[4] Paradiso, A. (2008) Energy Scavenging for Mobile and Wireless Electronics. Massachusetts Institute of Technology Media Laboratory Thad Starner, Georgia Institute of Technology, GVU Center.
[5] Mateu, L. and Moll, F. (2005) Review of energy harvesting techniques and applications for microelectronics. *Proceedings of SPIE*, **5837**, 359–373.
[6] Meindl, J. (1995) Low power microelectronics: Retrospect and prospect. *Proceedings of the IEEE*, **83**, 619–635.

[7] Rabaey, J., Ammer, J., Karalar, T., *et al.* (2002) Picoradios for wireless sensor networks: The next challenge in ultralow-power design. *IEEE International Solid-State Circuits Conference (ISSCC), Digest of Technical Papers,* San Francisco, CA, February 3–7, pp. 200–201.

[8] Calhoun, B., Daly, D., Verma, N., *et al.* (2005) Design considerations for ultra-low energy wireless microsensor nodes. *IEEE Transactions on Computers,* **54**(6), 727–740.

[9] Chapman, P. and Raju, M. (2008) Designing power systems to meet energy harvesting needs. *TechOnline India,* **8**(42).

[10] Paulo, J. and Gaspar, P.D. (2010) Review and future trend of energy harvesting methods for portable medical devices. *Proceedings of the World Congress on Engineering (WCE),* San Francisco, CA, October, Vol. **2**.

[11] Huang, T.-C., Hsieh, C.-Y., Yang, Y.-Y., *et al.* (2012) A battery-free 217 nW static control power buck converter for wireless RF energy harvesting with α-calibrated dynamic on/off time and adaptive phase lead control. *IEEE Journal of Solid-State Circuits,* **47**, 852–862.

[12] Zhang, X.Y., Jiang, H.J., Zhang, L.W., *et al.* (2010) An energy-efficient ASIC for wireless body sensor networks in medical applications. *IEEE Transactions on Biomedical Circuits and Systems,* **3**(1), 11–18.

[13] Moore, G.E. (1998) Cramming more components onto integrated circuits. *Proceedings of the IEEE,* **86**, 82–85.

[14] ZigBee Alliance (2006) ZigBee Specifications, Version 1.0 r13, December. http://www.zigbee.org/(accessed November 13, 2015).

[15] Institute of Electrical and Electronics Engineers, Inc. IEEE Std. 802.15.4- 2003 (2003) IEEE Standard for Information Technology—Telecommunications and Information Exchange between Systems—Local and Metropolitan Area Networks—Specific Requirements—Part 15.4: Wireless Medium Access Control (MAC) and Physical Layer (PHY) Specifications for Low Rate Wireless Personal Area Networks (WPANs). Institute of Electrical and Electronics Engineers, Inc., New York.

[16] Banerjee, K., Souri, S.J., Kapur, P., and Saraswat, K.C. (2001) 3-D ICs: A novel chip design for improving deep-submicrometer interconnect performance and systems-on-chip integration. *Proceedings of the IEEE,* **89**(5), 602–633.

[17] Pereyma, M. (2007) *Overview of the modern state of the vibration energy harvesting devices.* Proceedings of the International Conference on Perspective Technologies and Methods in MEMS Design, May, pp. 107–112.

[18] Cheng, S., Wang, N., and Arnold, D.P. (2007) Modeling of magnetic vibrational energy harvesters using equivalent circuit representations. *Journal of Micromechanics and Microengineering,* **17**, 2329–2335.

[19] Anderson, M.J., Cho, J. H., Richards, C.D., *et al.* (2005) A comparison of piezoelectric and electrostatic electromechanical coupling for ultrasonic transduction and power generation. *Proceedings of the IEEE Ultrasonics Symposium,* pp. 950–955.

[20] Flynn, A.M. and Sanders, S.R. (2002) Fundamental limits on energy transfer and circuit considerations for piezoelectric transformers. *IEEE Transactions on Power Electronics,* **17**, 8–14.

[21] Koutroulis, E. and Kalaitzakis, K. (2006) Design of a maximum power tracking system for wind-energy-conversion applications. *IEEE Transactions on Industrial Electronics,* **53**, 486–494.

[22] Tan, Y.-K. and Panda, S.K. (2011) Self-autonomous wireless sensor nodes with wind energy harvesting for remote sensing of wind-driven wildfire spread. *IEEE Transactions on Power Electronics,* **26**(4), 1367–1377.

[23] Tan, Y.-K. and Panda, S.K. (2011) Optimized wind energy harvesting system using resistance emulator and active rectifier for wireless sensor nodes. *IEEE Transactions on Power Electronics,* **26**(1), 38–50.

[24] Roundy, S., Steingart, D., Frechette, L., *et al.* (2004) Power sources for wireless sensor networks. Presented at the *Proceedings of 1st European Workshop on Wireless Sensor Networks (EWSN),* Berlin, Germany.

[25] Ramadass, Y.K. and Chandrakasan, A.P. (2011) A battery-less thermoelectric energy harvesting interface circuit with 35 mV startup voltage. *IEEE Journal of Solid-State Circuits,* **46**(1), 333–341.

[26] Carlson, E.J., Strunz, K., and Otis, B.P. (2009) A 20 mV input boost converter with efficient digital control for thermoelectric energy harvesting. *IEEE Journal of Solid-State Circuits,* **45**(4), 741–750.

[27] Tellurex (N.D.) Thermoelectric Generators. http://www.tellurex.com/(accessed November 14, 2015).

[28] Kishi, M., Nemoto, H., Hamao, T., *et al.* (1999) Micro thermoelectric modules and their application to wristwatches as an energy source. *Proceedings of the International Conference on Thermoelectrics,* pp. 301–307.

[29] Lineykin, S. and Ben-Yaakov, S. (2007) Modeling and analysis of thermoelectric modules. *IEEE Transactions on Industry Application,* **43**(2), 505–512.

[30] Bandyopadhyay, S. and Chandrakasan, A.P. (2012) Platform architecture for solar, thermal, and vibration energy combining with MPPT and single inductor. *IEEE Journal of Solid-State Circuits,* **47**(9), 2199–2215.

[31] Schoeman, J.J. and van Wyk, J.D. (1982) A simplified maximal power controller for terrestrial photovoltaic panel arrays. *IEEE Power Electronics Specialists Conference,* pp. 361–367.

[32] Sullivan, C.R. and Powers, M.J. (1993) A high-efficiency maximum power point tracker for photovoltaic arrays in a solar-powered race vehicle. *IEEE Power Electronics Specialists Conference*, pp. 574–580.

[33] Esram, T. and Chapman, P.L. (2007) Comparison of photovoltaic array maximum power point tracking techniques. *IEEE Transactions on Energy Conversion*, **22**(2), 439–449.

[34] Ziegler, S., Woodward, R.C., Iu, H.H.-C., and Borle, L.J. (2009) Current sensing techniques: A review. *IEEE Sensors Journal*, **9**(4), 354–376.

[35] Paing, T., Falkenstein, E., Zane, R., and Popovic, Z. (2009) Custom IC for ultra-low power RF energy harvesting. IEEE Applied Power Electronics Conference and Exposition (APEC 2009), pp. 1239–1245.

[36] Smith, A.A. (1998) *Radio Frequency Principles and Applications: The Generation, Propagation, and Reception of Signals and Noise*. IEEE Press, New York.

[37] Ungan, T. and Reindl, L. (2008) Harvesting low ambient RF-sources for autonomous measurement systems. *Proceedings of the IEEE International Instrumentation and Measurement Technology Conference (IMTC)*, pp. 62–65.

[38] Nintanavongsa, P., Muncuk, U., Lewis, D.R., and Chowdhury, K.R. (2012) Design optimization and implementation for RF energy harvesting circuits. *IEEE Journal on Emerging and Selected Topics in Circuits and Systems*, **2**(1), 24–33.

[39] Williams, C.B. and Yates, R.B. (1996) Analysis of a micro-electric generator for microsystems. *Sensors and Actuators*, **52**(1–3), 8–11.

[40] Maurath, D., Becker, P.F., Spreemann, D., and Manoli, Y. (2012) Efficient energy harvesting with electromagnetic energy transducers using active low-voltage rectification and maximum power point tracking. *IEEE Journal of Solid-State Circuits*, **47**(6), 1369–1380.

[41] Arroyo, E. and Badel, A. (2011) Electromagnetic vibration energy harvesting device optimization by synchronous energy extraction. *Sensors and Actuators A: Physical*, **171**(2), 266–273.

[42] Roundy, S., Wright, P.K., and Rabaey, J. (2002) Micro-electrostatic vibration-to-electricity converters. *Proceedings of the ASME 2002 International Mechanical Engineering Congress and Exposition*, pp. 487–496.

[43] Huang, T.-C., Lee, Y.-H., Du, M.-J., et al. (2012) A photovoltaic system with analog maximum power point tracking and grid-synchronous control. *Proceedings of the IEEE 15th International Power Electronics and Motion Control Conference (EPE/PEMC)*, September, pp. LS1d.3-1–LS1d.3-6.

[44] Kurs, A., Karalis, A., Moffatt, R., et al. (2007) Wireless power transfer via strongly coupled magnetic resonances. *Science Express*, **317**(5834), 83–86.

[45] Nakamoto, H., Yamazaki, D., Yamamotot, T., et al. (2006) A passive UHF RFID tag LSI with 36.5% efficiency CMOS-only rectifier and current-mode demodulator in 0.35 μm FeRAM technology. *IEEE International Solid-State Circuits Conference (ISSCC), Digest of Technical Papers*, San Francisco, CA, February 3–7, pp. 310–311.

[46] Yi, J., Ki, W.-H., Mok, P.K.T., and Tsui, C.-Y. (2009) *Dual-power-path RF-DC multi-output power management unit for RFID tags*. Proceedings of the IEEE Symposium on VLSI Circuits, June, pp. 200–201.

[47] Rao, Y. and Arnold, D.P. (2011) An input-powered vibrational energy harvesting interface circuit with zero standby power. *IEEE Transactions on Power Electronics*, **26**(12), 3524–3533.

[48] Mandal, S. and Sarpeshkar, R. (2007) Low-power CMOS rectifier design for RFID applications. *IEEE Transactions on Circuits and Systems I: Regular Papers*, **54**(6), 1177–1188.

[49] Guo, S. and Lee, H. (2009) An efficiency-enhanced CMOS rectifier with unbalanced-biased comparators for transcutaneous-powered high-current implants. *IEEE Journal of Solid-State Circuits*, **44**(6), 1796–1804.

[50] Lu, Y., Ki, W.-H., and Yi, J. (2011) A 13.56 MHz CMOS rectifier with switched-offset for reversion current control. *Proceedings of the IEEE Symposium on VLSI Circuits*, June, pp. 246–247.

[51] Dwari, S. and Parsa, L. (2010) An efficient AC–DC step-up converter for low-voltage energy harvesting. *IEEE Transactions on Power Electronics*, **25**(8), 2188–2199.

[52] Kwon, D. and Rincon-Mora, G.A. (2010) A single-inductor AC–DC piezoelectric energy-harvester/battery-charger IC converting ±(0.35 to 1.2V) to (2.7 to 4.5V). *IEEE International Solid-State Circuits Conference (ISSCC), Digest of Technical Papers*, San Francisco, CA, February 7–11, pp. 494–495.

[53] Huang, T.-C., Du, M.-J., Lin, K.-L., et al. (2014) A direct AC-DC and DC-DC cross-source energy harvesting circuit with analog iterating-based MPPT technique with 72.5% conversion efficiency and 94.6% tracking efficiency. *Proceedings of the IEEE Symposium on VLSI Circuits*, June, pp. 26–27.

[54] Erickson, R.W. and Maksimovic, D. (2001) *Fundamentals of Power Electronics*, 2nd edn. Kluwer Academic Publishers, Secaucus, NJ.

[55] Pylarinos, L. (2003) Charge Pumps: An Overview. http://www.eecg.utoronto.ca/~kphang/ece1371/chargepumps.pdf (accessed November 14, 2015).

[56] Favrat, P., Deval, P., and Declercq, M.J. (1998) High-efficiency CMOS voltage doubler. *IEEE Journal of Solid-State Circuits*, **33**(3), 410–416.

[57] Huang, T.-C., Du, M.-J., Yang, Y.-Y., *et al.* (2012) Non-invasion power monitoring with 120% harvesting energy improvement by maximum power extracting control for high sustainability power meter system. Proceedings of the IEEE Custom Integrated Circuits Conference (CICC), September, pp. 1–4.

[58] Paing, T., Shin, J., Zane, R., and Popovic, Z. (2008) Resistor emulation approach to low-power RF energy harvesting. *IEEE Transactions on Power Electronics*, **23**(3), 1494–1501.

[59] Enne, R., Nikolic, M., and Zimmermann, H. (2010) A maximum power-point tracker without digital signal processing in 0.35 μm CMOS for automotive applications. IEEE International Solid-State Circuits Conference (ISSCC), Digest of Technical Papers, February 7–11, pp. 494–495.

Index

active rectifier, 506–507, 511, 524
adaptive energy recovery control (AERC), 354–356, 406, 439
adaptive instruction-cycle control (AIC), 102–103, 106–110
adaptive voltage positioning (AVP), 226–227, 254
advanced reduced instruction set computing machines (ARMs), 100
amplifier (AMP), 287, 290
analog dynamic-voltage scaling (ADVS), 8, 45, 49, 119, 126–130, 134–137
analog-to-digital converter (ADC), 102, 298, 314–115, 500, 536
automatic energy bypass (AEB), 371–372, 390, 392, 394–395

bandwidth (BW), 28, 82, 260, 283, 339–341
bipolar transistor (BJT), 31–33, 37–38, 484
Bluetooth low energy (BLE), 483
boosted gate (BG), 399, 420–421, 423
buffer (BUF), 317

capacitor free *(C-free)*, 28–29, 50, 56, 58–63, 65–68, 70, 74–75, 81, 95–96, 113, 119
charge pump, 33, 506, 509, 513, 514
charge pump rectifier, 506
clustered voltage-scaling (CVS), 100–101
code-excited linear prediction (CELP), 103

common gate (CG), 41–42
common source (CS), 33, 40, 50, 68, 85, 212–213, 215–224, 226–227, 229, 231–233, 236, 243, 248, 253–255, 292–293
comparator (CMP), 304
compensation enhancement multi-stage amplifier (CEMA), 62–64, 66, 367–368, 373
complementary metal–oxide–semiconductor (CMOS), 7–8, 10–12, 15, 18, 19, 26–27, 37–38, 62, 75, 85, 101, 104, 106, 120–121, 169, 262, 269, 335, 343–344, 348, 373, 376, 387, 389, 396, 428, 441–442, 463, 484, 525–526
computer-aided design (CAD), 27, 102, 106, 108, 441
constant-charge auto-hopping (CCAH), 351–352
constant current/constant voltage (CC/CV), 443–444, 449, 452, 463
constant-current generator (CCG), 331–333
continuous built-in resistance detection (CBIRD), 449, 472–476, 480–482
continuous conduction mode (CCM), 170, 175, 178–182, 190, 264, 267, 271, 305, 327, 358, 365, 405–406, 408–410, 413, 415, 417, 427, 436, 441, 474
continuous conduction mode/green mode relative skip energy control (CCM/GM RSEC), 406, 408–410

Power Management Techniques for Integrated Circuit Design, First Edition. Ke-Horng Chen.
© 2016 John Wiley & Sons Singapore Pte Ltd. Published 2016 by John Wiley & Sons Singapore Pte Ltd.

current feedback compensation (CFC), 66–68, 70, 72, 74–75
current mode control (CMC), 270, 297–298
current-ripple based (CRB), 96–98, 295
current transformer (CT), 499–500

digital-dynamic voltage scaling (DDVS), 17
digital-low dropout (D-LDO), 28–29, 93–97, 99–100, 102, 106–108, 113
digital-to-analog converter (DAC), 102, 121, 298–299
discontinuous conduction mode (DCM), 170, 175, 178–182, 327, 351–354, 358, 362, 441, 517, 521–522
drain-induced barrier lowering (DIBL), 9, 13
dropout voltage, 29, 31–33, 42–43, 45–46, 75, 81, 85, 88–89, 110–114, 118–120, 442, 512
dual stress liner (DSL), 1, 16, 18
dynamic-voltage scaling (DVS), 29, 62, 95, 96, 100–102, 106, 109–113, 118–121, 392–394, 441

electrical oxide thickness (EOT), 7–8, 11, 18–19, 26–27
electrocardiogram (ECG), 483, 525
energy distribution regulation amplifier (EDRA), 367, 373, 385
error amplifier (EA), 29, 31, 33–36, 39, 44, 47, 50–51, 55–56, 81, 85–86, 91, 113–117, 119, 126, 136, 139, 142–145, 147, 150–151, 160, 162–165, 167–169, 272, 274–275, 361–362, 365–367, 369, 373, 385, 404, 425

graphics processing unit (GPU), 465

hardware description language (HDL), 107–108
high-threshold voltage (HVT), 15
high-voltage metal–oxide–semiconductor (HVMOS), 2, 5–7, 26

instruction cycle-based dynamic voltage scaling (iDVS), 96, 100, 102–110, 121
internal body control (IBC), 399–404, 420–422

Kirchhoff's current law (KCL), 57, 68, 258

leakage power dissipation, 9
left-half-plane (LHP), 40–42, 58–60, 69–70, 75–77, 135, 139, 142, 145
length of diffusion (LOD), 19
low-threshold voltage (LVT), 15

maximum power point (MPP), 493, 498, 505, 514–516, 518–519, 521–522
maximum power point tracking (MPPT), 498, 504, 514–515, 518–519, 521–524
micro-electro-mechanical system (MEMS), 484, 524
Moore's law, 1, 312, 483
multi-layer ceramic capacitor (MLCC), 197–201, 204, 214, 228, 256, 258, 262–265, 267
multiply-and-accumulate (MAC), 103–105, 524

normalize (NORM), 103

operating system (OS), 100–101, 103
ordered power-distributive control (OPDC), 406
output independent gate drive (OIGD), 399–405, 408, 415, 420–422, 427–429, 432, 436–440
over-current protection (OCP), 30

phase-lock-loop (PLL), 20, 212, 298, 299, 302–303, 321, 339, 343–344, 349, 360, 406, 441
photovoltaics (PV), 496–498, 514
poly–insulator–poly (PIP), 81
poly space effect (PSE), 20
power conversion efficiency (PCE), 28–29, 33, 44–45, 61, 93, 95–97, 111–114
power management unit (PMU), 118, 120, 345, 349–351, 358, 363–364, 371–372, 395, 397, 484
printed circuit board (PCB), 28. 33. 50, 81, 148, 226, 320, 345, 348, 350–351, 361, 397, 439, 470
pseudo-continuous-conduction mode (PCCM), 352–356, 358, 361–363
pulse width modulation (PWM), 47, 124, 126–127, 151, 155, 171–172, 174–175, 177, 180, 182, 185, 275, 289, 292, 297–298, 321, 344, 351–352, 357–358, 360–362, 364, 366, 371–372, 386, 388, 390, 402, 455, 458, 464, 517, 518, 520, 523

quadratic differential and integration (QDI), 285–287, 289–294

regular-threshold voltage (RVT), 15
resistor emulation, 516, 518, 526
right-half-plane (RHP), 40–42, 58–60, 69, 76, 78, 155, 173, 188–189, 205
ripple-based on-time control (RBOTC), 297–298, 321

Index

ripple-recovered compensator (RRC), 256–267
rising edge detector (RED), 97, 108

scaling effects, 7, 22
short-channel effect (SCE), 10
single inductor dual-output (SIDO), 118–121, 348–350, 356–357, 363–366, 368–369, 371–373, 377–378, 381–382, 385, 387, 389, 390. 392–396
single-instruction multiple-data (SIMD), 103
smooth switch technique (SST), 115
solar cell, 496, 498, 511, 514, 518
source follower (SF), 33, 37, 38, 40–41, 353–354, 362
state of charge (SOC), 444–445, 449, 458, 466
stress memorization technique (SMT), 1, 15, 26
switching power regulator (SWR), 28, 30, 44, 94, 101–102, 110–111, 114, 118, 122–126, 131, 135, 138, 150, 158, 186, 201, 299, 348, 363

temperature coefficient (TC), 43
terminal reflector (TR), 97
thermal design power (TDP), 465
through-silicon via (TSV), 484
timing control unit (TCON), 351, 363

ultra-low-threshold voltage (U-LVT), 15
unity gain frequency (UGF), 40, 42, 52–56, 58, 63, 66, 70, 71, 76, 78, 81, 86–87, 173, 175, 273, 275
universal asynchronous receiver/transmitter (UART), 102

voltage-control current-source (VCCS), 288, 291–293, 476
voltage-mode control (VMC), 270, 297–298
voltage-ripple-based (VRB), 294

well proximity effect (WPE), 19–20
wireless sensor network (WSN), 483, 502, 504, 524